General Geology

Second Edition

Robert J. Foster
California State University at San Jose

Charles E. Merrill Publishing Co.
A Bell & Howell Company
Columbus, Ohio

Published by
Charles E. Merrill Publishing Co.
A Bell & Howell Company
Columbus, Ohio 43216

International Standard Book Number—0-675-09061-X

Library of Congress Number—72-85136

6 7 8 9 10—77

Printed in the United States of America

Preface

Revising a book is generally a dull, tedious job, but the many recent advances in geology made the revision of this book a stimulating project. In only four years, sea-floor spreading and plate tectonics have changed geological thinking in a manner that can only be described as revolutionary. These new concepts required the rewriting of much of the sections on structural geology. The other main change from the first edition is the expansion of historical geology from four chapters to nine chapters. This expansion was made in response to the needs of many of the users of the first edition. The third change is the addition, at the end of the historical geology section, of a chapter on the geologic future and geologic aspects of environment. This addition was made at the request of many students who wanted more information of how geology directly affects their lives. Hopefully, then, this edition is updated and somewhat expanded so that it will more nearly meet the needs and desires of instructors and students.

As in the first edition, this revision begins with minerals and rocks so that they can be introduced in the early laboratory periods if the course has a required laboratory. Most laboratory courses begin with these topics, and this commonly forces the instructor to use the chapters in the text in a different sequence than the author intended. In some courses this forces an undue hardship on the student. In many laboratory courses the emphasis is on identification and naming. To this should be added the interpretation of rocks as stressed in this text.

The sequence of topics progresses from materials, such as minerals and rocks, to processes and appeals to students in laboratory and nonlaboratory courses. Every part of geology is related to every other part so there is no best sequence of topics. To show the interrelationships among the various topics frequent cross-references are in the text. These references help the student to see these relationships and keep his interest high on the initial reading. They also facilitate using the chapters in a different sequence, and they greatly aid in study and review for the final examination.

A number of features have been incorporated in this book to make it more useful to the general education student. The number of technical terms (jargon) used is kept to a minimum so that the student will concentrate on principles instead of names. A glossary has been added. The Supplementary Reading listed at the end of each chapter stresses *Scientific American* articles, especially those available separately, and books available in paperback editions. Such readily available, inexpensive materials may encourage more outside reading. The chapter end questions are mainly simple factual questions that will help the student to review and test his retention. Some, however, are thought questions that may help him to see beyond the text. In a few cases some points not specifically covered in the text are covered by the questions. The history of life stresses the vertebrates because they are of more interest to the nongeologist than the stratigraphically more important invertebrates.

Geosynclines and continental drift are covered in Part III on structural geology rather than in Part IV, historical geology, because these topics are more closely allied with structural geology. The concept of the geosyncline has been retained although many would include geosynclines as an aspect of sea-floor spreading and plate tectonics. The geosyncline is an important step in the development of our ideas of mountain building, and most instructors are comfortable with the concept in spite of the fact that the classical ideas about geosynclines must be modified as they are here in later chapters.

The final chapter has much information on geological hazards that could have been incorporated in earlier chapters. However, such additions would have made some chapters unduly long and would have strayed away from the main concepts. This would be awkward for those courses with limited time. Some instructors may want to cover parts of the last chapter when the related material is studied earlier in the course. Even if omitted in the course, many students will read the last chapter because of their interest in environment.

This revision owes much to those who have aided me in many ways, not only in this revision but with my other books. Dr. Leigh Mintz took time when his *Historical Geology* was in press to read Part IV critically. My wife Joan drafted, edited, and read proofs. I am indebted to many people and agencies for the illustrations.

Robert J. Foster

Contents

Chapter 1

Introduction

Why Study the Earth?

Geology is the study of the earth. That we live on the earth is reason enough to study it. The more that we know about our planet, especially its environment and resources, the better we can understand, use, and appreciate it. To man the earth is the most important body in the universe.

In the broader view, however, the importance of the earth shrinks. At least three other stars, and possibly as many as eight, are thought to have planets. Even with our largest telescopes we cannot see these other planets, but we infer their existence from the wavering motions of the stars. The earth is a medium-sized planet orbiting a medium-sized star. Thus the earth appears to be an average planet. However, the earth is unique in having abundant water, exemplified by the oceans, and an atmosphere that can support life. The earth's surface temperature, controlled largely by its distance from the sun, makes these features possible, and these features, in turn, make life possible on the earth. The development and history of life are important aspects of historical geology. The space program has also revealed that the earth is unique in having a magnetic field. The earth's magnetic field, for reasons that will be discussed later, is believed to be caused by a liquid iron core that is also believed to store the energy that causes the formation of such surface features as mountain ranges. These features, which, as far as we know, are unique to the earth, are central to the processes of erosion and deformation of the earth's surface that are the main aspects of physical geology. Thus geology is in large part the study of the consequences of the earth's unique features.

1

Geology has contributed a great deal to civilization both intellectually and economically. Among the great concepts gained from geological studies are an understanding of the great age of the earth, and the development of an absolute time scale. Geology differs from most other sciences in that it is concerned with absolute time. Time appears in the equations of physics and chemistry, but these sciences are generally concerned with rates of change, and the time is relative—not absolute. Geologic time extends back almost five billion years to when the earth formed. Thus geology is concerned with immense lengths of time when measured against human experience. It is difficult to comprehend the lengths of time involved in geologic processes, but this must be done to appreciate geology fully. (See Fig. 1-1.)

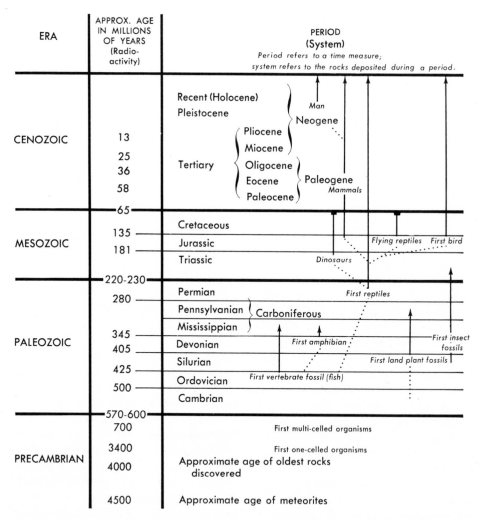

Figure 1-1. The geologic time scale. Shown to the right is a very simplified diagram showing the development of life. Not included on the diagram are many types of invertebrate fossils such as clams, brachiopods, corals, sponges, snails, etc., which first appeared in the Cambrian or Ordovician and have continued to the present. This figure is discussed in Part IV.

Another important point learned from geology is that constant change, both biological and physical, has been and is occurring on the earth.

The origin and development of life is part of geology. It is closely related to the history and development of the earth's surface and thus cannot be separated from the physical history of the earth. Geology shows that, in the broadest sense, all life is related. Biologists share this concern with life, but much of the evidence is geologic in nature. Thus geology and biology overlap in part.

As we shall see in Part IV, there are only a few fundamental laws in geology. The most important of these were recognized by the ancient Greeks, but because men were not yet ready to accept them, they were forgotten and had to be rediscovered. The fundamental principle that underlies most of geology is simply that the present processes occurring on the earth have occurred throughout geologic time. Thus ancient rocks can be interpreted in terms of present processes.

The economic contributions of geology to civilization show that in many ways, too, geology is a very practical science. Geologic knowledge is used to locate and to exploit our mineral resources. Except for water and soil, all mineral resources, such as sand and gravel, petroleum, coal, and metals, are non-renewable. Once mined they are gone, and new deposits must be found. Geologists have discovered the deposits of metal and energy-producing minerals on which our civilization is based. We take these things for granted now; but a hundred years ago, when the West was opening up and the industrial revolution was occurring, these mineral deposits were being discovered at a rapid rate and geologists were the most influential scientists of the day. At this time, too, the principles of geology were being formulated.

Today mineral resources still occupy many geologists, but geologists are also concerned with other economic problems, such as urbanization. The development of large cities has resulted in the building of large structures, such as tall buildings and dams. Geology helps in designing foundations for these structures. Examples of both large and small structures that have failed through neglect of simple geologic principles, easily understood by elementary students, are common. (See Fig. 1-2.) Dams fail because they are built near active faults or on porous foundations. During their first rainy season, new freeways are washed out or blocked by landslides. Homes built on hillsides are destroyed by landslides and mudflows. Geologists have also recognized the need for earthquake-resistant structures in some areas and have helped in their design.

Scope of Geology

Geology is concerned with both the processes operating in and on the earth, and with the history of the earth including the history of life. In the broadest sense, geology includes the study of the continents, the oceans, the atmosphere, and the earth's magnetic and radiation fields. Clearly, this scope is too broad for any one scientist, so geologists generally, but not exclusively, limit themselves to the solid earth that can be studied directly. Other specialties have developed to study the other aspects of the earth. Geophysicists study the deep parts of the earth and its fields, mainly by indirect methods; oceanographers study the hydrosphere; and meteorologists study the atmosphere.

Even with this restriction, geology is a very broad field. Most geologists specialize in one or more facets of geology, much as engineers specialize in various fields of

Figure 1-2. Landsliding at Point Fir-
min, California. Geologic studies can
prevent losses such as this. Spence air
photo.

physical science such as electronics or construction. However, geology is even broader
than engineering because it encompasses both physical and biological science.

Mention of a few of the specialties in geology will illustrate. Those who study
minerals and rocks need specialized training in chemistry and physics, as does the
geochemist who is concerned with chemical processes in the earth. Those who study
fossils must be trained in biology of plants and animals, both vertebrate and invertebrate, so that they can interpret the age and environment of fossils. Those who study
deformed rocks must know mechanics. Ground water and petroleum geologists must
be familiar with hydrodynamics. A complete listing would be very long, but these
examples will illustrate the point. All of these specialties overlap somewhat, and, no
matter what his specialty, a geologist must be familiar with all facets of geology.

Geologic Methods

Geology is based mainly on observations and seeks to determine the history of the earth
by explaining these observations logically, using other sciences such as physics, chemistry, and biology. Only a small part of geology can be approached experimentally. For
example, although the important use of fossils to date or establish contemporaneity of

rock strata is based on the simple, basic principle that life has changed during the history of the earth, this principle could not be established experimentally; it was the result of careful observations and analyses over a long period of time by many people of varied backgrounds.

Geologic problems are many, diverse, and complex; almost all must be approached indirectly, and in some cases, different approaches to the same problem lead to conflicting theories. It is generally difficult to test a theory rigorously for several reasons. The scale of most problems prevents laboratory study; that is, one cannot bring a volcano into the laboratory, although some facets of volcanoes can be studied indoors. It is also difficult to simulate geologic time in an experiment. All of this means that geology lacks exactness and that our ideas change as new data become available. This is not a basic weakness of geology as a science, but means only that much more remains to be discovered; this is a measure of the challenge of geology.

Reasoning ability and a broad background in all branches of science are the main tools of the geologist. The geologist uses the method of multiple working hypotheses to test his theories and to attempt to arrive at the best-reasoned theory. This thought process requires as many hypotheses as possible and the ability to devise ways to test each one. Not always is it possible to arrive at a unique solution—but this is the goal. In the sense that the geologist uses observation, attention to details, and reasoning, his methods are similar to those of fictional detectives.

The most important method used by the geologist is to plot on maps the locations of the rock types exposed at the earth's surface. The rocks are plotted according to their type and age on most maps. (See Fig. 1-3.) From such maps it is possible to interpret the history of the area. The early geologists had to make their own maps, and work in a remote area was very difficult in many cases. Now, excellent maps produced from

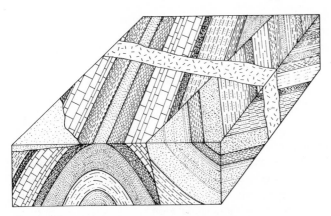

Figure 1-3.   Block diagram showing a geologic map with cross-sections on the sides. A geologic map shows the distribution of rock types on the surface. From such maps, the history of an area is interpreted.

aerial photographs are available for most areas. Much geologic mapping is done directly on either black and white or color aerial photographs, which have proved to be unexcelled for accurate location of rock units. In addition, the outcrop pattern of the rock units generally shows well on air photos. (See Fig. 1-4.) Radar images that show

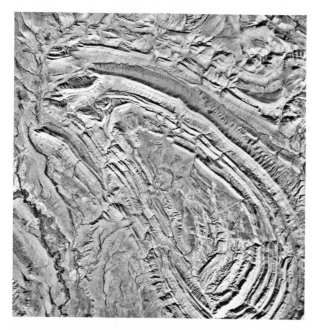

Figure 1-4. Vertical aerial photograph of folded rocks in Wyoming. Erosion has etched out the differences in resistance of the various rock layers. Such a photo is essentially a geologic map of the area. Photo from U.S. Geological Survey.

the surface beneath thick forest cover now extend the use of aerial photographs into such areas. (See Fig. 1-5.) Helicopters and jeeps have largely, but not entirely, replaced pack animals and back packing. (See Fig. 1-6.) However, in spite of all of these advances that have accelerated geologic mapping, much of the earth is not mapped geologically or at best has been mapped only in reconnaissance fashion.

Recent advances have extended classical geologic mapping by providing new things to map. As an example, it is possible to obtain, with some complex instrumentation, aerial views of areas in infrared or heat wavelengths rather than visible light. These are used in the study of active volcanic areas and ground water studies as well as other types of studies. (See Fig. 1-7.) Other examples of the types of data now mapped include the earth's magnetic field, radioactivity, heat flow, seismic properties, and many more.

Mapping this new data is giving geologists new insights in understanding the earth.

The other more obvious method used extensively by the geologist is the detailed study of rocks themselves. (See Figs. 1-8, 1-9, and 1-10.) This can be done at several levels, and Part I is largely concerned with this. The first methods applied were optical microscopy and chemical analysis, and these are both still in wide use. Newer methods such as the electron microscope; X-ray analysis; optical, infrared, and X-ray spectroscopy; and the electron microprobe now permit very detailed studies of rocks.

Recent Trends

Recent discoveries have expanded the interest of the geologist far beyond his classical realm, the rocks of the earth's surface. In addition, geology is becoming more quantitative, although it is still qualitative when compared to a science such as physics. The

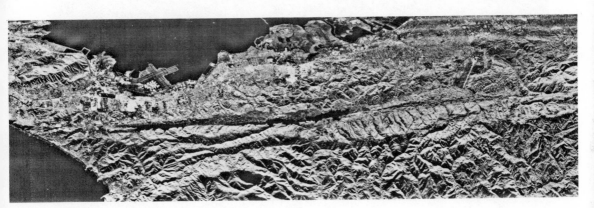

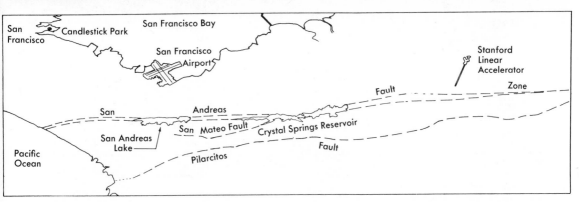

A.

Figure 1-5. Radar images. A. San Francisco Peninsula. Unlike conventional aerial photos, radar images can be obtained in cloudy weather or even at night. The radar penetrates the vegetation and reveals the actual surface. The bottom (west) part of this area has thick redwood forests. From U.S. Geological Survey in cooperation with NASA and Westinghouse Electric Co.

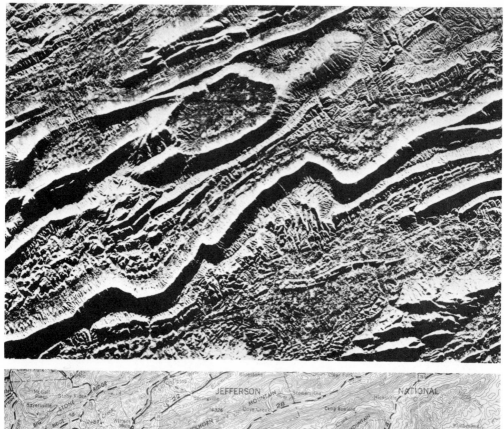
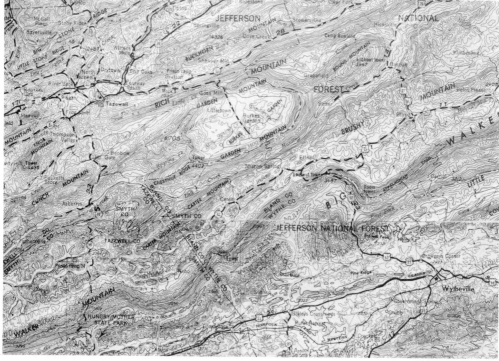

Figure 1-5. Radar images. (cont.)
B. Appalachian Mountains in western
Virginia. The structures of these folded
rocks are well shown. Courtesy of
Raytheon Company, Autometric Facility.

B.

Figure 1-6. Mapping on the northwest coast of Alaska. The helicopter is used to transport geologists to this remote area and for movement about this trackless region. Photo by Standard Oil Co. of California.

early geologists soon understood in a general way the surface processes such as erosion. A geologic time scale based on fossils was also soon developed, and this was used to decipher the history of the continents. In the early part of the twentieth century, radioactive dating enabled the geologist to work with the very old rocks, and the problem of the origin and early history of the earth began to take form. Since World War II oceanography has developed rapidly, and the features of the ocean bottoms have been mapped. These oceanographic discoveries have shaped geologic thinking profoundly, and have focused attention on causes of geologic structures. Study of the continents had presented these questions, but offered no solutions; the questions are still not answered, but the new data have made possible a better formulation of the problems.

At present, the space programs are giving us information about other parts of the solar system. (See Fig. 1-11.) Most astronomers believe that all of the solar system formed at the same time, so this new information gives much help to the geologist. This has given rise to a new science of "geology" of planets, termed astrogeology, or, perhaps better, planetology. The origin of the solar system is, of course, a classic problem of astronomy, but is also part of geology. These advances, plus the sharpening by geophysicists of the tools that probe the depths of the earth, have greatly expanded the viewpoint of the geologist. At the present time geology, astronomy, oceanography,

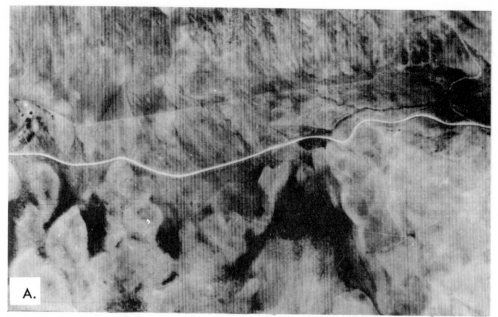

A.

Figure 1-7. Infrared image and aerial photograph of the same area. A. Infrared image shows warm areas in light tones and cool areas in dark tones. The curving light line is a blacktop road, and the straight lines are fences separating fields with different kinds of plant cover. The light pattern near the bottom is landslides, and the dark areas are poorly drained areas between the landslides.

B.

Figure 1-7. (cont.) B. Aerial photograph shows landslides at the bottom and hills near the top. The landslides moved from the bottom toward the top. From an unpublished report to NASA by U.S. Geological Survey, by R. E. Wallace and R. M. Moxham with J. Vedder.

geophysics, and meteorology are coming closer together; and it is this combination of sciences that must be used together to study the whole earth.

This Book

This book is about the solid rock earth, especially its history and the processes that formed its surface features. In this sense it covers classical geology, but in order to

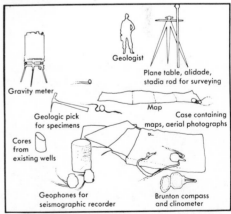

Figure 1-8. A geologist and some of his tools. The gravity meter measures the earth's gravity. The plane table, alidade, and stadia rod are used in surveying. Cores from wells, if available, reveal the rocks at depth. The geophones are used in seismic studies of the buried rocks. The Brunton compass and clinometer is used to measure the attitude of layered rocks. Photo by Standard Oil Co. of California.

Figure 1-9. Close study of rock outcrops exposed on the coast of northwestern Alaska. Photo by Standard Oil
Co. of California.

Figure 1-10. This geologist is examining a sample of a sedimentary rock that has abundant fossil fragments. The hand lens is used in the field and is commonly followed by microscopic, X-ray, or many other more detailed methods in the laboratory. Photo by Standard Oil Co. of California.

Figure 1-11. Oblique photo of the moon's surface from a height of 33.3 miles, taken by Lunar Orbiter III. This photo shows both impact and volcanic features. The moon is discussed in Chapter 18. Photo from NASA.

understand these features, it will be necessary to consider evidence from a wide range
of fields. Thus, the viewpoint will be wide, and many of the topics would not have
appeared in a survey of geology only a few years ago. This broadness, however, makes
modern geology a more valuable course for the non-science major.

We begin our study by considering the crust—the outer few miles—of the earth. We
study, first, the minerals and rocks of the crust, because they tell us much about the
composition of the crust and the processes that take place within the crust as well as
on its surface. Then we consider erosion—the dominant surface process. Next, we
consider the interactions of the processes in the crust and in the deeper parts of the earth
and learn how the present surface features of the earth formed. This leads us to a model
of the entire earth. Finally, the methods of dating and deciphering the earth's history
are described. The geologic history of North America is outlined as an example of
geologic interpretation.

SUPPLEMENTARY READING

Albritton, C. C. (ed.), *The Fabric of Geology.* Reading, Mass.: Addison-Wesley Publishing Co.,
    Inc., 1963, 372 pp.

American Geological Institute, *Geology: Science and Profession.* Washington, D.C.: American
    Geological Institute, 1965, 28 pp.

Chamberlin, T. C., "The Method of Multiple Working Hypotheses," *Journal of Geology,* 1897,
    Vol. 5, pp. 837-848.

Mather, K. F., and S. L. Mason, *Source Book in Geology.* New York: McGraw-Hill, 1939,
    702 pp.

Roy, C. J., *The Sphere of the Geological Scientist.* Washington, D.C.: American Geological
    Institute, 1962, 28 pp.

PART 1

COMPOSITION OF THE CRUST – MINERALS AND ROCKS

Chapter 2

Atoms, Elements, and Minerals

Which was which he could never make out
  Despite his best endeavour.
Of *that* there is no manner of doubt—
No probable, possible shadow of doubt—
  No possible doubt whatever.

The Gondoliers
Sir W. S. Gilbert and Sir Arthur Sullivan

Abundance of Elements

On casual examination, the surface of the earth appears to be quite complex. Many different-appearing rocks and soils are to be seen. The complexity is much more apparent than real, and close examination reveals a very simple composition. The chemical composition of the crust of the earth will be considered first.

Over 100 elements are known, and of these 80 are stable. The crust of the earth is composed almost entirely of only eight elements, and all of the other elements make up less than one per cent, as shown in Table 2-1. Thus, in spite of its heterogeneous aspect, the crust of the earth has a very simple composition. Why the crust is composed of these elements will be discussed in Chapter 18. Notice that many of our common, cheap metals, such as chrome, nickel, zinc, copper, lead, and tin, are really rather rare and form only about 0.04 per cent of the crust. How these elements are concentrated into mineable deposits will be discussed in Chapters 3 and 6.

Before discussing the important implications of these data, we should look at their origin. All abundance data are calculated from estimates of the amounts of the different rock types and their chemical analyses. Obviously, the main problem is estimating the amounts of the rock types; however, geologic maps showing the distribution of rock types have been made for many parts of the earth's surface. In spite of this, such estimates are difficult, and different workers have made different estimates. The

Table 2-1. The average composition
of continental crustal rocks. After Brian
Mason, 1966, and Konrad Krauskopf,
1968. (A few very rare elements and
shortlived radioactive elements are
omitted.)

Symbol	Name	Weight (per cent)	Symbol	Name	Weight (per cent)	Symbol	Name	Weight (per cent)
O	Oxygen	46.4	Nd	Neodymium	0.0028	Mo	Molybdenum	0.00015
Si	Silicon	28.15	La	Lanthanum	0.0025	Ho	Holmium	0.00015
Al	Aluminum	8.23	Co	Cobalt	0.0025	Eu	Europium	0.00012
Fe	Iron	5.63	Sc	Scandium	0.0022	Tb	Terbium	0.00011
Ca	Calcium	4.15	N	Nitrogen	0.0020	Lu	Lutetium	0.00008
Na	Sodium	2.36	Li	Lithium	0.0020	I	Iodine	0.00005
Mg	Magnesium	2.33	Nb	Niobium	0.0020	Tl	Thallium	0.000045
K	Potassium	2.09	Ga	Gallium	0.0015	Tm	Thulium	0.000025
Ti	Titanium	0.57	Pb	Lead	0.00125	Sb	Antimony	0.00002
H	Hydrogen	0.14	B	Boron	0.0010	Cd	Cadmium	0.00002
P	Phosphorus	0.105	Th	Thorium	0.00096	Bi	Bismuth	0.000017
Mn	Manganese	0.095	Sm	Samarium	0.00073	In	Indium	0.00001
F	Fluorine	0.0625	Gd	Gadolinium	0.00073	Hg	Mercury	0.000008
Ba	Barium	0.0425	Pr	Praseodymium	0.00065	Ag	Silver	0.000007
Sr	Strontium	0.0375	Dy	Dysprosium	0.00052	Se	Selenium	0.000005
S	Sulphur	0.026	Yb	Ytterbium	0.0003	A(r)	Argon	0.000004
C	Carbon	0.020	Hf	Hafnium	0.0003	Pd	Palladium	0.000001
Zr	Zirconium	0.0165	Cs	Cesium	0.0003	Pt	Platinum	0.000001
V	Vanadium	0.0135	Er	Erbium	0.00028	Te	Tellurium	0.000001
Cl	Chlorine	0.013	Be	Beryllium	0.00028	Ru	Ruthenium	0.000001
Cr	Chrome	0.010	U	Uranium	0.00027	Rh	Rhodium	0.0000005
Rb	Rubidium	0.009	Br	Bromine	0.00025	Os	Osmium	0.0000005
Ni	Nickel	0.0075	Ta	Tantalum	0.0002	Au	Gold	0.0000004
Zn	Zinc	0.0070	Sn	Tin	0.0002	He	Helium	0.0000003
Ce	Cerium	0.0067	As	Arsenic	0.00018	Re	Rhenium	0.0000001
Cu	Copper	0.0055	Ge	Germanium	0.00015	Ir	Iridium	0.0000001
Y	Yttrium	0.0033	W	Tungsten	0.00015			

amounts shown in the tables are subject to some revision, but they show the general
features of the composition of the crust. Estimates of the composition of the whole earth
or of the solar system are quite different.

Definitions

So far we have discussed *elements,* which are the smallest units in nature that cannot
be subdivided by ordinary chemical methods. The smallest unit of an element is the
atom. The atom can be broken down into smaller particles such as electrons, neutrons,

protons, etc., but except in the case of radioactive elements, this requires the large amounts of energy available in atom-smashing machines such as the cyclotron.

In nature, elements are combined to form *minerals,* which can be defined as

1. naturally occurring, crystalline,
2. inorganic substances with
3. a definite small range in chemical composition and physical properties.

Note that all three conditions must be met by a mineral.

Minerals are *crystalline* substances; that is, they have an orderly internal structure, as opposed to such things as glasses which are super-cooled liquids. A *crystal* is a solid form bounded by smooth planes which give an outward manifestation of the orderly internal structure. (See Fig. 2-18.) Note the distinction between crystal and crystalline. Although all minerals are crystalline, they do not necessarily occur as geometric crystals.

The minerals group themselves naturally to form *rocks,* which can be defined as

1. a natural aggregate of one or more minerals, *or*
2. any essential and appreciable part of the solid portion of the earth (or any other part of the solar system).

Note that this definition does not require all rocks to be composed entirely of minerals. Some non-crystalline material is present in many quickly chilled volcanic rocks.

Oxygen and Silicon, the Building Blocks of the Rock-forming Minerals

Because the rocks of the crust are composed predominantly of eight elements, we would suspect at first glance that these eight elements, which must form most minerals, would occur in many different combinations. However, most rocks are composed of only a few combinations of elements—or minerals. The reason for this happy situation, which means that a beginner needs to learn only a few minerals to understand most rocks, can be seen in the volume data in Table 2-2. Rocks are composed of over 90 per cent

Table 2-2. Main elements in the continental crust.

Element	Weight per cent	Atom per cent	Volume per cent
Oxygen	46.40	62.17	94.05
Silicon	28.15	21.51	.88
Aluminum	8.23	6.54	.48
Iron	5.63	2.16	.48
Calcium	4.15	2.22	1.19
Sodium	2.36	2.20	1.11
Magnesium	2.33	2.05	.32
Potassium	2.09	1.15	1.49
Totals	99.34	100.00	100.00

oxygen, by volume, because oxygen atoms (ions) are relatively large. Thus, the number of possible combinations is controlled by the ways in which oxygen atoms can be arranged to build minerals. This is especially true because relatively few other atoms are readily available to form minerals. Therefore, although about 2,000 minerals have been discovered, only about 20 are common, and fewer than 10 form well over 90 per cent of all rocks.

Thus the key to understanding the structure of minerals lies in how the atoms (more correctly called *ions*, because they have gained or lost one or more electrons) in a mineral are arranged, i.e., how the other atoms are fitted among the oxygen atoms. Clearly, the size of the atoms in question, compared to the size of the oxygen atom, will determine the structure of any mineral. This relationship is shown in two dimensions in Fig. 2-1. The number of oxygen ions that surround an element can be calculated from simple geometric relationships.

Small atom fits in space
between 3 large atoms

Larger atom fits in space
between 4 large atoms

Figure 2-1. The relative sizes of atoms determine how they can fit together. Shown here in two dimensions.

As can be predicted from the abundance data, most of the rock-forming minerals are composed largely of oxygen and silicon, generally with aluminum and at least one more of the "big eight." The sizes of oxygen and silicon are such that the small silicon ion fits inside four oxygen ions. This unit, which is the building block of the silicate minerals, is called the SiO_4 *tetrahedron,* because the four oxygen ions can be considered to be situated on the four corners of a tetrahedron with the silicon ion at the center. (See Fig. 2-2.)

 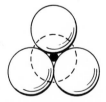 

Figure 2-2. Three views of an SiO_4 tetrahedron. The large spheres are oxygen, and the small spheres are silicon.

The structures of the silicate minerals are determined by the way these SiO_4 tetrahedra are arranged. Two of these tetrahedra, if joined at a corner, can share a single oxygen ion. This forms a very strong bond. With this type of linkage, through sharing of oxygen, the structure of the silicate minerals is formed. The possible structures are shown in Fig. 2-3.

		Formula of silicon-oxygen unit	Number of oxygen shared	Example
1. Single tetrahedron		(SiO_4)	0	Olivine
2. Double tetrahedron	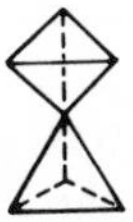	(Si_2O_7)	1	Epidote
3. Ring	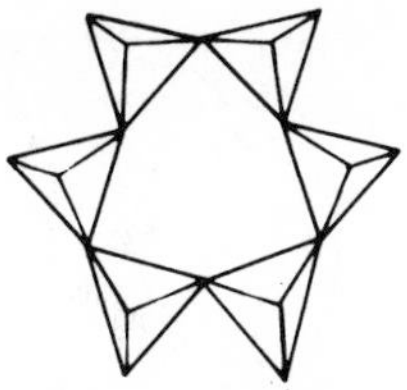	(Si_6O_{18})	2	Tourmaline
4. Chains	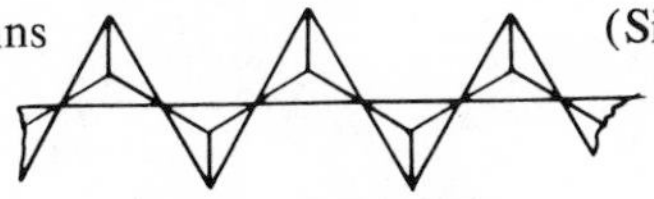	(SiO_3)	2	Pyroxene (augite)
5. Double chains	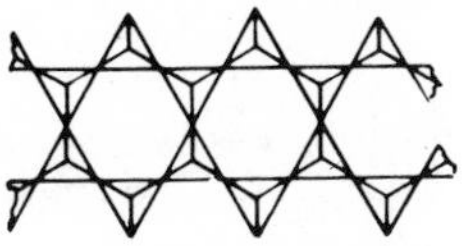	(Si_4O_{11})	2 & 3	Amphibole (hornblende)
6. Sheets	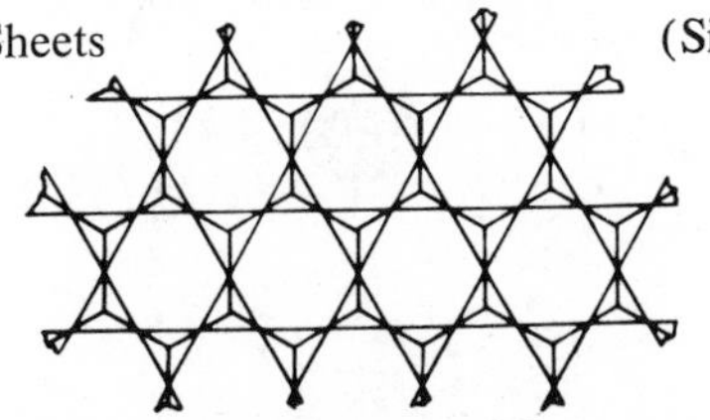	(Si_2O_5)	3	Micas (muscovite, biotite)
7. Three-dimensional networks	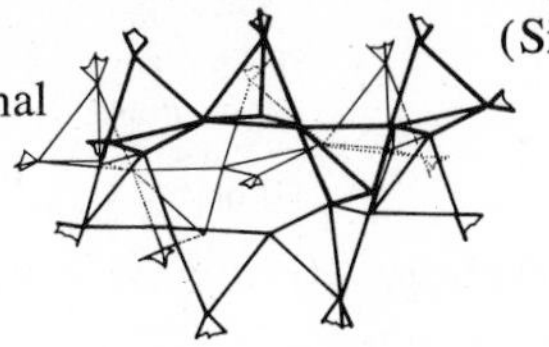	(SiO_2)	4	Quartz, feldspars

Figure 2-3. Structures of silicate minerals.

Atomic Structure

The sizes of ions determine the internal geometry of minerals and other crystalline substances; however, we must learn some features of atomic structure to understand how minerals are held together. This study will also provide an understanding of natural radioactivity, which is an important source of energy in the earth and will be discussed in later chapters.

Atoms are composed of smaller particles, and although physicists have discovered many such particles, only a few need be discussed here. Most of the mass of an atom is in its nucleus. The nucleus is composed of protons and neutrons (except in the case of the hydrogen nucleus that is composed of only a single proton). Protons are heavy and have a positive electrical charge. The nucleus is surrounded by a cloud of the much lighter, negatively charged electrons. (See Fig. 2-4.) An atom is mostly empty space because the diameter of the outer electron orbits, which is the size of the atom and averages nearly 0.00000002 centimeters, is between 20,000 and 200,000 times the diameter of the nucleus. The number of electrons equals the number of protons; hence,

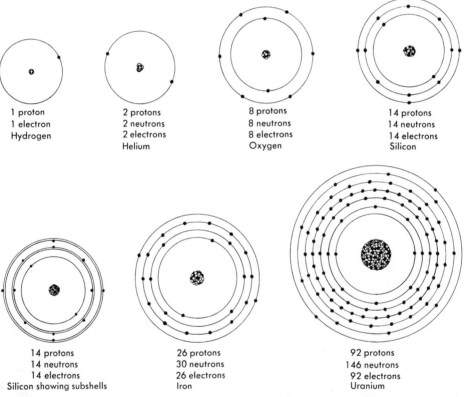

1 proton
1 electron
Hydrogen

2 protons
2 neutrons
2 electrons
Helium

8 protons
8 neutrons
8 electrons
Oxygen

14 protons
14 neutrons
14 electrons
Silicon

14 protons
14 neutrons
14 electrons
Silicon showing subshells

26 protons
30 neutrons
26 electrons
Iron

92 protons
146 neutrons
92 electrons
Uranium

Figure 2-4. Diagrammatic representation of some common atoms. Although the electron shells are shown in only two dimensions here, the electrons actually move in spherical shells. The electron shells are complicated by subshells shown only for silicon. The nuclei contain almost all of the mass of the atoms, but the atomic radii are between 20,000 and 200,000 times the radii of the nuclei.

an atom is electrically neutral. The number of protons and electrons determines the chemical properties and so determines the identity of an element. The number of protons (or electrons) is called the *atomic number* of an element. In addition, an atom may have neutrons in its nucleus. Neutrons have a mass slightly greater than that of a proton plus an electron, and are electrically neutral. The number of neutrons in the atom of an element may vary, giving rise to *isotopes.* Changing the number of neutrons in the atom of an element does not change the chemical properties, but it does change the mass of the atom. The *atomic mass number* is the number of neutrons and protons in the nucleus. Most elements have naturally occurring isotopes, as shown in Table 2-3.

Table 2-3. Examples of naturally occurring isotopes.

Element	Symbol*	Atomic number (= number of protons)	Nucleus protons	Nucleus neutrons	Atomic mass number (= number of protons and neutrons)	Electrons (= number of protons)	Remarks
Hydrogen	$_1H^1$	1	1	0	1	1	Common form
	$_1H^2$	1	1	1	2	1	Deuterium
Helium	$_2He^4$	2	2	2	4	2	Common form
	$_2He^3$	2	2	1	3	2	
Oxygen	$_8O^{16}$	8	8	8	16	8	Common form
	$_8O^{18}$	8	8	10	18	8	
Carbon	$_6C^{12}$	6	6	6	12	6	Common form
	$_6C^{14}$	6	6	8	14	6	
Potassium	$_{19}K^{39}$	19	19	20	39	19	Common form
	$_{19}K^{40}$	19	19	21	40	19	
Lead	$_{82}Pb^{206}$	82	82	124	206	82	
	$_{82}Pb^{207}$	82	82	125	207	82	
	$_{82}Pb^{208}$	82	82	126	208	82	
Uranium	$_{92}U^{235}$	92	92	143	235	92	
	$_{92}U^{238}$	92	92	146	238	92	

* The subscript is the atomic number and the superscript is the atomic mass number. Sometimes only the superscript is used because only the mass number can vary for a given element.

Bonding

The bonding of atoms to form minerals is done in several ways, most of which involve the electrons surrounding the nucleus. The electrons are not randomly distributed around the nucleus, but move in distinct orbits. Each orbit or shell contains a fixed number of electrons, as shown diagrammatically in two dimensions in Fig. 2-4. Most of the electron shells have subshells as indicated in Fig. 2-4 and Table 2-4. Up to seven shells are possible to accommodate the number of electrons necessary to balance the number of protons in the nucleus. As the number of protons in the nucleus increases, the electron shells are filled. The first shell contains 2 electrons when filled, and the second, 8. The third shell can contain up to 18 electrons; the fourth, 32; the fifth, 32;

the sixth, 9; and the seventh, 2. The electrons are distributed among the outer shells of any element in such a manner that the outer shell always contains eight or fewer electrons. Larger numbers can be accommodated only in interior shells.

The main principle involved in mineral bonding is to achieve a stable configuration with the outer shell filled. This can be accomplished in different ways. One possibility is for the atom to gain or lose electrons. An example of this is an atom with one electron in its outer shell losing this electron to another atom with seven electrons in its outer shell. The resulting *ions*, as atoms that have gained or lost electrons are called, have opposite charges and so are mutually attracted. (See Fig. 2-5.) This is called *ionic bonding*, and most minerals have ionic bonds.

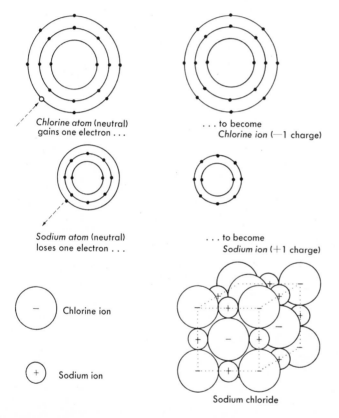

Chlorine atom (neutral) gains one electron . . .

. . . to become
Chlorine ion (—1 charge)

Sodium atom (neutral) loses one electron . . .

. . . to become
Sodium ion (+1 charge)

Chlorine ion

Sodium ion

Sodium chloride

Figure 2-5. Ionic bonding. Ions are formed by the completion of an electron ring by the gain or loss of electrons. The resulting ions are held together (bonded) by the attraction of unlike electric charges.

Another way that the outer shell of an atom can be filled is by two or more atoms sharing electrons. In this way the outer shells of the atoms are filled even though the total number of electrons is not enough to satisfy both atoms individually. This is called *covalent bonding* and is shown in Fig. 2-6. Diamond is one of the few minerals with covalent bonding, and this accounts for the hardness of the diamond. Bonding of the silicate minerals is midway between ionic and covalent.

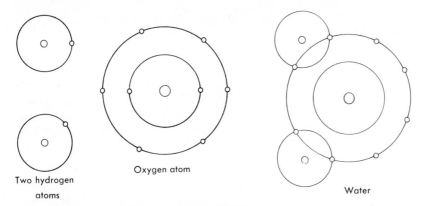

Two hydrogen atoms

Oxygen atom

Water

Figure 2-6. Covalent bonding, shown diagrammatically, results when electrons are shared by more than one atom.

    A third common type of bonding is the *metallic bond.* In a metal, the atoms are closely packed so that the electrons are not owned by any particular nucleus. This allows the electrons to move rather freely through the metal and so accounts for the good thermal and electrical conductivity of metals. In minerals, metallic bonds are found in some sulfides and in native elements, such as gold.

    The size of atoms or ions is very important in determining the structure of minerals. Ions that are within about ten per cent of the same size may mutually substitute in many minerals. Some examples of ion pairs that can mutually substitute are silicon and aluminum, sodium and calcium, iron and magnesium, and gold and silver. See Table 2-4 and Fig. 2-7. The type of bond also affects possible substitution, and although

Table 2-4. Atomic and ionic radii of some elements. Note the general relationship between gain or loss of electrons and the difference between the atomic and ionic radii.

Element	Atomic number	Number of electrons per shell and subshell						Ion formed	Number of electrons gained or lost to form ion	Radii (Angstroms)	
		1st	2nd	3rd	4th	5th	6th			Ionic	Atomic
Oxygen	8	2	2•4					O^{-2}	2 gained	1.40	0.60
Sodium	11	2	2•6	1				Na^+	1 lost	0.97	1.86
Magnesium	12	2	2•6	2				Mg^{+2}	2 lost	0.66	1.60
Aluminum	13	2	2•6	2•1				Al^{+3}	3 lost	0.51	1.43
Silicon	14	2	2•6	2•2				Si^{+4}	4 lost	0.42	1.17
Sulfur	16	2	2•6	2•4				S^{-2}	2 gained	1.85	1.04
Chlorine	17	2	2•6	2•5				Cl^-	1 gained	1.81	1.07
Potassium	19	2	2•6	2•6	1			K^+	1 lost	1.33	2.31
Calcium	20	2	2•6	2•6	2			Ca^{+2}	2 lost	0.99	1.96
Titanium	22	2	2•6	2•6•2	2			Ti^{+3}	3 lost	0.76	1.46
								Ti^{+4}	4 lost	0.68	1.46
Iron	26	2	2•6	2•6•6	2			Fe^{+2}	2 lost	0.74	1.24
								Fe^{+3}	3 lost	0.64	1.24
Silver	47	2	2•6	2•6•10	2•6•10	1		Ag^+	1 lost	1.26	1.44
Gold	79	2	2•6	2•6•10	2•6•10	2•6•9	1	Au^+	1 lost	1.37	1.44

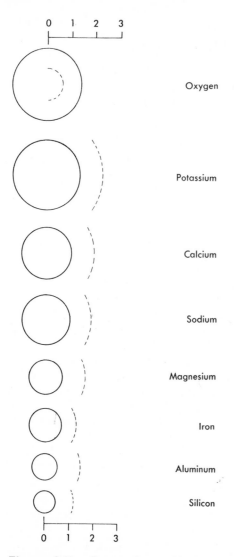

Figure 2-7. Comparison of ionic sizes of some common elements. Their atomic sizes are indicated by the dashed lines. The scale is angstroms (1 angstrom is one ten-millionth of a millimeter).

potassium and gold are nearly the same size, they do not substitute for each other. Potassium forms ionic bonds, and gold forms covalent bonds with oxygen.

Our knowledge of the internal structure of minerals comes from X-ray studies. X rays have wavelengths in the same size range as atoms, and so are deflected by the atoms in a crystal. This deflection occurs because the atoms in a crystal are packed in an orderly way. If the packing were not orderly, the X rays would be deflected in a random way. Instead, the regular pattern of the deflected X rays can be used to determine the internal structure of a mineral. (See Fig. 2-8.)

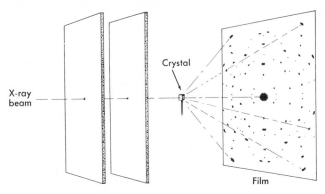

Figure 2-8. A beam of X rays is deflected by the atoms in a crystal into a regular pattern, thus revealing the internal structure of the crystal.

The material in the rest of this section is not directly related to mineralogy—the main topic of this chapter. The geological application of the rest of this section appears in later chapters. It is included here because it is directly related to atomic structure.

Natural Radioactivity

Not all nuclei are stable. Some nuclei are radioactive and disintegrate spontaneously, releasing energy in the process. Understanding why and how this takes place was one of the great advances in physics during the first half of the twentieth century. We will only look at the geologically important aspects of radioactivity. In natural radioactivity, the unstable nucleus emits several types of high energy particles and also releases energy in the form of electromagnetic waves similar to light energy.

The more important particles released in natural radioactivity are called alpha and beta particles. Alpha particles are helium nuclei (ions) and so consist of two protons and two neutrons. They are therefore positively charged. Beta particles are electrons moving at velocities near the speed of light and are negatively charged, as well as much smaller and less massive than alpha particles. The amount of energy released by emission of a particle depends on the mass and the velocity of the particle.

The other type of radiation in natural radioactivity is gamma radiation. Gamma rays are energy in the form of electromagnetic waves. Visible light, radio waves, and X rays are also energy in the form of electromagnetic waves and differ only in their frequency or wavelength. Gamma rays differ from X rays only in frequency; gamma rays have a somewhat higher frequency.

In radioactive decay, an atom is changed to an atom of another element. This means that the number of atoms of a radioactive element decreases with the passage of time. If the rate of disintegration is known, then by measuring the amounts of the parent and daughter elements, the age of crystallization of the mineral containing the parent element can be found. This is the principle of radioactive dating, and works only because the rate of disintegration is not affected by temperature, pressure, etc.

The rate of radioactive decay is expressed in terms of the half-life. In one half-life, one-half of the radioactive atoms present will decay to the daughter element. In the next half-life, one-half of the remaining radioactive atoms will decay, and so on. (See Fig. 2-9.)

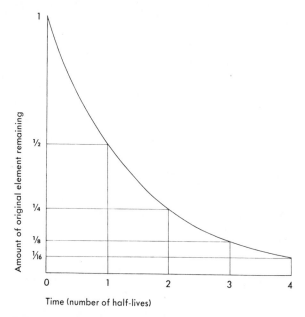

Figure 2-9. Rate of radioactive decay. During each half-life, one-half of the remaining amount of the radioactive element decays to its daughter element.

The products of radioactive decay can be understood rather easily from the foregoing descriptions of nuclei and radiation particles. Every element has a certain equal number of protons and electrons, and a variable number of neutrons that determines the isotope of that element. If the element decays by alpha emission, it loses the two protons and two neutrons that make up the alpha particle. Thus, the atomic number is reduced by two, and the atomic mass is reduced by four as shown in the following equation:

$$_{62}Sm^{147} \rightarrow {}_{60}Nd^{143} + {}_{2}\alpha^{4}$$

Alpha particles can be seen on a luminous dial watch. In a dark room the flash of light that is produced each time an alpha particle hits the zinc sulfide on the dial of the watch can be seen with a magnifiying lens.

In beta decay an electron is lost. This can be thought of as occurring because a neutron separates into a proton and an electron; the electron is emitted at high velocity, and the proton is retained by the nucleus. Thus the atomic number is increased by one, and the atomic mass remains the same. The additional proton in the nucleus is balanced by gaining an electron for the outer shell. Free electrons are generally available and can be captured by the atom. In equation form

$$_{37}Rb^{87} \rightarrow {}_{38}Sr^{87} + \beta^{-}$$

The third geologically important type of radioactivity occurs by electron capture. In this case, the nucleus takes an electron from one of the inner shells. This electron unites with a proton to form a neutron. Thus the atomic number is reduced by one, and the mass remains the same, as in

$$_{19}K^{40} + e^- \rightarrow {}_{18}A^{40}$$

The only changes in rate of radioactive decay occur in decay by electron capture, in which the rate changes slightly as the abundance of nearby electrons changes.

In most radioactivity, gamma rays are also emitted. Emission of a particle may leave the nucleus in an excited state, and emission of gamma rays can allow the atom to settle down. In any case, mass must be conserved in a radioactive event, and this can explain why gamma rays are emitted. Note in Table 2-5 that the mass of a neutron is slightly greater than the combined masses of a proton and an electron. The question then, using beta decay as an example, is: Where did the excess mass go when the neutron changed into a proton and an electron? The answer is that the mass was changed into energy in the form of gamma rays. The equivalence of mass and energy was postulated by Einstein, and experiments have so far shown this to be correct. His famous equation is

$$E = mc^2$$

or, in words, energy equals mass times the speed of light squared. However, in beta decay the energy of the gamma rays is less than the amount of mass lost. This led Fermi to postulate the emission of a small neutral particle he called the neutrino. Many years

Table 2-5. Mass and charge of some particles and atoms. Masses are in atomic mass units (amu) based on oxygen atom = 16.0000 amu (old unit).* Note that the mass of an electron plus a proton is less than the mass of a neutron.

	Charge	Mass		Charge	Mass
Electron	−1	0.000549	$_8O^{16}$	0	16.000000 (standard)
Proton	+1	1.007593	$_8O^{18}$	0	18.004855
Neutron	0	1.008982	$_{82}Pb^{204}$	0	204.0361
α-particle	+2	4.002675	$_{82}Pb^{206}$	0	206.0386
$_1H^1$	0	1.008142	$_{82}Pb^{207}$	0	207.0403
$_1H^2$	0	2.014735	$_{82}Pb^{208}$	0	208.0414
$_1H^3$	0	3.016997	$_{90}Th^{232}$	0	232.11034
$_2He^3$	0	3.016977	$_{92}U^{235}$	0	235.11704
$_2He^4$	0	4.003873	$_{92}U^{238}$	0	238.12493

* 1 amu = 1/16 of the weight of an oxygen atom. A molecular weight of oxygen is 16 grams and there are 6.02×10^{23} atoms in a molecular weight (Avogadro's number). Therefore

$$1 \text{ amu} = 1/16 \times \frac{16 \text{ grams}}{6.02 \times 10^{23}} = 1.66 \times 10^{-24} \text{ grams}$$

10^{23} means 1 followed by 23 zeros
10^{-24} means 1/1 followed by 24 zeros

later the existence of neutrinos was proved. The neutrino is only one of a few dozen tiny particles found or postulated by atomic scientists. How they are formed and whether they exist in the nucleus as particles or in the form of energy is not known. With this in mind, it is clear that the descriptions above of how radioactive decay takes place are only convenient fictions to account for the main results.

Radioactivity in Geology

The naturally radioactive isotopes are shown in Table 2-6 and Fig. 2-10. Note that there are three radioactive series that go through many steps before a stable isotope is formed, as well as a number of isotopes that decay to a stable product in a single step. Also some isotopes, such as potassium 40, decay by two processes to two different daughter products. In addition to the three series shown in the figures is the neptunium series; however, the half-life of this series is much shorter than that of the others, so only very small amounts of some of the members of this series are found in nature. Only a few of the natural radioactive isotopes are of geologic importance. Some of these are useful in age determination, and others are sources of radioactive heating of the earth. Those that have proved most useful in dating are carbon 14, potassium 40, rubidium 87, uranium 235, uranium 238, and thorium 232. The requirements for dating are a reasonable rate of decay (half-life), the retention of daughter isotopes, and the existence of common minerals containing the parent element. Those most important as a source of radioactive heating can be determined from consideration of abundance data and amount of energy released, both of which are shown in the tables. They are uranium 235, uranium 238, thorium 232, and potassium 40.

Calculation of the amount of energy released by radioactivity is simple. One needs only to know the masses of the parent and daughter elements and the masses of the particles released. (See Table 2-5.) These data and Einstein's famous equation, $E = mc^2$ (energy equals loss of mass multiplied by the speed of light squared), tell the amount of energy released by the decay of one atom of the parent isotope. The rate of decay (half-life) is then used to determine how often an atom of the parent decays. The data for the main heat-producing isotopes are shown in Table 2-7. Notice that because all of these isotopes have been decaying throughout geologic time, more of them were present in the early history of the earth and so produced more heat at that time. Note also that potassium 40 and uranium 235 both have much shorter half-lives than uranium 238 and thorium 232 and so are assumed to have been much more abundant at that time and thus produced much more heat during the early stages of the earth.

In the earth the energy released by radioactivity goes mainly into heat. Each kind of radiation behaves differently. The heavy, slow-moving alpha particles are stopped by a sheet of paper, but their impact is very destructive. They impart some of their kinetic energy to electrons and, being heavy, may disrupt whole atoms. Beta particles are much lighter and faster. They are about one hundred times more penetrating than alpha particles, but do about one hundred times less damage. Because they are too light to affect nuclei, they lose their kinetic energy by interaction with electrons. Gamma rays move with the velocity of light and have no mass. They are about one hundred times more penetrating than beta particles but do much less damage. Like beta rays, gamma rays only disrupt electrons. Heat is molecular motion, so the main effect of radioactivity is to heat the surroundings by imparting motion to electrons and atoms. Other effects are damage to a crystal lattice and increase of potential energy by moving electrons from one shell to another. These less important features are discussed below.

Table 2-6. Summary of naturally occurring radioactive isotopes. A few isotopes with very long half-lives and of very small abundances are not included.

A. Single-step decay

Isotope	Decay	Half-life (in years unless noted)	Per cent of total amount of element in crust
Neutron	$_1n \longrightarrow {}^1H^+ + \beta^-$	12 minutes	————
Tritium ($_1H^3$)	$_1H^3 \longrightarrow He^3 + \beta^-$	12.26	0.0001—0.00001
Carbon 14	$_6C^{14} \longrightarrow {}_7N^{14} + \beta^-$	5660 ± 30	(in biologic material)
Potassium 40	$_{19}K^{40}$ $\nearrow^{89\%} {}_{20}Ca^{40} + \beta^-$ $\searrow_{11\%} {}_{18}A^{40}$ (electron capture)	1.3×10^9	0.0118
Rubidium 87	$_{37}Rb^{87} \longrightarrow {}_{38}Sr^{87} + \beta^-$	4.7×10^{10}	27.85
Indium 115	$_{49}In^{115} \longrightarrow {}_{50}Sn^{115} + \beta^-$	6×10^{14}	95.72
Lanthanum 138	$_{57}La^{138}$ $\nearrow^{85\%} {}_{56}Ce^{138} + \beta^-$ $\searrow_{15\%} {}_{58}Ba^{138}$ (electron capture)	1.1×10^{11}	0.089
Samarium 147	$_{62}Sm^{147} \longrightarrow {}_{60}Nd^{143} + {}_2\alpha^4$	1.06×10^{11}	15.1
Lutetium 176	$_{71}Lu^{176}$ $\nearrow^{33\%} {}_{72}Hf^{176} + \beta^-$ $\searrow_{67\%} {}_{70}Yb^{176}$ (electron capture)	2.1×10^{10}	2.59
Rhenium 187	$_{75}Re^{187} \longrightarrow {}_{76}Os^{187} + \beta^-$	4×10^{10}	62.93

B. Multiple-step decay

All isotopes with atomic number greater than 83 are radioactive, and all of these that occur naturally are members of the three radioactive families listed below. Some isotopes with atomic number less than 83 also occur in these families. The complete decay of these families is shown in Fig. 2-10.

Isotope	Decay	Half-life (in years unless noted)	Per cent of total amount of element in crust
Uranium 235	$_{92}U^{235} \rightarrow {}_{82}Pb^{207} + 7_2\alpha^4 + 4\beta$	7.1×10^8	0.72
Uranium 238	$_{92}U^{238} \rightarrow {}_{82}Pb^{206} + 8_2\alpha^4 + 6\beta$	4.5×10^9	99.28
Thorium 232	$_{90}Th^{232} \rightarrow {}_{82}Pb^{208} + 6_2\alpha^4 + 4\beta$	1.4×10^{10}	100

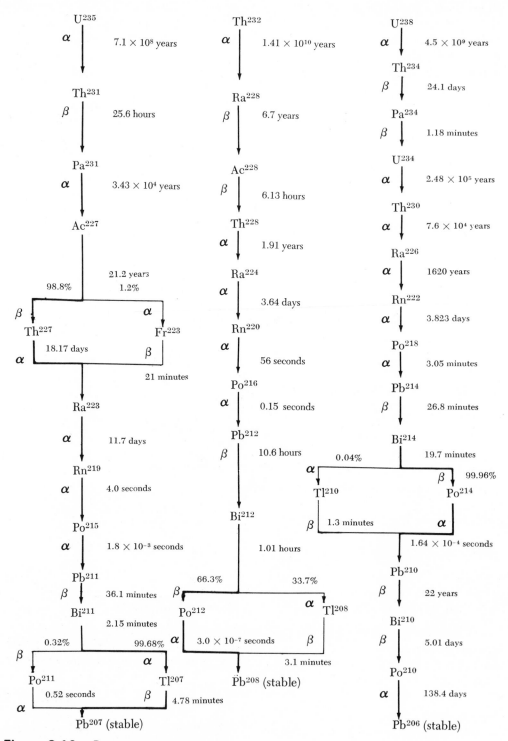

Figure 2-10. Decay schemes for uranium 235, uranium 238, and thorium 232. Type of decay, branching ratios, and half-lives are indicated. This figure is an expansion of Table 2-6B.

Table 2-7. Radioactive heat production in rocks. At the time the earth formed, 4.5 billion years ago, the shorter-lived radioactive elements were more abundant and so produced more heat than at present. The units used are 10^{-14} calories per second per gram of rock. From J. W. Winchester, "Terrestrial heat flow, radioactivity, and the chemical composition of the earth's interior," *Journal of Geological Education,* Vol. 14, No. 5, December, 1966, pp. 200-204.

Radioactive isotope	Granitic rocks		Basalt	
	Present	*4.5 billion years ago*	*Present*	*4.5 billion years ago*
Uranium 238	10.60	22.2	1.34	2.68
Uranium 235	0.46	37.5	0.06	4.74
Thorium 232	11.47	14.4	1.67	2.10
Potassium 40	3.30	40.1	0.73	8.89
Total	25.83	114.2	3.80	18.41

It should also be clear that only gamma rays have high energy and can penetrate very far, so most radiation detectors work mainly on gamma rays. Alpha rays are absorbed by about four inches of air, and beta rays by about a yard of air, but gamma rays can travel over a half-mile in air. Less than an inch of rock will absorb alpha and beta radiation, but gamma rays can penetrate over a foot of rock. These points should be taken into account in prospecting for radioactive mineral deposits. Figure 2-11 shows one means of detecting radioactivity.

The crystal structure of minerals containing alpha-emitting elements can be damaged or destroyed by the alpha particles, especially if the mineral has weak bonding. In other cases alpha radiation produces discoloration, and this can cause gray (smoky) quartz,

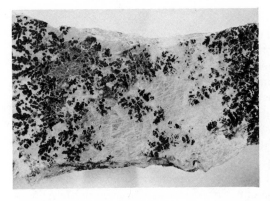

A.

B.

Figure 2-11. The mineral uraninite. A. An ordinary photograph. B. An autoradiograph made by placing photographic film next to the mineral in a darkroom. The radiation from the uraninite exposes the film even though no visible light reaches the film. This can be done without a darkroom by wrapping the film in opaque paper and laying the flat mineral on the film. Exposure times are generally a few weeks. Photo from Ward's Natural Science Establishment, Inc. Rochester, N.Y.

violet fluorite, and blue or yellow halite. Beta particles can also cause discoloration in some minerals. Another common example of alpha damage is the halos or rings that commonly surround tiny crystals of radioactive minerals. These small crystals generally occur as inclusions in later-formed minerals, and dark halos are common in biotite. Uranium and thorium minerals emit alpha particles with various amounts of energy, and because the alpha particles do most of their damage near the end of their movement, traveling a distance determined by the energy of each particle, they form a series of concentric halos. (See Fig. 2-12.) The amount of damage done to a mineral (the

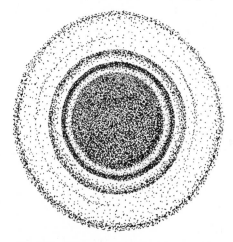

Figure 2-12. Radiation halo. Such halos are commonly seen in biotite crystals where they surround tiny crystals of radioactive minerals such as uraninite. Typical size is less than one-tenth of a millimeter.

thickness and darkness of the halo) can be used to calculate the age of the mineral, but so far such dating methods have met with only limited success. Uranium 238 also decays by spontaneous fission at a very much slower rate than it does by alpha decay. The high-energy fission particles released in this way do much damage to the enclosing crystal. Under high magnification, the paths of such particles can be seen in some minerals. The path is a small tube, not unlike a bullet hole. (See Fig. 2-13.) By counting the tubes, the number of fission disintegrations that have occurred can be determined. The number of remaining uranium atoms, the source of the particles, can be determined by instrumental analysis, and thus the age of the mineral can be calculated.

Radiation can increase the energy of an atom by moving its electrons into different orbital shells and by trapping within the atom other electrons from beta particles or from other atoms. Remember that each electron belongs in a certain shell. This energy can be released by freeing the electrons so that they can escape or move back to the original shell. One way to do this is to excite the atom by heating it. The energy is released in the form of light. This phenomenon can be seen easily by heating a mineral such as fluorite in a frying pan in a dark room. The complete explanation of this

Figure 2-13. Fossil fission tracks. The tubes are caused by high-energy radioactive particles escaping from the mineral at the center. The tubes are much like bullet holes. The longer tubes are about one one-hundredth of a millimeter in length.

phenomenon would require much deeper knowledge of atomic structure. Because the amount of energy stored depends on the length of time a mineral has been exposed to radiation, the age of some minerals can be determined by measuring the amount of stored energy. So far this has not proved to be an accurate method of dating minerals.

Neutrons—Cosmic Rays

To complete our understanding of natural radiation, it is necessary to consider another particle. That particle is the neutron, and it is particularly interesting because of its central role in atomic and nuclear bombs. The main source of natural neutrons is cosmic rays, and only a few of the possible types of neutron reactions will be considered here.

Primary cosmic rays are nuclei (atoms that have lost their electrons) moving at nearly the speed of light. They are mainly hydrogen and helium, but most other elements are also present in small quantities. That they come from outer space is shown by their increase with height. The number does not change appreciably from night to day, so they do not come from the sun. The atoms are believed to come from exploding stars (supernova), and are accelerated to very high energy levels by the associated magnetic fields.

The primary cosmic rays collide with the atoms of gas that form the earth's atmosphere, producing secondary cosmic rays. These secondary cosmic rays are particles knocked out of the atoms; they consist of neutrons, protons, and other particles. Neutrons outside a nucleus are not stable, but decay, with a half-life of 12.8 minutes, to a proton and a beta particle. Each primary cosmic ray can produce several secondary cosmic rays, and each of the secondaries can produce more cosmic rays, and so on. Each successive generation has less energy; however, primary cosmic rays and certain of the early-formed particles have great penetrating power. Neutrons have great penetrating power because they have no charge, and thus are not deflected by protons or electrons.

Some of the less energetic nuclei of the primary cosmic rays form the van Allen radiation belts around the earth. Because these nuclei have electric charges, they are deflected by the earth's magnetic field. The interaction is such that some of the less energetic charged particles move in a spiral course around lines of force of the earth's magnetic field. Following such a path, these particles move from pole to pole and back. (See Fig. 2-14.) The more energetic particles are able to penetrate the magnetic field; these are the primary cosmic rays discussed above.

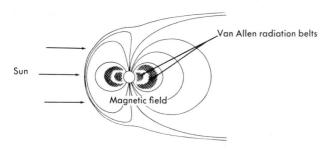

Figure 2-14. The earth's magnetic field and radiation belts.

In addition to cosmic rays, there are also particles from the sun that affect the earth. These particles are electrons and protons, and though they are constantly expelled from the sun, at times many more are expelled. These times of intense solar activity correspond with intense sun spot activity and solar flares. Because these charged particles disrupt the earth's magnetic field, these events are called magnetic storms. Such changes, of course, affect the earth's radiation belts, as well as radio communications. These particles also follow the lines of force of the earth's magnetic field and so fall on the earth near the magnetic poles. It is the interaction between these particles and the atoms of oxygen and nitrogen in the upper atmosphere that causes the aurora, or northern and southern lights. The processes may be likened to the effect of an electrical current on neon gas that produces the red light in a neon sign.

Of the cosmic rays, only the neutron is important geologically. A neutron colliding with a nitrogen atom knocks a proton out, converting the nitrogen to carbon 14:

$$_7N^{14} + n^1 \rightarrow {}_6C^{14} + {}_1H^1$$

Carbon 14 is radioactive and decays by beta emission to nitrogen 14:

$$_6C^{14} \rightarrow {}_7N^{14} + \beta^-$$

These reactions are utilized in the carbon 14 method of dating.

Cosmic rays also make possible the dating of meteorites. When cosmic rays hit iron atoms, helium 3 is produced. Thus the amount of helium 3 reveals how long the iron meteorite has undergone cosmic-ray bombardment. This effect was first noticed when attempts were made to date meteorites by measuring the amount of helium produced by alpha decay. More helium than expected was found, and analysis of the helium revealed both helium 4 from alpha activity and helium 3 from cosmic-ray activity. It was also discovered that the helium 3 concentration was greatest near the surface of a meteorite and became less dense in the interior, as would be expected from cosmic-ray activity. Similar effects were noted with other isotopes formed by cosmic-ray bombardment. Thus ordinary radioactive dating can tell the time of crystallization of a meteorite, and cosmic-ray dating tells when the body broke up to form a meteorite.

The principle involved in these cosmic-ray transformations is used commonly by nuclear physicists. By bombarding atoms with high-speed particles, alpha particles and electrons, as well as neutrons, they have succeeded in making many isotopes. Radioactive isotopes of almost every element have been made artificially.

The energy used in geologic processes ultimately comes from the atom. It is neutrons that cause the splitting or fission of atoms. The large, unstable uranium 235 atom releases neutrons when it undergoes fission, making a continuous, or chain, reaction possible. Natural uranium consists of only 0.7 per cent uranium 235. The 99.3 per cent uranium 238 absorbs neutrons, preventing natural chain reactions. Thus to produce an atomic bomb, uranium 235 must be concentrated in a large enough mass (the critical mass) that the neutrons cannot escape.

The opposite of fission is fusion, the making of a more massive atom from lighter atoms. This is the process that releases energy in the sun and other stars. An example is

$$_1H^1 + {}_1H^1 \rightarrow {}_1H^2$$
$$_1H^2 + {}_1H^1 \rightarrow {}_2He^3$$
$$_2He^3 + {}_2He^3 \rightarrow {}_2He^4 + 2{}_1H^1$$

or, in summary, the net reaction is

$$4{}_1H^1 \rightarrow {}_2He^4 + 2e^-$$

These reactions take place at very high temperatures of the order of several billion degrees. Such temperatures occur in uranium atomic bombs. The hydrogen bomb, which is a fusion device, is made by adding deuterium ($_1H^2$) and tritium ($_1H^3$) to a uranium bomb, producing what is called a thermonuclear device.

Identification of Minerals—Physical Properties

With this knowledge of atoms and their structure, minerals, the natural materials formed by atoms, can be considered.

The definition of mineral suggests that minerals can be identified in several ways. The fact that minerals have reasonably fixed chemical compositions suggests that ordinary methods of chemical analysis can be used, and, historically, this approach was important in the development of both mineralogy and chemistry. The drawbacks to its current use are that a laboratory, as well as much skill and knowledge, is needed, and the relatively recent discovery that certain minerals having identical chemical compositions have very different internal (crystal) structures. In this latter case, the polymorphs (many forms) generally have very different characteristics, such as graphite and diamond, which are both composed of pure carbon. Proof that certain minerals were indeed polymorphs had to wait until the use of X rays to probe crystal structure was perfected. Another problem in the use of chemical analysis is that analysis of the all-important silicate minerals is an extremely difficult task, even in the best-equipped laboratories. Fortunately, this problem can be bypassed by using the easily determined physical properties to identify minerals, and this can be supplemented where necessary with spot chemical tests.

The physical properties include such things as color, luster, hardness, weight (specific gravity), crystal form, fluorescence, taste, solubility, cleavage, magnetism, radioactivity, and many more. The use of the more important physical properties follows.

Cleavage and Fracture. These terms describe the way a mineral breaks. Any irregular break is termed a fracture. Many terms have been coined to describe fracture, but most, such as even, fibrous, splintery or hackly, are self-explanatory. A common type of fracture, called *conchoidal,* is the hollow, rounded type of break that occurs in glass. (See Fig. 3-12.)

A mineral has a cleavage if it has a direction of weakness that, when broken, produces a smooth plane that reflects light (Figs. 2-15 and 2-16). The common minerals may have up to six such directions. Note that a cleavage is a direction; therefore, one cleavage direction will, in general, produce two cleavage surfaces. Cleavages can be recognized by their smoothness and by their tendency to form pairs or steps, as suggested in the sketches. A series of steps is more common in minerals with poor cleavage, that is, minerals in which the cleavage direction is only slightly weaker than any other direction in the mineral. Even in this case, each of the cleavage surfaces will be smooth and parallel and, so, will reflect light as a unit.

The planes of weakness that are the cleavage directions in a mineral are caused by the atomic structure of the mineral. Cleavages form along directions of weak bonding as shown in Fig. 2-17.

Crystal Form. All crystalline substances crystallize in one of six crystal systems (Figs. 2-18, 2-19, and 2-20). If the mineral grows in unrestricted space, it develops the external shape of its crystal form; if it cannot grow its external shape, its crystalline nature can be determined only under the microscope or by X-ray analysis. Many external forms

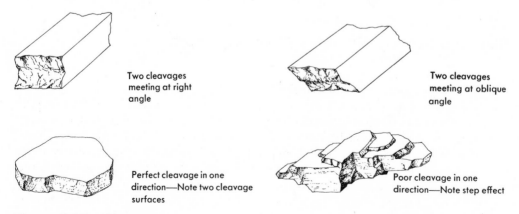

Two cleavages meeting at right angle

Two cleavages meeting at oblique angle

Perfect cleavage in one direction—Note two cleavage surfaces

Poor cleavage in one direction—Note step effect

Figure 2-15. Cleavage.

A.   B.

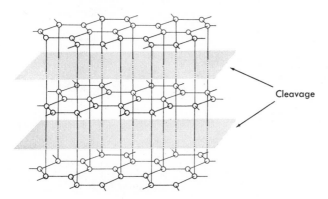

C.

Figure 2-16. Minerals showing cleavage. A. Calcite. Note the double refraction of the string behind the crystal. B. Serpentine asbestos with fibrous cleavage. C. Cleavage surface of perthite showing the intergrowth of two types of feldspar. Parts A. and B. are from Ward's Natural Science Establishment, Inc., Rochester, N.Y.; Part C. is from the Smithsonian Institution.

![graphite structure diagram] Cleavage

Figure 2-17. Structure of graphite. The bonds between the carbon sheets are weak, causing the cleavage in graphite. Graphite and diamond are both composed of carbon only; the differences between them are caused by the internal structure.

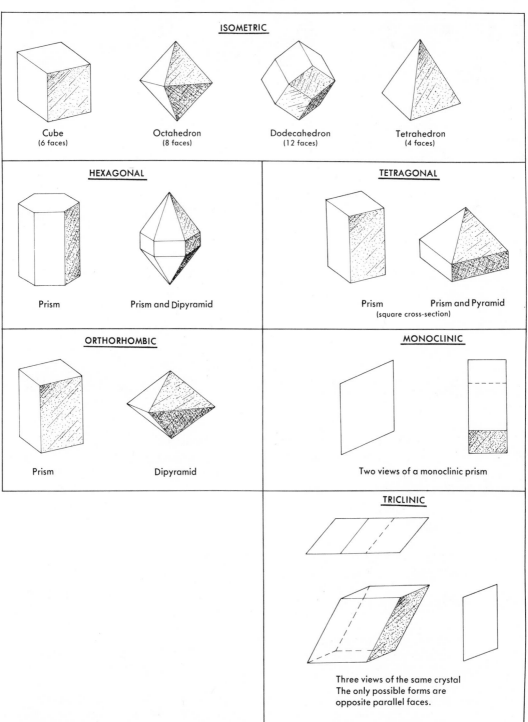

Figure 2-18. Examples of some of the more common crystal forms. They are shown and named for reference only.

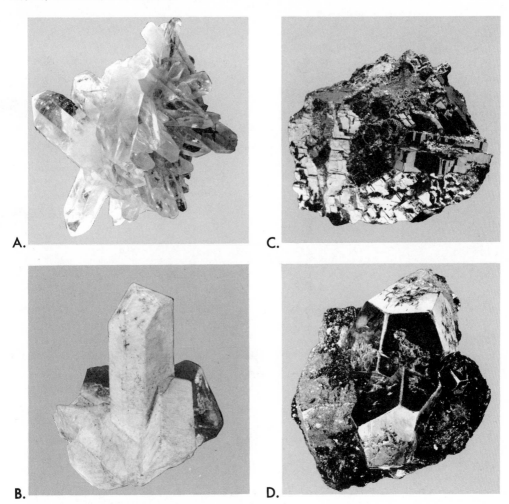

A.

C.

B.

D.

Figure 2-19. Mineral specimens showing crystal forms. Parts A., Quartz, and C., Galena, from Ward's Natural Science Establishment, Inc., Rochester, N.Y.; Parts B., Orthoclase (potassium feldspar), and D., Pyrite, are from the Smithsonian Institution.

are possible in each of the systems; however, the system can be determined by the symmetry of the crystal. Crystallography is a fascinating subject, and more information can be found in the mineralogy references listed at the end of this chapter.

Hardness. The hardness of a mineral is its resistance to scratching. As might be guessed, it is a difficult property to measure exactly because the amount of force, the shape of the scratches, and the relation of the surface tested to the crystal structure will all affect the measured hardness. In spite of this, the gross hardness is easily measured and is of great help in identification. Because surface alteration may change the apparent hardness, fresh surfaces only must be tested.

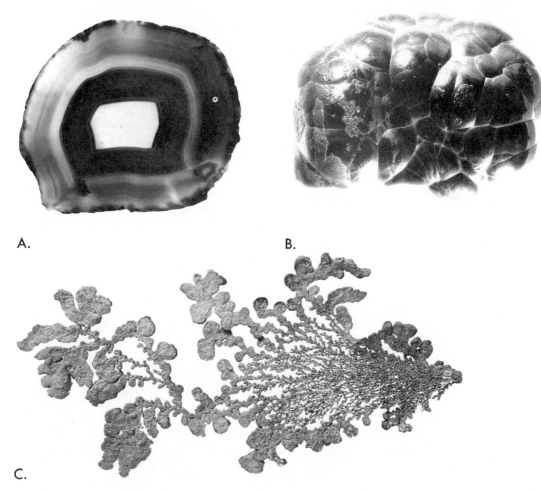

A.

B.

C.

Figure 2-20. Minerals whose crystalline nature is not readily apparent. In most of the specimens shown, the individual crystals are too small to be seen with the naked eye. A. Agate. B. Hematite. C. Native copper. Part A. is from Ward's Natural Science Establishment, Inc., Rochester, N.Y., Parts B. and C. are from the Smithsonian Institution.

The hardness scale (Mohs) follows.

1. Talc   Softest mineral
2. Gypsum
3. Calcite
4. Fluorite   (The steps are not of equal value but are
5. Apatite   arbitrarily defined. The steps are reasonably
6. Orthoclase   equally spaced except that the step between
7. Quartz   corundum and diamond is very large.)
8. Topaz
9. Corundum
10. Diamond   Hardest mineral

A useful hardness scale is:

2.
   —Fingernail
3.
   —Copper cent (Use a bright shiny one, or you will only test the hardness of the
      tarnish)
4.
5.
   —Knife blade or glass plate
6. —File
7. —Quartz
8.

To determine the hardness of a mineral, you must find the softest mineral on the hardness scale that will scratch it. With a little practice, only a knife blade and your fingernail will be necessary to estimate the hardness. Note that two substances of the same hardness can scratch each other. A pitfall that can trap the neophyte is accepting any mark on the mineral as a scratch, because just as soft chalk leaves a mark on a hard blackboard, a soft mineral may leave a mark on a hard one. You can check this by trying to remove the mark with your moistened finger, by feeling the mark with your fingernail (even a tiny depression can be detected), and by looking very closely at the mark with a lens. The final test should always be to reverse the test and see whether the mineral can scratch the test mineral.

Color. Strictly speaking, color means the color of a fresh, unaltered surface, although in some cases the tarnished or weathered color may help in identification. For some minerals color is diagnostic, but many, such as quartz, may have almost any color, due to slight impurities.

Streak is the color of the powdered mineral. To see the streak, the mineral is rubbed on a piece of unglazed porcelain called a streak plate. The color of the streak in many minerals is a more constant property than the color. The streak may be a very different color from the mineral color, and this, too, is of help in identification. Care should be taken, especially when working with small disseminated crystals, that only the mineral, and not the matrix, is rubbed on the streak plate. Many powdered minerals are white and are said to have no streak. Minerals with a hardness greater than the streak plate will, of course, have no streak.

Luster is the way a mineral reflects light. The two main types of luster are metallic and nonmetallic. The distinction is difficult to describe in words; either it looks like metal, or it does not. A few fall in between and are sometimes called sub-metallic. Dozens of terms have been proposed to describe all of the types of luster. A simplified outline of the most commonly used terms follows. Most terms are self-explanatory.

Metallic
   Bright
   Dull
(Sub-metallic) best to avoid if possible.

Nonmetallic
   Adamantine (brilliant luster, like a diamond)
   Vitreous (glassy)
   Greasy
   Resinous
   Waxy
   Pearly
   Silky
   Dull or earthy

Specific Gravity is a measure of the relative weight of a substance. It is the ratio of the mass of a substance to the mass of an equal volume of water. It is measured by weighing the substance in air and in water,

$$\text{Specific gravity} = \frac{\text{weight in air}}{\text{weight in air} - \text{weight in water}}$$

This is a difficult measurement to make with most specimens. Estimations of specific gravity can be made fairly accurately by one who is experienced in handling minerals. It is of help in identification, especially with the very heavy minerals.

The *density* of a substance is its mass per unit volume; when expressed in grams per cubic centimeter, density is numerically equal to specific gravity.

Taste and Solubility can be determined by touching the specimen with the tongue. A soluble mineral will have the feel of a lump of sugar, and an insoluble mineral may feel like glass.

Reaction with dilute hydrochloric acid is another solubility test. Calcite will react, producing effervescence. Dolomite (see Tables A-1 and A-2 in Appendix A) must be powdered to increase the surface area before effervescence occurs. Dolomite will also react with hot dilute acid or with strong acid, both of which also increase the rate of chemical reaction.

Other physical properties may require special equipment to test. Examples are radioactivity (Geiger counter), magnetism (magnet), fluorescence (ultraviolet light).

Rock-forming Minerals

The main rock-forming minerals are silicates, minerals in which elements such as iron, magnesium, sodium, calcium, and potassium are combined with aluminum, silicon, and oxygen. Their identification will be mentioned here and summarized in Tables A-1 and A-2 in Appendix A. Their structures are shown in Fig. 2-3.

Feldspars are the most abundant minerals. They are composed of sodium, calcium, and potassium, combined with silicon, aluminum, and oxygen. There are two main types, and they are complicated by mixing.

Plagioclase feldspar is an example of a continuous mineral series. The high-temperature plagioclase is *calcic,* $CaAl_2Si_2O_8$; at low temperature *sodic plagioclase,*

$NaAlSi_3O_8$, forms. A continuous series of intermediate plagioclase containing both sodium and calcium can form, depending on the temperature and composition of the melt. This is possible because sodium and calcium ions are about the same size. Aluminum and silicon, which are also nearly the same size, also mutually substitute. The microscope must be used to determine the relative amounts of sodium and calcium in plagioclase.

Potassium feldspar, $KAlSi_3O_8$, is the third, more complicated type of feldspar. A number of varieties have been recognized, but their identification requires microscopic or X-ray techniques. The variety that generally forms in granitic rocks is commonly, although generally incorrectly, called *orthoclase;* and the variety in volcanic rocks, which crystallize at higher temperatures than granitic rocks, is called *sanidine.*

Limited amounts of sodium can be accommodated in potassium feldspar in spite of the differences in ionic sizes. This limited mixing is further complicated in that the temperature of formation further controls the amount of mixing. (See Fig. 2-21.)

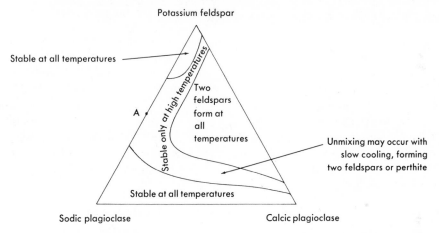

Figure 2-21. This triangular composition diagram shows the possible mixtures of the three feldspars and their temperature stability. This diagram is based on experimental data. Compositions on the left side of the diagram are common in the granitic rocks discussed in the next chapter. A feldspar composed of 50 per cent sodic plagioclase and 50 per cent potassium feldspar would fall on point A on the diagram.

At high temperatures, such as in volcanic rocks, all mixtures of sodium and potassium are possible and can be preserved by rapid chilling. Slow cooling will allow, in the intermediate mixture, the potassium feldspar and the sodic plagioclase to unmix. In some rocks the unmixing is complete and two distinct crystals form; but in many, the two feldspar crystals are intergrown and form *perthite.*

Feldspars are identified by their light color, hardness (6), and two cleavages at nearly right angles. Striations on one cleavage direction will distinguish plagioclase (Fig. 2-22). Perthite will appear as mottled, irregular veining of different color (Fig. 2-16C).

Olivine, $(Mg,Fe)_2SiO_4$, is the simplest of the ferromagnesian or dark-colored minerals. The parentheses in the formula mean that magnesium and iron may be present in

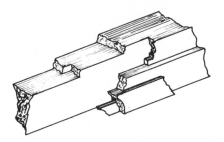

Figure 2-22. Plagioclase striations appear on only one of the two cleavage directions.

varying amounts because they are nearly the same size. Magnesium olivine forms at higher temperatures than the iron-rich types. Olivine generally forms small, rounded crystals that, because of alteration, may give deceptively low hardness tests. It can be recognized by its distinctive olive-green color.

Pyroxene is a family name of a very large and complex group of minerals. We will consider only *augite*, $(Ca,Na)(Mg,Fe^{II},Fe^{III},Al)(Si,Al)_2O_6$, the main rock-forming mineral of the family. The possible range in composition is suggested by its complex formula. It is generally recognized by its dark color, hardness (5-6), and two cleavages that meet at nearly right angles.

Amphibole is a family name of another very large and complex group of minerals. We will consider only *hornblende*, $Ca_2Na(Mg,Fe^{II})_4 (Al,Fe^{III},Ti)(Al,Si)_8O_{22}(O,OH)_2$, the main rock-former in the group. Its composition is even more complex than that of augite. It is black, has a hardness of 5-6, and has two cleavages that meet at oblique angles, distinguishing it from augite.

Mica. The micas are easily recognized by their colors and perfect cleavage. The common micas and their formulas are

Biotite—black mica, $K(Mg,Fe)_3(AlSi_3O_{10})(OH)_2$
Muscovite—white mica, $KAl_2(AlSi_3O_{10})(OH)_2$
Chlorite—green mica, $(Mg,Fe,Al)_6(Al,Si)_4O_{10}(OH)_8$

Quartz (SiO_2) is generally clear or white but may have any color. It is recognized by its hardness (7) and lack of cleavage.

As in many other minerals, the color of quartz can be caused by very minor amounts of impurities. How this can occur is shown in the following diagrams of crystal defects (Fig. 2-23). In other cases color is caused by radiation as described earlier.

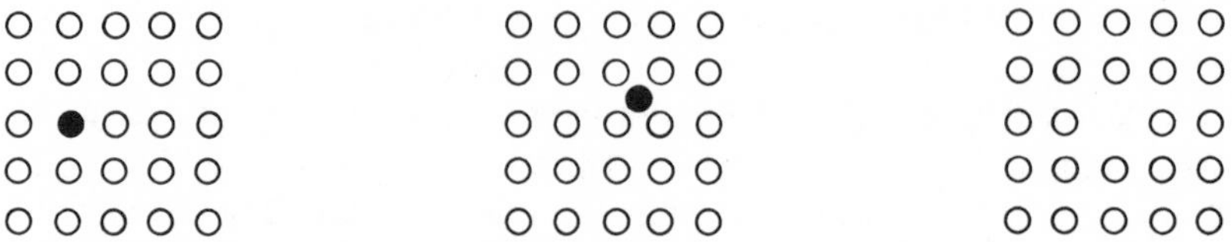

Substitutional impurity   Interstitial impurity   Vacancy

Figure 2-23. Some types of crystal defects shown diagrammatically.

QUESTIONS

2-1. The four most abundant elements (weight per cent) in order are ————,
     ————, ————, and ————.
2-2. Define mineral.
2-3. Distinguish between crystal and crystalline.
2-4. Why are there only eight important rock-forming minerals?
2-5. Can you relate the perfect cleavage in mica to its internal structure?
2-6. How are minerals distinguished from rocks?
2-7. Distinguish between silicon and silicate.
2-8. Why are the physical properties of graphite and diamond so different?
2-9. What factors determine whether one element can substitute for another in a
     mineral?
2-10. Why are silicon-oxygen tetrahedra the building blocks of most minerals?
2-11. Describe the units of which atoms are made.
2-12. What are ions, and what is ionic bonding?
2-13. What is an isotope?
2-14. Which are the geologically more important radioactive elements?
2-15. Define half-life. Why was radioactive heating more important in the early history
     of the earth than it is now?
2-16. How is radioactivity used to determine the age of a mineral by decay products?
     By crystal damage?

SUPPLEMENTARY READING

GENERAL

Akasofu, Syun-Ichi, "The Aurora," *Scientific American* (December, 1965), Vol. 213, No. 6,
     pp. 54-62.
Fleischer, R. L., P. B. Price, and R. M. Walker, "Nuclear Tracks in Solids," *Scientific American*
     (June, 1969), Vol. 220, No. 6, pp. 30-39.
Hurley, P. M., *How Old Is the Earth?* (Science Study Series). Garden City, New York: Double-
     day & Co., Inc., 1959, 153 pp. (paperback).
O'Brien, B. J., "Radiation Belts," *Scientific American* (May, 1963), Vol. 208, No. 5, pp. 84-96.
Rankama, Kalervo, *Isotope Geology.* London: Pergamon Press, Inc., 1954, 477 pp.
Reynolds, J. H., "The Age of the Elements in the Solar System," *Scientific American* (November,
     1960), Vol. 203, No. 5, pp. 171-182. Reprint 253, W. H. Freeman and Co., San Francisco.

MINERALS

Berry, L. G., and Brian Mason, *Mineralogy.* San Francisco: W. H. Freeman and Co., 1959,
     612 pp.
Hurlbut, C. S., Jr., *Dana's Manual of Mineralogy.* New York: John Wiley & Sons, Inc., 1959,
     609 pp.
Hurlbut, C. S., Jr., and H. E. Wenden, *Changing Science of Mineralogy.* Boston: D. C. Heath
     and Co., 1964, 128 pp. (paperback).
Pough, F. H., *A Field Guide to Rocks and Minerals.* Boston: Houghton Mifflin Co., 1955,
     349 pp.
Rapp, George, Jr., "Color of Minerals," *ESCP Pamphlet Series,* PS-6, Boston: Houghton Mifflin
     Co., 1971, 30 pp.

Chapter 3

Igneous Rocks and Volcanoes

Rock Cycle

The rocks of the crust are classified into three types according to their origin. Two of these types are formed by processes deep in the earth and, therefore, tell us something about conditions within the crust. They are

1. *Igneous rocks,* which solidify from a melt or *magma;* and
2. *Metamorphic rocks,* which are rocks that have been changed—generally by high temperature and pressure within the crust.

The third type, which records the conditions at the surface is

3. *Sedimentary rocks,* which are deposited in layers near the earth's surface by water, wind, and ice.

Much of geology is concerned with the interactions among the forces that produce these three rock types. The relationships are quite involved, but can be illustrated by the rock cycle (Fig. 3-1).

Crystallization

Igneous rocks are formed by the crystallization of magma. A magma is a natural, hot melt composed of a mutual solution of rock-forming materials (mainly silicates) and

49

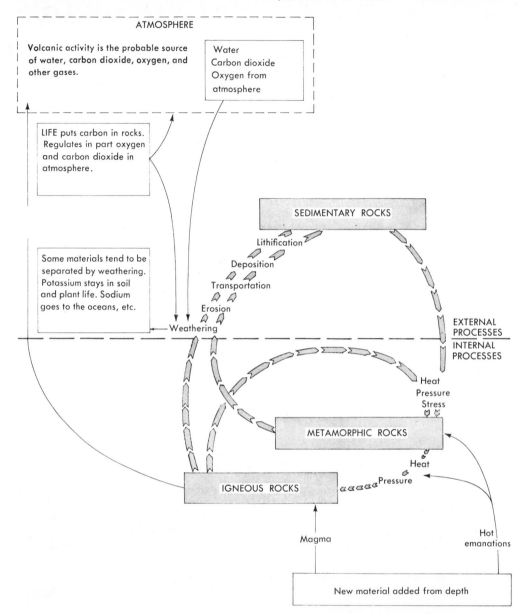

Figure 3-1 The middle of this diagram, showing the relationships among igneous, sedimentary, and metamorphic rocks, is what is generally considered the rock cycle. The upper and lower parts of the figure show how material is added to and subtracted from the rock cycle. The external processes of the upper half of the diagram are discussed in Part II of this book, and the internal processes in the lower half are discussed in Part III.

some volatiles (mainly steam) that are held in solution by pressure. Magmas may contain some solid material, but are mobile. Magma probably originates near the bottom of the crust, but its origin is an important problem.

One of the most important concepts in igneous geology is the reaction series first described by Bowen, an American physical chemist who studied silicate melts in the laboratory. This series shows the sequence in the crystallization of a basaltic melt, which is the most common type of volcanic rock. It also suggests how the earliest formed crystals might be separated from the melt or magma, forming rocks with a composition different from that of the original melt. Such a process is called *differentiation,* and we will discuss its role when we consider the origin of magmas and of the various types of igneous rocks. Although Bowen originated the idea of a reaction series from studies of experimental melts, the general sequence of crystallization of most rocks was already known as a result of intensive microscopic studies. The reaction series is an oversimplification but is very useful. (See Fig. 3-2.)

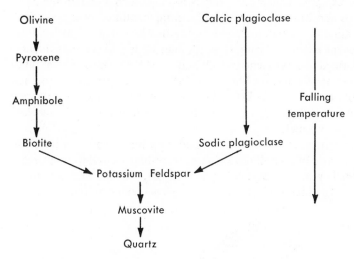

Figure 3-2. Bowen's reaction series. Shows the general sequence of mineral formation with falling temperature in a basaltic melt.

The reaction series really consists of three series. The right-hand side is a *continuous series;* that is, all compositions of plagioclase from entirely calcic to entirely sodic and all compositions in between occur. The left-hand side is a *discontinuous series;* that is, the changes from one mineral to the next occur in discrete steps. Each of the minerals named in the left-hand series is actually a continuous series in composition of that mineral group. The lower series is only a sequence of crystallization. The arrangement shows the relative, but not exact, sequence of crystallization; calcic plagioclase and olivine tend to crystallize at the same time, and the lower series is last to crystallize. *It should be emphasized that Bowen's series holds for only some basaltic magmas,* but, as we shall see, basalt magma is very important in igneous geology.

The interpretation of the crystallization of a basaltic magma in terms of Bowen's series is as follows. In general, the first mineral to crystallize from the melt is olivine. In most cases, at the same time, or a little later, calcic plagioclase begins to crystallize. The discontinuous series continues with falling temperature; the olivine reacts with the melt to form pyroxene, which at the same time also starts to crystallize directly from

the melt. Plagioclase continues to crystallize during falling temperature; the composition, however, continuously becomes more sodic. If cooling is slow enough, these two
processes go on: the minerals on the left side react with the melt to form the next lower
mineral, while that mineral is crystallizing from the melt, and the plagioclase continuously reacts with the melt to form more plagioclase that is more sodic in composition.
This process ends when all the ferromagnesian minerals and plagioclase are formed.
Then the potassium feldspar-muscovite-quartz series begins. This series may in part
overlap the sodic part of the plagioclase series.

Another way to explain the discontinuous series is to consider what happens on
heating one of the minerals. When one of the discontinuous series minerals is heated,
it does not melt at a single temperature the way ice or similar substances do. Instead,
when an amphibole, for example, is heated, it begins to melt when a certain temperature, depending on its composition, is reached, and melting continues over a range of
temperature. If we examine the remaining solid during this first stage of melting, by
quick quenching to prevent reaction during slow cooling, we would see that the amphibole has been transformed to a pyroxene and a liquid—just the reverse of what takes
place during the slow cooling of a magma. On further heating, a temperature is reached
at which further melting would occur. As before, melting would take place over a range
in temperature, resulting in a melt and olivine. Finally, a temperature would be reached
at which all the olivine would melt.

The evidence for the reaction series is abundant. Under the microscope, for instance,
pyroxene crystals with jackets of amphibole are seen, recording a discontinuous reaction that was interrupted, probably by cooling too rapidly for the reaction to be
completed. (See Fig. 3-3.) In differentiated rock bodies, slow cooling and low viscosity

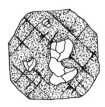

A.   B.

Figure 3-3. Evidence for the Bowen reaction series seen in microscopic study of igneous rocks. A. Augite crystal with remnants of olivine inside it. B. Hornblende crystal with augite remnants. In both cases the reaction was incomplete, thus preserving the remnants of the earlier mineral. Note the cleavage angles in augite and hornblende. The minerals shown here are typically about one-eighth inch in diameter, but may be much larger or smaller.

have combined to preserve some of the main points of the reaction principle. In such
rock bodies, the early-crystallized olivine sank to the bottom and so was removed from
contact with the main mass of the magma, allowing progressively lighter-colored rocks
to form in the upper parts of the body. (See Fig. 3-4.)

Some believe that such differentiation explains the great variety of igneous rocks, but
this is probably an oversimplification. The reaction series applies to basaltic magmas,
and the vast areas of basalt lava certainly indicate that basalt magmas are common in

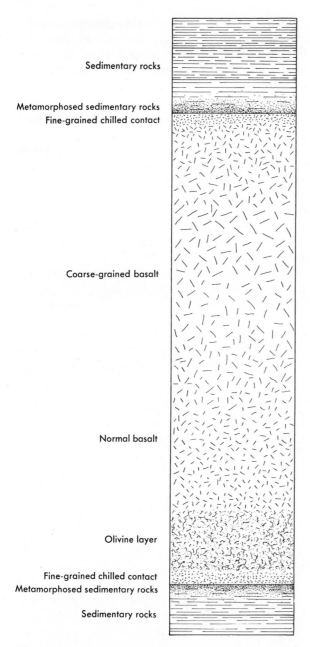

Sedimentary rocks

Metamorphosed sedimentary rocks
Fine-grained chilled contact

Coarse-grained basalt

Normal basalt

Olivine layer

Fine-grained chilled contact
Metamorphosed sedimentary rocks

Sedimentary rocks

Figure 3-4. Differentiation in a basalt sill. A narrow column through a sill is shown here. The olivine layer develops by the sinking of the early-formed, heavy olivine crystals.

the earth's crust. But, as we will see later, rocks of granitic composition are also very abundant. The vast amount of granitic rocks and their volcanic equivalents suggests that these rocks also represent important magma types in the crust, because they are far too abundant to be derived from basalt by differentiation. Complete differentiation of basalt can produce, at the most, 15 per cent granitic-composition rocks. This points out a major problem of geology—the origin of magmas. Understanding the crystallization of basalt gives some insight into the genesis of rock types as shown in the next section.

Classification

Igneous rocks are classified on the bases of composition and texture. The classification is an attempt to show the relationships among the various rock types; hence, it is an attempt at natural or genetic classification.

Composition, in the classification of igneous rocks, follows the reaction series. Most rock names came into use long before Bowen described the reaction series. These names had been applied to the more common rocks. It is of more than passing interest that these common groups of rocks fall into the various steps in the differentiation of a basaltic magma according to Bowen's reaction series. This correlation suggests that the compositional part of igneous rock terminology may be genetic. This is shown in Fig. 3-5.

Because almost all coarse-grained rocks contain feldspar, the most common mineral family, the *type* of feldspar is the most important factor in the composition. Of secondary importance is the type of dark mineral which generally accompanies each type of feldspar. See classification diagram, Fig. 3-38.

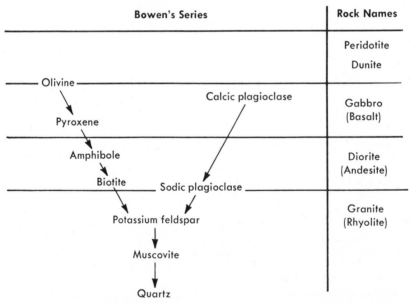

Figure 3-5. This diagram shows in a very general way how differentiation by separation of successively formed crystals could result in several rock types originating from a single basalt magma. (Volcanic rock types are in parentheses.) It also shows why minerals widely separated on the reaction series, such as quartz and olivine, rarely appear in the same rock.

In a similar way, the texture tells much about the cooling history of an igneous rock. In igneous rocks the texture refers mainly to the grain size. Rocks that cool slowly are able to grow large crystals; but quickly chilled rocks, such as volcanic rocks, are fine grained. To see the significance, we must consider the mode of occurrence of igneous rocks.

Texture and Mode of Occurrence

Igneous rocks occur in two ways, either as *intrusive* (below the surface) bodies or as *extrusive* (surface) rocks. The ultimate source of igneous magma is probably deep in the crust or in the upper part of the mantle, and the terms intrusive and extrusive refer to the place where the rock solidified. Intrusive igneous rocks can be seen only where erosion has uncovered them. They are described as *concordant* if the contacts of the intrusive body are more or less parallel to the bedding of the intruded rocks, and *discordant* if the intrusive body cuts across the older rocks. (See Fig. 3-6.)

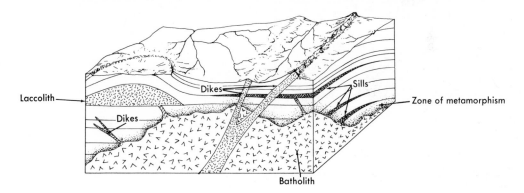

Figure 3-6. Intrusive igneous rock bodies. The laccolith and sills are concordant with the enclosing sedimentary beds, and the batholith and dikes are discordant. The heat from the batholith metamorphoses the surrounding rocks.

The largest discordant bodies are called *batholiths.* These are very large features, the size of mountain ranges such as the Sierra Nevada in California, which is nearly 400 miles long by about 50 miles wide. Many other ranges, both larger and smaller, also contain batholiths. (See Fig. 3-7.) Because batholiths are large and also because they probably were emplaced at least several thousand feet below the surface, they cooled very slowly. This slow cooling permitted large mineral grains to form; therefore, it is not surprising that batholiths are composed mainly of granitic rocks with crystals large enough to be easily seen. (See Fig. 3-8.) Smaller bodies of coarse-grained igneous rocks are termed *stocks* or *plutons.*

Dikes are tabular, discordant intrusive bodies. They range in thickness from a few inches to several thousand feet, but generally are of the order of a few feet to tens of

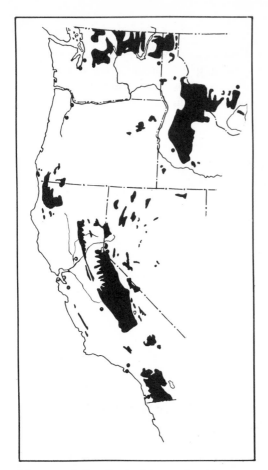

Figure 3-7. Batholiths of western United States.

Figure 3-8. Granite has a coarse, interlocking texture formed by crystallization from a melt. Photo from Ward's Natural Science Establishment, Inc., Rochester, N.Y.

feet in thickness. (See Figs. 3-9 and 3-10.) They are generally much longer than they are wide, and many have been traced for miles. Most dikes occupy cracks and have straight, parallel walls. Because the intrusive rock is commonly resistant, dikes generally form ridges when exposed by erosion.

The corcordant intrusive bodies are *sills* and *laccoliths.* They are very similar and are intruded between sedimentary beds. Sills are thin and do not noticeably deform sedimentary beds. Laccoliths are thicker bodies and up-arch the overlying sediments, in some cases forming mountains. (See Fig. 3-11.)

Dikes and sills are small bodies compared to batholiths and have much more surface area for their volume; thus, these bodies cool much more rapidly and are commonly fine grained. Laccoliths are generally fine grained but, depending on their size, may be coarse grained or intermediate.

The extrusive rocks result from lava flows and other types of volcanic activity. These rocks commonly cool rapidly and are fine grained, or glassy (see Fig. 3-12) if cooled so rapidly that no crystallization occurs. Other types of extrusive rocks will be discussed below when volcanoes are described.

Many igneous rocks are a mixture of coarse and fine crystals, and this texture is called *porphyritic.* (See Fig. 3-13.) Such a texture, in general, records a two-step history of the rock. The large crystals form as a result of slow cooling, perhaps in a deep magma chamber. Then the magma is moved higher in the crust to form a dike, sill, or flow. The remainder of the rock cools rapidly, resulting in a fine-grained or even glassy matrix surrounding the early-formed larger crystals. A rock with a coarse matrix and some much larger crystals may also be described as porphyritic. It is doubtful whether all such rocks have a two-step history as outlined here.

The temperature of basalt lava is in the range from 1,000°C to 1,200°C, felsite lava is about 900°C, and granitic rocks are believed to be emplaced at about 820°C. Advancing basalt flows at Hawaii have been measured at temperatures as low as 745°C. The

Figure 3-9. Dikes cutting through horizontal sedimentary rocks. The dikes are more resistant to erosion than are the sedimentary rocks. West Spanish Peak, Colorado, is in the background. Photo by G. W. Stose, U.S. Geological Survey.

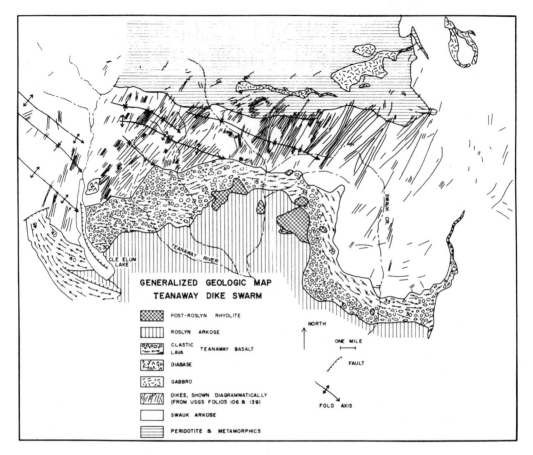

Figure 3-10. Teanaway dike swarm, Cascade Mountains, Washington. At places the dikes, shown diagrammatically here, comprise over 90 per cent of the surface, recording much dilation. From R. J. Foster, *American Journal of Science,* Vol. 256, 1958.

cooling time of igneous rocks is not well known. Batholiths are believed to solidify in one to a few million years. Dikes, depending on size, probably solidify in several tens of thousands of years. Basalt lava is known to have hardened to a depth of over ten feet in two years. Rocks are poor conductors of heat; hence, intrusive rocks take a very long time to cool because the heat must be conducted away by the surrounding rocks. Lava flows lose their heat to the atmosphere much more quickly.

A summary of textural terms will be of help in learning to recognize igneous rock types:

Glassy—no crystals
Fine-grained—grains can be seen in sunlight with a good hand lens
Coarse-grained—crystals easily seen
Pegmatitic—crystals over one inch (This texture will be discussed below.)
Porphyritic rocks—generally are named for the texture of the groundmass, giving such terms as porphyritic basalt for a dark, fine-grained rock with some larger crystals, and prophyritic granite for a granite with some much larger crystals.

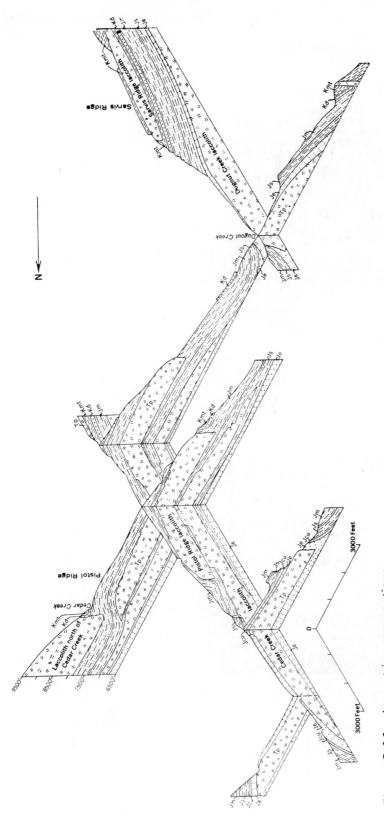

Figure 3-11. Isometric cross-sections of laccoliths in the Henry Mountains, Utah. The letter symbols designate the rock units. From C. B. Hunt, U.S. Geological Survey, *Professional Paper 228.* 1953.

Figure 3-12. Obsidian or volcanic glass. The round, shell-like fractures are called conchoidal fractures. Photo from Ward's Natural Science Establishment, Inc., Rochester, N.Y.

Figure 3-13. Prophyritic texture. The light-colored feldspar crystals grew during slow cooling, and subsequent chilling probably produced the fine-grained dark matrix. Photo from Ward's Natural Science Establishment, Inc., Rochester, N.Y.

Volcanoes

A great deal can be learned about igneous processes from study of volcanoes. They are, after all, the only direct evidence we have for the existence of magma in the crust.

The active volcanoes today are of several different types, and the differences among them seem to depend on the composition of their magmas, more particularly on the behavior of the volatiles in the magmas. The main volatile in magma is water, which escapes from volcanoes in the form of steam. Carbon dioxide is a common volcanic gas, but the sulfur gases (hydrogen sulfide and the oxides of sulfur), because of their strong odors, are the gases most easily noted near volcanoes. In addition, minor amounts of other gases, such as carbon monoxide, hydrochloric and hydrofluoric acid, ammonia, hydrogen, and hydrocarbons, are also released. Ordinary air also escapes from some volcanoes, especially those with porous rocks. Volcanic emanations are believed to have played a major role in the formation of the oceans and the atmosphere. (See Table 3-1.) Recent eruptions at Hawaii are estimated to release one to two per cent gas during the early stages of eruption and about one-half per cent during later stages. Laboratory studies suggest that, at depth, magmas may contain up to five to eight per cent dissolved water.

A simple analogy will illustrate the behavior of the water and other gases in volcanic magma. Most magmas are believed to contain at least a few per cent water in solution. The water is dissolved in the magma just as carbon dioxide gas is dissolved in a bottle of soda pop. Most magma is able to dissolve this water because the magma is under considerable pressure due to the weight of the overlying rocks. The gas is dissolved in the soda pop as a result of the pressure inside the bottle. When the bottle is opened, the pressure is reduced and the gas begins to bubble out of the liquid. In the same way, when magma is erupted to the surface, the confining pressure is reduced and the gas, in this case mainly steam, is released.

Table 3-1. Composition of volcanic gas at Hawaii.

	(Volume per cent)
Water	70.75
Carbon dioxide	14.07
Hydrogen	0.33
Nitrogen	5.45
Argon	0.18
Sulfur dioxide	6.40
Sulfur trioxide	1.92
Sulfur	0.10
Chlorine	0.05
	99.65%

After Shepherd, 1921 in Macdonald, G. A., "Physical Properties of Erupting Hawaiian Magmas," *Geological Society of America Bulletin* (August, 1963), Vol. 74, No. 8, pp. 1071-1078.

If the magma has a low viscosity, the gas merely escapes or forms bubbles in the resulting rock. The tops of lava flows can commonly be distinguished from the bottoms by the presence of these bubbles at the top. (See Fig. 3-14.) In some cases, the bubble holes are filled by material deposited by the fluids, or are filled at a later time. The fillings are commonly silica materials such as opal or chalcedony, but many other minerals also occur here. Large filled holes of this kind are actively sought by rock collectors who cut and polish them.

If a fluid magma is erupted very suddenly, it may form a froth of tiny bubbles which quickly solidifies to form the light, glassy rock *pumice.* As a result of the many bubble holes, pumice is light enough to float. The formation of pumice can be likened to the opening of a bottle of warm soda pop or champagne, which results in the production of a froth. A rock with coarser bubbles is called *scoria.*

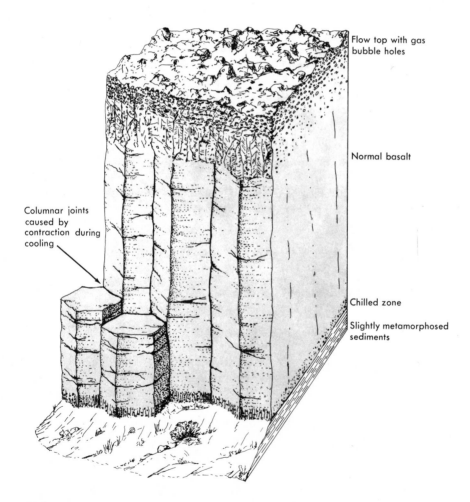

Figure 3-14.   A section through a lava flow.

More violent events occur when more viscous magmas are erupted to the surface and the gas attempts to escape. In this case, the plastic rock is shattered by the pressure of the expanding gas. The gas may expand so rapidly that the rock explodes, blowing the fragments high in the air. The fragments produced in this way are called *volcanic ash* if sandsize or finer, and *volcanic blocks* if larger. The term ash implies burning, but, of course, these rocks do not result from burning. The term was first applied when volcanoes were thought to result from subterranean burning and so is a historical mistake. The rocks formed by this process are called *pyroclastic rocks* (fire-broken rocks). Pyroclastic rocks are named according to the size of the fragments without reference to composition. Consolidated ash is called *tuff;* if the rock contains larger fragments as well, it is called *tuff-breccia,* and if composed of mainly large fragments, *volcanic breccia.* (See Fig. 3-15.) The main problem with these rocks is distinguishing them from normal sedimentary rocks formed by erosion of a volcanic area. The latter rocks are called volcanic sandstone and volcanic conglomerate. In general, field study is needed to distinguish between these two types of accumulation.

Figure 3-15. An outcrop of volcanic breccia. This rock is composed of coarse, angular fragments of volcanic rocks in a fine-grained matrix. The dark fragments are obsidian, and the light are pumice. Photo by M. H. Staatz, U.S. Geological Survey.

Volcanic bombs are formed by pyroclastic material that is still molten when thrown from the volcano. Again, the name is not really appropriate. Bombs are recognized by their shapes. A liquid assumes a streamlined shape while in flight. Most volcanic bombs are somewhat viscous when thrown from a volcano and solidify during flight. Most spin during flight and so have an elongate shape. (See Fig. 3-16.) Some are still fluid when they hit the ground, and these flatten out.

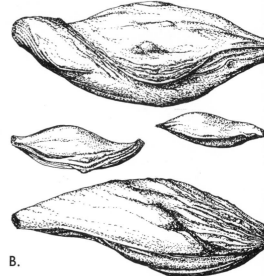

A.

B.

Figure 3-16. Volcanic bombs. A. An unusual volcanic bomb 13 feet long. Photo taken in 1902 in southern Idaho by I. C. Russell, U.S. Geological Survey. B. Typical volcanic bombs from Hawaii. Sizes range from under an inch to several feet.

Volcanic eruptions range from quiet to violent, as will be evident from the following descriptions. The examples used are currently active volcanoes and those that are recent enough that erosion has not greatly modified them. A volcano that has erupted in historical time is generally considered an active volcano; but only at one or two closely studied, well-instrumented volcanoes is it possible to predict eruptions, and even at these, warning comes just before eruption, and the place of eruption is known only generally. Thus, designating active volcanoes is difficult.

The rocks covering the surface of much of western United States are volcanic, and many of these are tens of millions of years old. This may seem old but is relatively young by geologic standards. Many of the original features of these rocks have been removed by erosion, but they are readily recognizable as volcanic. Less easily recognized are buried volcanic terranes. Such rocks are exposed by uplift and erosion in all parts of the earth, and the age of these rocks is hundreds of millions of years in many cases. Thus study of present volcanoes will help in interpretation of ancient rocks—an important principle in geology.

Present-day volcanoes occur in relatively few areas, such as the rim of the Pacific Ocean, the Mediterranean Sea, and the Mid-Atlantic Ridge. These are all geologically active areas and will be discussed in Chapter 15. (See Fig. 3-17.)

The quietest eruptions are lava flows that come from dikes or dike swarms. The lava is basalt and apparently is very fluid. These quiet flows can produce great thicknesses and cover huge areas. An example is the Columbia River Basalt of southeast Washington and northeast Oregon that covers an area of over 30,000 square miles and is over

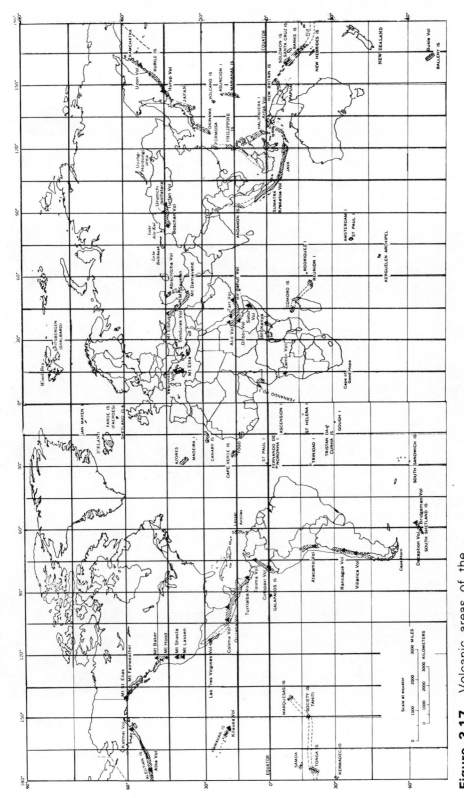

Figure 3-17. Volcanic areas of the world. From G. A. Waring, U.S. Geological Survey. *Professional Paper 492.* 1965.

65

5,000 feet thick at places. Thus this unit comprises at least 30,000 cubic miles of basalt. The source of this basalt field was four widely separated dike swarms. The very fluid basalt must have flowed like water because individual flows can be followed for over one hundred miles. (See Fig. 3-18.) Similar basalt fields are found at several places on the earth, such as the Deccan Plateau of India.

At other places similar basalt erupts from central vents, and large *shield volcanoes* form. They are called shield volcanoes because the very fluid magma builds gentle slopes that resemble shields. (See Fig. 3-19.) The Hawaiian Islands are examples of this type of quiet eruption whose most violent displays are fountains of incandescent lava thrown up from lava pools in the craters. (See Figs. 3-20 and 3-21.) Despite the gentle

Figure 3-18. Oblique aerial view looking east near the confluence of the Snake and Imnaha Rivers on the Oregon-Idaho border. The bedded rocks on the valley sides are basalt flows of the Columbia River Basalt. The lower parts of the valleys are cut in metamorphic rocks that underlie the basalt. Photo from U.S. Geological Survey.

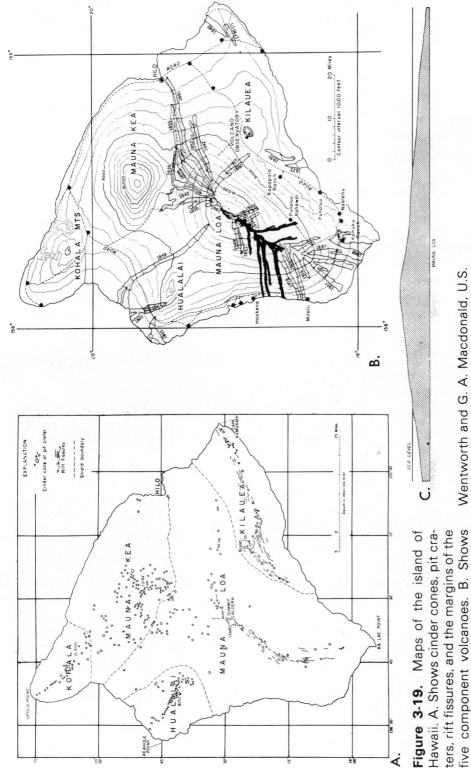

Figure 3-19. Maps of the island of Hawaii. A. Shows cinder cones, pit craters, rift fissures, and the margins of the five component volcanoes. B. Shows the historic lava flows thru 1950 (black). C. Cross-section of Mauna Loa, a shield volcano, without vertical exaggeration. Parts A. and B. are from C. K. Wentworth and G. A. Macdonald, U.S. Geological Survey. *Bulletin* 994, 1953. Part C. is from K. O. Emery and others, U.S. Geological Survey. *Professional Paper* 260A. 1954.

67

Figure 3-20. Lava fountains 300 feet high on Kilauea, September 23, 1961. Note the trees burning around the active margin of the flow fed by the fountains. Photo by D. H. Richter, U.S. Geological Survey.

A.

B.

Figure 3-21. A. Oblique aerial view of the summit of Kilauea. The craters are believed to be collapse features. The inner crater, Halemaumau, has been the site of eruptions with lava fountaining and, at times, has been partially filled by lava forming a lava lake. This view is toward the north and shows the profile of 13,784 foot Mauna Kea. Photo from U.S. Geological Survey. B. Looking southwest into Halemaumau at the collapse features formed during February 1960. An eruptive phase followed. The deep pit is about 750 feet below the rim. Photo by D. H. Richter, U.S. Geological Survey.

slopes, the Hawaiian Islands rise a total of 28,000 feet from the floor of the Pacific to 13,784 feet above sea level.

The fluid basalts discussed so far are easily recognized by their ropy surfaces. (See Fig. 3-22.) However, even in Hawaii, not all flows are of this type. Some flows are more

Figure 3-22. Ropy or pahoehoe lava.

viscous; their surfaces harden during flowage and are broken into blocks by the traction of the fluid lava a few inches below the crust. These are the blocky lava flows. (See Fig. 3-23.) The terms *blocky* and *ropy,* although somewhat descriptive, are not too good; and many geologists use the Hawaiian words *aa* and *pahoehoe* for blocky and ropy lava respectively. At some places the top of a flow may solidify, and the melted interior flows on, forming a lava tube or tunnel. Examples can be seen at many places, including the Modoc lavas in northeastern California and Craters of the Moon in Idaho.

Cinder cones probably result if the lava is more viscous than that in blocky flows. Pyroclastic material thrown out of the vent builds up a cone around the vent. A central depression around the vent remains at the summit. The cone is composed of tuff, breccia, ash, cinders, blocks, and bombs. (See Fig. 3-24.) The slope is generally near 30 degrees, which is the steepest slope that can be made of loose material. Cinder cones are very common features and, because the loose material is easily eroded, are generally young features. They are rarely more than a few hundred feet high because these slopes are so unstable; such loose slopes are commonly tiring and frustrating to climb. Cinder cones are commonly made of basalt or andesite.

The very high volcanic mountains are generally *composite volcanoes,* so called because they are generally made of a combination of pyroclastic material and lava. Such volcanoes are generally made of andesite (intermediate composition) and combine explosive eruptions with outpouring of lava. Mt. Shasta and Mt. Rainier (Fig. 3-25) are examples. The lavas cover the pyroclastic material and form the skeleton that enables the mountain to grow large. The lava is more viscous here, and the vents are blocked from time to time by congealing lava. These plugs cause the gas pressure to build up until the pressure exceeds the strength of the plug, and a greater or lesser explosion results. Under these conditions new vents may form on the sides of the volcano. Obviously, as the magma becomes more viscous, the explosive eruptions become more numerous and/or more destructive. Crater Lake, Oregon, (Fig. 3-26) was formed by the explosive destruction of the top of a composite volcano similar to Mt. Rainier or Mt. Shasta. Interestingly, the volume of material blown out of Crater Lake is much

Figure 3-23. Blocky or aa lava. A. Blocky lava flow and cinder cone. B. Steep front of a flow of viscous, blocky obsidian. Photos from U.S. Forest Service.

Figure 3-24. Aerial view of Paricutin, a large cinder cone that began in a cornfield in 1943 and was active for about nine years. The lava flow spread from vents beneath the cone. Photo from Chicago Natural History Museum.

Figure 3-25. Mt. Rainier, a massive composite volcano in the Cascade Mountains, Washington. Photo from Washington State Department of Commerce and Economic Development.

smaller than the volume of the missing top of the mountain, suggesting that after explosive discharge of the magmatic gases the magma subsided. Most craters and calderas, as large craters are called, are formed by subsidence.

Krakatoa, an island between Java and Sumatra, has a somewhat similar history. It was a volcano 2,600 feet above sea level that formed within a ring of islands marking the rim of an older volcano that lost its summit, probably in much the same way as Crater Lake formed. In 1883, the island of Krakatoa erupted explosively, destroying the island. The ash from this great explosion was the cause of beautiful sunsets all over the world long after the eruption.

Lavas ranging in composition from andesite to rhyolite tend to be viscous, and they cause explosive eruptions. Plugs tend to block the vents, resulting in violent explosions of various kinds. One type has just been described. In 1902, Mt. Pelée on Martinique Island was in eruption, and a large spine formed in the vent, plugging it. This spine was pushed up about 1,500 feet, forming quite a spectacle. Finally, the pressure of the magmatic gases was enough to enable the gas-charged magma to break through. The resulting eruption was one of the most destructive ever witnessed, and led to the recognition of a very important type of pyroclastic rock. (Recall the earlier description of the formation of pumice.) At Mt. Pelée, the magmatic gas expanded so violently that instead of forming a frothy pumice, the expanding bubbles completely shattered the pumice, producing a mixture of hot gas and small glass fragments. Such a mixture is quite fluid because of the gas and can move very rapidly down even a gentle slope. When it comes to rest, the hot glass fragments weld together, forming a *welded tuff.* Eruptions of this type are called glowing clouds *(nuées ardentes)* or glowing avalanches. The 1902 eruption of Mt. Pelée wiped out the city of St. Pierre, killing all of its about 30,000 inhabitants except a prisoner who although badly burned was saved by the thick walls of his cell and another person who was on the outskirts of the town. It was an eruption of this kind from Vesuvius that buried Pompeii in 79 A.D.

Many welded tuffs closely resemble obsidian or felsite, and only recently has the abundance of welded tuffs been recognized. Their composition is similar to that of batholiths, and their abundance on the continents may be as great as that of basalt.

Figure 3-26. Crater Lake, Oregon, fills a depression formed partly by explosive eruptions and mainly by subsidence of the magma. Wizard Island is a small volcanic cone with a crater at its summit. Photo from Oregon State Highway Department.

Recent studies also indicate that not all welded tuffs are associated with dome volcanoes. Much of Yellowstone National Park is covered by welded tuffs, and the geysers there probably get their heat from the cooling lava.

Lava plugs or domes, such as the spine at Mt. Pelée, are believed to form during the waning phase of some composite volcanoes. The only recently active volcano in the United States, except those in Alaska and Hawaii, is Mt. Lassen and is of this type. The plug or dome is composed of very viscous lava, and its extrusion is probably very similar to the squeezing of toothpaste from a tube. At Mt. Lassen the plug dome where exposed has steep, smooth cliffs with vertical grooves that developed when the dome was squeezed up. The eruptions of 1914–1915 were small; the most violent events were mudflows caused by melting snow. (See Fig. 3-27.)

Underwater volcanic eruptions are common, and lavas sometimes flow into water bodies. This causes rapid chilling of the lava, and causes the lava to subdivide into ellipsoidal bodies. The term *pillow lava* is applied to the resulting rocks. (See Fig. 3-28.)

Erosion may remove the fragmental rocks from a volcano, leaving the lava that congealed in the central conduit as shown in Fig. 3-29.

Volcanic eruptions cannot yet be predicted; however, recent studies have detected the movement of magma at a few very closely studied, well-instrumented volcanoes. The movement is shown by many thousands of shallow, very weak earthquakes. The location of these earthquakes may ultimately reveal the magma plumbing system. It

Figure 3-27. Mount Lassen in eruption. This is the only volcano that has erupted in recent times in conterminous United States. Photo by B. F. Loomis, from National Park Service.

Figure 3-28. Pillow lava in the Alaska Gulf region. Photo by F. H. Moffit, U.S. Geological Survey.

also appears that the magma moves upward to shallow magma chambers just before eruption. The latter hypothesis is suggested by the swelling of a volcano just before

Figure 3-29. Shiprock, New Mexico, is probably part of an ancient volcano. Dikes radiating from it can be seen on the left skyline and the right foreground. Photo from New Mexico Department of Development.

eruption. This is detected by very sensitive tilt meters. After an eruption the magma apparently returns to the depths. This is indicated by the tilt meters and, at Hawaii, by the draining of the lava lakes that form in the craters there. (See Fig. 3-30.)

Mt. Rainier, Washington, is a volcano with some steam activity in its crater. It is being monitored near the summit by several temperature sensors that transmit their data via Nimbus satellite several times each week. An eruption of Mt. Rainier would endanger nearby cities. Volcanic hazards are discussed in Chapter 25.

Origin and Emplacement

The origin of magma is a problem about which only some suggestions can be made. More will be said in the structure chapters, where volcanoes and other igneous bodies will be considered in relation to the other features with which they are associated. Seismic data suggest that volcanic magma forms near the base of the crust and moves upward to a shallow magma chamber before erupting to the surface. Why such a local hot spot forms is not known, nor is it known, certainly, whether the magma is melted, or partially melted, crust or mantle. (See Fig. 3-31.)

Once the magma is generated, the next problem is: why does it rise in the crust? Several suggestions have been made. Rocks, like most substances, expand when they are melted. Thus, magma is less dense than the surrounding rocks and tends to rise, much as oil mixed with water rises and floats on the top.

Salt domes may illustrate intrusion caused by gravitational rise of lighter material. Salt domes occur in many parts of the world, and because some are associated with

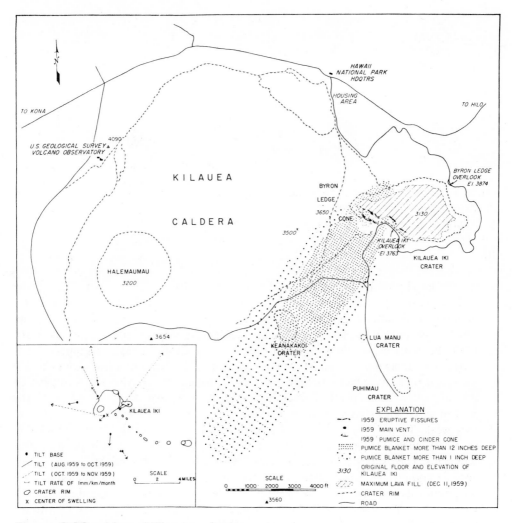

Figure 3-30. Map of Kilauea volcano, Hawaii, showing principal features of the 1959 eruption of Kilauea Iki. Small-scale inset map shows tilt pattern on the summit of Kilauea prior to the eruption. From J. P. Eaton and D. H. Richter, *Geotimes*, Vol. 5, 1960.

oil fields, they have been closely studied. Salt domes occur in sedimentary rocks; the salt originates from salt beds formed by the evaporation of sea water. These sedimentary rocks will be described more fully in Chapter 4. Salt, or halite as the mineral is called, has a specific gravity of 2.2. The specific gravity of sedimentary rocks increases with depth of burial because the weight of the overlying rocks compresses the deeper rocks. At the Gulf Coast of the United States, the near-surface sedimentary rocks have a

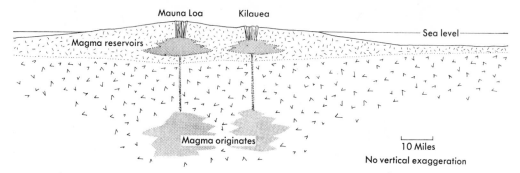

Figure 3-31. Hypothetical cross-section through Hawaii, showing possible source and accumulation zones. After G. A. Macdonald, *Science,* Vol. 133, 1961.

specific gravity of 1.9, at 2,000 feet the specific gravity has increased to 2.2, and at 10,000 feet it is 2.4. In this area the salt beds are very deep and thus are much lighter than the surrounding rocks. The salt of lesser density tends to rise and float to the top. This seems unlikely because we are used to considering rocks as strong, rigid materials. It can be easily shown, however, that rocks are not strong enough to support themselves. This topic will be pursued further in Chapter 14, but a moment's thought will show that rocks have finite strength and that the weight of overlying rocks will exceed this strength at some depth. Below this depth rocks behave as viscous fluids. In the present case, salt behaves as a viscous fluid and the overlying sedimentary rocks are very weak. To start the formation of a salt dome, some initial irregularity is probably necessary, such as a high point on the salt bed, or a weakness or break in the overlying sediments. The rising salt may remain connected to the salt bed or might, in some cases, become completely detached from the salt bed. The tops of many salt domes are mushroom shaped. (See Fig. 3-32.) Rises up to eight miles have occurred in the Gulf Coast. The mechanics of salt dome formation are complex, and some downward movement of the surrounding sedimentary rocks may have occurred.

Magmas are less viscous than salt and, like salt, are less dense than the rocks at depth. Thus, the same mechanism, which may account for the rise of salt domes, may also account for the rise of some magmas. The internal structure of salt domes is somewhat similar to the internal structures of some intrusive igneous rocks, further suggesting a similarity. The heat of igneous rocks could weaken the surrounding rocks and thus assist the rise.

A possible source of pressure within the magma is the expansion of the volatiles. This was discussed earlier in connection with the surface manifestations of volcanism. In the present context, the volatiles in the newly formed magma would tend to expand as the magma rises because the confining pressure (weight of overlying rocks) is reduced. This might crack the enclosing rocks, allowing the magma to rise further.

Expanding gas bubbles might also be able to erode the enclosing rocks. This principle is used in a type of drilling called *jet piercing,* and may be the mechanism by which some volcanic rocks reach the surface. Possible examples are the cylindrical, pyroclastic-filled conduits found in some volcanic areas.

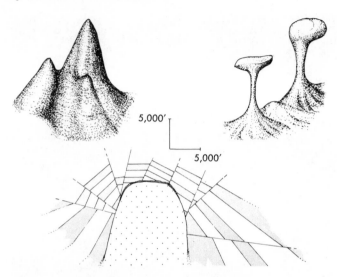

Figure 3-32. Salt domes showing some common shapes and the fractures caused in the surrounding rocks by their rise.

The classic explanation of the emplacement of batholiths in the crust is *stoping.* The name comes from the mining term for the process of removing ore overhead. The hot magma causes blocks of the overlying enclosing rock to be broken off and fall through the melt. Such a process requires a low-viscosity magma and would allow much contamination of the magma by reaction with the stoped blocks. Most batholiths contain many inclusions, probably of the enclosing rocks, giving some credence to this hypothesis. (See Figs. 3-33 and 3-34.)

Another possible way that magma can work its way through the crust is to react with the enclosing rocks. The magma could, by chemical reaction, dissolve or partially dissolve the rocks above the magma. Some melting or partial melting is also possible. For melting to occur, a similar amount of crystallization would have to occur elsewhere in the magma because magmas do not have enough excess heat energy to do much melting. Excess heat energy would have to be in the form of high temperature, but lava at the surface is generally very close to its melting point, indicating that, in general, magmas do not have excessively high temperatures. All of these processes would tend to change the composition of the magma by assimilation of parts of the enclosing rocks. Differentiation would probably also be going on in the magma.

Emplacement of Batholiths

The emplacement of batholiths is a major problem of geology. The great size of a batholith like the Sierra Nevada immediately raises the question: What happened to the large volume of rocks that the batholith displaced? Traditionally this question has been answered by two opposing points of view. One school of thought holds that the batholith made room for itself by a combination of stoping and pushing aside the pre-existing rocks. The many included blocks, presumably of prebatholith rocks, found in some batholiths, and the zone of the deformed earlier rocks that surrounds most

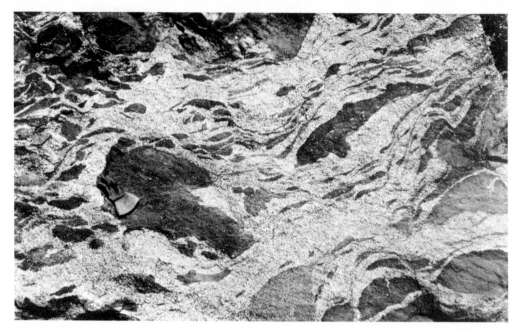

Figure 3-33. Granite with many dark inclusions. The glove shows the scale. Photo by F. C. Calkins, U.S. Geological Survey.

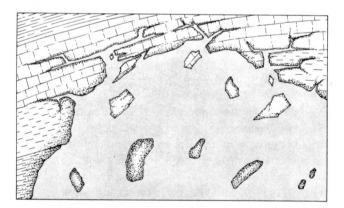

Figure 3-34. Stoping. Forceful injection of the magma breaks fragments from the roof that sink through the magma, reacting with it.

batholiths are the main evidence for this viewpoint. The opposing viewpoint is that the batholith was created more or less in place by the metamorphism of the pre-existing rocks that accompanied the mountain-building event during which the batholith was emplaced. This possibility is included in the rock cycle, Fig. 3-1. The evidence for this theory is the metamorphic rocks that surround most batholiths. In some cases these zones of high-grade metamorphic rocks are very wide, and in other cases the heat from the crystallizing magma is enough to account for the smaller halo of metamorphism. However, recent studies show that most batholiths have different isotope compositions than upper crustal rocks, and so could not have formed by the metamorphism of such rocks.

It seems unreasonable to think that a huge batholith the size of the Sierra Nevada was all emplaced at the same time. Recent studies have shown that this batholith is composed of a number of much smaller plutons. The finer-grained margins of these plutons that resulted from more rapid cooling indicate that at no time was the entire batholith molten. Radioactive dates indicate that the emplacement of the batholith took place over a few tens of millions of years. (See Fig. 3-35.)

Because of their large surface area, batholiths have traditionally been considered to be thick bodies extending deep into the crust. Some recent seismic and gravity studies suggest that batholiths may be relatively thin features. Figures 3-6 and 3-36 show these two interpretations of batholiths.

All of these data have been incorporated into one recent theory. This theory holds that the batholith is formed by the bubble-like rise of the detached pods of magma generated in the upper mantle or deep crust. These pods of magma form a thin batholith near the surface. At some places the magma reaches the surface and forms extensive volcanic rocks, and some batholiths intrude volcanic rocks that are similar in composition to the main batholith. These magma pods tend to spread out near the surface and so might account for the deformation that surrounds many batholiths. In passing through the crustal rocks, the magma would react with those rocks and could assimilate some of them by reaction and partial melting. In any case, the rocks through which the magma passes would be metamorphosed. Erosion of the thin layer of granitic rocks would expose an extensive metamorphic terrane similar to that found in many mountain ranges. Thus this theory would explain most of the data of batholiths and also explain the origin of extensive areas of metamorphic rocks. Like batholiths, these metamorphic terranes are difficult to explain because the known rise in temperature with depth is not enough to account for the amount of metamorphism.

Many aspects of the batholith problem are related to the structural significance of batholiths and will be discussed further in Chapter 14.

Pegmatites and Quartz Veins

The water and other volatiles (gases) in magma may also be involved in the formation of pegmatite and ore veins. When a deep intrusive body, such as a batholith, is emplaced, the rim is quickly chilled and solidifies first. This is commonly recorded by the fine-grained margins of an intrusive body. The volatiles are thus trapped inside the intrusive body; and as it solidifies, presumably from the margin inward, the volatiles are confined to a smaller and smaller volume in the center of the body. The residual fluid is rich in the elements that form the last minerals on the Bowen series, especially quartz, feldspar, and mica, as well as less common elements that do not fit into the rock-forming minerals. This water-rich residual fluid eventually escapes, perhaps as a result of the increase in pressure that can result when the volatiles are concentrated in

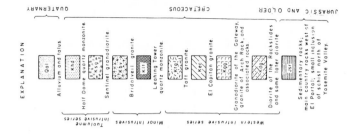

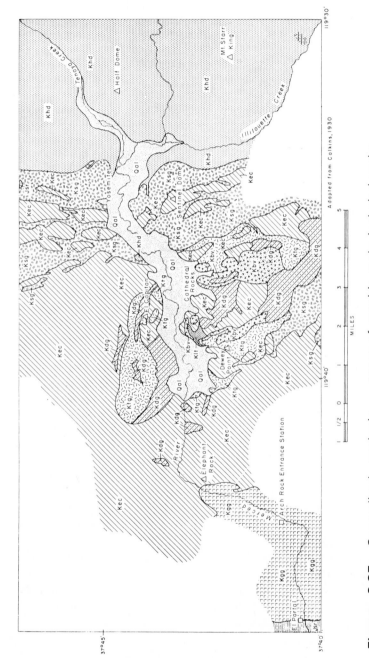

Figure 3-35. Generalized geologic map of Yosemite Valley in the Sierra Nevada, California. The Sierra Nevada batholith is made of many smaller intrusive bodies as shown on this map. Granodiorite and quartz monzonite are types of granitic rocks included under the term granite in this text. From F. C. Calkins and D. L. Peck, California Division of Mines and Geology, *Bulletin* 182, 1962.

80

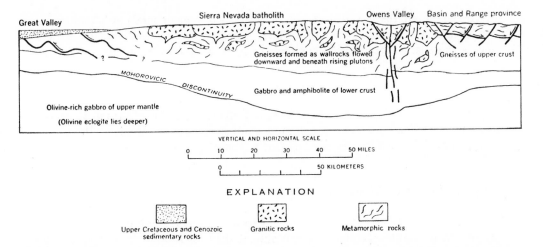

Figure 3-36. Hypothetical cross-section through the Sierra Nevada, based on the idea that the granitic rocks of the batholith moved through the metamorphic rocks to their present positions. This cross-section is based in part on seismic data. Compare with Figure 3-6 that shows a thicker batholith. From Warren Hamilton and W. B. Myers, U.S. Geological Survey, *Professional Paper 554C, 1967.*

a small volume. The increase in pressure may cause cracks that are filled by this material (Fig. 3-37). Contraction due to cooling may also help to cause the cracks. The fluid has a low viscosity because of the volatiles, and this increases the mobility of the ions. As a result, very large crystals form. The resulting very-coarse-grained rocks are called *pegmatites,* and they may have the same composition as the enclosing magma or may contain unusual minerals. In the latter case the fluids are believed to concentrate the minor elements to form the unusual minerals. Pegmatites commonly occur as dike-like masses in or near intrusive bodies. They are actively sought because they contain large and, often, beautiful minerals, such as beryl, tourmaline, topaz, fluorite, all of the rock-forming minerals, and many more.

The residual fluids are probably also responsible for the formation of quartz-metal veins, but the process is not yet understood. The last fluid in a batholith would be rich in quartz and in the elements such as copper, lead, zinc, gold, silver, and perhaps sulfur,

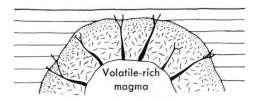

Figure 3-37. Cross-section of a crystallizing batholith showing origin of veins. Crystallization has proceeded inward from the fine-grained, quickly chilled margin. The volatiles are concentrated in the still molten center. The increased pressure causes cracks to form that are filled by the metal-bearing quartz-rich volatiles.

that do not fit into the crystal structures of the rock-forming minerals. A batholith may contain only a very tiny percentage of these elements, but the total amount of metal may be quite large. Quartz veins containing metals in the form of sulfide minerals are found around some intrusive bodies. (See Fig. 3-37.) These veins do not grade into pegmatites, so they must form by a somewhat different, although probably related, process.

In still other cases these fluids may cause widespread alteration of the intrusive body and/or the surrounding rocks. Much more study is needed to understand the actions of the magmatic volatiles that are believed to cause all of these diverse effects.

Identification and Interpretation

Identification of the coarse-grained igneous rocks will present no serious problems if one is able to identify the minerals and estimate their relative amounts. The first clue comes from the color of the rock which, following the chart, will suggest which minerals should be looked for. In the use of the term *light color,* it should be understood that any shade of red or pink, no matter how dark, must be considered a light color. In the same way, any shade of green should be considered a dark color. Overall color should not be used in place of good mineral determination, because exceptions are common enough to be troublesome (especially on examinations).

In a sense, the fine-grained, or volcanic, rocks are more difficult to identify than are the coarse-grained rocks because less data, in the form of identifiable minerals, are available; but because of this, the classification is less rigid. The procedure is similar in that color is used as the first step and, if no porphyritic crystals are present, the only step. If the porphyritic crystals can be identified, they are used in the same way as are the identifiable minerals in coarse-grained rocks; but because all potential minerals may not be present, color must be considered in deciding on a name. Care must be used not to mistake filled gas holes for porphyritic crystals. In many cases, it is necessary to have a chemical analysis in order to name a fine-grained rock accurately. Also, the volcanic rocks are not simply fine-grained equivalents of the coarse-grained rocks, as the chart might lead one to believe, but contain some small but distinct variations in composition, which chemical analyses will detect.

On the chart shown as Fig. 3-38, granite is used for most of the coarse-grained, quartz-bearing rocks. A number of other names are also in use for these rocks, depending on whether they contain one feldspar, two feldspars, or mixed feldspar (perthite). Diorite generally has more dark minerals than does granite, and the feldspar is plagioclase. Quartz diorite is distinguished from granite in that it contains only plagioclase feldspar. The latter distinction, however, is not always easy to make.

Understanding the formation of a rock is much more important than simply naming it. It is necessary to discuss classification so that the meaning of rock names and the relationships among the various rock types are understood. Therefore, when studying a rock, try to interpret its origin, following mainly the discussion in the section on texture and mode of origin. Thus the coarse texture of a granite implies that the specimen came from a large, slowly cooled, intrusive body, but a pumice sample must have come from a volcano, perhaps from an explosive eruption. When actual outcrops are studied, all of the information discussed here can be used in deciphering the origin and history of the rocks.

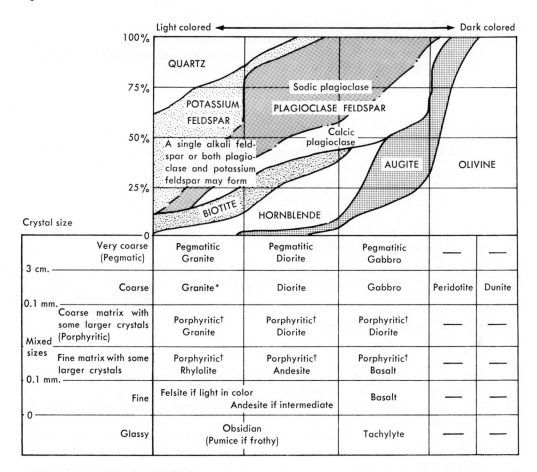

Figure 3-38. Classification of the igneous rocks. The diagram shows the range in mineral composition within the various rock types.

Minor accessory minerals are omitted.

*Granite is used in the widest sense. Several names are in common use, depending on the type of feldspar.

†Modifier porphyritic is generally omitted unless crystals are very prominent.

°Coarse and fine are used in the sense that phaneritic and aphanitic are used in petrography.

QUESTIONS

3-1. Define igneous rock.

3-2. Describe the crystallization of a basaltic melt.

3-3. Describe the crystallization of plagioclase in a magma.

3-4. What happens when a crystal of calcic plagioclase is melted?

3-5. Is the classification of most igneous rocks genetic? Discuss.

3-6. How can a sill be distinguished from a flow?

3-7. What cooling history does a porphyritic texture imply?

3-8. What is a batholith; what is the usual composition and what problems do batholiths present?

3-9. Name a concordant igneous body.

3-10. In what type of occurrence (sill, flow, batholith, etc.) would you expect to find glassy, fine-grained, porphyritic, and coarse-grained igneous rocks?

3-11. Describe the formation of pumice.

3-12. Describe the types of volcanoes. Include how they are formed and which is steepest and why.

3-13. Describe the formation of ore veins.

3-14. Igneous rocks are classified on the bases of _____ and _____.

3-15. What minerals are most likely to be visible in a felsite porphyry?

3-16. How does pegmatite form?

3-17. What are the main volcanic gases?

SUPPLEMENTARY READING

GENERAL

Fenton, C. L., and M. A. Fenton, *The Rock Book.* New York: Doubleday & Co., Inc., 1940, 357 pp.

Pough, F. H., *A Field Guide to Rocks and Minerals.* Boston: Houghton Mifflin Co., 1955, 349 pp.

Spock, L. E., *Guide to the Study of Rocks.* New York: Harper & Row, Pubs., 1962, 298 pp.

IGNEOUS PRINCIPLES

Mason, Brian, *Principles of Geochemistry* (3rd ed.), Chapter 5, "Magmatism and igneous rocks," pp. 96-148. New York: John Wiley & Sons, Inc., 1966.

Tuttle, O. F., "The Origin of Granite," *Scientific American* (April, 1955), Vol. 192, No. 4, pp. 77-82. Reprint 819, W. H. Freeman and Co., San Francisco.

VOLCANOES

Bullard, F. M., *Volcanoes: In History, in Theory, in Eruption.* Austin, Texas: University of Texas Press, 1962, 441 pp.

Eaton, J. P., and K. J. Murata, "How Volcanoes Grow," *Science* (October 7, 1960), Vol. 132, No. 3432, pp. 925-938.

Macdonald, G. A., "Volcanology," *Science* (March 10, 1961), Vol. 133, No. 3454, pp. 673-679.

Williams, Howel, "Volcanoes," *Scientific American* (November, 1951), Vol. 185, No. 5, pp. 45-53. Reprint 822, W. H. Freeman and Co., San Francisco.

Chapter 4

Weathering and Sedimentary Rocks

Sedimentary rocks are formed at or near the surface of the earth. They are composed of rock fragments, weathering products, organic material, or precipitates. Most are deposited in beds or layers. They comprise a very small volume of the earth, only about 5 per cent of the crust. In spite of their small volume, however, they cover about 75 per cent of the surface. They are formed in a number of different ways, but because the raw materials come from weathering, we will begin there.

Weathering

Mechanical weathering breaks up a pre-existing rock into smaller fragments, and *chemical weathering,* acting on these small fragments, rearranges the elements into new minerals.

Mechanical weathering is done mainly by the forces produced by the expansion of water due to freezing, and to a lesser extent by such things as forces of growing roots (see Fig. 4-1), worms and other burrowing animals, lightning, and work of man. The most important of these methods is the result of the 9 per cent expansion that water undergoes when it freezes. Tremendous forces are produced in this way, and their efficacy is shown by the thick layers of broken rock that form in areas that go through a daily freeze-thaw cycle, many times a year. (See Fig. 4-2.)

Figure 4-1. Sidewalk and curb have been moved by tree roots.

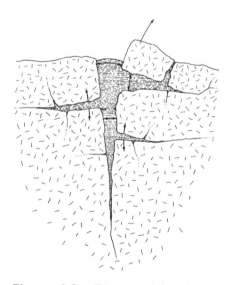

Figure 4-2. Frost wedging. Water in crack freezes from top down because ice is less dense than water (floats) and because the water in a crack is cooled most at the surface. This restricts the water in the crack. Because ice occupies more volume than water, further freezing creates forces due to this expansion. These forces cause further cracking of the rock.

Another type of mechanical weathering is caused by the expansion of rock when erosion removes the weight of the overlying rocks. The outermost layer of the rock may expand enough so that it breaks from the main mass, forming cracks or joints parallel to the surface. (See Fig. 4-3.) This may also occur in boulders released by frost action.

Figure 4-3.   Cracks or joints parallel to the surface are believed to be caused by expansion. The expansion occurs as erosion removes the overlying rocks. The granitic rocks in this view crystal-lized at least a mile beneath the surface. Jointing of this type on a larger scale is conspicuous in many areas of granitic rocks. Photo by G. K. Gilbert, U.S. Geological Survey.

In this case layers like onion skin may form. (See Fig. 4-4.) However, most weathering of boulders in this way is caused by chemical weathering, as noted below.

The products of mechanical weathering include everything from the huge boulders found beneath cliffs to particles the size of silt. The erosion and transportation of these fragments will be discussed in Chapters 7 through 11. These fragments can form rocks such as sandstones, and the shape of the fragments in such a rock can reveal some of the history of the rock. Thus if the fragments are sharp and angular, they probably were buried quickly; but if the fragments are rounded, they were abraded during long transportation. (See Fig. 4-5.)

Water is important in most types of weathering. The role of water is largely determined by its structure. The water molecule is composed of an oxygen atom and two smaller hydrogen atoms that are attached to the oxygen about 105 degrees apart, as shown in Fig. 2-6. The hydrogen has a slight residual positive charge, and the oxygen a similar weak negative charge. Because of these charges the molecules tend to form in groups of two to eight. More groups are formed near the freezing point than in liquid water, and because the groups take up more space, water expands on freezing. These residual charges also make water an excellent solvent.

The rate of chemical weathering depends in general on the temperature, the surface area, and the amount of water available. Except in cold or very dry climates, it generally

Figure 4-4. Expansion of surface layers can produce spherical masses with onion skin layers. Such weathering probably begins on joint cracks; and although splitting-off of the layers is mechanical weathering, the expansion of the layers is probably caused largely by chemical weathering. This photo of basalt in Puerto Rico by C. A. Kaye, U.S. Geological Survey.

Figure 4-5. Sand grains showing different degrees of roundness. The rounded sand grains were probably shaped by abrasion during longer transportation than the angular grains. If the fragments are of different materials, the softer will be rounded first.

keeps up with mechanical weathering, and the two can be separated only in concept. Mechanical weathering provides the large surface area necessary for chemical activity to take place at the surface of the earth. (See Fig. 4-6.)

In following chemical weathering, we will see what happens to the elements in the main rock-forming minerals. The main reactions involved are *oxidation, hydrolysis,* and *carbonation.* Oxidation is reaction with oxygen in the air to form an oxide.

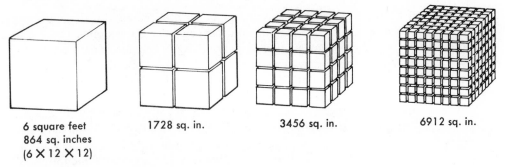

6 square feet
864 sq. inches
(6 × 12 × 12)

1728 sq. in.

3456 sq. in.

6912 sq. in.

Figure 4-6. Surface area versus particle size. Shows how mechanical weathering increases the surface area and thus helps to increase chemical weathering. A cubic foot is shown in each case.

Hydrolysis is reaction with water, and carbonation is reaction with carbon dioxide in the air to form a carbonate. The carbonation reaction begins with the uniting of carbon dioxide (CO_2) and water (H_2O) to form carbonic acid (H_2CO_3), or

$$CO_2 + H_2O \rightarrow H_2CO_3$$

This acid plays an important role in many weathering reactions. In all of these reactions, substances are added so that the total volume is increased. The details of how these reactions take place are not well understood and are an important area of current research in soil science, a relatively new branch of science. The reactions take place slowly and are probably influenced by the composition of the water films that surround the particles and by the organic materials present (particularly the acids). The importance of organic materials in weathering and soil formation is covered in the next section. One of the processes is *leaching,* which is the removal of soluble materials by ground water.

Air pollutants such as carbon dioxide and sulfates, released by burning fossil fuels and industrial processes, cause accelerated weathering of building stone and other materials in cities. The ultimate fate of many pollutants is not known, but it appears that some are absorbed and decomposed by soil.

The rain water involved in most weathering is not pure water. In addition to carbon dioxide, it contains dissolved salts. Atmospheric water vapor, in general, will not condense to form clouds or rain without a nucleus to condense around. Dust particles provide some of these nuclei, but most are tiny crystals of salts from the ocean. Spray from ocean waves evaporates, and the tiny crystals of salts formed in this way are carried away by the wind. The salts formed when sea water is evaporated are described in a later section. Near the ocean, rain water has the same chemical composition as sea water, but is very much diluted. Farther inland, the composition is variable because the winds transport the various salts from the ocean differently. The chemical composition of rain water does have an effect on the clays produced by weathering, but little more can be said on the complex subject of clay mineralogy here. However, the salts in rain water accumulate in lakes that have no outlet to the sea, such as Great Salt Lake.

Chemical weathering is most intense on grain surfaces; because of the increase in volume during chemical weathering, this trait can cause disaggregation of a rock. This process may account for fragmentation of rocks in areas where mechanical weathering by frost action does not occur. Laboratory experiments show that expansion and contraction from daily heating and cooling are not enough to cause disintegration, even on a desert.

Climate affects the topography produced by weathering. Dry areas where mechanical weathering predominates generally have bold, angular cliffs (see Fig. 4-7), and humid areas have rounded topography. Limestone, which is made of the soluble mineral calcite, is readily dissolved by carbonic acid. Thus it is a resistant rock and forms bold outcrops in dry areas, but it is rapidly weathered and tends to form valleys in humid regions. In warm, moist climates, weathering may extend to depths of over 400 feet. Maximum weathering probably takes place just below the ground surface—the depth at which fence posts tend to break.

Figure 4-7. Angular topography in an arid area. Weathering has accentuated minor differences among the sedimentary layers. Bryce Canyon National Park, Utah. Photo from Union Pacific Railroad.

Chemical weathering can be defined also as changes taking place near the surface that tend to restore the minerals to equilibrium with their surroundings. The primary minerals form, in most cases, at elevated temperature and pressure and are in equilibrium under those conditions. When these minerals are exposed at the surface, they are in conditions far from those under which they formed; therefore, they react to form new minerals which will be in equilibrium. One can then predict that the highest-temperature minerals will probably be affected first in chemical weathering. Thus it turns out that Bowen's reaction series (Fig. 3-2), which shows the sequence of mineral formation with falling temperature, also shows the relative stability to weathering, with olivine the first to be affected and quartz the least affected. Organic soil acids can in some cases upset this relationship.

The weathering of potassium feldspar will illustrate chemical weathering. An overall reaction can be written as

2 potassium feldspar + carbonic acid + water $\longrightarrow$

$$2KAlSi_3O_8 \quad + \quad H_2CO_3 \quad + H_2O \longrightarrow$$

clay + potassium carbonate + 4 silica

$$Al_2Si_2O_5(OH)_4 + K_2CO_3 \qquad + 4\ SiO_2$$

The feldspar is broken down by a process in which hydrogen replaces potassium. The hydrogen can come from water, from carbonic acid, or from the hydrogen ions that surround plant roots. The first two sources are inorganic, and the last is an example of biological weathering. The potassium released in the latter way is taken up by the plant. The clay mineral shown in the illustrative reaction is not the only clay mineral that can form. Potassium and magnesium clays are also common and form in some weathering environments. Weathering reactions of other feldspars are similar, differing only in that sodium and calcium are involved in place of potassium.

The weathering of the ferromagnesian minerals is also similar, and magnesium-bearing clays may form in this way. The ferromagnesian minerals are more easily weathered because the iron is easily oxidized, breaking down the mineral.

The products of chemical weathering of the common minerals can be summarized as follows:

> *Quartz*—very little chemical weathering, sand grains produced.
> *Feldspars*—
>> Potassium feldspar ($KAlSi_3O_8$).
>> K by carbonation produces K_2CO_3.
>> Al,Si,O by hydrolysis produce clay ($Al_2Si_2O_5(OH)_4$).
>> SiO_2 is released as soluble silica and colloidal silica.
>> Plagioclase feldspars—produce the same end products except that $CaCO_3$ and Na_2CO_3 are produced instead of K_2CO_3.
> *Muscovite*—produces the same products as potassium feldspar.
> *Ferromagnesian minerals*—depending on composition, produce the same products plus iron oxides—hematite Fe_2O_3 (red iron oxide) or limonite $FeO\,(OH)$ (yellow iron oxide).
>> Magnesium forms soluble magnesium carbonate, and some is used in producing clay minerals.

In summary, the products of weathering and their ultimate disposition are as follows: (See also Fig. 4-8)

Weathering product	*Disposition*
Clay	Forms shale.
Silica	Forms chert and similar rocks.
Potassium carbonate (K_2CO_3)	Some is transported to ocean, some is used by plant life, and some is adsorbed on or taken into certain clays.
Calcium carbonate ($CaCO_3$)	Forms limestone.
Sodium carbonate (Na_2CO_3)	Dissolves in ocean.
Iron oxides	Form sedimentary deposits of iron oxide.
Magnesium	Some replaces calcium in limestone to form dolostone, and some is transported to ocean.

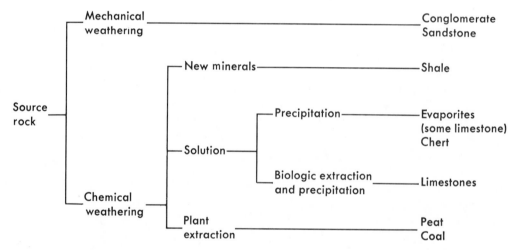

Figure 4-8. Simplified chart showing origins of sedimentary rocks.

From this brief outline of chemical weathering it is possible to make some interpretation of sedimentary rocks. As might be expected, most sandstones are composed largely of quartz because it is the mineral most resistant to weathering. Some sandstones contain much feldspar, and such rocks formed under conditions that did not permit chemical weathering to complete its job. This situation could have developed several ways. Accumulation might have been too rapid to permit complete weathering, or the climate may have been too dry or too cold. Certainly there are other possibilities, but this serves as an example of how sedimentary rocks can be interpreted.

Soils

Soils are of great economic importance. They can be said to be the bridge between life and the inanimate world. Soil, as the term is used here, is the material that supports plant life, and so supports all life. Only lichen and primitive plants, such as silver sword in Hawaii, can get nutrition directly from rock without soil. The development of soil is a complex interplay of weathering and biologic processes. The study of soils is a rather new science, and many of the processes by which soils develop are not fully understood.

The importance of soil to life is shown by the composition of some foods. Steak contains about 12½ per cent phosphorus, iron, and calcium by weight. Thus each pound of steak contains about two ounces of rock material obtained from soil by plants. A much larger amount of rock had to be weathered to release these two ounces. Plants use much weathered rock material. For example, an acre of alfalfa yields about four tons of alfalfa per year. To produce this alfalfa, over two tons of rock must be weathered to obtain the phosphorus, calcium, magnesium, potassium, and other elements in the alfalfa.[1] This corresponds to about 0.007 inches of rock weathered. The average rate of erosion is estimated to be about one inch per 9,000 years, so this is about 5.5 times the annual erosion rate. This "uses" soil at a very high rate, so fertilizers must be added. Thus soils can be used faster than they form, a fact not always taken into account in agricultural economics.

There are many different types of soils. At first it was thought that the parent rock determined the soil type. We now know that climate is more important than parent rock; but the situation is complex, and other factors, especially the type of vegetation, are involved. Soils can be described as mature and immature. On hillslopes, erosion may prevent soils from reaching maturity and the transported soil may accumulate in valleys. (See Fig. 4-9B.) It apparently takes at least a few hundred years for a soil to develop. For this reason soil conservation is very important. The range in rate of soil formation is great. In 45 years, 14 inches of soil formed on pumice from the eruption of the volcano Krakatoa. At the other extreme, in many places no soil has formed on glacially polished surfaces in about 10,000 years. (See Fig. 10-9.)

In most soils three layers, or *horizons,* can be recognized. They are the topsoil, the subsoil, and the partly decomposed bedrock. In addition, in some areas, especially forests, a layer of organic material may form at the surface. (See Figs. 4-9 and 4-10.) The material that gives the upper parts of many soils their dark color is *humus.* Humus

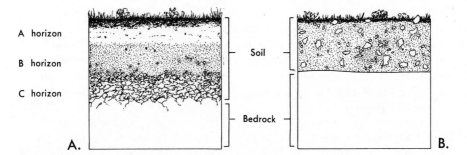

Figure 4-9. Soil profiles. A. Residual soil with well-developed horizons or layers. B. Transported soil without layering.

[1]W. D. Keller, "Geochemical weathering of rocks: source of raw materials for good living," *Journal of Geological Education,* Vol. 14, No. 1, February, 1966, pp. 17-22.

Figure 4-10. Soil profile revealed in a pit. Photo by H. E. Malde, U.S. Geological Survey.

is formed by the action of bacteria and molds on the plant material in the soil. Chemical weathering releases some of the materials used by plants, and humus also provides food for plants.

Soils form as a result of the weathering processes just discussed and the effects of the organic material present, mainly humus. One of the most important processes in the formation of soils is the leaching of material from the top soil and the deposition of this material in the subsoil. Water reacts with humus to form acid that is very effective both in leaching and in causing chemical weathering. Earlier, the similar role of carbonic acid, formed by carbon dioxide and water, was discussed. Carbon dioxide is also formed in soil from the decay of organic material.

We can now consider a few of the more important types of soils. In the United States the soil of the humid east is different from that of the drier west. The dividing line is approximately at the 25-inch annual rainfall line and runs almost north-south near 97° longitude. (See Fig. 4-11.) The characteristics of the soils about to be discussed are summarized in Table 4-1.

The soil that predominates in eastern United States is called *pedalfer* (*pedon,* Greek for soil, and Al, Fe, the symbols for aluminum and iron). Humus is well developed in the temperate, moist climate. The abundant rain water becomes strongly acid by reaction with the humus and leaches the topsoil. Aluminum and iron leached from the topsoil are deposited in the subsoil, giving it a brown (limonite) color. Soluble materials such as calcite or dolomite are leached from this type of soil and are generally removed by the ground water. These soils are quite acid, and clays are well developed. Ground water carries the clay to the subsoil so that topsoil may be somewhat sandy, due to

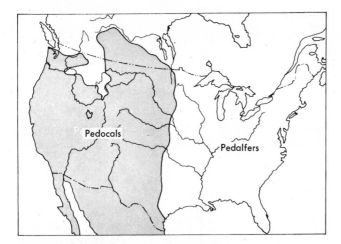

Figure 4-11. Map of major soil types in the United States. The dividing line is approximately at the 25-inch annual rainfall line except in the northern Rocky Mountains where low temperature affects the soil type.

resistant quartz fragments. Pedalfers generally are best developed under forests. Conifer forests produce a thick layer of litter on the forest floor and so develop more humus. Therefore, the soil is very acid and its development is accentuated; the resulting soil is light gray in color. It is in the pedalfers that calcite, generally called agricultural lime or limestone, is spread on the soil to combat its acidity.

In the drier west, less humus develops so the limited amounts of ground water are much less active chemically. Chemical weathering is also much slower in this climate so there is less clay than in pedalfers. Under these conditions leaching is not complete, and the soluble materials, especially calcite, are deposited in the subsoil because the ground water evaporates before it can remove them. Such soils are called *pedocals* (*pedon* for soil, and *cal* for calcite). The calcite forms as whitish material in the soil, called *caliche* (Spanish for lime). Caliche may accumulate in the topsoil as well as in the subsoil. This can occur when the ground water moves upward through capillary openings because of the evaporation at the surface. Pedocals tend to develop under grass- or brush-covered areas.

Soil formation in tropical climates is not fully understood, in part because so few areas have been well studied. The soil in rain forests is somewhat similar to the pedalfers. The soil of the grass- and tree-covered savannas, however, is much different. Apparently because of the seasonal heavy rainfall, leaching is the dominant process. Everything, including silica and clays, is leached from the soil and carried away by ground water except iron and aluminum, which form hydrated oxides. Apparently, at the high temperatures, bacteria destroy the humus so that the ground water is not acid and cannot remove iron and aluminum. The soils produced in this way are called *laterites.* Laterite is from the Latin word for brick, and its reddish-brown color is similar to that of a brick. Laterites do not form distinct layers as other soils do. As

Table 4-1. Generalized summary of soil types. See text for details.

Climate	Temperate humid (>25″ rainfall)	Temperate dry (<25″ rainfall)	Tropical Savanna (heavy seasonal rainfall)	Extreme Arctic Desert (hot or cold)
Vegetation	Forest	Grass-brush	Grass and tree	Almost none, so no humus develops
Typical area	Eastern U.S.	Western U.S.	———	———
Soil type	Pedalfer	Pedocal	Laterite	No real soil forms because there is no organic material. Chemical weathering is very slow.
Topsoil	Sandy; light colored; acid.	Commonly enriched in calcite; whitish color.	Enriched in iron (and aluminum); brick red color. *(Zones not developed)*	
Subsoil	Enriched in aluminum, iron, clay; brown color (limonite).	Enriched in calcite; whitish color.	All other elements removed by leaching. *(Zones not developed)*	
Remarks	Extreme development in conifer forests because abundant humus makes ground water very acid. Produces a light gray soil because of removal of iron.	*Caliche* is the name applied to the calcite-rich soils.	Apparently bacteria destroy the humus so no acid is available to remove iron. Tropical rain forests develop soils similar to pedalfers.	

might be expected, they are poor soils on which to grow crops, as shown by the many unsuccessful attempts to develop new farmlands in emerging tropical countries. Most laterites are iron-rich; however, in some areas they are mainly aluminum. Such a mixture of hydrated aluminum oxides is called *bauxite* and is a valuable ore. Bauxite probably forms only from the weathering of aluminum-rich rocks that have little iron, such as some feldspar-rich igneous rocks. At some places laterites are mined for iron ore and, at a few places, for their nickel or manganese; here the parent rock probably was rich in these elements.

In the extreme climates, such as arctic and desert, very little chemical weathering occurs and there is almost no organic material; therefore, no real soil can develop.

Lithification

The transformation of a sediment into a rock is called *lithification*. Several processes are involved. *Cementation* is the deposition by ground water of soluble material between the grains and, as might be expected, is most effective in coarse-grained, well-sorted, permeable rocks. The main cementing agents are

Calcite—recognized by acid test. (Do not mistake the effervescence of cement for the overall effervescence of a calcite-bearing rock.)

Silica—generally produces the toughest rocks.

Iron oxides—color the rock red or yellow.

Other cements, such as dolomite, are possible. In certain poorly sorted rocks, clay, which generally colors the rock gray or green-gray, may be thought of as forming the cement. *Compaction* is an effective lithifier of fine-grained rocks such as shale and siltstone. Compaction generally comes about from the weight of overburden during burial, and the reduction in volume due to compaction and squeezing out of water may amount to over 50 per cent in some shales. *Recrystallization* probably is important in producing chert and in transforming calcite-bearing mud into limestone. Recrystallization produces an interlocking texture.

Classification and Formation

As with the igneous rocks, it is more important to interpret the formation of these rocks than merely to name them. Knowing the origin of the materials that compose a sedimentary rock and understanding the origin of its sedimentary features will permit such interpretation.

According to the way they were formed, sedimentary rocks can be classified as marine, lacustrine (lake deposited), glacial, eolian (wind deposited), fluvial (river deposited), etc. They can be classified also by composition as limestone, chert, quartzose sandstone, etc., or by their mode of origin as clastic, chemical precipitate, or organic. In practice, these are blended to produce a practical classification as follows:

Clastic rocks—composed of rock fragments or mineral grains broken from any type of pre-existing rock. (See Fig. 4-12.) They are subdivided according to fragment size. Commonly, sizes are mixed, requiring intermediate names such as sandy siltstone. They are recognized by their clastic texture. The fragments originate from mechanical weathering. The agents of erosion, transportation, and deposition that bring these fragments together and so form the clastic sedimentary rocks are described in Chapters 7 through 11.

Clastic texture

Interlocking texture. Developed in chemical precipitates such as some limestones and in many igneous and metamorphic rocks.

Figure 4-12. Recognition of textures is very important in the identification and interpretation of rocks.

Composition is not used in the classification of clastic sedimentary rocks because the composition may be affected by several factors that cannot be satisfactorily evaluated from study of the rock alone. The examples of this are almost endless, but a few simple ones will illustrate. A quartz sandstone may be formed if the source area is mainly quartz (perhaps an older quartz sandstone), or a quartz sandstone may be the result of prolonged weathering, leaving only quartz fragments. Large amounts of feldspar in a sandstone may imply rapid deposition and burial before chemical weathering could decompose the feldspar, or it might imply a cold climate in which chemical weathering is very slow.

Size	Sediment	Rock
Over 2 mm	Gravel	*Conglomerate* — generally has a sandy matrix. May be subdivided into roundstone and sharpstone conglomerate (*breccia* is a synonym) depending on the fragment shape. (See Fig. 4-13.)

$1/_{16}$-2 mm	Sand	*Sandstone* — recognized by gritty feel. (See Figs. 4-14 and 4-15.) Generally designated as coarse, medium, or fine if well sorted. They have also been subdivided on the basis of composition. The more important types are
		  Quartzose sandstone (mainly quartz)
		  Arkosic sandstone (arkose) — over 20 per cent feldspar.
		  Graywacke — poorly sorted, with clay or chloritic matrix.

$1/_{256}$-$1/_{16}$ mm	Silt	*Siltstone (mudstone)* — may be necessary to rub on teeth to detect grittiness, thus distinguishing it from shale.
Less than $1/_{256}$ mm	Clay	*Shale* — distinguished from siltstone by its lack of grittiness and its *fissility* (ability to split very easily on bedding planes). Rocks composed of clay that lack fissility are called *claystone*.

The actual formation of shale is somewhat more complex than indicated above. Much of the clay that gives a shale its fissility apparently develops after deposition. This is suggested by the lack of geologically young shale. The young fine-grained clastic rocks are mainly mudstones.

Non-clastic rocks—formed by chemical precipitation, by biologic precipitation, and by accumulation of organic material. These processes extract specific materials from their surroundings, generally sea water, and precipitate these substances, forming rocks. The rocks are classified mainly by composition. As with the clastic rocks, these rocks are commonly mixed, both among themselves and with the clastic rocks.

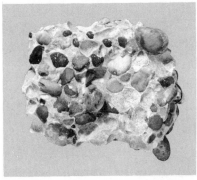

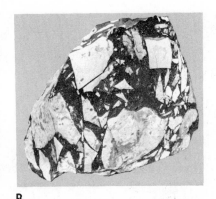

A.   B.

Figure   4-13. Conglomerate.   A.
Rounded pebbles in a sandstone matrix.
B. Sharpstone conglomerate (breccia).
The angular pebbles in this rock were

probably not transported far compared
with the pebbles in A. Photos from
Ward's Natural Science Establishment,
Inc., Rochester, N.Y.

Figure 4-14. Sandstone. The bedding is too thick to show in this specimen. Photo from Ward's Natural
Science Establishment, Inc., Rochester,
N.Y.

Limestone is composed of calcite. It is recognized by effervescence with dilute
hydrochloric acid. It is generally of biologic origin and may contain fossils. (See
Fig. 4-16A.) A rock composed mainly of fossils or fossil fragments is called
coquina. (See Fig. 4-16B.)

Limestone can form in many ways as shown in Table 4-2. Most limestone
probably originates from organisms that remove calcium carbonate from sea
water. The remains of these animals may accumulate to form the limestone
directly, or they may be broken and redeposited. Calcite can also be precipitated

Figure 4-15. Sandstone outcrop. Photo from New Mexico State Tourist Bureau.

Table 4-2. Origins of the more important non-clastic sedimentary rocks.

| | Biologic | Chemical | | Remarks |
		Evaporite	Other	
Limestone	×	× rare	×	
Dolostone		× rare		Forms mainly by replacement.
Chert	×		× precipitation from volcanic silica	Most biologic chert recrystallizes from opal.
Gypsum		×		
Rock salt		×		
Chalk	×			
Diatomite	×			
Coal	×			

chemically. The amount of carbon dioxide in sea water controls the amount of calcite that can remain in solution. If the amount of carbon dioxide is reduced by warming the water, as would occur in shallow tropical water, calcite may be precipitated. Marine plants also remove carbon dioxide from sea water and aid in calcite precipitation. Bacteria can also aid chemical precipitation of calcite

A.

B.

Figure 4-16. Limestone. A. Fossilifer- stone composed almost entirely of
ous limestone. A fine-grained white fossil fragments. Photos from Ward's
limestone with abundant fossil frag- Natural Science Establishment, Inc.,
ments. B. Coquina, a variety of lime- Rochester, N.Y.

by making the water more alkaline. Decaying organic material can increase the
amount of carbon dioxide and so retard precipitation of calcite. Limestone is
presently being formed by chemical precipitation on the shallow Bahama Banks
where the factors just discussed are favorable.

Dolostone (dolomite) is composed of dolomite. It is recognized by effervescence
after scratching (to produce powder) with dilute hydrochloric acid; it will also
react (without scratching) with concentrated or with warm dilute hydrochloric
acid. Dolostone is generally formed by replacement of calcite, presumably very
soon after burial. The reduction in volume in this replacement may produce
irregular voids and generally obliterates fossils.

Chert is microscopically fine-grained silica (SiO_2). It is equivalent to chalcedony
(see Mineral Table, Appendix A) unless microscopic or other detailed methods
of study are used. It is unfortunate that many names have been applied to the
fine-grained varieties of silica. These names were based on minor differences
such as color and luster that cannot be substantiated by modern methods, and
there is little agreement among mineralogists on their naming. Both chert and
chalcedony may contain opal and fine-grained quartz. Dozens of names have
been applied to chalcedonic silica; of these, a few may be useful, e.g., agate for
banded types, and flint for dark gray or black chert.

Chert originates in several ways. Some may precipitate directly from sea
water in areas where volcanism releases abundant silica. Most comes from the
accumulation of silica shells of organisms. These organisms remove dissolved
silica from sea water to form their shells or skeletons. These silica remains come
from diatoms, radiolaria, and sponge spicules, and are composed of opal. Opal
is easily recrystallized to form chert. Thus much chert is recrystallized, making
the origin difficult to discern. In recrystallization the silica may replace other
materials, and fossils replaced by silica are common. If this occurs in limestone,
beautifully preserved fossils with delicate features intact can be recovered by
dissolving the limestone with acid. Chert may form either beds or nodules.

Gypsum forms from the evaporation of sea water. Rocks formed in this way are
termed *evaporites;* they are discussed below. The identification of gypsum is
described in Appendix A.

Rock salt is composed of halite. It is recognized by its cubic cleavage and its taste.
(See Appendix A.) It is deposited when restricted parts of the sea are evaporated.

Chalk is soft, white limestone formed by the accumulation of the shells of microscopic animals. The shells are composed of calcium carbonate, and chalk is
recognized by effervescence with acid.

Diatomite is a soft, white rock composed of the remains of microscopic plants.
Because the remains are composed of silica, it is distinguished from chalk by
lack of effervescence.

Coal is formed by the accumulation of plant material.

Table 4-3. Composition of sea water.

Element	Per cent	Origin	Rock formed
Chlorine	0.019	volcanoes	rock salt (halite)
Sodium	0.011	chemical weathering	
Magnesium	0.001	chemical weathering	dolomite
Sulfur	0.0009	volcanoes	gypsum
Calcium	0.0004	chemical weathering	limestone (calcite)
Potassium	0.0004	chemical weathering	potassium salts

Evaporite deposits are formed by the evaporation of sea water. Gypsum and rock
salt are the main rocks formed in this way. When sea water is evaporated at surface
temperatures, such as in a restricted basin, the first mineral precipitated is calcite.
Dolomite is the next mineral precipitated, but only very small amounts of limestone
and dolomite can be formed in this way. Evaporation of a half-mile column of sea water
would only produce an inch or two of limestone and dolomite. After about two-thirds
of the water is evaporated, gypsum is precipitated; and when nine-tenths of the water
is removed, halite forms. During the last stages of precipitation, potassium and magnesium salts form. Thick beds of rock salt imply evaporation of large amounts of sea
water. This is discussed in Chapter 6.

The rocks described above are the most important, but by no means the only,
sedimentary rocks. Any mixture of types is possible and, because any weathering
product may form a sedimentary rock, endless variety is possible. Some less abundant
types include economically important deposits, described in Chapter 6, such as iron
oxides, phosphorous rocks, bauxite (aluminum ore), and potash.

Shale, sandstone, and limestone make up about 99 per cent of all sedimentary rocks.
Because feldspars are the most abundant minerals, their main weathering product,
shale, is the most abundant sedimentary rock. Sandstone is next in abundance, and
limestone is last. The abundance of shale is somewhat less than is predicted from the
abundance of clay-forming silicate minerals, suggesting that some clay is deposited in
the deep sea basins. Clay particles are very small and sink slowly; they can be carried
thousands of miles by gentle currents.

Features of Sedimentary Rocks

The most noticeable feature of an outcrop of sedimentary rocks is the bedding, which records the layers in the order of deposition with the oldest at the bottom. In some instances the beds are too thick to show in a small outcrop, and, for the same reason, many hand specimens do not show obvious bedding.

Study of the type of fossils and of sedimentary features can tell much about the environment of deposition and, in some cases, the post-depositional history. Mud cracks on bedding planes record periodic drying and suggest shallow water and, perhaps, seasonal drying (Fig. 4-17). Ripple-marks also suggest shallow water with some

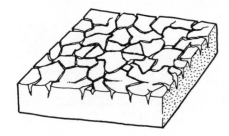

Figure 4-17. Mud cracks, which form as a result of drying mud flats, can be used to indicate top and bottom.

current action, but are also known to form in deep water. Detailed study of ripple-marks can show type and direction of current. (See Figs. 4-18 and 4-19.) Current action can also cause localized scouring or cutting to produce cut-and-fill features. (See Fig. 4-20.) Another type of current feature is cross-bedding. (See Figs. 4-21 and 4-22.) These features can also be used to recognize beds that have been overturned by folding or faulting. Currents can also align mineral grains and fossils and so impart linear internal structures to rocks that can be used to determine the current direction. An example

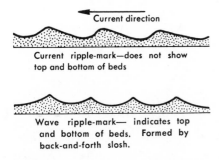

Current direction

Current ripple-mark—does not show top and bottom of beds

Wave ripple-mark— indicates top and bottom of beds. Formed by back-and-forth slosh.

Figure 4-18. Ripple-marks form where currents can act on sediments. Ripple-marks form at right angles to the current direction, and asymmetrical ripple-marks indicate which way the current flowed. To see why wave ripple-marks only can be used to determine top and bottom of beds, turn the page upside-down.

Figure 4-19. Oscillation ripple-marks. Note the presence of ripple-marks in successive layers. Photo by D. F. Demarest, U.S. Geological Survey.

Figure 4-20. Scouring produced by localized currents. A. Initial deposition. B. Channel cut by current. C. Deposition is resumed, filling the channel. Such a feature records an interruption in deposition and can be used to distinguish top and bottom.

Figure 4-21. Development of crossbeds. A. Initial deposition. B. Channel cut by current. C. Channel filled by deposition from one side. D. Normal deposition after channel is filled. Cross-beds can also be used to distinguish top and bottom.

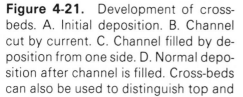

A.   B.

Figure 4-22. Cross-bedding. A. In river-deposited coarse sandstone. Photo by H. E. Malde, U.S. Geological Survey. B. In wind-deposited sandstone in Zion National Park, Utah. Photo by H. E. Gregory, U.S. Geological Survey.

of the reconstruction of a sedimentary basin from interpretation of the features of scattered outcrops is shown in Fig. 4-23.

If a mixture of sediments is suddenly deposited into a sedimentary basin, then the large fragments sink faster than the small ones. This does not violate Galileo's famous experiment, because here we are concerned with falling bodies in a viscous medium in which velocity of descent is controlled mainly by the size. The type of bedding produced by this kind of sedimentation is called *graded bedding* (Fig. 4-24); we can demonstrate this experimentally by putting sediments of various sizes in a jar of water, shaking, and then allowing the contents to settle. Graded beds can form in many ways, such as the sudden influx of sediment by a storm or seasonal flow of rivers.

The sorting of a sediment determines other of its textural features. The *porosity* of a sediment is a measure of its empty or void space, and its *permeability* is a measure of the interconnections of the pore spaces. These properties become quite important when one is concerned with recovering oil or ground water from an underground reservoir. The amount of pore space in sedimentary rocks is quite varied but surprisingly large. If we consider packing of uniform spheres (and most sand grains are rounded), we find that the closest possible packing contains 26 per cent void space, and the loosest, 48 per cent. (See Fig. 4-25.) The permeability will vary depending on the amount of pore space, the sorting, the cementation, and the size of the particles. A shale may have over 30 per cent porosity, but the surface tension of water will prevent flowage through the tiny openings.

Slump features as shown in Fig. 4-26 form if sediments are deposited on a slope or if they are tilted while still soft. Another type of deformation of soft rock is the vertical movement of sediment shown in Fig. 4-27.

Outcrop of Morrison formation   Boundary of physiographic province

A.

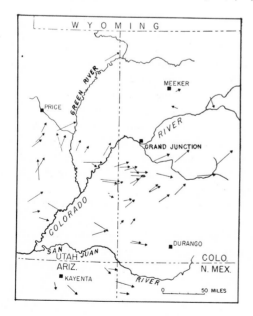

B.

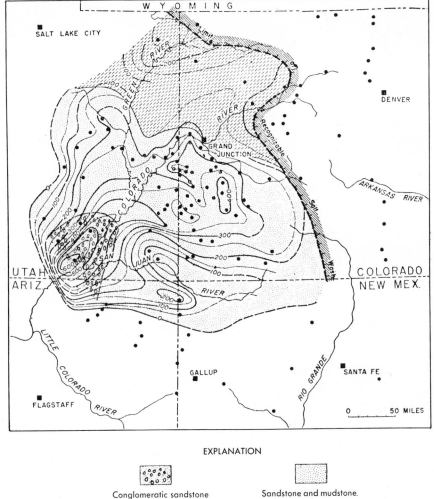

C.

EXPLANATION

Conglomeratic sandstone

Sandstone and mudstone.

Claystone and lenticular
sandstone

Claystone and limestone

• Location of measured section

————100————
Thickness, dashed where inferred,
interval 50 feet

Figure 4-23. Reconstruction of an ancient sedimentary basin from interpretation of the features in scattered outcrops. A. Map of outcrops of the Morrison Formation. B. Current direction inferred from cross-bedding in the Salt Wash Member of the Morrison Formation. The arrows show the downslope direction of cross-beds. Length of the arrows is proportional to the consistency. C. Thickness and sediment type in the Salt Wash Member. Note that the coarser sediments are closer to the source area inferred from both thickness and cross-bedding. From L. C. Craig and others, U.S. Geological Survey, *Bulletin* 1009E, 1955.

Figure 4-24. Graded bedding. Produced by the more rapid settling of coarse material than fine material.

Close packing

Open packing

Figure 4-25. The packing controls the porosity, as illustrated here in the idealized case of spheres.

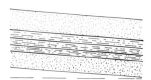

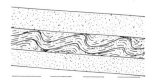

Figure 4-26. Soft sediments can slide down even very gentle slopes, developing slump.

Deposition

Most sedimentary rocks are deposited in oceans, and these marine rocks, easily dated by their contained fossils, record much of the earth's history. The dating and interpretation of such rocks is the topic of Chapter 17 of this book. In the present section a few of the more important marine environments will be considered. The geologic importance of these depositional sites will be covered in later chapters.

One of the interesting observations of marine sedimentary rocks is that at many places individual sedimentary units, such as sandstones, can be traced for hundreds of miles. At other places the beds are irregular and have limited horizontal extent. To see how this comes about we will consider how marine sedimentary rocks are formed. For simplicity, we will limit the discussion to clastic rocks, sandstones and shales. As will be seen, the origin of the fragments that compose these rocks is erosion. These fragments are delivered to the ocean by rivers.

Because most marine sediments consist of material delivered by rivers, most sedimentary rocks form near the continents. Limited sampling and drilling in the oceans show this to be true. If we wish to interpret the thick accumulations of old sedimentary rocks exposed in today's mountain ranges, we should study the continental shelves and the continental slopes. The structure of these areas is described in Chapter 15. (See Fig. 4-28.)

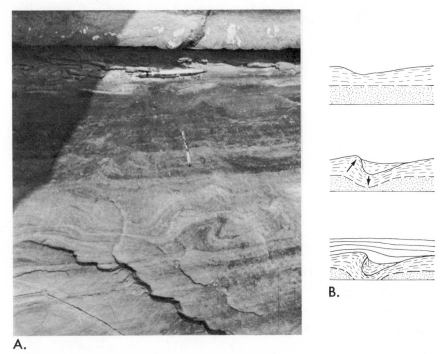

A.

B.

Figure 4-27. A. Photo of deformed bedding. B. Diagram showing how deformation of soft rock forms the structure shown in photo.

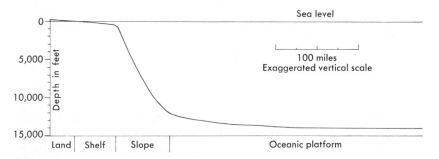

Figure 4-28. Continental margin showing typical continental shelf, continental slope and oceanic platform. The vertical scale is exaggerated.

The continental shelf is part of the continent, and its extent is determined by seismic studies of the crustal structure. However, the 100 fathom, or 600 foot, depth line is generally close to the edge of the continental shelf. The continental shelf is almost flat; it has an average slope of about 10 feet per mile. The break in slope between the shelf and the continental slope generally begins at a depth of about 400 to 500 feet, and the range is from near sea level to 1,500 feet. The average depth of the shelves is near 200 feet. The continental slope extends on the average to a depth of two miles or more, with a slope of two to six degrees. This is a very steep slope compared to the mountain ranges of the earth.

The continental shelves are covered with clastic sediments for the most part. One would expect that, in general, the coarsest material should be found near shore or near river mouths and the finer silt and clay should be farther from the source. (See Fig. 4-29.) This is probably true in a very general way, but the many departures suggest that

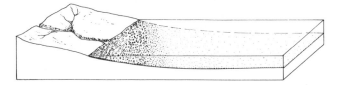

Figure 4-29.   Idealized diagram showing coarse material being deposited
near the shore and finer material further
out.

the situation is not simple. (See Fig. 4-30.) One very obvious reason is that during the recent glacial time the oceans were lower because much water was held in the glacial ice. Apparently, too, tidal and other currents are fairly effective in moving and sorting clastic material on the relatively shallow shelves, as will be shown in Chapter 11. This type of reworking of the sediments may possibly account for the uniform units extending over large areas that are found on the present continents. In any event, pebbles and conglomerate are not uncommon far out on the shelf.

The present coastal plains on the Atlantic and Gulf coasts are underlain by what are apparently continental shelf deposits. Uplift has exposed the landward parts of these sedimentary accumulations, and seismic studies and deep drilling indicate that they extend onto and thicken on the present continental shelf. On the Gulf Coast these deposits are several miles thick, and their features and fossils indicate shallow-water deposition. The latter suggests slow subsidence.

Much less is known about the structure and formation of the continental slopes. They will be discussed more fully in Chapter 15. Limited sampling shows the surface, at least, to be covered mainly with fine sediments, presumably swept off the shelves. A more important process in the building of the slopes may be submarine landslides or mud-slides. Following earthquakes, submarine telephone cables may be broken. The cables do not all break at the time of the earthquake but at later times, farther down the slope. In some cases many cables have been broken and the time of breaking is accurately known. This, plus the burial of some of the cables reported by the repair ships, strongly suggests that submarine landslides move down the continental slopes.

Another feature of continental shelves and slopes is the submarine canyons. These submarine canyons have many of the features of river canyons, and some are larger than such river canyons as the Grand Canyon. The origin of such features has been the subject of lively speculation, and not everyone is yet satisfied with the explanations offered. At first they were thought to have formed by ordinary river erosion at a time when sea level was lower than it is at present. Many of the submarine canyons have heads near the mouths of large rivers, making this a reasonable hypothesis. However,

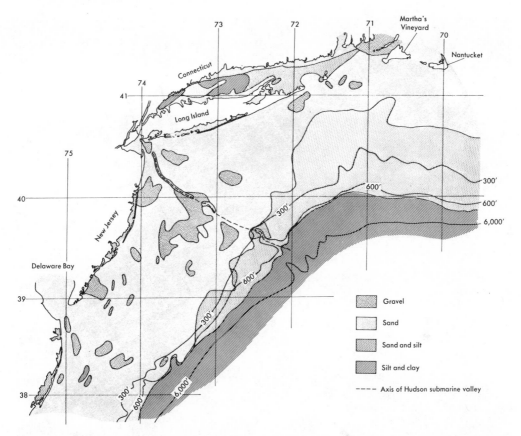

Figure 4-30. Bottom sediments on part of the continental shelf in northeastern United States. The many departures from the idealized case shown in Figure 4-29 are probably caused by bottom currents and by deposition when sea level was lowered during glacial times. Adapted from F. P. Shepard and G. V. Cohee, *Bulletin,* Geological Society of America, Vol. 47, 1936.

many of them extend to depths of over two miles, and sea level changes of this magnitude in the recent past seem impossible. The submarine landslides mentioned above offer the most likely explanation. Such landslides consist of sediment-laden water. Because they are heavier than the surrounding water, these landslides or mudflows move along the bottom and are capable of eroding the soft bottom material. These muddy waters are called *turbidity currents.* They can move down very gentle slopes and can transport fairly large pebbles. Landsliding of unconsolidated material on the continental slopes can account for the canyons there; and turbidity currents initiated by stream discharge, storms, or other currents can account for the submarine canyons on the continental shelves. As well as explaining the canyons, these processes can also account for the building of the continental slopes, for seismic studies show the slopes to be composed of sedimentary rocks. Fig. 4-31 shows some features of deep ocean sediments.

A.

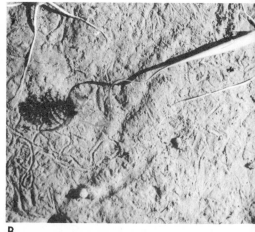
B.

C.

Figure 4-31. Ocean bottom photographs. A. Hydrographer Canyon off the east coast of the United States. Soft coral and other bottom organisms at a depth of 2,400 feet. B. Hydrographer Canyon showing brittle stars and other organisms. C. Deep sea bottom in Romanche Trench. Photos courtesy of R. M. Pratt.

Turbidity currents and submarine landslides can also account for the smooth ocean bottoms at the foot of the continental slope at many places. (See Fig. 4-32.) These smooth surfaces are apparently similar to the fans that develop where rivers enter broad valleys. These submarine fans contain pebbles and shallow-water shells at some places. These fans do not occur everywhere at the foot of the continental slope.

Figure 4-32. Turbidity current in Lake Ontario at mouth of Niagara River. The turbidity current is caused by the sediment-bearing river water. Note that part of the river water is carried by long-shore currents.

To further evaluate the possible existence and importance of turbidity currents in the geologic past, we will look at their depositional features. Turbidity currents, because of their density and speed, can transport large fragments. These larger fragments and the main mass of fine material are deposited fairly rapidly. The larger, heavier fragments settle first, forming graded beds. The eroding ability of turbidity currents suggests that they may erode or otherwise deform the top of the underlying bed. This is shown in Fig. 4-33.

A.   **B.**

Figure 4-33. Depositional features of turbidity currents. A. Photo. The bottom of a bed is shown. B. Diagram. The lobe-like casts point in the opposite direction to current movement. The feature that crosses the specimen is the cast of a groove, probably caused by a boulder carried by the turbidity current. The specimen shown is about two feet in diameter.

QUESTIONS

4-1. Describe two methods of mechanical weathering.

4-2. Describe two methods of chemical weathering.

4-3. List the common minerals in the order of their stability to chemical weathering.

4-4. In what climate is chemical weathering most effective and why?

4-5. In what climate is mechanical weathering most effective and why?

4-6. Why is quartz a common mineral in sandstones?

4-7. What effects would you expect the increased carbon dioxide in industrial areas to have on stone buildings?

4-8. Why is mature soil zoned? Describe the zones.

4-9. Discuss the importance of climate and rock type on the soil produced.

4-10. What is the origin of chert?

4-11. What are the common cements in sandstones?

4-12. Clastic sedimentary rocks are classified on the basis of _____.

4-13. What rocks would form from the weathering of a granite in a cold climate? In a warm, moist climate?

4-14. How can top and bottom of the beds be determined in a group of vertical sedimentary beds?

4-15. How might sediments deposited in shallow water differ from those deposited in deep water?

4-16. What features could be used to distinguish turbidity-current-deposited sediments?

4-17. What geologic history might be inferred from each of the following rocks?
    a. A feldspar-rich sandstone.
    b. A well-rounded conglomerate.
    c. A sandstone composed almost entirely of quartz.
    d. A limestone with abundant fossils.

4-18. What kinds of fossils would you expect to find in rocks deposited in the following environments?
   a. Tropical shallow water.
   b. Deep water.
   c. Temperate near shore.
   d. Marsh.

SUPPLEMENTARY READING

GENERAL

Fenton, C. L., and M. A. Fenton, *The Rock Book.* New York: Doubleday & Co., Inc., 1940, 357 pp.

Pough, F. H., *A Field Guide to Rocks and Minerals.* Boston: Houghton Mifflin Co., 1955, 349 pp.

Spock, L. E., *Guide to the Study of Rocks.* New York: Harper & Row, Publishers, 1962, 298 pp.

WEATHERING

Boyer, R. E., "Field Guide to Rock Weathering," *ESCP Pamphlet Series, PS-1.* Boston: Houghton Mifflin Co., 1971, 37 pp.

Foth, Henry, and H. S. Jacobs, "Field Guide to Soils," *ESCP Pamphlet Series, PS-2.* Boston: Houghton Mifflin Co., 1971, 38 pp.

Keller, W. D., "Geochemical Weathering of Rocks: Source of Raw Materials for Good Living," *Journal of Geological Education* (February, 1966), Vol. 14, No. 1, pp. 17-22.

Keller, W. D., *Principles of Chemical Weathering.* Columbia, Mo.: Lucas Brothers, 1957 (revision), 111 pp.

Kellogg, C. E., "Soil," *Scientific American* (July, 1950), Vol. 183, No. 1, pp. 30-39. Reprint 821, W. H. Freeman and Co., San Francisco.

Reiche, Parry, *A Survey of Weathering Processes and Products.* Albuquerque, N.M.: University of New Mexico Press, 1950, 95 pp.

SEDIMENTARY FEATURES

Freeman, Tom, "Field Guide to Layered Rocks," *ESCP Pamphlet Series, PS-3.* Boston: Houghton Mifflin Co., 1971, 43 pp.

Heezen, B. C., "The Origin of Submarine Canyons," *Scientific American* (August, 1956), Vol. 195, No. 2, pp. 36-41. Reprint 807, W. H. Freeman and Co., San Francisco.

Kuenen, P. H., "Sand," *Scientific American* (April, 1960), Vol. 202, No. 4, pp. 94-106. Reprint 803, W. H. Freeman and Co., San Francisco.

Pettijohn, F. J., and P. E. Potter, *Atlas and Glossary of Primary Sedimentary Structures.* New York: Springer-Verlag, 1964, 370 pp.

Potter, P. E., and F. J. Pettijohn, *Paleocurrents and Basin Analysis.* Berlin: Springer-Verlag, published in United States by Academic Press, Inc., New York, 1963, 296 pp.

Shrock, R. R., *Sequence in Layered Rocks.* New York: McGraw-Hill Book Co., 1948, 507 pp.

Chapter 5

Metamorphic Rocks

Agents of Metamorphism

Metamorphic rocks are rocks that have been changed while in the solid state either in texture or in mineral composition by any of the following—heat, pressure, directed pressure (stress), shear, or chemically active solutions. Because any rock of any type or composition can be subjected to any or all of the above agents, the variety of metamorphic rocks is endless. Note that the changes that take place in the lithification of a sedimentary rock are, according to the definition, metamorphic. Such changes, however, are not considered metamorphic, but are arbitrarily excluded.

The changes that occur during metamorphism are the result of an attempt to reestablish equilibrium with the new conditions to which the rock is now subjected. Again note the similarity to weathering. Weathering occurs when rocks that formed deep in the earth are subjected to surface conditions. However, when surface-formed rocks, such as shales, are metamorphosed, the changes that occur proceed in the opposite direction and high-temperature minerals are formed.

Many factors promote or retard metamorphic reactions. Large surface area promotes chemical reaction so fine-grained rocks react faster than coarse-grained rocks. Glass is less stable than crystalline material and so also reacts faster.

Heat is an important agent in causing metamorphism. It is well known that as one goes deeper into the earth in mines or drill holes, the temperature increases. Thus, deep burial alone can cause the temperature to rise, but most metamorphic rocks probably form at places where the temperature rise with depth is greater than normal.

119

Pressure also is the result of burial. The pressure is due to the weight of the overlying rocks. At the increased temperature and pressure deep in the earth, rocks lose much of their strength and so behave plastically. This causes the pressure due to the weight of overlying rocks to act uniformly in all directions, just as the pressure due to the weight of the water in a swimming pool acts in all directions.

Directed pressure (stress) is also commonly present during metamorphism. The effect of this is to fold the more or less plastic rocks. Stress may deform rocks, and flattened or stretched pebbles or fossils record some of this deformation.

Shear results when the rocks are broken and moved by directed pressure, as in faults. In metamorphic rocks the movement is commonly distributed on many closely spaced shear surfaces.

Chemically active solutions are very important in metamorphic reactions. Even if temperature and pressure are high enough to cause a metamorphic reaction to occur, some water is generally necessary to allow the reaction to take place. Without water to act as a catalyst, the reaction may be so slow that it almost does not occur. Water is generally present in most rocks, and only a thin film on the grain boundaries is necessary. The water may contain other materials that cause or promote the reactions. It can also bring in or remove elements from the rock. Small amounts of water may also greatly reduce the strength of rocks undergoing metamorphism.

The hot waters that cause low-grade metamorphism have apparently been found in some areas in the quest for geothermal power. Hot springs are relatively common and are spectacular at places such as Yellowstone National Park. Recent study of the geothermal area near the Salton Sea in southern California shows that the hot waters have caused metamorphism in the surrounding sediments. Geothermal waters are described in the next chapter.

Types of Metamorphism

Metamorphic rocks can be subdivided into three genetic groups, although there is gradation among the groups.

1. *Thermal* or *contact metamorphic rocks.* These rocks generally are found at the margins of intrusive igneous bodies such as batholiths. The rocks are termed *hornfels.*
2. *Regional metamorphic rocks* are so-called because they generally occupy large areas. These rocks form deep in the crust, and their presence at the surface reveals much uplift and erosion. The rocks generally have recrystallized under stress so that the new minerals grow in preferred orientation. These rocks are said to be *foliated,* and the most common are *gneiss* (pronounced "nice") and *schist.*
3. *Dynamic metamorphic rocks* result from breaking and grinding without much recrystallization.

Formation and Identification of Metamorphic Rocks

For classification and identification, metamorphic rocks can be subdivided into two textural groups:

1. Foliated—having a directional or layered aspect.
2. Non-foliated—homogeneous or massive rocks.

Non-foliated Metamorphic Rocks

Because the non-foliated rocks are the simpler, we will consider them first. There are two types of non-foliated metamorphic rocks. The first type consists of thermal or contact metamorphic rocks called *hornfels* (singular). They generally occur in narrow belts around intrusive bodies and may originate from any type of parent rock. Sometimes these are called *baked rocks;* the name is appropriate, because their formation is quite similar to the baking of clay pottery in a kiln. They are generally fine-grained, tough rocks that are difficult to identify without microscopic study unless the field relations are clear. They range from being completely recrystallized with none of their original features preserved to being slightly modified rocks with most of the original features preserved. (See Fig. 5-1.) In some of these rocks new minerals grow. At the early stages of growth the irregular growths produce what are called spotted rocks. Calcite-bearing contact rocks are generally quite spectacular, containing large crystals of garnet and other minerals.

Figure 5-1. Hornfels, showing bedding and graded bedding. This polished specimen is just over two inches in diameter. Photo by F. C. Armstrong, U.S. Geological Survey.

The second type of non-foliated metamorphic rock develops if the newly formed metamorphic minerals are equidimensional and so do not grow in any preferred orientation. The best examples of this are the rocks that result when monomineralic rocks such as limestone, quartz sandstone, and dunite are metamorphosed. In the case of limestone, which is composed of calcite, no new mineral can form; thus, the calcite crystals, which are small in limestones, grow bigger, develop an interlocking texture, and become *marble*. (See Fig. 5-2.) Marble can be distinguished from limestone only by its larger crystals and lack of fossils; being composed of calcite, both effervesce in acid. Marble may have any color; pure white calcite rocks are more apt to be marble than limestone. Impure limestones, as noted above, develop new minerals, particularly garnet. Dolomite marble also occurs.

Because quartz sandstone similarly can form no new minerals, the quartz crystals enlarge and intergrow to form *quartzite*. Quartzite is distinguished from chalcedony, chert, and opal by its peculiar sugary luster. The term quartzite is best reserved for the metamorphic rock, although quartz sandstone is called quartzite by some and the metamorphic rock is called metaquartzite.

Both marble and quartzite may form under stress, and so the crystals may be aligned. However, because the crystals are equidimensional, this preferred orientation can be

Figure 5-2. Marble, showing coarse
interlocking texture. Photo from Ward's
Natural Science Establishment, Inc.,
Rochester, N.Y.

detected only by microscopic examination. Thus, although we call these rocks non-foliated, some may have a hidden foliation.

So far we have been concerned with *progressive* metamorphism, which is metamorphism caused by increased temperature, pressure, and other agents of metamorphism. The next example is one of *retrogressive* metamorphism, which is metamorphism at a temperature lower than that of the rock's original formation. Dunite is composed mainly of olivine, which is the highest-temperature mineral in Bowen's series. When it is subjected to the same metamorphic conditions that produce the rocks discussed earlier, it too is metamorphosed—but at a temperature lower than that of its formation. The olivine of dunite is changed to the mineral *serpentine,* and the metamorphic rock produced is called *serpentinite.* Its identification is discussed in Appendix A. In other instances, instead of serpentine, *talc* is the metamorphic mineral that forms. If non-foliated, the talc rock is called *soapstone;* if foliated, it is called *talc schist.* Identification of talc is discussed in Appendix A. Slight differences in the composition of the parent rock or of the water solutions that cause this type of metamorphism probably determine which mineral is formed.

Foliated Metamorphic Rocks

Regional metamorphism generally takes place in areas undergoing deformation. These very active areas are generally the cores of mountain ranges. Regional metamorphic rocks are produced by the same type of forces that make mountain structures. The forces that fold and fault the shallow rocks exposed on the flanks of mountain ranges can provide the stress fields under which the deep-seated metamorphic rocks, now exposed in the uplifted cores of mountain ranges, were recrystallized.

One example of the rocks produced by increasing metamorphic grade will show most of the foliated metamorphic rock types. These same rock types will form when many other parent rocks are metamorphosed; the differences in original chemical composition will change the relative amounts of the metamorphic minerals only somewhat.

Shale is composed mainly of clay. During lower-grade metamorphism, the clay is transformed to mica, and at higher grades of metamorphism, to feldspar. Which micas and which feldspars form depends on the bulk chemical composition. Compare this with chemical weathering and note that weathering is, in a sense, retrogressive metamorphism. Because mica is a flat, platy mineral, it tends to grow with its leaves perpendicular to the maximum stress, forming a preferred orientation. (See Figs. 5-3, 5-4, and 5-5.) Microscopic examination of some foliated rocks shows that the bedding

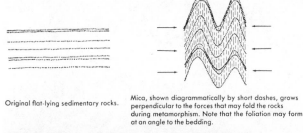

Original flat-lying sedimentary rocks.

Mica, shown diagrammatically by short dashes, grows perpendicular to the forces that may fold the rocks during metamorphism. Note that the foliation may form at an angle to the bedding.

Figure 5-3.   Development of foliation.

Figure 5-4.   Foliation and bedding in folded rocks. The rock type is slate. Compare with Figure 5-3. Photo by A. Keith, U.S. Geological Survey.

Figure 5-5. Typical exposure of slate. Bedding is nearly vertical and is shown by color differences. Foliation is perpendicular to the bedding. Photo by J. B. Woodworth, U.S. Geological Survey.

planes are slightly displaced in the plane of the foliation. This suggests that foliation, at least at some places, forms in shear directions. The shear direction is in general at about 45 degrees to the maximum stress in uniform materials. Thus foliation may form perpendicular to the direction of maximum stress, or at about 45 degrees to that direction.

The sequence produced in the metamorphism of a shale is as follows:

Sedimentary rock	Low-grade metamorphism	Medium-grade metamorphism	High-grade metamorphism
SHALE ⟶	SLATE ⟶	SCHIST ⟶	GNEISS
Clay	Clay begins to be transformed into mica. Mica crystals are too small to see, but impart a foliation to the rock. May also form by mechanical rearrangement of clay alone.	Mica grains are larger so that the rock has a conspicuous foliation.	Mica has transformed largely to feldspar, giving the rock a banded or layered aspect.

The rock names are applied for the overall texture of the rock and not strictly for the mineral transformations which occur over a range in temperature and pressure. (See Figs. 5-6, 5-7, and 5-8.) Most schists contain feldspar, but the name gneiss is reserved for rocks with much more feldspar than mica. Other rocks that have the same general bulk chemical composition, thus producing the same metamorphic rocks, are certain pyroclastic and volcanic sedimentary rocks, many sandstones, arkose, granite, and rhyolite. The coarse-grained rocks in this list are rarely the parents of slate or fine-grained schist but, in general, remain more or less unaffected until medium-grade

Figure 5-6. Outcrop of foliated rock. The rock is an impure marble, and the foliation is parallel to the bedding. The light-colored pods above the knife are quartzite, and the pods formed when a sandstone bed broke brittlely when the rocks were folded. Photo by J. A. Donaldson, Geological Survey of Canada, Ottawa.

Figure 5-7. Garnet-mica schist. The large garnet crystals grew during metamorphism. Foliation does not show clearly in the orientation pictured. Photo from Smithsonian Institution.

A.   B.

Figure 5-8. Gneiss. A. The foliation is
shown by discontinuous layers of dark
and light colored minerals. B. The layers
in this gneiss are more irregular. Photos
from Ward's Natural Science Establishment, Inc., Rochester, N.Y.

metamorphism is reached. Occurrence of slate in areas where the non-shale rocks are
not metamorphosed suggests that slate may be formed by a mechanical reorientation
of clay particles during folding, together with limited recrystallization.

Only one other group of regional metamorphic rocks must be considered among the
most common metamorphic rocks. These rocks are formed by the metamorphism of
basalt, certain pyroclastic and volcanic sedimentary rocks, gabbro, graywacke, and
some calcite-bearing or dolomitic sedimentary rocks. The sequence is

Parent rock	Low-grade metamorphism	Medium- and high-grade metamorphism
BASALT ⟶	GREENSCHIST ⟶	AMPHIBOLITE (Amphibole schist)
	Fine-grained, foliated Chlorite (green mica) Green amphibole Quartz	Coarse-grained, foliated Plagioclase Dark amphibole

Dynamic metamorphism, the third type of metamorphism, results when rocks are
broken, sheared, and ground near the surface where the temperature and pressure are
too low to cause any significant recrystallization. Thus, these rocks are commonly
associated with fault zones, but there are all gradations between these rocks and
ordinary schist, especially lower-grade schist formed from coarse-grained rocks. The
fine-grained, ground rock in a fault zone, sometimes called gouge, is *mylonite,* and the
rocks gradational with schist are called *semischist* or cataclastic schist or gneiss. The
origin of these rocks is clear when they are encountered in the field, but they are very
difficult to identify in hand specimen.

An unusual type of dynamic metamorphism can occur at impact craters caused by
meteorites, as described in Chapter 18.

The origin, texture, and mineralogy of the common metamorphic rocks are summarized in Figs. 5-9 and 5-10.

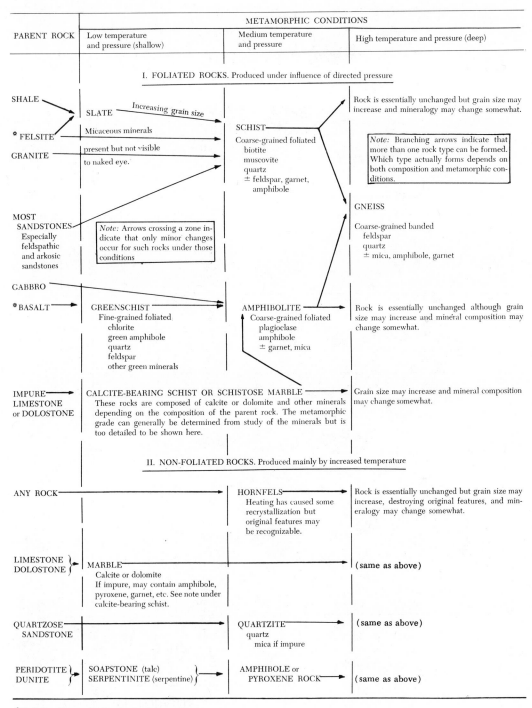

Figure 5-9. A generalized chart showing the origin of common metamorphic rocks.

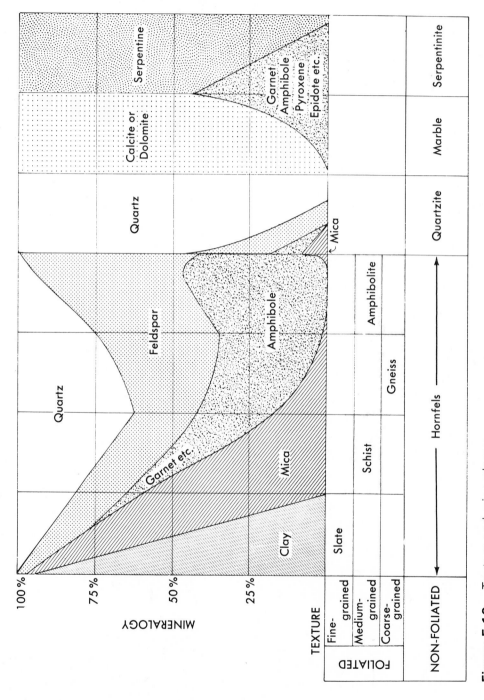

Figure 5-10. Texture and mineralogy of common metamorphic rocks.

128

Interpretation

Study of metamorphic rocks, coupled with laboratory experiments, yields much information about conditions and processes deep in the crust. That is, if we know the conditions of formation of a mineral or, better, an assemblage of minerals, when we find rocks composed of such minerals, we know the conditions under which the rocks formed. The problem is complicated by the diversity of natural rocks in that they contain many more elements, which are the variables in experiments, than can be easily studied in experiments. It is also not possible to reproduce the effects of geologic time in the laboratory, nor is it generally possible to detect the effects of solutions or volatiles that may have catalyzed the reactions in natural rocks. Thus, although we are very rapidly gaining an understanding of metamorphic processes, many problems remain.

The purpose of this chapter is to give the student some ability in identifying metamorphic rocks. He will not, except in a very general way, be able to interpret the detailed geologic history of these rocks without knowledge of chemistry, mineralogy, and field and microscope study. To interpret metamorphic rocks fully, one must know whether they contain equilibrium mineral assemblages or whether conditions changed before the metamorphic reactions were complete; whether material was added or subtracted during metamorphism; and whether the rocks were subjected to one or more periods of metamorphism. One can think of many more possibilities that must be considered: for example, if the reactions take place during deep burial, will the rocks revert to something like their original state during slow uplift? No doubt some such effect may operate in some instances, but most of the chemical reactions involved go very slowly in the reverse direction. If this were not so, nothing but weathering products would be visible at the surface.

Metamorphic Grade

The concept of metamorphic grade was introduced above in discussing the formation of metamorphic rocks. Metamorphic grade is an interpretation based on field and laboratory evidence and should be further examined. The first attempts to map and study grade of metamorphic rocks consisted of mapping mineral zones. This was done by restricting the study to a single parent rock type, such as shale, and recording the first occurrence of certain minerals as the metamorphic grade increased. (See Fig. 5-11.) This method was limited to rocks of a single composition and did not take into account nonequilibrium minerals. This simple method has been improved by considering assemblages of several minerals and by comparing such equilibrium assemblages in rocks of many compositions. Many problems are encountered in recognizing equilibrium assemblages, and much field and laboratory work has been required to find widespread equilibrium assemblages that can be used in this way.

Problems

The origin of metamorphic rocks presents many geologic problems. The regional metamorphic rocks are the most difficult to explain. These rocks formed under high temperature and pressure. The depths at which these temperatures and pressures can occur, assuming reasonable temperature gradients, are in the range of 10 to 20 miles. (See Fig. 5-12.) The amount of erosion in areas where high-grade regional metamorphic rocks are exposed is probably less than this, indicating that the temperature gradient

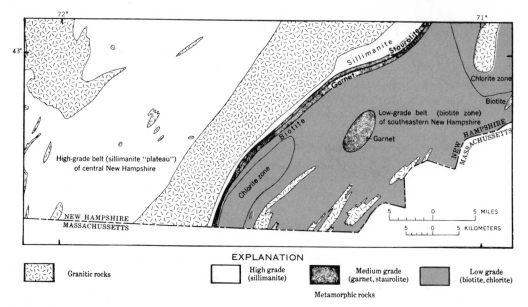

Figure 5-11. Metamorphic grade as interpreted from metamorphic minerals in shaly rocks in southeastern New Hampshire. From Warren Hamilton and W. B. Myers, U.S. Geological Survey, *Professional Paper* 554C, 1967.

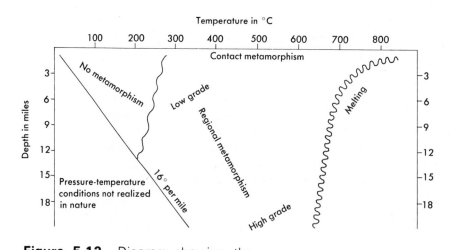

Figure 5-12. Diagram showing the conditions under which metamorphism occurs. Pressure is proportional to depth. Biotite is believed to crystallize between 200 and 300°C and sillimanite between 600 and 700°C, or very near the melting point of granite.

must have been greater. A higher temperature gradient is not unreasonable because such rocks form in geologically active areas. The origin of such high-temperature gradients is a problem. A possible solution to this problem is the passage of magma through the rocks, as suggested in Chapter 3 in connection with the origin of batholiths. When viewed on a large scale, as in Fig. 5-11, the narrow transition zone between high-grade metamorphic rocks (sillimanite-bearing rocks) and low-grade rocks (chlorite-bearing rocks) is also difficult to explain. That is, if the heat comes from deep burial, the heat conduction of rocks is such that the surrounding rocks should have been much hotter.

Figure 5-13. The small, irregular occurrence of massive granitic rock that cut across the foliation may have formed by partial melting. The granitic rock is localized in a minor fault zone where the foliation is disturbed. Photo by A. W. Postel, U.S. Geological Survey.

Another problem is the origin of granite. As shown in Fig. 5-12, at the highest grades of metamorphism some melting may occur, and so metamorphic processes may merge with igneous activity. Many of the small lenses of granitic rocks found in high-grade gneiss probably originate from such partial melting. (See Fig. 5-13.) The origin of larger bodies of granite, such as batholiths, in high-grade metamorphic terranes is a problem. Some show chilled, fine-grained margins and other features that are clearly igneous. Others have completely gradational contacts and could be metamorphic rocks or could have intruded the gneiss just after the gneiss formed and was still hot. Some of the latter rocks are probably the result of metamorphism, but, as pointed out in Chapter 3, the isotope composition of most batholiths indicates an origin in the lower crust or upper mantle.

QUESTIONS

5-1. Define metamorphic rock.

5-2. What are the types of metamorphism, and which agents are most important in each?

5-3. What is foliation, and how does it form?

5-4. What is contact metamorphism, and what types of rocks are formed?

5-5. How does a shale differ from a slate?

5-6. What is the low-grade metamorphic equivalent of a basalt?

5-7. Serpentinite forms from metamorphism of _____.

5-8. Soapstone forms from metamorphism of _____.

5-9. How is the metamorphic grade established?

5-10. How might the field relationships differ between an igneous and a metamorphic granite?

QUESTIONS
COVERING ALL THE ROCKS AND MINERALS
IN CHAPTERS 2, 3, 4, and 5

1. Rockhounds collect many different rocks and minerals that they cut and polish. What physical properties must such rocks and minerals have? List the various minerals and rocks that have these qualities.

2. Name and interpret as completely as possible the significance of the following rocks:

    a. A coarse-grained rock with interlocking texture composed mainly of quartz and feldspar with some dark minerals.

    b. A fine-grained, dark-gray rock with a few well-formed plagioclase and olivine crystals.

    c. A foliated rock with much muscovite.

    d. A fine-grained, white rock composed of calcite with some clam shells.

    e. A clastic rock composed of well-rounded quartz grains about 1 mm in diameter.

    f. A clastic rock composed of quartz and feldspar with some hornblende and mica. The grains are mainly between 1 and 2 mm in diameter and are angular.

    [*Note:* After reading Chapters 7 through 11 you may want to review your interpretations of the last two rocks.]

3. Before starting to identify rock specimens, it may help to list all of the rocks mentioned in the text in a form like the one suggested below. This will show the importance of texture in recognition and interpretation of rocks.

Check which type:

	Igneous	Sedimentary	Metamorphic

Coarse-grained
Clastic texture

_____

_____

Interlocking texture
  Massive

_____

_____

  Foliated

_____

Fine-grained

_____

_____

Glass

_____

Frothy

_____

SUPPLEMENTARY READING

GENERAL

Fenton, C. L., and M. A. Fenton, *The Rock Book.* New York: Doubleday & Co., Inc., 1940, 357 pp.

Pough, F. H., *A Field Guide to Rocks and Minerals.* Boston: Houghton Mifflin Co., 1955, 349 pp.

Spock, L. E., *Guide to the Study of Rocks.* New York: Harper & Row, Publishers, 1962, 298 pp.

METAMORPHIC ROCKS

Fyfe, W. S., F. J. Turner, and John Verhoogen, *Metamorphic Reactions and Metamorphic Facies,* Chapter 1, "Concept of metamorphic facies," pp. 3-20; Chapter 7, "Mineral assemblages of individual metamorphic facies," pp. 199-239. Geol. Soc. of America, 1958, Memoir 73.

Mason, Brian, *Principles of Geochemistry,* (3rd ed.), Chapter 10, "Metamorphism and metamorphic rocks." New York: John Wiley & Sons, Inc., 1966, pp. 248-282.

Chapter 6

Mineral Resources

Many of the materials that form the crust of the earth are useful to man. The term *mineral resources* is used to mean natural resources that come from the earth. Note that the term "mineral" used in this sense does not follow the definition of mineral used earlier because many of the materials discussed in this chapter, strictly speaking, are not minerals. The most important mineral resources are metals and energy-producing materials, and this chapter will be limited mainly to these. Many other mineral products are mined, however, such as sand, gravel, building stone, clay, diatomite, and limestone.

This chapter will be concerned mainly with the origin of mineral resources. Few new principles are involved, so it will, in a way, serve as a review of minerals and rocks. Before discussing formation of mineral deposits, we will consider some of their unique features.

Distribution and Importance

Mineral resources are not spread evenly around the continents, but are scattered. (See Fig. 6-1.) This creates the have and the have-not nations. A mine or other mineral deposit is a rare feature, geologically speaking.

The importance of mineral deposits to man cannot be overemphasized. The wealth of nations is dependent on their natural resources. Although not always emphasized, mineral deposits have played a very important role in history. In ancient times, wars

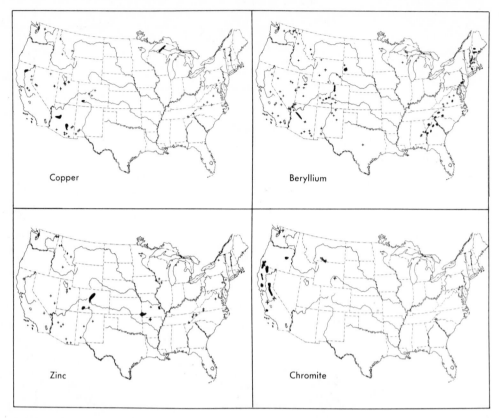

Figure 6-1. Maps showing the occurrence of economic deposits of some metals in the United States. Note the similarities between gold and silver, and among silver, lead, and zinc. Most of these deposits are in the geologically active areas that will be discussed in Chapter 14. Data from Mineral Resource Maps of the U.S. Geological Survey.

were fought and civilizations prospered or failed because of gold and silver deposits. In recent times, metals and energy-producing materials, such as iron, coal, and petroleum, have determined the course of history.

Exhaustibility

Every mineral deposit has finite dimensions. When the ore is mined out, the deposit has been used. Thus all mineral deposits are exhaustible, and there is no way to replace them. This is not news to anyone who has traveled through the west and seen the many ghost towns. The importance of the exhaustibility of mineral deposits is seen when rates of mining are considered. In the last fifty years more metal has been mined than in all preceding time. Thus, we are using our natural resources at an unprecedented rate. Unless new sources are found, we face the prospect of nations decaying like the ghost towns of the west. This drastic possibility will be averted by increased exploration

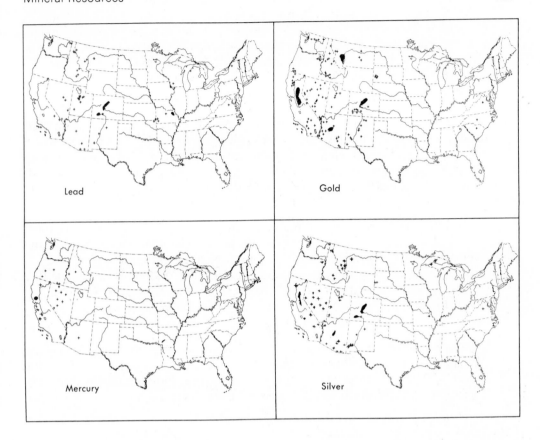

Lead

Gold

Mercury

Silver

(prospecting), using new, sensitive chemical, physical, and biologic methods; by finding substitutes, such as plastics; and by developing new refining methods that are effective with very low-grade ores. In any case, it seems quite evident that we will have to pay more for our metals in the future.

An ore is anything that can be mined at a profit. Therefore, what is ore in the United States may not be ore in the Canadian Arctic or Central Australia because of the cost of mining and the distance to market. It also follows that what is worthless today may be ore tomorrow because of changing economic conditions or technological changes.

Water, both surface and underground, and soil are the most important mineral resources. Life would not be possible without both. These two basic resources are different from all other mineral resources in that they are replenishable, although the time required to replenish them may be very long by human standards.

Formation of Mineral Deposits

Mineral deposits occur at places where unusual conditions caused the concentration of some elements far in excess of their normal abundances. Table 2-1 shows the average amounts of all elements in the earth's crust. Copper, for instance, has an average abundance of 0.0055 per cent, but copper ore in the United States must average 0.5 per cent copper. Thus copper must be concentrated by at least 90 times its average amount to form copper ore. Many mines produce more than one metal, thus reducing the percentage of any one element that must be present to make the deposit economic. In any case, to form a metal mine, an element must be concentrated at least several to a few thousand times its average abundance. We will now see how this can be done. (See Tables 6-1 and 6-2.)

Many mines are associated with igneous rocks, and we will consider them first. Because we do not know how igneous rocks form, we have little knowledge of where the metals come from, except that they are part of the magma. The different magma types, such as granite and gabbro, each have typical metals that are associated with them. The distribution of ore metals is probably inherited from the original distribution of metals that developed when the earth formed.

Disseminated Magmatic Deposits are the simplest of the magmatic deposits. The valuable mineral is disseminated or scattered throughout the igneous body. In the diamond deposits of South Africa, for example, the diamonds are disseminated in an unusual rock, somewhat similar to peridotite. (See Fig. 6-2.) In this example, the valuable mineral is unique to this rock type and, because of its great value, can be mined. In the other types of deposits, the ore mineral or minerals are concentrated rather than disseminated within the body.

Crystal Settling was already described in the discussion of differentiation of magma. (See Figs. 3-4 and 6-3.) Heavy, early-crystallized minerals sink to the bottom of the igneous body. Many chromite and/or magnetite deposits form in this way, generally in gabbro bodies. (Magnetite crystallizes early under some conditions and late under

Table 6-1. Genetic classification of mineral deposits.

Type	Example
Magmatic	
Disseminated	Diamonds — South Africa (Fig. 6-2.)
Crystal settling	Chromite — Stillwater, Montana (Fig. 6-3.)
Late magmatic	Magnetite — Adirondack Mountains, New York (Fig. 6-4.)
Pegmatite	Beryl and lithium — Black Hills, South Dakota (Fig. 6-5.)
Hydrothermal	Copper — Butte, Montana (Fig. 6-6.)
Contact-metamorphic	Lead and silver — Leadville, Colorado (Fig. 6-8.)
Sedimentary	
Evaporite	Potassium — Carlsbad, New Mexico
Placer	Gold — Sierra Nevada foothills, California (Fig. 6-10.)
Weathering	Bauxite — Arkansas (Fig. 6-11.)

Table 6-2. Geologic occurrence of some metals.

Metal	Main ore minerals	Principal type occurrence	Example
Aluminum	Bauxite $Al_2O_3 \cdot nH_2O$	Weathering	Arkansas
Antimony	Stibnite Sb_2S_3	Hydrothermal	Yellow Pine, Idaho
Beryllium	Beryl $Be_3Al_2Si_6O_{18}$	Pegmatite	Black Hills, S.D.
Cadmium	By-product of zinc ores	Hydrothermal (in sphalerite)	Missouri (Tri-State Region)
Chromium	Chromite $Fe(Cr,Fe)_2O_4$	Magmatic	Stillwater, Montana
Cobalt	Cobaltite $CoAsS$ Smaltite $CoAs_3$	Hydrothermal	Republic of the Congo
Copper	Bornite Cu_5FeS_4 Enargite Cu_3AsS_4 Chalcocite Cu_2S	Hydrothermal	Bingham, Utah Butte, Montana
Gold	Native gold Au Tellurides $AuTe_2$	Hydrothermal Placer	Homestake, S.D. Motherlode, Calif.
Iron	Hematite Fe_2O_3 Magnetite Fe_3O_4	Sedimentary and Weathering	Lake Superior Region
Lead	Galena PbS	Hydrothermal	Missouri (Tri-State Region) Coeur d'Alene, Id.
Manganese	Oxides MnO_2	Weathering	Chamberlain, S.D.
Mercury	Cinnabar HgS	Hydrothermal	New Almaden, Calif.
Molybdenum	Molybdenite MoS_2	Hydrothermal	Climax, Colo.
Nickel	Sulfides NiS	Magmatic	Sudbury, Ontario
Platinum	Native platinum Pt	Magmatic Placer	Ural Mountains, Russia
Silver	Sulfides Ag_2S By-product of zinc-lead mine	Hydrothermal	Comstock, Nev. Coeur d'Alene, Id.
Tin	Cassiterite SnO_2	Hydrothermal Placer	Malaya
Titanium	Ilmenite $FeTiO_3$ Rutile TiO_2	Magmatic	Adirondack Mtns., New York
Tungsten	Scheelite $CaWO_4$ Wolframite $(Fe,Mn)WO_4$	Contact-metamorphic	Small deposits in Nev. and Calif.
Uranium	Pitchblende U_3O_8 Uraninite UO_2 Carnotite $K_2(UO_2)_2(VO_4)_2 \cdot 3H_2O$	Sedimentary (Carnotite)	Colorado Plateau
Vanadium	Carnotite $K_2(UO_2)_2(VO_4)_2 \cdot 3H_2O$ Roscoelite $K(V,Al)_2(OH)_2AlSi_3O_{10}$	Sedimentary (Carnotite)	Colorado Plateau
Zinc	Sphalerite ZnS	Hydrothermal	Missouri (Tri-State Region)

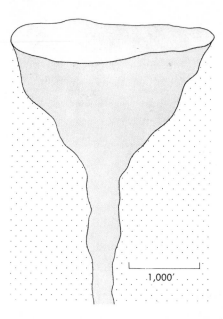

Figure 6-2. South African diamond pipe. Diamonds are scattered through more or less vertical, cylindrical bodies of an unusual igneous rock somewhat similar to peridotite.

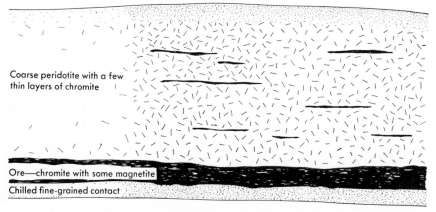

Coarse peridotite with a few thin layers of chromite

Ore—chromite with some magnetite

Chilled fine-grained contact

Figure 6-3. Chromite layers form by settling of early-formed chromite crystals. The parent igneous rock is peridotite, dunite, or gabbro. Commonly the igneous body has a more irregular shape than shown.

others.) In some cases, the segregation of the early-formed ore minerals is effected by squeezing the liquid magma away, leaving the early-formed crystals.

Late Magmatic Deposits form when the ore-forming constituents of the magma are among the last to crystallize. The metalliferous material is heavy; it tends to settle to the bottom of the igneous body and to crystallize there and in the interstices between the earlier-formed minerals. (See Fig. 6-4.) At the late stage when these deposits form, as pointed out earlier, the magma may be water-rich.

Pegmatite bodies also form late in the crystallization of a magma. These extremely coarse-grained rocks may form as dikes or irregular pods. (See Fig. 6-5.) Just how the large crystals form is not known, but the water- and volatile-rich solutions must permit the growth of large crystals. Not all pegmatites are igneous, for many are found in metamorphic terranes far from any igneous body. They are sought for their large crystals of mica, which are used in industry, and for feldspar, quartz crystals, and gems. They are also the main source of columbium, tantalum, the rare earths, beryl (the main beryllium mineral), and lithium. Many other minerals occur in pegmatites, including some high-temperature sulfides.

Hydrothermal, or hot water, deposits are the most common type of metal deposit. Their possible origin from the late magmatic fluids was discussed earlier. They grade from high-temperature to low-temperature deposits. Typical hydrothermal deposits contain sulfide minerals, such as pyrite, chalcopyrite, and galena, in quartz veins. (See Figs. 6-6, and 6-7.) Generally they are associated with batholiths, and the veins are both in the batholith and the surrounding rocks. However, many hydrothermal deposits have no nearby outcropping igneous rocks and the origin of the depositing solutions is not known. In such cases the ores may have been deposited by circulating ground

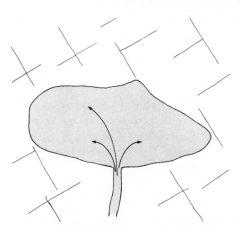

A. Injection of magma

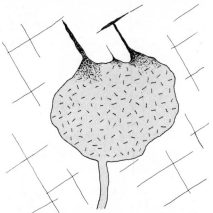

B. After partial crystallization, the magma body is deformed and the late fluid is squeezed into newly-formed cracks.

Figure 6-4. An example of segregation of late magmatic fluid. In this case, the late fluid is separated from the rest of the magma; in other cases, the late fluid may sink and fill the spaces between the early crystals near the bottom of the igneous body.

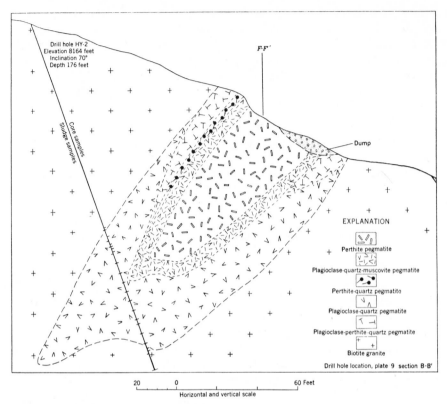

Figure 6-5. Cross-section of the
Hyatt pegmatite, Larimer County, Colorado. From W. R. Thurston, U.S. Geological Survey, *Bulletin* 1011, 1955.

water, as mentioned later in connection with geothermal power. In these instances, the
metals might come from deep magmas or might be leached from other rocks by the
circulating ground water. Examples of hydrothermal deposits include copper at Butte,
silver at the Comstock Lode, lead-zinc at Missouri, and gold at the Mother Lode in
California.

Contact-metamorphic Deposits form in the zone of metamorphism that surrounds an intrusive body. The magma and its fluids react with the wall rock and the
ground water and replace the wall rock with ore minerals. These deposits are commonly
found in limestone near intrusive bodies because the limestone is especially reactive
with the fluids. (See Fig. 6-8.) Many deposits formed in this way contain high-temperature sulfide minerals similar to those in high-temperature hydrothermal deposits. Some
iron deposits, a few fairly large, formed in this way: Iron Springs, Utah, is an example.

Sedimentary processes can also concentrate valuable elements.

Evaporites, or chemical precipitates, such as salt beds, especially those containing
potassium salts, are sought actively. Such beds come from the evaporation of sea water
or of a salt lake. Because sea water contains only 3.5 per cent dissolved material, mostly

Figure 6-6. Underground mining, Kelley Mine, Butte, Montana. Some of the veins in the granitic rock being drilled can be seen. Photo from Anaconda Co.

common salt, several thousand feet of sea water must evaporate to produce 100 feet of salt. Many salt deposits, however, are much thicker than this. The evaporation is thought to occur in shallow, slowly sinking basins, partially dammed by sand bars, and fresh brine from the ocean flowing over the sand bars, perhaps during storms. Recently it has been suggested that salts could accumulate in deep basins with high evaporation and stagnant bottoms. The chemistry of these brines is complex, but the general sequence of precipitation is calcium and magnesium carbonate, gypsum, salt (halite), and finally, potassium and bromine salts. Salt beds are mined in the area of an ancient basin in Michigan, Ohio, and New York. Potassium salts are mined near Carlsbad, New Mexico.

At other times and places in the geologic past, generally in restricted basins, deposits of phosphate and iron-rich rocks have occurred. Rather specialized conditions are probably necessary to form such rocks. The phosphate deposits of Idaho, Wyoming, and Montana formed in such a restricted sea.

Placer Deposits form by the concentration of heavy mineral fragments by rivers. A heavy, or more properly, a dense mineral fragment is moved less easily than a lighter fragment of the same size. For this reason a river tends to separate, or sort, its transported load. The denser fragments are the first to be deposited and tend to be concentrated on the insides of curves, at places where the gradient is less, and in similar places. These processes will be described in Chapter 8.

Figure 6-7. Aerial view of the open pit mine at Chuquicamata, Chile. The veins here are too narrow to mine economically underground. The ore averages 1⅓ per cent copper, and about 142 tons of ore and waste rock must be moved for each ton of copper recovered. This low-grade ore body has been in production over 50 years. Photo from Anaconda Co.

Minerals concentrated in this way must be dense and resistant to weathering. Examples include gold, diamond, and platinum. Gold placers were very important in the opening of the west, especially California in 1849. (See Fig. 6-10.) Placer deposits are "poor man's deposits" because very little equipment is necessary to prospect for them or to exploit them. This easily-obtained wealth started the westward movement. Ocean waves can also produce placers, and beach sands containing valuable heavy minerals are mined at some places.

Weathering Processes also contribute to mineral wealth. Soils are surely our most valuable mineral commodity. In addition, some tropical soils (laterites) are mined for their aluminum (bauxite), iron, or other elements, as was described earlier. (See Fig. 6-11.)

Weathering processes are also active on outcropping mineral deposits. Their effect on many deposits is great and, unless understood, can be very misleading. If an ore or its weathering products are insoluble, the ore will be enriched by weathering unless the weathered material is removed by erosion. This is especially true in mineral deposits in soluble rocks such as limestone. (See Fig. 6-12.) This kind of enrichment has fooled many prospectors and investors who thought that they had a good mine on the basis of near-surface assays, but found the ore too low-grade to be economical when a shaft was sunk. Gold, lead, copper, silver, and mercury enrichments of this type are very common.

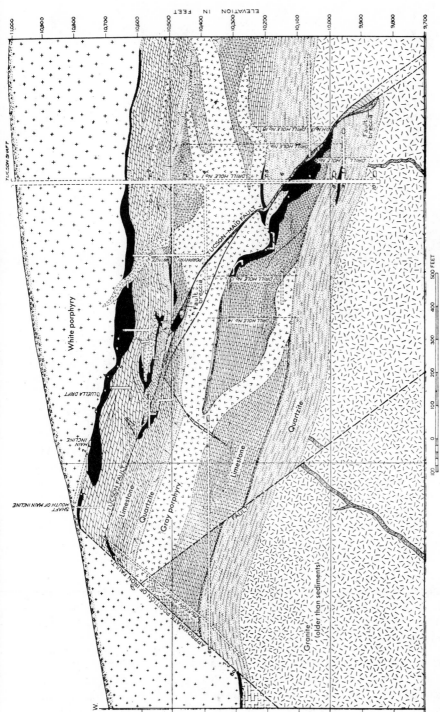

Figure 6-8. Cross-section of part of the Leadville district, Colorado. Most of the ore bodies replace limestone and are believed to have originated from the porphyry. Some veins are present, and the ore bodies along the fault show that it acted as a passage for some of the ore fluids. Replacement ore bodies generally have complex shapes. See Figure 6-9. From F. A. Aicher, U.S. Geological Survey, *Professional Paper* 148. 1927.

145

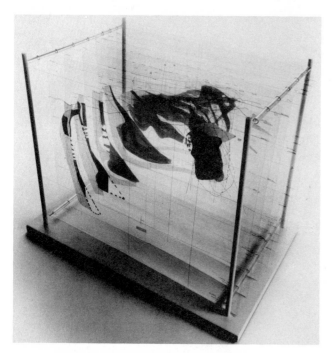

Figure 6-9. A model of a mine made by drafting successive cross-sections on transparent material. The complex shapes of ore bodies can be studied by this technique. Photo from Anaconda Co.

At sulfide deposits, the weathering processes can become very complex. The sulfide minerals are oxidized to become soluble sulfates. Pyrite, which is present in almost every case, weathers to sulfuric acid and soluble iron sulfates, all of which aid in the solution of other sulfide minerals. Iron, silver, and copper sulfides are dissolved in this way. The soluble sulfates trickle slowly downward, enriching the lower part of the deposit. The main zone of enrichment occurs at the water table. Here the chemical environment differs, mainly in its lack of oxygen, and precipitation as sulfides occurs. Again, the shrewd prospector should not be fooled by the near-surface shows unless they, by themselves, are rich and large enough to mine.

Weathering of mineral deposits does help the prospector. The oxidation of the ever-present pyrite leaves a surface residue of iron oxides, especially hematite and limonite. These red-brown caps clearly mark the outcrops of sulfide veins, and so they are the prime target of the prospector. (See Fig. 6-13.)

Deep Ocean Deposits have recently attracted much attention. Bottom photographs and dredge samples have revealed that parts of the floor of the Pacific Ocean are covered by manganese nodules. (See Fig. 6-14.) These interesting rocks have formed by the precipitation of manganese and iron dissolved in sea water onto sand grains and pebbles on the sea bottom. This precipitation is so slow that 1000 to 100,000 years are required to deposit a layer one millimeter thick. The deposition occurs in the form of oxides and hydroxides. These nodules average about 24 per cent manganese, 14 per cent iron, and

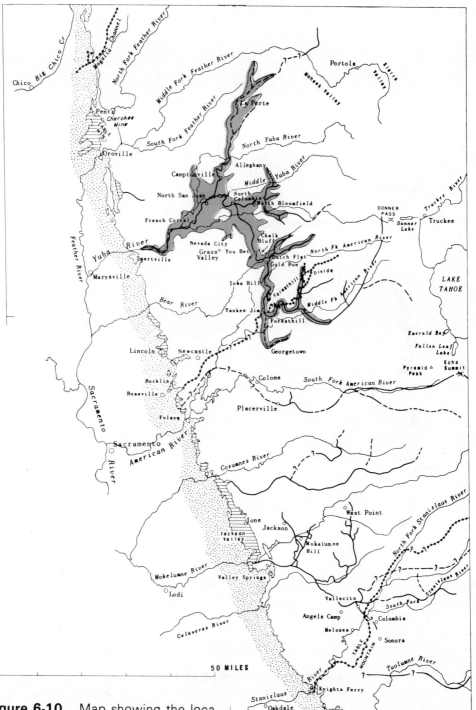

Figure 6-10. Map showing the location of placer gold deposits in the Sierra foothills in California. The ancient rivers along which these placers accumulated were in different locations than the present rivers. The older of the rivers are shown by heavy lines and the younger by dotted lines. These rivers flowed into a sea indicated by the stippled area. From P. C. Bateman and Clyde Wahrhaftig, California Division of Mines and Geology, *Bulletin* 190, 1966.

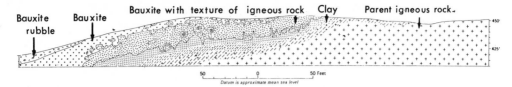

Figure 6-11. Cross-section of the Pruden bauxite mine, Arkansas. The bauxite is the weathering product of the aluminum-rich igneous rock. From Mackenzie Gordon, Jr. and others, U.S. Geological Survey, *Professional Paper* 299, 1958.

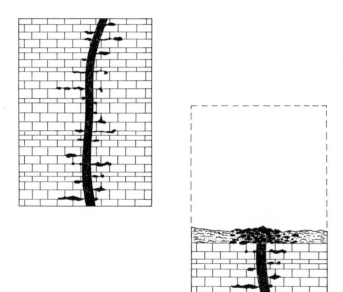

Figure 6-12. Residual concentration by solution of the limestone that encloses the deposit. The ore minerals must be insoluble.

small amounts of nickel, cobalt, copper, and other elements. Some contain two per cent or more copper, nickel, or cobalt. The elements come from chemical weathering of the continents, by leaching of submarine basalt, and from waters associated with volcanic activity. Because of the large areas involved, these nodules constitute a very valuable resource, and techniques for their mining are being developed. However, the nodules also contain large amounts of silica (SiO_2), and current metallurgical techniques do not permit the separation of manganese from silica profitably.

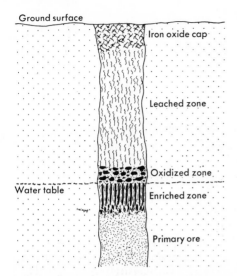

Figure 6-13. Weathering of a sulfide vein deposit. Rain water percolating to the water table leaches metals from the vein and redeposits them near the water table, creating an enriched zone. The reddish iron oxide cap helps to attract the attention of prospectors.

Figure 6-14. Manganese nodules on the floor of the Atlantic Ocean. Photo courtesy of R. M. Pratt.

Energy Resources

The main mineral energy sources are coal and petroleum. Doubtless in the future atomic and nuclear power will overshadow these mineral fuels. Geothermal power is currently being produced and also will be more important in the future.

Coal and petroleum are sometimes called fossil fuels because they originate from once-living material. Coal comes from plant material. To form coal, plant material must accumulate; but under most conditions, even with abundant growth of vegetation as in a tropical rain forest, plant material does not accumulate. Instead, it decays and is attacked by bacteria as soon as it falls. Most coal beds seem to have developed in swamps, where, as soon as a fallen tree becomes waterlogged, it sinks to the bottom and ceases, or nearly ceases, to rot, especially if the water is stagnant as it is in many swamps. The plant material at the bottom of such a swamp becomes *peat.* The peat ultimately becomes coal, largely due to compaction from the weight of overlying sediments. The first stage in this process produces *lignite,* and further pressure produces *bituminous coal* and finally, in rare cases, *anthracite.* (See Fig. 6-15.) In this process the original peat may be compressed to less than one twenty-fifth of its original thickness. To produce a reasonably pure coal, little other sediment can be accumulating in the swamp at the same time.

Petroleum also comes from organic material, but the process is not so clear as it is with coal. Oil and gas come from marine sediments. Dark organic-rich shale is the probable source. Such sediments are accumulating today in the Black Sea where the bottom muds contain up to 35 per cent organic material compared with about 2.5 per cent in ordinary marine sediments. In normal marine sediments the organic material is oxidized, but in stagnant bottom water it can accumulate. The situation is very similar to the formation of coal. As with coal, apparently, deep burial, some heat, and much time are necessary to produce gas and oil. Once formed, the petroleum must migrate to a permeable rock from which it can be extracted. Oil and gas reservoirs and traps will be discussed in Chapter 12 in connection with ground water. Not all oil forms in marine rocks. The oil shales of the Rocky Mountain area, which contain a large percentage of our petroleum reserves, formed in fresh-water lakes, probably by the same process.

The elements useful for producing atomic power are uranium and thorium. Thorium occurs mainly in hydrothermal veins. Uranium occurs in hydrothermal veins, pegmatites, and in some sedimentary rocks.

Geothermal power is relatively new in this country but has been used in Italy for some years. Basically, it is simply the use of steam escaping from the earth to generate power. Hot springs and geysers are fairly common, but at only a few places does steam escape from the earth continuously. Yellowstone National Park is a thermal area where super-heated water or steam could be obtained by drilling. Electric power is being generated in this way at The Geysers in north central California. (See Fig. 6-16.) The water escaping from these thermal areas is mostly rain water that has been heated, probably by contact with a cooling magma. Isotope studies show that very little of the water is primary water from a magma. In most areas, these thermal waters contain much dissolved material, a fact that is both an asset and a major problem. After being used to generate power, the waters can be evaporated by solar heat to harvest some of the salts. This is a clear gain. The problems are the disposal of the raw or partially harvested waters, and the corrosion and deposition that these waters cause in pipes. It

Lignite

Bituminous—cannel coal

Coke

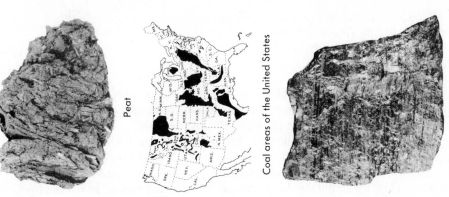

Peat

Coal areas of the United States

Bituminous—splint coal

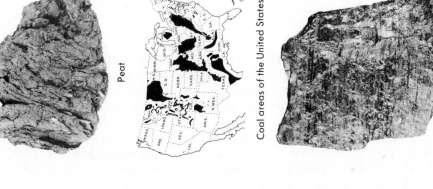

Peat bog

Bituminous—bright coal

Anthracite

Figure 6-15. Types of coal. Courtesy U.S. Bureau of Mines.

Figure 6-16. Geothermal steam escaping from wells at The Geysers, California. Photo by Pacific Gas and
Electric Co.

is interesting to note that at some places these waters contain appreciable amounts of
metals such as lead, copper, and iron. Thus these hot waters may closely resemble the
hydrothermal solutions that form mineral veins. As we learn more about such hot
waters and how to harness them, they may become an important power source. As will
be discussed later, we know that the earth's interior is hot, and perhaps some day we
can use deep holes to obtain heat from which substantial power can be generated.

Hot brines (133°F), containing large amounts of dissolved metals, have been found
in the Red Sea. The sea water that forms these brines is believed to have moved large
distances through rocks and to have leached the metals during this movement.

QUESTIONS

6-1. What are the similarities and differences between mineral resources and other
types of natural resources?

6-2. What types of mineral deposits are directly associated with igneous rocks? Might
some of these be found in sedimentary or metamorphic rocks?

6-3. Is the gold pan still a useful tool for the modern prospector?

6-4. What types of mineral deposits are associated directly with sedimentary rocks?

6-5. What parts of the earth are relatively unexplored for mineral resources?

6-6. Name an ore of lead_____.

zinc_____.

mercury_____.

iron_____.

copper_____.

SUPPLEMENTARY READING

Bachmann, H. G., "The Origin of Ores," *Scientific American* (June, 1960), Vol. 202, No. 6, pp. 146-156.

Barnea, Joseph, "Geothermal Power," *Scientific American* (January, 1972), Vol. 226, No. 1, pp. 70-77.

Committee on Resources and Man, National Academy of Science—National Research Council, *Resources and Man.* San Francisco: W. H. Freeman and Co., 1969, 259 pp. (paperback).

Degens, E. T. and D. A. Ross, "The Red Sea Hot Brines," *Scientific American* (April, 1970), Vol. 222, No. 4, pp. 32-42.

De Nevers, Noel, "Tar Sands and Oil Shales," *Scientific American* (February, 1966), Vol. 211, No. 2, pp. 21-29.

Flawn, P. T., *Mineral Resources: Geology, Engineering, Economics, Politics, Law.* Chicago: Rand McNally, 1966, 406 pp.

Hibbard, W. R., Jr., "Mineral Resources: Challenge or Threat?," *Science* (April 12, 1968), Vol. 160, No. 3824, pp. 143-149.

Lang, A. H., "Prospecting in Canada," Geological Survey of Canada, Economic Geology Series No. 7, 1956, 401 pp.

Lovering, T. S., *Minerals in World Affairs.* New York: Prentice-Hall, Inc., 1943, 394 pp.

Park, C. F., Jr., *Affluence in Jeopardy: Minerals and the Political Economy.* San Francisco: Freeman, Cooper and Company, 1968, 368 pp.

Riley, C. M., *Our Mineral Resources.* New York: John Wiley & Sons, Inc., 1959, 338 pp.

Skinner, B. J., *Earth Resources.* Englewood Cliffs, New Jersey: Prentice-Hall, Inc., 1969, 150 pp. (paperback).

U. S. Bureau of Mines, *Minerals Yearbook.* Washington, D. C.: Government Printing Office, published annually.

Wenk, Edward, Jr., "The Physical Resources of the Ocean," *Scientific American* (September, 1969), Vol. 221, No. 3, pp. 167-176.

PART 2

THE EXTERNAL PROCESSES – EROSION

The outer few miles of the earth are called the crust. We will define the term later. Although the crust is the only part of the earth accessible to direct observation, studying the earth by observations in the crust is like trying to read a book by its covers. Actually, the problem is even greater than this because 71 per cent of the surface is covered by oceans.

Classical geology developed from observations made on the continents. One of the most fundamental ideas or laws to have been developed is that of *uniformitarianism,* which says that *the present is the key to the past.* In other words, any structure in old rocks must have been formed by processes now going on upon the earth. Although this simple statement of uniformitarianism has limitations, it was very important to the development of geology. Uniformitarianism was not always accepted. We see erosion going on around us, note the material carried by muddy rivers, and conclude from this that rivers eroded the valleys in which they flow. Not too many years ago some geologists believed that valleys, especially deep valleys such as Yosemite, were formed by great earthquakes which split the earth open and that the rivers then flowed into them simply because they were the lower areas. This theory is called *catastrophism.*

Uniformitarianism implies slow processes going on over great lengths of time. Rapid processes, such as earthquakes, landslides, and volcanic eruptions, do occur, but, in general, these occur at widely separated times so that their average rate is slow.

The main process going on today on the earth's surface is clearly erosion and is shown by gullies on hillsides, landslides, and similar examples. The agents of erosion, of which rivers are far and away the most important, are constantly wearing down the continents; nevertheless, we still have mountains. The present rate of erosion is such that, theoretically, all topography will be removed in about 12 million years, but we know that erosion and deposition have been going on for a few billion years. Something must interrupt the work of the rivers and uplift the continents. Thus, much of geology is concerned with the struggle

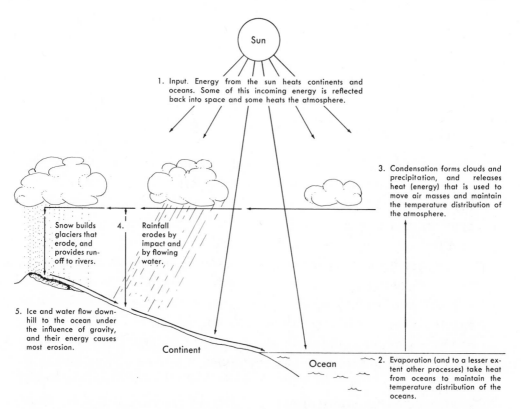

Figure II-1. The water cycle and the earth's energy balance. Incoming energy from the sun is distributed around the earth by the circulation of the oceans and the atmosphere. The water cycle is part of the energy exchange between the oceans and the atmosphere, and it is the immediate source of the energy used in most erosion.

between the external forces of erosion and the internal forces which cause uplift.

Some familiar demonstrations of the internal forces which cause uplift are earthquakes, volcanoes, and folded and tilted sedimentary rocks that are now high on mountains, but were originally formed as flat beds at or below sea level. These internal processes will be discussed later. Physical or dynamic geology is the study of this battle between the internal force of uplift and the external force of erosion. We will consider these two forces separately in order to eliminate confusion, but remember that they go hand in hand.

The underlying cause of erosion is downslope movements under the influence of gravity; the agents that cause such movements are water (rivers, etc.), ice (glaciers), and wind. The material that is moved by these agents comes from the weathering processes discussed earlier.

The sun's energy is the ultimate source of the energy used in erosion, as shown in Fig. II-1. This diagram also shows that erosion is a necessary by-product of the distribution of the sun's energy around the earth. Without this distribution of energy by the oceans and the atmosphere, the equatorial regions would become hotter and the poles colder because most of the sun's energy falls near the equator.

Chapter 7

Downslope Movement of Surface Material

Importance of Downslope Movement

All of the surface material of the earth tends to move downslope under the influence of gravity. The surface material is more or less weathered and is generally termed *overburden* to differentiate it from bedrock. These downslope movements shape the earth's surface and so must be understood to interpret landscapes. The downslope movement generally ends at a river or stream that carries the material away. Thus, rivers and downslope movements work together to shape the landscape. A small example of this, discussed in the following chapter on rivers, shows that although a river downcuts to form a valley, the shape of the valley sides is determined by downslope movements, which widen the narrow cut made by the river. (See Fig. 8-10.) The efficiency of downslope movements is easily shown by analysis of slopes. Such analysis shows that most of the earth's surface slopes less than 5 degrees and that very little of the earth's surface slopes more than 45 degrees. Many forms of downslope movement are slow processes. However, when slopes are disturbed by buildings, roads, or irrigation, more rapid movements may be triggered, causing much damage. (See Fig. 7-1.) Thus, knowledge of these processes is important when any type of engineering project is considered on even moderately sloping ground. (See Fig. 7-2.)

Figure 7-1. Urban landslide in Oakland, California. This slide developed after heavy winter rains. Construction on the top of the hill may have altered the infiltration of rain water. Only a small part of the slide is shown in the photo.

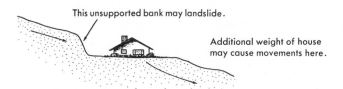

Figure 7-2. Hillside construction can cause earth movements if not properly planned.

Types

Downslope movements may be rapid or very slow; and they may involve only the very topmost surface material, or they may be massive movements that involve the total thickness of overburden or even bedrock. The processes at work on any slope depend on many factors. On steep slopes no overburden generally can develop; therefore, the dominant process on such slopes is *rockfall,* and the slope that develops at the foot is called a *talus.* (See Fig. 7-3.) On less steep slopes—how much less depends on the climate which controls the weathering processes—overburden may develop, and this material may move downslope in a number of ways discussed below. Thus weathering,

Figure 7-3. A talus develops at the foot of a steep slope as a result of mechanical weathering as shown in this view of Bryce Canyon National Park, Utah. Photo from Union Pacific Railroad.

which is controlled by the climate, determines the type of downslope movement, thereby causing the different aspects of the topography in a humid area and a dry area. (See Fig. 7-4.) The shapes of slopes and the theories of how these shapes are developed will be discussed in Chapter 8. The structure and rock types on a hill may also control the erosion processes. Differences in rate or kind of weathering can determine the amount of material available for downslope movement, and so the underlying bedrock structure may be etched in relief.

Figure 7-4. The Appalachian Mountains in North Carolina. Compare the rounded hillsides shown here with the angular topography of arid areas shown in Figure 7-3. Photo by A. Keith, U.S. Geological Survey.

Erosion by Rain

Falling rain erodes surface material in two ways, by impact and by runoff. The impact of raindrops loosens material and splashes it into the air (see Fig. 7-5); and on hillsides, some of this material falls back at a point lower down the hill. Raindrops hit the ground with a velocity of about 30 feet per second. The resistance of the air controls this velocity. One-tenth of an inch of rain amounts to about 11 tons of water on each acre,

Figure 7-5. Erosion by raindrop. From U.S. Department of Agriculture, 1955 Yearbook, *Water.*

and the energy of this water is about 55,000 foot-pounds per second. Ninety per cent of this great amount of energy is dissipated in the impact. Ninety per cent of the impact splashes are to a height of one foot or less, and the lateral splash movement is about four times the height. Because impact erosion is most effective in regions that have little or no grass or forest cover, it is most effective in desert areas (or newly plowed fields) that are subjected to sudden downpours such as thunderstorms. Splash erosion accounts for the otherwise puzzling removal of soil from hilltops where there is little runoff. It can also ruin soil by splashing up the light clay particles that are then carried away by the runoff, leaving the silt behind.

The rain water that does not infiltrate into the ground runs down the hillside and erodes in the same ways that river water erodes. This water may gather in lower places and cut gullies. The amount of erosion caused in these ways depends on the steepness of the slope and on the type and amount of vegetative cover. The effects of such erosion are best seen in areas where the vegetation has been removed by fires or man.

Massive Downslope Movements

The massive types of downslope movement that involve appreciable thicknesses of material are creep, earthflow, mudflow, and landslide.

A number of names are applied to these movements as described in the next few pages. They are termed *flows* if they move as viscous fluids, *slides* if they move over the surface, and *falls* if they travel in the air in free fall.

Creep is the slow downslope movement of overburden and, in some cases, bedrock. It is recognized by tilted poles and fence posts if the movement is great enough. Commonly, in such cases, trees will be unable to root themselves, and only shrubs and grass will grow on the slope. In other cases, where the creep is slower, the trunks of trees will be bent as a result of the slow movement. (See Figs. 7-6 and 7-7.)

If the amount of water in the overburden increases, an *earthflow* results. In this case a tongue of overburden breaks away and flows a short distance. An earthflow differs from creep in that a distinct, curved scarp is formed at the breakaway point. (See Fig. 7-8.)

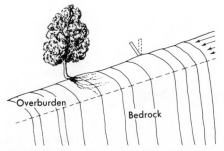

Figure 7-6. The effect of creep on fence posts and trees. This is caused by the slow downslope movement of the overburden.

Figure 7-7. Creep in slate. Photo by
F. H. Moffit, U.S. Geological Survey.

Figure 7-8. An earthflow, showing
the curved scarp at the breakaway
point.

With increasing water, an earthflow may grade into a *mudflow*. The behavior of
mudflows is similar to that of fluids. Rain falling on the loose pyroclastic material on
the sides of some types of volcanoes produces mudflows, and this is a very important
mode of transportation of volcanic material. Another type of mudflow is common in
arid areas. In this case, a heavy thunderstorm produces large amounts of runoff in the
drainage area of a stream. The runoff takes the form of rapidly moving sheets of water
that pick up much of the loose surface material. Because these sheet floods flow into
the main stream, all of this muddy material is concentrated in the main stream course.
As a result, a dry stream bed is transformed very rapidly into a flood. This flood of
muddy material is a mudflow, and it moves very swiftly, in some cases with a steep,
wall-like front. Such a mudflow can cause much damage as it flows out of the mountains. Eventually, loss of water (generally by percolation into the ground) thickens the
mudflow to the point that it can no longer flow. (See Fig. 7-9.)

Figure 7-9. Mudflow in the San Juan Mountains, Colorado. Its source can be seen on the skyline, and it ends in a lake. The trees growing on the mudflow show its age. Photo by W. Cross, U.S. Geological Survey.

Another type of mudflow, called *solifluction,* occurs when frozen ground melts from the top down, as during warm spring days in temperate regions or during the summer in areas of permafrost. (See Fig. 7-10.) In this case the surface mud flows downslope. Solifluction causes many problems in construction, especially in the far north areas of permafrost. Permafrost areas, as the name implies, are areas where the deeply frozen ground does not completely thaw during the summer.

Another type of movement of soil material, called *frost heaving,* is associated with freezing and thawing. In this case the expansion of water as it freezes pushes pebbles upward through the soil or overburden. This phenomenon is well known to farmers in regions such as New England where each spring their fields are covered with pebbles that were not visible in the fall. In the Middle Ages peasants thought that the stones grew in the soil just as crops do. Frost heaving is a complex process, not yet fully understood. The small expansion of water during freezing is not enough to account for the several inches of uplift of pebbles or fence posts common in areas of frost heaving. Most of the soil water is in the pore spaces of the fine material. When this water freezes, it expands and so pushes nearby pebbles upward. During the process of freezing, soil moisture migrates to these pockets of fine material and forms lenses of ice; it is this additional volume of ice that causes the several inches of uplift. When melting occurs, the surface tension of the water in the fine sediments pulls the fine material together, leaving the larger pebbles at nearly the height to which the expansion pushed them.

In very cold areas, frost heaving causes stone rings or polygons. (See Fig. 7-11.) In such areas, the extreme cold causes contraction cracks to form at the surface. In summer, melt water fills the cracks, and frost heaving proceeds as outlined above. Pebbles rise to the surface, forming the polygons.

Figure 7-10. Solifluction probably caused the smooth slopes in the middleground in this view of the Sierra Nevada in California. Photo from U.S. Forest Service.

Figure 7-11. Stone polygons formed by frost action. A. Close-up view in Alaska. Photo by H. B. Allen, U.S. Geological Survey.

A.

Landslides are rapid slides of bedrock and/or overburden. Two types involve bedrock: the *landslide* or *rockslide,* and the *slump.* Rockslides develop when a mass of bedrock breaks loose and slides down the slope (Fig. 7-12). In many cases the bedrock is broken into many fragments during the fall, and this material behaves as a fluid, spreading out in the valley. It may even flow some distance uphill on the opposite side of the valley if the valley into which it falls is narrow enough. Such landslides are sometimes called *avalanches,* but this term is best reserved for snow slides. Landslides are generally large and destructive, involving millions of tons of rock. As suggested in the diagrams, landslides are apt to develop if planes of weakness, such as bedding or jointing, are parallel to a slope, especially if the slope has been undercut by a river, a glacier, or men. Thus, landslide danger can be evaluated by geologic study. Slides involving overburden are called *debris slides.*

Slumps tend to develop in cases where a strong, resistant rock overlies weak rocks. Note in Fig. 7-13 the curved plane of slippage, the reverse tilt of the resistant unit that may provide a basin for a pond to develop, and the rise of the toe of the slump, which is most pronounced where the underlying rock is very weak and can flow plastically. Unlike rockfalls, slumps develop new cliffs nearly as high as those previous to the slump, setting the stage for a new slump. Thus slumping is a continual process, and generally, in areas of slumping, many previous generations can be seen far in front of the present cliffs.

B.

Figure 7-11. (cont.) Stone polygons formed by frost action. B. Oblique aerial view near Churchill, Manitoba. Photo from Department of Energy, Mines and Resources, Ottawa, Canadian Government Copyright (A 1426-66).

A.

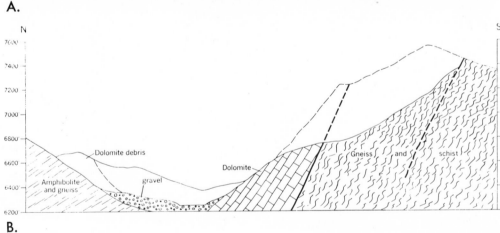

B.

Figure 7-12. Landslide near Hebgen Lake, Montana. A. The slide was triggered by an earthquake and came from the left, leaving a large scar on the mountain side. The debris traveled uphill on the right side of the valley. The slide dammed the Madison River, creating the lake in the foreground. The light-colored area on the slide is a cut made to drain the lake partially. Photo by J. R. Stacy, U.S. Geological Survey. B. This cross-section of the slide was drawn looking in the opposite direction from A. From J. B. Hadley, U.S. Geological Survey, *Professional Paper* 435, 1964.

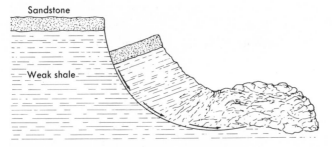

Figure 7-13. Slump block showing curved plane of slippage and reverse tilt of the slump block.

Triggering

Massive downslope movements may be triggered in many ways. The most common are (1) undercutting of the slope, (2) overloading of the slope so that it cannot support its new weight and, hence, must flow or slide, (3) vibrations from earthquakes or explosions that break the bond holding the slope in place (see Figs. 7-12 and 7-15), and (4) additional water. The addition of water is generally seasonal, and this is why many newly made cuts stand until the following spring. The effect of the water is twofold: it adds to the weight of the slope, and it lessens the internal cohesion of the overburden. Although its effect as a lubricant is commonly considered to be its main role, this effect actually is very slight. The main effect is lessening the cohesion of the material by filling the spaces between the grains with water. A simple experiment with sand will illustrate this point. Dry sand can be piled only in cones with slopes slightly over 30 degrees. Damp sand, however, can stand nearly vertically because the small amounts of moisture between the grains tend to hold the grains together by surface tension. (See Fig. 7-14.) Additional water completely fills the intergrain voids; thus, there is no surface tension to hold the grains together, and a mud forms that flows outward as a fluid.

Slides associated with the 1964 Alaskan earthquake show two other factors. (See Figs. 7-15, 7-16, and 7-17.) Slides may occur under water and create destructive waves, commonly called tidal waves, but better called seismic sea waves because they have nothing to do with tides. (See Chapter 13.) The Alaskan slides also illustrate sliding by a peculiar type of clay that behaves as a fluid when disturbed. This type of clay is

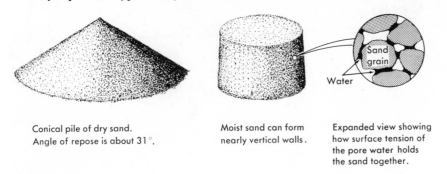

Conical pile of dry sand.
Angle of repose is about 31°.

Moist sand can form
nearly vertical walls.

Expanded view showing
how surface tension of
the pore water holds
the sand together.

Figure 7-14. The effect of water on the cohesion of sand.

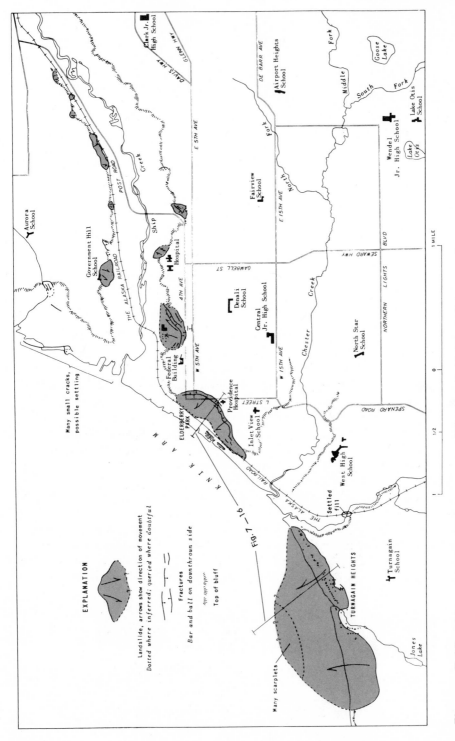

Figure 7-15. Map of Anchorage, Alaska, showing the location of slides caused by the Good Friday Earthquake of 1964. Cross-sections of the L Street and Turnagain Heights slides are shown in Figure 7-16. From Arthur Grantz and others. U.S. Geological Survey, *Circular* 491, 1964.

170

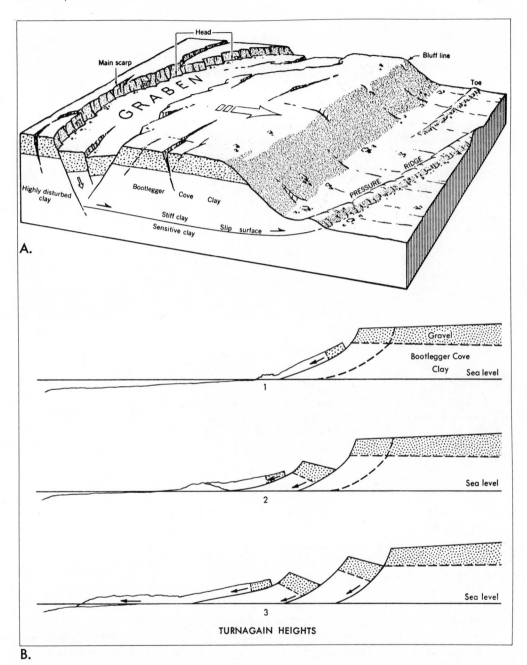

Figure 7-16. Diagrams showing L Street slide (A.) and probable development of Turnagain Heights landslide (B.). See Figure 7-15 for locations. A graben is a down-dropped block. A. From W. R. Hansen, U.S. Geological Survey, *Professional Paper* 542-A, 1965. B. After Arthur Grantz and others, U.S. Geological Survey, *Circular* 491, 1964.

A.

B.

Figure 7-17. Landslides at Anchorage, Alaska, caused by the 1964 earthquake. See Figure 7-15 for locations. A. Looking west at Government Hill School. Note the slumping. B. Aerial view looking south at the hospital on Fourth Avenue. Photos by W. R. Hansen, U.S. Geological Survey, *Professional Paper* 542-A, 1965.

termed *quick clay;* it is generally formed from the very fine products of glacial erosion and so is found mainly at northern latitudes. These clays contain much water, commonly 50 per cent or more by weight, and the tiny clay platelets are randomly oriented. Because most quick clays were deposited in the ocean, the interstitial water is saline. The salt ions in this water help to keep the clay particles stuck together, but if fresh ground water moves through the clay, flushing out the salt, the clay becomes unstable. Vibrations may then squeeze the water out, bringing the clay platelets into contact and making the mixture a fluid. (See Fig. 7-18.)

Recognition of any of the types of mass wasting can be of great economic importance when contemplating any construction. The history of many housing developments

Figure 7-18. A dramatic example of quick clay. On the left an undisturbed sample of quick clay supports 11 kilograms (24 pounds). On the right another sample of the same quick clay pours like a liquid after being stirred. Photo from C. B. Crawford, National Research Council of Canada.

suggests that even house or lot buyers could benefit by the ability to recognize them. Generally, mass movements are best detected by observation from a distance followed by detailed close study. Generally, a slope in motion will have few trees and may have a bumpy, hummocky appearance. Because some types of glacial deposits also have this appearance, one should know how to recognize glacial topography, which will be discussed in Chapter 10.

QUESTIONS

7-1. Erosion tends to level the land surfaces. Why aren't the continents worn flat?

7-2. Describe the slow means of downslope movement.

7-3. Describe the rapid means of downslope movement.

7-4. Of what importance is raindrop impact in erosion?

7-5. How might rapid mass movements be triggered? What is the role of water in triggering?

SUPPLEMENTARY READING

Ellison, W. D., "Erosion by Raindrop," *Scientific American* (November, 1948), Vol. 179, No. 5, pp. 40-45. Reprint 817, W. H. Freeman and Co., San Francisco.

Kerr, P. F., "Quick Clay," *Scientific American* (November, 1963), Vol. 209, No. 5, pp. 132-142.

Sharp, R. P., and L. H. Nobles, "Mudflow of 1941 at Wrightwood, Southern California," *Bulletin,* Geological Society of America (May, 1953), Vol. 64, No. 5, pp. 547-560.

Sharpe, C. F. S., *Landslides and Related Phenomena.* New York: Columbia University Press, 1938, 137 pp.

Chapter 8

Geologic Work of Rivers and Development of Landscapes

In the space of one hundred and seventy-six years the lower Mississippi has shortened itself two hundred and forty-two miles. This is an average of a trifle over one mile and a third per year. Therefore, any calm person, who is not blind or idiotic, can see that in the Old Oölitic Silurian Period, just a million years ago next November, the Lower Mississippi River was upwards of one million three hundred thousand miles long, and stuck out over the Gulf of Mexico like a fishing rod. And by the same token any person can see that seven hundred and forty-two years from now the Lower Mississippi will be only a mile and three quarters long, and Cairo and New Orleans will have joined their streets together, and be plodding comfortably along under a single mayor and a mutual board of aldermen. There is something fascinating about science. One gets such wholesale returns of conjecture out of such a trifling investment of fact.

> Mark Twain,
> *Life on the Mississippi,*
> Chapter 17

Rivers are agents of erosion, transportation, and deposition. That is, they carve their own valleys and carry the eroded material downstream where it is either deposited by the river or delivered into a lake or the ocean. Rivers are the most important agents in transporting the products of erosion: they drain vastly larger areas than do glaciers, and, even in their infrequent times of flowage on deserts, they are able to carry more material than the wind can. Thus it is the rivers that receive and transport the products of weathering that are fed into them by runoff from precipitation as well as by downslope movements. Weathering, downslope movement, and rivers work together to shape the landscape. Although the geologic role of rivers has been known for a long time, rivers are complex systems with many variables and much remains to be learned about them. Even more factors are involved in the development of landscape, and this is one of the active fields of research in geology.

Flow of Rivers

Rainfall and other forms of precipitation are the source of river water. Only a small fraction of the total precipitation forms the rivers, but the amount of water is great. The fate of most of the rest of the precipitation is described in Chapter 12. The energy of rivers comes from the downward pull of gravity acting on the river water. In physics this is called *potential energy* and is equal to the mass times the acceleration of gravity times the height (mgh). Table 8-1 shows the tremendous amount of energy theoretically available from rivers. Only a small fraction of this energy is actually used to do geologic work. If this were not so, the landscapes of the earth would be far different. To see where the potential energy of rivers goes, and to gain some insight into river processes, the velocity of river flow must be considered.

Table 8-1. Calculation of approximate amount of energy available from rivers and the theoretically possible average velocity of rivers.

Potential energy = mass × acceleration of gravity × height

P.E. = mgh

mg = weight

P.E. = weight × height

Average elevation of continents = 2755 feet (See Chapter 14)
Average runoff of continental U.S. = 1,200 billion gallons per day
(See Chapter 12)

1 gallon of water weighs 8.35 pounds
P.E. = 1,200,000,000,000 × 8.35 × 2755
= 2.76 × 10^{16} foot-pounds per day

(The total energy used in the U.S. in 1960 was about
3.1 × 10^{19} foot-pounds, or about 1,000 times more.)
If all of the river's energy was used to heat the water,
it would only rise about 3.5 °F.

The velocity can also be calculated:
Potential energy = Kinetic energy

$mgh = \frac{1}{2}mv^2$

$v^2 = 2gh$

= 2 × 32 × 2755

= 176,000

v = 418 feet per second = 286 mph

The average velocity of river flow is less than five miles per hour, and most rivers are in the range from a fraction of a mile per hour to about 20 miles per hour. (Fifteen miles per hour is the fastest ever measured.) These velocities are far less than the 286 miles per hour average velocity (Table 8-1) predicted on the basis of the potential energy of rivers. Apparently the energy of rivers goes largely into heating the water by friction along the bed and banks and by turbulence. It is this friction and turbulence, however, that enable a river to erode and transport, as will be described in the next section. (See Fig. 8-1.)

The flow of a river is determined by many factors. The discharge, or flow, can be defined as the amount of water passing a given cross-section of the river in a time unit (usually given in cubic feet per second, c.f.s.). The discharge, then is equal to the area of the cross-section times the velocity. The cross-section is determined by the shape of

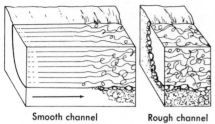

Smooth channel   Rough channel

Figure 8-1. Idealized diagram showing flow of water. The lines represent
paths of water particles. Where the
channel is smooth, the water flows in
straight paths; but where the channel is
rough, the flow is turbulent. Turbulence
increases with increasing velocity.

the river's channel, and the velocity is determined by a number of factors, the chief of
which are the roughness, shape, and curving of the channel and the slope of the river.
The amount of water available for runoff is also a factor, and this is determined by such
climatic factors as precipitation, evaporation, vegetation, and permeability of soil and
bedrock. On any given river, all of the factors that control the flow vary from place
to place and with season.

The channel is probably the most important feature of a river; however, it is difficult
to study because it is generally obscured by the water. In recent years, many soundings
have been taken, and much has been learned about channel shape and, as will be seen
in the next section, the changes in shape that occur with changes in flow. The shape
of a channel greatly influences drag or friction. In general, the smaller the wetted
perimeter, the less the drag. This suggests that if this were the only factor, rivers in
eroding their channels might tend to form semicircular cross-sections because this
shape has the least perimeter for its area. However, many other factors influence
channel shape and actual channels are quite irregular. (See Fig. 8-2.) At flood stage,
rivers are higher than their channels and spill over into their valleys. This greatly
increases the wetted perimeter, and the resulting drag slows the flow in the flooded
valley, reducing the potential damage to the inundated area. The size and amount of
the material transported by the river and the erodability of the banks have a great
influence on channel shape. (These factors will be described in the next section.) Figure
8-3 shows the distribution of velocity and turbulence in an idealized river channel both
on a straight reach and on curves. A river changes in a systematic way from its
headwaters to its mouth. Downstream the discharge increases because a larger area is
drained; and to accommodate this increased flow, the width and depth both increase,
as does the velocity. (See Fig. 8-4.)

The slope of a river also changes systematically from headwaters to mouth as shown
in Fig. 8-5. The vertical scale in this figure is greatly exaggerated. A typical river has
a slope of several hundred feet per mile near its headwaters and just a few feet per mile
near its mouth. The lower 600 miles of the Mississippi River has a slope of less than
one half foot per mile. A steep, rapidly flowing river generally has a slope of 30 to 60
feet per mile. A slope of one degree is 92.4 feet per mile. Notice that the profile (Fig.

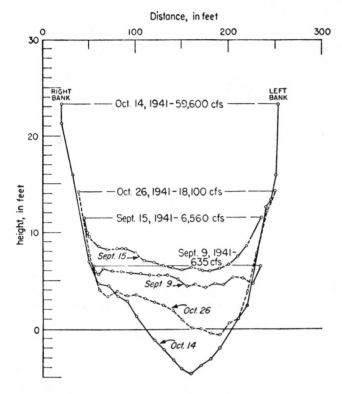

Figure 8-2. Changes in channel of San Juan River near Bluff, Utah, during flood in 1941. Note the changes in elevation of both the water surface and the channel depth. From L. B. Leopold and Thomas Maddock, Jr., U.S. Geological Survey, *Professional Paper* 252, 1953.

8-5) is not smooth but has irregularities, especially where tributary streams join. These irregularities are caused by abrupt changes in discharge and load. This suggests that the overall concave upward shape of the long profile is also the result of a balance between discharge, load, and slope, as well as other factors. A brief study of river processes in the next section will give insight into this balance.

Erosion and Transportation

The main geologic work of rivers is transportation. The load is delivered to the river by downslope movement of material on the valley sides, by erosion by the river itself, and by tributary streams. The load of a river can be considered in three parts: dissolved load, suspended load, and bed load. The total amount of material delivered by the rivers of the world to the oceans is estimated to be about 40 billion tons per year. Man's activities, especially farming and construction, contribute much of this load.

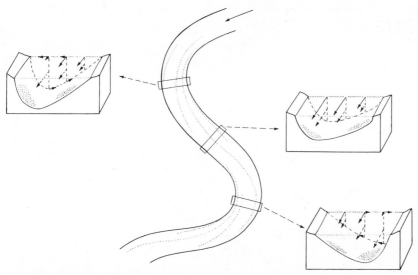

Figure 8-3. Distribution of velocity and turbulence in a river channel. In the cross-sections, the arrows show the direction and magnitude of the velocity. The surface velocity is indicated by two components, and the path of the surface water is shown by the dotted lines in the map view. The areas of most turbulence are indicated by stippling in the cross-sections.

The dissolved load comes from chemical weathering and from solution by the river itself. Unlike the rest of the stream's load, it has no effect on the flow of the stream. The dissolved load of the world's rivers is estimated at 3.9 billion tons per year, just under ten per cent of the total load. Table 8-2 shows the average dissolved load. About half of the dissolved load is bicarbonate ion (HCO_3^-) that results from the solution of carbon dioxide in water, as described in weathering in Chapter 4.

The suspended load consists of the fine material that, once in the water, settles so slowly that it is carried long distances. As the flow of a river increases, some of the finer bed load becomes suspended load so that the distinction can be made only at a given moment. Suspended material is the cause of muddy river water. Because clay particles settle very slowly, they are carried relatively long distances. Silt settles much more rapidly; it is carried as suspended load only by turbulent 'water which has enough upward currents to keep the silt from settling. Thus clay is carried farther than silt, resulting in sorting of sediment by rivers. The suspended load is estimated to be almost two-thirds of the total load carried by rivers and amounts to over 25 billion tons per year for all of the world's rivers.

The bed load is the material that is moved by rolling and sliding along the bottom. It, too, is somewhat sorted by size; also the particles are worn by abrasion, making the pebbles and boulders in a typical stream bed subspherical. In addition to abrasion and solution, the bed-load pebbles are at times cracked and broken by impacts during movement. Almost all of the movement of most bed loads occurs during the generally very limited times of high flow. Thus the floods that occur seasonally, or even less commonly, mark times that bed loads are moved. In some rivers the ratio of maximum to minimum flow is a few hundred to one. In the summer or fall, it is difficult to believe that the large boulders in the stream bed, particularly a small stream, are part of the

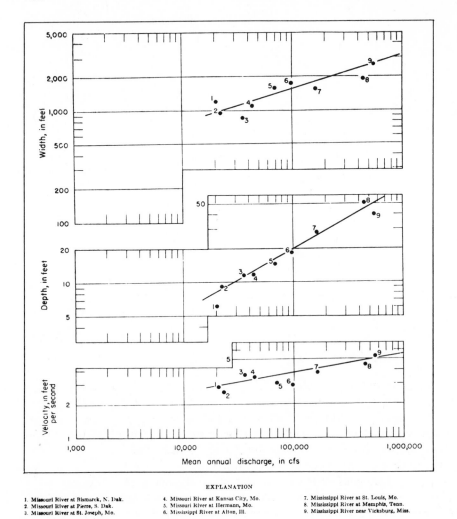

EXPLANATION

1. Missouri River at Bismarck, N. Dak.   4. Missouri River at Kansas City, Mo.   7. Mississippi River at St. Louis, Mo.
2. Missouri River at Pierre, S. Dak.   5. Missouri River at Hermann, Mo.   8. Mississippi River at Memphis, Tenn.
3. Missouri River at St. Joseph, Mo.   6. Mississippi River at Alton, Ill.   9. Mississippi River near Vicksburg, Miss.

Figure 8-4. Graphs showing the increase in width, depth, velocity, and discharge downstream in the Missouri River and lower Mississippi River. From L. B. Leopold and Thomas Maddock, Jr., U.S. Geological Survey, *Professional Paper* 252, 1953.

bed load and are being moved by the stream. (See Fig. 8-6.) One must remember that the largest boulders may move only during periods of very great flow that may occur a few times each century. However, if one visits such a stream during the spring when it is filled with melt water, the audible clicks when a large boulder rolls or slides into its neighbor can be heard. The bed load of most streams is about a quarter or less of the total load. The maximum size particle that can be moved by a river is proportional to the square of the velocity. Thus doubling the velocity increases by four times the size

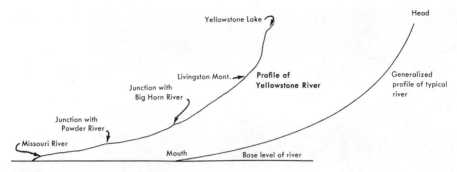

Figure 8-5. Longitudinal river profiles. Vertical scale exaggerated. Yellowstone River profile from U.S. Geological Survey, *Water Supply Paper* 41, 1901.

Table 8-2. Average composition of dissolved material in river waters of the world. From *Data of Geochemistry,* sixth edition, Chapter G, "Chemical Composition of Rivers and Lakes," D. A. Livingstone, U.S. Geological Survey, *Professional Paper* 440-G, 1963.

Bicarbonate (HCO$_3$)	58.4 parts per million	
Calcium	15.0 " " "	
Silica (SiO$_2$)	13.1 " " "	
Sulfate (SO$_4$)	11.2 " " "	
Chlorine	7.8 " " "	
Sodium	6.3 " " "	
Magnesium	4.1 " " "	
Potassium	2.3 " " "	
Nitrate	1.0 " " "	
Iron	0.7 " " "	
	119.9 " " "	

of particle that can be moved. In general, doubling the velocity increases the amount of material in transport between four and eight times. The movement of bed load during times of high flow is shown by the changes in size and shape of the channel. (See Fig. 8-2.)

Rivers erode by several processes, some of which were necessarily discussed in describing transportation. If parts of the channel are soluble, erosion by solution will occur. Abrasion occurs as a result of the transported material rubbing against the sides and bottom of the channel. A common type of abrasion results in cylindrical holes, called potholes, worn in bedrock, apparently by stones that are spun by turbulent eddy currents. (See Fig. 8-7.) Potholes can also be caused by *cavitation,* or bubble erosion.

Figure 8-6. A stream at low water level, showing part of the bed load that is moved during times of high discharge. Photo from California Division of Beaches and Parks.

This type of erosion is a problem on ship propellers and similar devices. It occurs when rapid increase in velocity reduces the internal pressure of the water, resulting in the formation of bubbles. These bubbles implode when the velocity decreases, and the resulting shock waves cause the erosion. Cavitation may also occur in plunge pools below waterfalls, although much of the erosion in such places is caused by the moving water itself. The impact and drag of the water erode and transport material. As noted earlier, the size of particle that can be moved in this way is proportional to the square of the velocity; however, velocities great enough to move sand, and even coarser material, are necessary to cause erosion of the smooth surface of consolidated clay.

The rate of erosion of the continents is determined by measuring the amount of material carried by rivers. The average rate of erosion in the United States, determined in this way, is 2.4 inches per thousand years. The lowest rate of erosion is 1.5 inches per thousand years, in the basin of the Columbia River, and the largest is 6.5 inches per thousand years, in the Colorado River. The rate of erosion depends on a complex

Figure 8-7. Potholes in a river bed cut in granitic rock. The potholes are formed during times of much greater flow. Photo by Mary Hill, courtesy California Division of Mines and Geology.

relationship of many factors, such as rainfall, evaporation, and vegetation, but note that the maximum rate of erosion occurs in areas with about ten inches of rainfall per year. (See Fig. 8-8.)

Erosion tends to deepen, lengthen, and widen a river valley. The ways that a river can deepen its channel were just discussed. At the head of a stream any downcutting lengthens the valley. This process is called headward erosion and is the way that streams eat into the land masses. (See Fig. 8-9.) A stream valley is widened by two methods. The valley side processes, such as creep and landsliding, widen the valley. (See Fig. 8-10.) Lateral cutting by the stream itself is the other process.

Lateral cutting is most pronounced on the outsides of curves where the valley sides may be undercut. Thus a curving stream widens its valley, and straight stretches of streams longer than ten times the width are not common. Clearly, any irregularity on a stream bottom or bank can deflect the flow and cause a curve to develop. Once formed, a curve tends to migrate, thus widening the valley. (See Fig. 8-11.) Many rivers have distinctive symmetrical curves called *meanders.* (See Fig. 8-12.)

The migration of meanders is shown in Fig. 8-13, which also shows how the meanders may intersect, causing ox-bow lakes. Meanders are not formed in the same way as are the irregular curves on a river. The curves that form meanders are not common curves, such as half circles. The meander-forming curve is the one that distributes the river's loss of energy most uniformly. Any other curve would tend to concentrate erosion and deposition at some point on the curve. (The development of river valleys by all of these methods of erosion will be described later.)

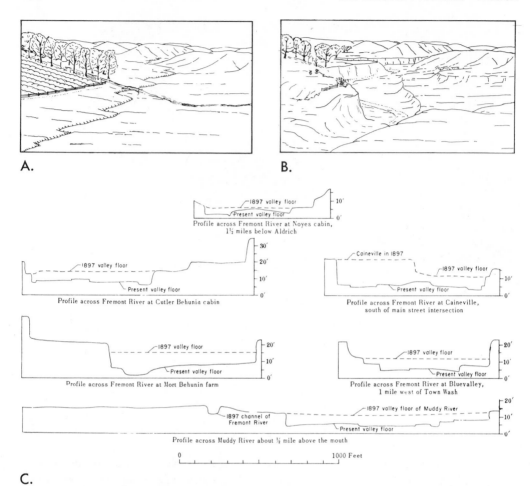

Figure 8-8. Examples of river erosion in Henry Mountains region, Utah, since 1897. Climatic change is the probable reason for this erosion. Parts A. and B. show Pleasant Creek at Notom. The creek in part B (present day) is 20 feet deep. Part C. shows a number of cross-sections of the Fremont River. From C. B. Hunt, U.S. Geological Survey, *Professional Paper* 228, 1953.

The River as a System

Rivers do not occur singly in nature. Each has tributaries and each tributary has smaller tributaries and so on, down to the smallest rill. All of these parts function together to form a river system, or drainage basin. A change in any part of the system generally affects all of the other parts. One obvious illustration of this balance is that all the tributary streams meet the main stream at the level of the main stream. This accordance of junctions is too universal to be due to chance. (This accordance is not found in glaciated areas where the stream levels have been changed by glacial erosion.)

Figure 8-9. The two streams in the foreground of this oblique aerial view are lengthening by headward erosion. The bigger one may ultimately capture the stream in the middleground. Photo by J. R. Balsley, U.S. Geological Survey.

Figure 8-10. Widening of a stream valley by a small slide. In this case a forest fire has reduced the vegetation and thus caused the slide. Photo from U.S. Forest Service.

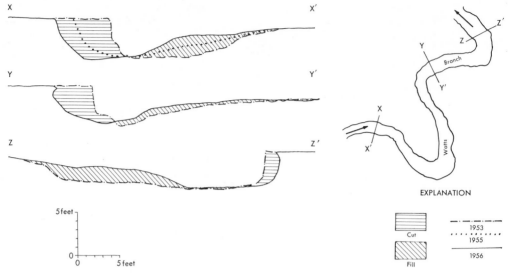

Figure 8-11. Erosion and deposition on Watts Branch near Rockville, Maryland, from 1953 to 1956. From M. G. Wolman and L. B. Leopold, U.S. Geological Survey, *Professional Paper* 282C, 1957.

Figure 8-12. Meanders on the Green River in Utah. Cut-off meanders and old meander scars show the movement of the river on the flood plain. Looking north (upstream) toward the Book Cliffs. Photo from U.S. Geological Survey.

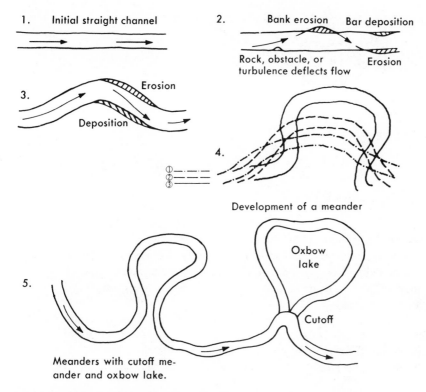

1. Initial straight channel

2. Bank erosion   Bar deposition
   Rock, obstacle, or   Erosion
   turbulence deflects flow

3. Erosion
   Deposition

4. ①
   ②
   ③

Development of a meander

5. Oxbow lake
   Cutoff

Meanders with cutoff me-
ander and oxbow lake.

Figure 8-13. Development of meanders.

    The main work of a river is to transport debris, and river systems regulate themselves so that they are able to transport the debris delivered to them. This is generally accomplished by changing the slope, by either erosion or deposition, so that at any given point for the discharge and channel characteristics the velocity is such that the debris is transported. Thus a change in any part of the system will affect the balance of the rest of the system. A stream that has just the right gradient to be able to transport the debris delivered to it by its own erosion, its tributaries, downslope movement of material, and rain wash is called a *graded stream.* This ability is a long-term average so that a given stretch of river at one season may be eroding its bed, and at another time it may be building up its bed. These changes in regimen vary from year to year with such climatic influences as amount and intensity of precipitation.

    The long profile of a river (see Fig. 8-5) was mentioned earlier, and the reason for its shape can now be appreciated. The slope near the head must be steep for the river to be able to do its work of eroding and transporting with the small discharge it has available. Farther downstream the velocity and discharge both increase and particle size decreases as does the number of banks (see Fig. 8-14), all of which permit the river to do its work with a lesser slope, even though the total load is increased. Thus the concave upward slope of a river enables the most uniform rate of work per unit of length and at the same time minimizes the work done in the whole system.

    In most river systems the profile is more complicated because the ability of a river to erode and transport depends largely on the velocity, the amount of flow, the gradient,

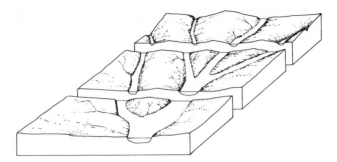

Figure 8-14. The number of banks decreases downstream, reducing the friction.

and the shape and roughness of the channel. At the point where a tributary joins the main stream, the regimen of the main stream is abruptly changed by the additional flow and the increased debris from the tributary. The main stream reacts to this change by altering its gradient so that it can handle the new material. These changes are superimposed on the simple profile.

The limit to which the river can erode is, of course, the elevation of its mouth. This limit is called the *base level* of the river. (See Fig. 8-5.) For most big rivers the base level is sea level, but for some that flow into lakes, it is the elevation of the lake. For tributary streams, the base level is the elevation at which they join the main streams.

Deposition

River-deposited sedimentary rocks (fluvial rocks) are relatively rare because rivers deliver most of their loads to lakes or to the ocean. River deposits in the form of deltas develop where rivers enter standing water; even in this case, however, much of the river's load may be reworked by offshore currents, thus becoming marine or lacustrine (lake-deposited) rocks. Deltas form because the velocity of river water is slowed so abruptly on entering standing water that the bed load is dropped. The dissolved and suspended loads may be carried farther out. (See Figs. 8-15 and 8-16.)

Alluvial fans are similar to deltas and are formed where streams flow out of mountains into broad, relatively flat valleys. In this case, the flatness of the valley slows the stream, causing deposition. In addition, especially in arid areas, the mountain stream may seep into the ground in the broad valley. Because the stream constantly shifts its position on the fan, the fan develops a cone shape (Fig. 8-17).

When the flow of a river fills its bed up to the top of the banks, any increase in flow will cause flooding. The floodwater spills over the river banks, filling part or all of the valley. A river is doing its maximum erosion and transportation at the time of flood, and as the floodwater spreads out, its velocity is quickly reduced, causing deposition near the banks of the river. Such deposits are termed *levees* (or natural levees, to distinguish them from man-made levees built for flood protection). (See Fig. 8-18.) Although the levees help to keep the river within its banks, the area between the levee and the side of the river valley may be lower than the river at flood stage; this area is

Figure 8-15. The Nile delta photographed from Gemini IV. The dark area is cultivated land. Mediterranean Sea to the left and Sinai Peninsula in the middleground. The Nile has a number of distributaries, and this accounts for the large cultivated area. Photo from NASA.

subject to inundation when the flooding river crests over the levee top. Because the river valleys are prime agricultural areas and because the mechanism of levee building provides a continuing supply of new soil, many people accept periodic flooding and live in the valleys of the large rivers. These people should expect spring flooding, for it is the normal, not the abnormal, behavior of the river. To some extent they can be protected by upstream flood-control dams, which regulate somewhat the flow during flood stage, but this system is expensive and not always effective.

Rivers may also build up their beds by deposition. In some cases, a river may deposit material during a long period of time so that its meandering causes deposition covering an entire valley. Such a process produces a smooth surface. Later, the river may, because of some change in its regimen, cut into this surface and so form one or more lower terraces (Fig. 8-19). Terraces can also be formed by lateral cutting of bedrock by the river. One common cause of terrace building is associated with the recent glaciation. In areas where glaciers used to be active in the mountains, much debris of glacial erosion was rather suddenly released to the rivers when the glaciers melted. Because the rivers could not transport the material as rapidly as it was delivered, they deposited it in their valleys. As conditions returned to normal, they downcut into these deposits and formed terraces, most of which, in time, will be removed by erosion. As is shown in a later section, the study of terraces can tell much about the history of a river.

Braided streams are an unusual type of stream that forms under some conditions, especially if the load is large and coarse for the slope and discharge, and the banks are easily eroded. In a braided stream the channel divides and rejoins numerous times. (See Fig. 8-20.) Apparently the stream deposits the coarser part of its abundant load in order

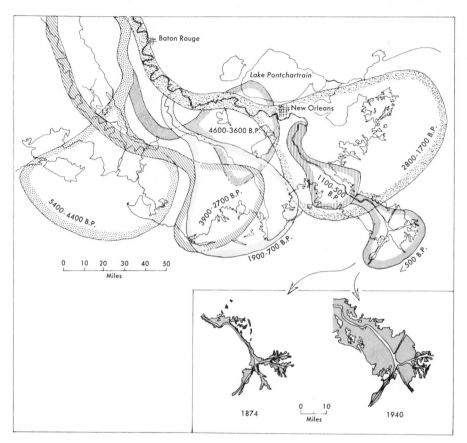

Figure 8-16. Development of the Mississippi River delta. The upper drawing shows the ages of subdeltas in years before present (B.P.). Upper drawing after Kolb and van Lopik, 1966, and lower drawings after Scruton, 1960.

Figure 8-17. Alluvial fans at canyon mouths. Photo by J. R. Balsley, U.S. Geological Survey.

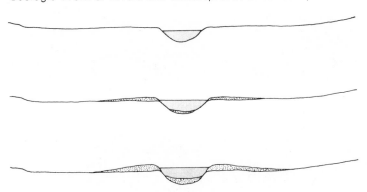

Figure 8-18. The development of levees. Levees form when silt is deposited beside the river during floods. If, as is generally the case, deposition also occurs on the river bed, the seriousness of later floods increases. The vertical scale is greatly exaggerated.

Figure 8-19. Terraces cut in river-deposited material by the Snake River. Teton Mountains on the skyline. Photo from Union Pacific Railroad.

to attain a steep enough slope to transport the remaining load. This deposition forces the stream to broaden and so erode its banks, adding to the abundant load, perhaps choking the channel. The places where two channels join have increased flow and reduced drag. At such places some of the bed load is moved, but this reduces the slope and deposition begins again; however, the load is moved in the process, albeit only a short distance.

Figure 8-20. Braided river. Aerial
view south up Nelchina River, Alaska,
toward glacier. Photo by J. R. Williams,
U.S. Geological Survey.

River Valleys

River valleys can be described as narrow or broad, although all intermediate steps
between the two exist. Narrow valleys are not much wider than the river channel, and
broad valleys are much wider than the river.

Narrow valleys, which have V-shaped cross-sections, are generally found where
rivers are actively downcutting. On many rivers they occur near the headwaters, but
narrow valleys can form anywhere. Many, but far from all, are downcutting, non-
graded rivers. Some narrow valleys occur in resistant rocks that slow the lateral cutting
of a river; they commonly have rapids and waterfalls where resistant rocks slow the
downcutting. Lakes may also form in narrow valleys, but, like falls and rapids, they
are only temporary features. (See Figs. 8-21, 8-22, 8-23, 8-24, and 8-25.)

Valley widening begins, in general, after a stream has reached grade and is no longer
rapidly downcutting. This occurs first, on most rivers, near the mouth where the
increased flow allows the river to erode more efficiently. Wide flood plains characterize
the lower parts of many rivers. (See Fig. 8-12.) In areas of weak rock, widening is also
common. Wide valleys are generally fairly flat in cross-section. Meanders are common
features of wide valleys, especially in areas with uniform banks composed of fine
sediment. Meanders may alternate with areas of irregular curves. Another feature,
familiar to fishermen, is the alternation of riffles and pools on most rivers. The riffles
form on alternate sides of the river and are composed of the coarser part of the bed
load. (See Fig. 8-26.)

Figure 8-21. An extremely narrow canyon with nearly vertical walls cut in sandstone. Downcutting was apparently faster than downslope movements could shape the valley sides. Photo from California Division of Beaches and Parks.

River valleys can be used in the interpretation of geologic history. Evidence of uplift can be found in individual river valleys because river valleys widen as erosion progresses. Thus, if a river is cutting a V-shaped, youthful valley within a much wider valley, a change has occurred causing the river to downcut actively. (See Fig. 8-27.)

Figure 8-22. The Yellowstone River in Yellowstone National Park, with falls and rapids. Photo from National Park Service.

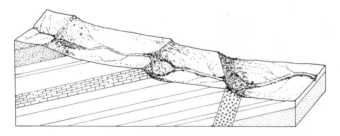

Figure 8-23. Block diagram showing how resistant rocks cause falls and rapids.

This change could be due to an uplift, a climatic change, or an increase in the flow of the river from a stream capture (discussed below) or other cause. If no further changes occur, the new valley will widen as erosion progresses, removing the evidence. If a new uplift occurs, a new V-shaped valley will be formed. This is one way river terraces, as these features are called, are produced (Fig. 8-28). If depression rather than uplift occurs, a river valley will be the site of deposition instead of erosion.

Figure 8-24. Rogue River, Oregon, has cut a narrow, V-shaped canyon in resistant rocks at this place. Photo from Oregon State Highway Department.

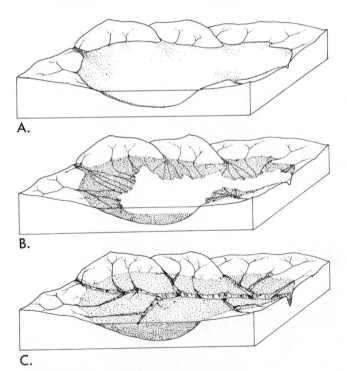

A.

B.

C.

Figure 8-25. Three stages in the life of a lake. A. Initial lake. B. Lake has been partially filled by stream deposition. C. Lake is completely filled, and stream has cut a channel across the lake fill.

Figure 8-26. A narrow, V-shaped valley in which the river has alternating pools and riffles. Photo from U.S. Forest Service.

Figure 8-27. A new valley is being cut into an already existing broad valley in this oblique aerial view of the Snake River on the Oregon-Idaho boundary. Photo from U.S. Geological Survey.

A.

B.

C.

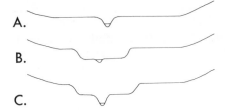

Figure 8-28. Successive uplifts have formed a terraced valley. A. An uplift caused the river initially in a broad valley to downcut, forming a V-shaped valley. B. Further erosion widened the new valley. C. A new uplift caused renewed downcutting. Ultimately erosion will destroy the older terraces, and it will not be possible to decipher this history.

At some places uplift is shown by rivers that have narrow, V-shaped valleys and a meandering map pattern. This happens most commonly when uplift occurs slowly enough that the river that was meandering in a broad valley maintains its meandering course, although downcutting. (See Fig. 8-29.)

Figure 8-29. Intrenched meanders on the San Juan River in Utah. Broad valley features like meanders in narrow, V-shaped valleys as shown here strongly suggest a change in the regimen of the river, such as a slow uplift. Photo from Utah Travel Council.

Drainage Patterns

The development of the drainage pattern is an important part of the erosional history of an area. A few of the ways that streams develop can be observed in nature, but because geologic processes are slow, most of the development must be imagined. Initially, the runoff flows down the irregularities in the slope in many small rills. The flowing water erodes these rills; the ones with greater flow are eroded most. These larger rills become deeper gullies, and ultimately one of them will become the master stream. The process of gully development can be seen in many excavations (see Fig. 8-30) and

Figure 8-30. Rills developed in an excavation in glacial deposits in Iowa. Photo by W. C. Alden, U.S. Geological Survey.

in fields where plowing has removed the natural vegetation that prevented erosion. (See Fig. 8-31.) The rills develop by downcutting and by headward erosion. The rill with the most water will downcut fastest, and this will accelerate headward erosion by this rill and all of its tributary rills. The lengthening by headward erosion enables this rill to capture the flow of the rills it intersects. This increases the area drained by the main rill and accelerates its erosion. The process of stream capture, called *stream piracy,* continues, and the rill develops first into a gully and finally into a master stream. (See Figs. 8-32 and 8-33.)

Figure 8-31. Gullies cut in a sandy slope in a two-year period in Wisconsin. Photo by W. C. Alden, U.S. Geological Survey.

Stream patterns develop in many ways. Most stream patterns are probably in part inherited from an earlier development and in part dependent on the underlying rock structure and changes in elevation. Other factors that influence drainage development include climate, runoff, and vegetation. The preceding description tacitly assumes an initially fairly smooth surface that was rapidly uplifted. These conditions, however, are probably not commonly found in nature. One can also envision an uplift slow enough that the initial streams are merely intrenched. Thus uplift may be slow or rapid, or may accelerate or slow down, and each case will affect the erosional development of an area in a different way. The landscapes developed by these differing conditions will be described in the next section.

The effect of bed rock structure on stream pattern can be seen in many places. Areas of uniform bed rock commonly have branching streams in a pattern similar to the veins in a leaf. (See Fig. 8-34.) In other areas, the joints, or cracks in the rocks, may influence

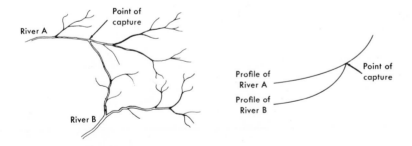

Figure 8-32. A common type of stream piracy. Rapidly downcutting minor tributary of River B captures the headwaters of River A. The gradients of the two rivers near the point of capture are shown in the lower diagram.

Figure 8-33. An example of stream capture is shown in this photo of the southern part of the Arabian Peninsula taken from Gemini IV. The Gulf of Aden is in the background. Wadi Hadramaut in the foreground clearly flows toward the right because it is wider there. Many of the tributary streams flow toward the left, and their flow is abruptly reversed where they join Hadramaut. Such reversals are common where rivers have been captured. Photo from NASA.

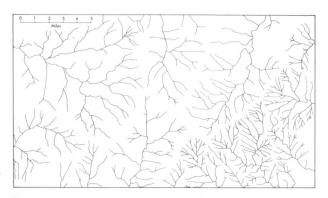

Figure 8-34. Branching stream pattern developed in massive rocks. From U.S. Geological Survey, Mount Mitchell quadrangle, North Carolina-Tennessee.

the stream pattern, and rectangular patterns commonly develop in this way. At other places, alternating hard and soft rocks may produce a trellis pattern. (See Fig. 8-35.) The initial slope of an area also affects the stream pattern, as shown in Fig. 8-36.

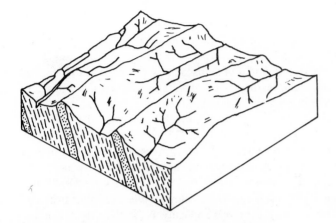

Figure 8-35. Alternating hard and soft rocks produce a trellis drainage pattern.

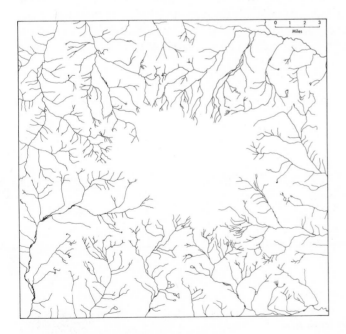

Figure 8-36. Radial stream pattern. From U.S. Geological Survey, Mount Rainier quadrangle, Washington.

Development of Landscape

Development of landscapes is even more complex than the development of river valleys and drainage patterns, and all of the factors discussed above are involved. Thus it is no wonder that much controversy exists. We will begin with the simplest case and use it to illustrate other possibilities, which, hopefully, will show why controversy exists. This is an area of active research; and as more data are obtained, much of the present controversy may be resolved.

The simplest viewpoint on the development of landscape was an American idea, and it had a pronounced influence on the history of geology. The cornerstone of modern geology is the doctrine of *uniformitarianism,* which, in the present context, states that the present processes at work on the earth, given enough time, caused the present landforms. This principle, although recognized by a few earlier men, was widely adopted near the end of the eighteenth century. Geology, however, developed mainly in Western Europe during the nineteenth century; and most geologists, especially the island-dwelling English, thought that ocean waves rather than rivers were responsible for most erosion. The opening of the American West following the Civil War provided a vast new area for geologic study. The great rivers and canyons, such as the Colorado, and the many mountain ranges at various stages of dissection, together with the dry climate that bared great vistas, made recognition of huge amounts of erosion simple. (See Fig. 8-37.)

Late in the nineteenth century, William Morris Davis of Harvard codified the earlier studies into what he termed the *cycle of erosion.* Davis visualized every landscape going through a series of changes from initial uplift to complete leveling by weathering and erosion. His emphasis was on the forms produced, and he paid little attention to the processes involved. He clearly recognized that an area rarely goes through the complete

Figure 8-37. A typical scene in the arid Southwest. The lack of vegetation and the fact that the flat-lying sedimentary rocks can easily be traced for many miles make recognition of the immense amount of erosion that has occurred easy.

cycle, but, rather, the cycle generally is restarted by new uplifts. This accounts for the fact that there are very few examples of areas that have gone through the complete erosion cycle. In the youthful stage, a newly uplifted area is relatively flat with deeply entrenched streams (Figs. 8-38 and 8-39). As erosion proceeds, the interstream areas are reduced in size as a result of downcutting by the rivers and widening of the valleys by downslope movement of material. When no flat interstream areas exist, the stage has progressed to maturity. (See Figs. 8-38 and 8-40.) This is the stage of maximum relief. As the interstream areas are rounded and lowered, the area passes into old age (Fig. 8-38). In the ultimate stage, when even these gentle interstream areas are worn almost flat, the area is called a *peneplain* (almost a plain, sometimes spelled peneplane). Flat eroded areas such as peneplains are rare features on the earth's surface, and this fact constitutes one of the weaknesses of the cycle-of-erosion hypothesis. However, many events, such as renewed uplift, can occur to prevent the complete development of a peneplain.

Perhaps the main objection to the Davis cycle is that it requires rapid uplift and then a stillstand during which erosion occurs. If the uplift of an area is slow enough, perhaps erosion keeps pace with the uplift and no change in form occurs. However, any type of uplift from rapid to slow, or from accelerating to slowing, is geologically possible, so at best the Davis cycle describes a special case.

Another objection is that the hills might not be reduced in the manner envisioned by Davis. Instead, a balance might result when the hill slopes and stream gradients

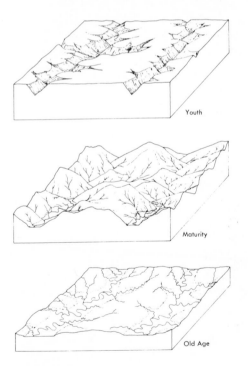

Figure 8-38. The cycle of erosion as envisioned by Davis.

Figure 8-39. An example of youthful topography. Oblique aerial view looking west at the Grand Canyon in Arizona. The flat surface is being attacked by headward erosion of the tributary streams. The inner canyon of the Colorado River is being cut in metamorphic rocks. Photo from U.S. Geological Survey.

Figure 8-40. An example of a mature area in the Sierra Nevada of California. Glacial erosion has modified this area. Photo from U.S. Geological Survey.

become adjusted. Such an equilibrium can, in a way, be thought of as an extension of the graded-stream concept discussed earlier. After this stage is reached, the shape of the area remains substantially the same as erosion progresses. Thus there is no great change in shape, as there is in the Davis cycle of erosion, because once an equilibrium shape develops, no new shape can form unless the balance is upset by uplift or some other change. (See Fig. 8-41.)

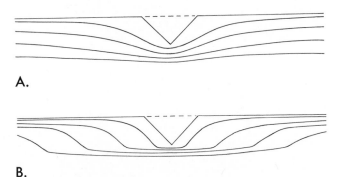

A.

B.

Figure 8-41. Successive profiles of the erosion of a river valley contrasting two theories of the development of slopes. A. The Davis concept of down-wearing of hillsides. B. Backwearing of hillsides, preserving approximately the same slope.

In an effort to resolve this latter objection, a number of recent studies have been made of hill slopes. The studies have been made in areas where the climate, vegetation, relief, and rock type are uniform. The slope angle of such hillsides is related to both stream gradient and density of streams. Where direct measurement of slope erosion was made, it was found that during retreat of slopes the slope angle remains constant if the debris is removed from the foot of the slope. (See Fig. 8-42.) If the erosional debris is not removed from the foot of the hill, the slope angle is reduced. The latter case is probably more common in humid areas where weathering may be deep and creep may be active. All of this suggests that the Davis cycle is not only a special case, but an oversimplification as well.

The length of time and the size of the area considered in the study of landscape may also influence one's ideas. A single hillside or stretch of river may be in equilibrium during the time of a short study of the type described above; however, if the whole drainage area is considered over a geologically long time interval, erosion must remove appreciable amounts of mass and so change the system. Thus, in the short term, landforms may not seem to change with time; but over a long interval they might actually do so. The dynamic earth may not remain still long enough, however, for these changes to occur, for the earth's surface is the result of the constant struggle between erosion and uplift.

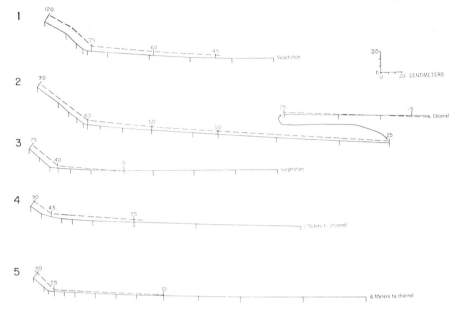

Figure 8-42. Measurement of erosion in the Badlands of South Dakota. The dashed profiles are July, 1953, and the solid lines are May, 1961. The numbers are the erosion depth in millimeters. From S. A. Schumm, *Bulletin,* Geological Society of America, Vol. 73, 1962.

QUESTIONS

8-1. How is most of the energy of a river used?

8-2. How do you account for the fact that the velocity of a river increases downstream?

8-3. What are the three ways by which a river transports material?
   1. _____ 2. _____ 3. _____

8-4. What are the processes of river erosion?

8-5. Describe the sequence of events in the formation and migration of a curve in a river course. (Use diagrams.)

8-6. A placer deposit is a place where a river has deposited fragments of heavy minerals, such as gold. Where would you look in a river for such deposits?

8-7. Can you think of some reasons why doubling the velocity of river increases the amount of material transported four to eight times but only increases the maximum size of material transported four times?

8-8. Show with a series of sketches the erosion of a river valley.

8-9. Describe the erosion of an upland area. How does the topography change as erosion proceeds? Your answer depends on several assumptions you must make.

8-10. What is the final surface produced by long erosion called?

8-11. What are some characteristics of narrow river valleys?

8-12. Describe the evidence showing rapid uplift of an area.

8-13. To what was Twain referring in the quotation at the beginning of this chapter?

SUPPLEMENTARY READING

Bloom, A. L., *The Surface of the Earth.* Englewood Cliffs, New Jersey: Prentice-Hall, Inc., 1969, 152 pp. (paperback).

Davis, W. M., *Geographical Essays,* especially Part II, "Physiographic Essays." New York: Dover Publications, Inc., (1909), 1954, 777 pp. (paperback).

Dury, G., *Face of the Earth.* Baltimore: Penguin Books, Inc., 1959, 223 pp. (paperback).

Hunt, C. B., *Physiography of the United States.* San Francisco: W. H. Freeman and Co., 1967, 480 pp.

Janssen, R. E., "The History of a River," *Scientific American* (June, 1952), Vol. 186, No. 6, pp. 74-83. Reprint 826, W. H. Freeman and Co., San Francisco.

Judson, Sheldon, "Erosion of the Land," *American Scientist,* (Winter, 1968), Vol. 56, No. 4, pp. 356-374. Reviews earlier estimates of erosion rates and estimates the effects of man. Concludes that the amount of erosion is much lower than stated in this chapter.

Leopold, L. B., and W. B. Langbein, "River Meanders," *Scientific American* (June, 1966), Vol. 214, No. 6, pp. 60-70.

Leopold, L. B., and Thomas Maddock, Jr., *The Hydraulic Geometry of Stream Channels and Some Physiographic Implications.* U.S. Geological Survey, Professional Paper 252, 1953, 57 pp.

Leopold, L. B., M. G. Wolman, and J. P. Miller, *Fluvial Processes in Geomorphology.* San Francisco: W. H. Freeman and Co., 1964, 504 pp.

Matthes, G. H., "Paradoxes of the Mississippi," *Scientific American* (April, 1951), Vol. 184, No. 4, pp. 18-23. Reprint 836, W. H. Freeman and Co., San Francisco.

Morgan, J. P., "Deltas—A Resume," *Journal of Geological Education* (May, 1970), Vol. 18, No. 3, pp. 107-117.

Morisawa, Marie, *Streams: Their Dynamics and Morphology.* New York: McGraw-Hill Book Co., 1968, 175 pp. (paperback).

Schumm, S. A., "The Development and Evolution of Hillslopes," *Journal of Geological Education* (June, 1966), Vol. 14, No. 3, pp. 98-104.

Shimer, J. A., *This Sculptured Earth: The Landscape of America.* New York: Columbia University Press, 1959, 256 pp.

Thornbury, W. D., *Principles of Geomorphology* (2nd ed.), Chapters 5, 6, 7, 8, pp. 99-208. New York: John Wiley & Sons, Inc., 1969, 594 pp.

Chapter 9

Geologic Processes in Arid Regions and the Work of Wind

Geologic Processes in Arid Regions

Dry regions form over one-quarter of the total land surface of the earth. Most of these arid areas are hot deserts or near-deserts, but some of the dry areas are cold. The main desert areas are in the subtropics near thirty degrees north and south latitude. They are caused by the primary atmospheric circulation of the earth that causes dry air to descend at these latitudes, and the resulting compression heats the already dry air. Other arid areas occur in the rain shadows of high mountain ranges. Southwestern United States is an example of such an area. The moist winds from the Pacific Ocean are cooled and precipitate their moisture when they are forced to rise over the Sierra Nevada and other mountain ranges in California. Here, too, the dry descending air is heated by compression.

Compared to more humid areas, the lack of moisture in deserts changes the emphasis but not the types of geologic processes. Rainfall, although infrequent and irregular, is the main agent of erosion. Rain may fall only once in many years at any one place, but is generally in the form of an intense cloudburst or thunderstorm. The lack of vegetation enables such storms to erode very efficiently by splash and runoff. Flash floods in normally dry stream channels many miles from the storm area may trap unwary travelers. The runoff carries much loose surface material to the stream channels, and the resulting flow may be a thick mudflow rather than a normal stream. Such mudflows can carry huge boulders and can do great damage to anything in their path. Weathering

211

is slow, and because mechanical weathering is much more important in arid regions than chemical weathering, little soil is formed. The fertile soils of Egypt are transported from more humid regions by the Nile River. Limestone that is rapidly dissolved in humid climates forms bold cliffs in the desert.

Desert Landscapes

Infrequent rain and the resulting lack of moisture cause desert landscapes to develop slowly. Because of the lack of rain, throughgoing rivers rarely develop. Exceptions are the Colorado and Nile Rivers that begin in high, well-watered mountains and then cross deserts. Most deserts, however, are areas of internal drainage, and this profoundly affects their erosional development.

The development of drainage patterns in deserts is well shown by the Great Basin of southwestern United States. This is an area of many mountain ranges formed by relatively recent faulting. The Basin and Range country is another name sometimes applied to much of this region because basins separate the ranges. The basins are low areas, and lakes may form in them during times of heavy rainfall, such as the ice age described in the next chapter. Lake-deposited sediments are fairly common in these basins, and dry lake beds form the surfaces of many. The lakes generally contain salt and other dissolved materials that come from the rain water and chemical weathering. Thus many of them are what are called alkali lakes, and the dry lake beds are alkali flats. Great Salt Lake and Bonneville Salt Flats are examples. Because these lakes are

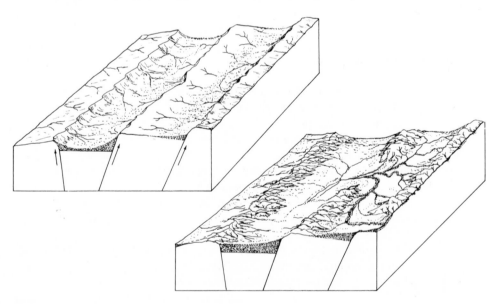

Figure 9-1. Development of drainage systems in the basin and range area. As each basin is filled, the drainage spills over the lowest divide into the next ba-sin. This process continues, slowly integrating the drainage of many basins until the ephemeral streams reach the ocean—a goal rarely reached.

not connected to the ocean, they develop in much the same way as do the oceans. They are, in effect, small oceans.

The usual development of drainage patterns in a desert begins with erosion in the mountains and deposition in the basins. The deposition may occur well out in the basin or as alluvial fans near the mountain front. These processes continue until the basin is filled to the level that the ephemeral streams can flow over the lowest divide separating it from an adjoining basin. (See Fig. 9-1.) The processes now continue in the new basin until it too is filled. While the second basin is being filled, the divide between the two basins is generally lowered by erosion; this initiates a stage of erosion in the first basin because of a lowered base level. (See Fig. 9-2.) This complicated series of events will continue until an outlet to the ocean is established; however, a connection to the ocean is rarely developed because lack of rain makes these processes very slow.

The erosional reduction of desert mountains is somewhat different from that of more humid climates. The differences may be basic, but are more likely the result of different emphasis of the various erosional processes. Also, as just noted, local base levels and

Figure 9-2. A complex drainage system that resulted from integration of basins. The low range in the foreground and middleground is being vigorously eroded, probably because its base level was lowered by the integration of basin drainage. Some possible future captures can be seen. In the far middle-ground, several basins that have integrated are visible. Note the lack of water in the stream courses and the almost complete lack of vegetation in this view near the southern border of California. Photo from U.S. Geological Survey.

small drainage areas are involved, rather than large watersheds, so that each desert range or part of a range develops at its own rate, making the developmental sequence easy to see.

The first step is generally the development of fans at the mountain front. After this, the wearing back of mountains begins. The initial steep mountain front retreats and a fairly smooth surface in the bedrock is eroded. This surface is called a *pediment* and is generally continuous with the fan. The pediment is commonly covered by a thin layer of gravel, making it very difficult to distinguish from a fan unless the generally present gullies cut through the gravel. The method by which pediments are formed is not clear, but streams issuing from the mountains may change their courses back and forth across the pediment in a process not unlike the way alluvial fans are formed. Runoff is probably also important in pediment formation. Pediments have concave-upward slopes, much like river profiles. The slope of most pediments is between one-half and seven degrees and is determined by the particle size of the gravel that covers it, and probably the amount of runoff. The steep mountain front behind the pediment seems to maintain its slope angle during retreat. The various stages in the erosion of a desert range are shown in Figs. 9-3, 9-4, and 9-5. The examples shown are faulted mountains

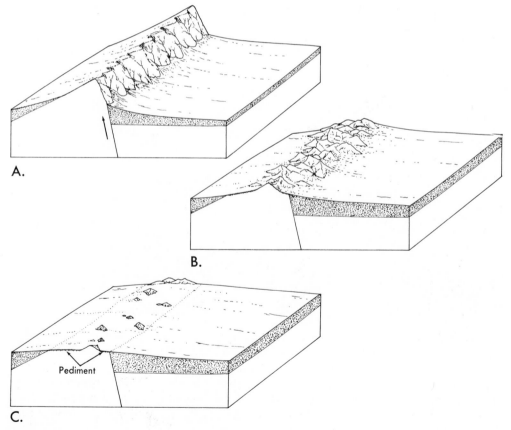

A.

B.

C.

Figure 9-3. Formation of pediments. A. Shortly after uplift. B. Pediments have begun to form. C. Late stage. The mountain range is mainly pediment.

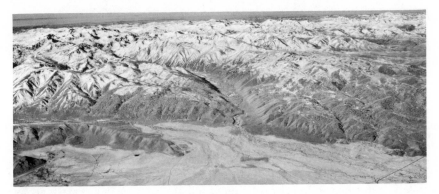

Figure 9-4. The faulted eastern face of the Sierra Nevada is only slightly modified by erosion. Photo from U.S. Geological Survey.

Figure 9-5. The main range shown here has been embayed by erosion with the development of pediments and fans. In the right foreground, the mountains have been reduced to low pediments. Photo from U.S. Geological Survey.

similar to those in the arid areas of southwestern United States. Other types of mountains are found in other arid regions. The formation of faulted mountains is described in Chapters 14 and 24.

Geologic Work of Wind

Only in dry areas is wind an active agent of erosion, transportation, and deposition. The rapid heating and cooling of deserts leads to strong winds. The behavior of wind is similar to that of water in some ways, but it is much less effective. Wind erosion occurs by two processes, deflation and sandblasting (or abrasion). *Deflation* is simply the removal by wind of sand- and dust-sized particles. Typical places where deflation may take place are dry, unvegetated areas such as deserts, dry lake or stream beds, and actively forming, glacier-outwash plains in dry seasons. Deflation may, in some areas, produce hollows, sometimes called blowouts, that are recognized by their concave shape, which could not have been produced by water erosion. (The only other common way that closed depressions form is by solution of soluble rocks, such as limestone, by ground water.) At other places deflation removes the fine material from the surface, leaving behind pebbles, to produce a surface armored by a pebble layer against further deflation. Alternate wetting and drying can also cause pebbles to rise by a process not unlike frost heaving discussed in Chapter 7. *Sandblasting* by wind-driven sand grains can cause some erosion near the base of a cliff or on a boulder, but probably is much less effective than weathering. (See Fig. 9-6.) Sandblasting does produce some interestingly shaped pebbles called *ventifacts,* or *dreikanter* (German for three edges). Sand

Figure 9-6. Double Arch in Arches National Monument, Utah. The arches are believed to be partly the result of wind erosion of thick sandstone beds. Water from infrequent rains probably loosens the sand near the surface, and wind removes the loose sand grains and causes some abrasion. Photo from National Park Service.

hitting a pebble on the side facing the prevailing wind wears a round surface that may be polished and slightly pitted. Most ventifacts have several such surfaces, three being common. (See Fig. 9-7.) The other surfaces may be produced by other common wind directions or, probably more commonly, by the turning of the pebble, perhaps due to the undermining of the pebble by deflation of its supporting sand. (See Fig. 9-8.)

Wind transportation is similar to water transportation in that both a suspended load and a bed load are involved. The suspended load produces dust storms, and the bed

Figure 9-7. Ventifacts. Photo by M.
R. Campbell, U.S. Geological Survey.

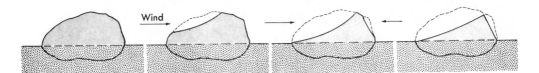

Figure 9-8. Formation of ventifacts
by wind abrasion. The main facet faces
the prevailing wind direction. Other
faces are formed by other common
wind directions or the pebble is turned,
exposing another side to the wind.

load forms sandstorms. Most windblown sand is less than one millimeter (mm) in
diameter, and the average size is about ¼ mm. Sand begins to move when the wind
reaches 11 miles per hour. All material finer than 1/16 mm (silt) is considered dust,
and most wind-blown dust is much smaller than this. The separation into these two
size ranges is very sharp, much more so than in the case of water transportation, so
that wind-deposited sediments are generally much better sorted by size than water-laid
deposits. The reason for this better sorting is that wind can raise sand particles only
a few feet above the ground, although it can raise dust thousands of feet. At one place
where measurements are made, 50 per cent, by weight, of the sand was raised less than
five inches above the ground and 90 per cent was raised less than 25 inches, although

some large fragments rose over 10 feet. At this place, maximum abrasion occurred nine inches above the ground. Such abrasive effects (only a few inches above ground) are common on fence posts and power poles in areas of sandstorms. As might be imagined, sandstorms are associated with dust storms, but dust storms may occur without sandstorms; so, in general, dust is moved much farther than sand. (See Fig. 9-9.)

Figure 9-9. A dust storm caused by a cold front at Manteer, Kansas, in April, 1935. The dust is in the cold air mass and rises to great heights. Photo from Environmental Science Services Administration.

The mechanism for moving sand is similar, in part, to the action of water. The process produces a bouncing motion of the sand grains. It begins when a grain rolls into or over another grain. When the first grain hits, its impact transfers its kinetic energy to one or more grains which, in turn, also bounce up and, on falling, continue the process by hitting other grains (Fig. 9-10). Except in large downdrafts, dust cannot be lifted by wind alone because at ground level there is a thin zone of no wind about one thirtieth of the diameter of the average particle. The impact of falling sand grains can put the dust into motion, and the turbulence of the wind will tend to keep it in motion.

The amount of wind-blown material raised by a storm can be very high. For example, during some of the intense dust storms in the United States in the 1930's, the load in

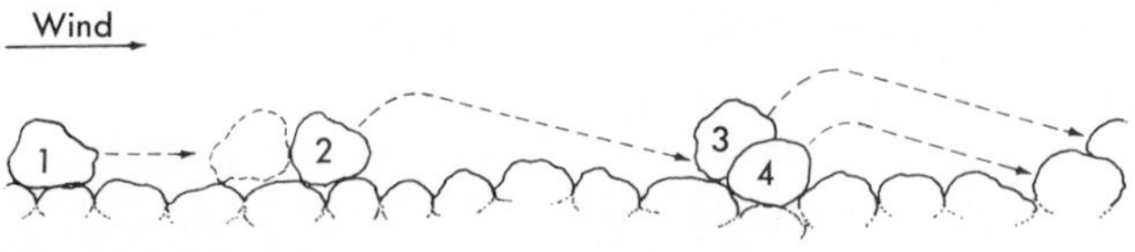

Wind

Figure 9-10. Movement of bed load. Sand grain 1 rolls over and hits grain 2 which bounces up and hits grains 3 and 4, both of which also bounce up and hit more sand grains.

the lower mile of the atmosphere in some areas was estimated at over 150,000 tons per cubic mile of air.

Wind deposits are of two types—loess composed of dust, and drifts and dunes composed of sand. *Loess* forms thick, sheet-like deposits that are relatively unstratified because of their rather uniform grain size, generally in the silt size range. (See Fig. 9-11.) Thick loess deposits in China are believed to have originated from the deserts

Figure 9-11. Loess near Vicksburg, Mississippi. The upper layer of loess is somewhat different from the lower. Although loess can stand in vertical cliffs, it is very easily eroded where a little water spills over the cliffs. Photo by E. W. Shaw, U.S. Geological Survey.

of Asia. Most American loess deposits are associated with glacial deposits and are believed to have formed when glacial retreat left large unvegetated areas on which winds could act. Wind-blown sand, unlike loess, does not generally form blanket deposits. *Dunes* are naturally formed accumulations of wind-blown sand. Most dunes probably start in the lee of an obstacle (Fig. 9-12). Unless they have become fixed by vegetation, most sand dunes move slowly across the desert (Fig.9-13). As might be expected, many dunes and dune areas are not easily classified; however, several distinct types of dunes are of wide occurrence (Figs. 9-14, 9-15, 9-16, 9-17, and 9-18). The sand grains in dune deposits are commonly frosted by the abrasion during transport.

Wind →

Figure 9-12. Accumulation of sand in lee of fence post.

Figure 9-13. The movement of a sand dune. The wind moves individual sand grains along the surface of the dune until they fall off the steep face.

Dune type	Sketch	Cross-section	Remarks		
Barchan	 Wind →		Max. size 100′ high, 1000′ point to point	25-50′ per year movement	Commonest. Generally in groups in areas of constant wind direction.
Transverse	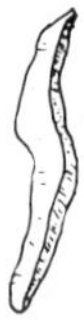 →		Similar to barchans but not curved. They form in areas with strong winds where more sand is available.		
Parabolic	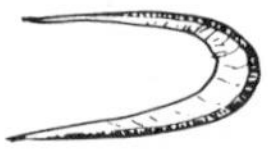 →		Max. size 100′ high	Form in areas with moderate winds and some vegetation.	Extreme curved types called *hairpin*. Common on sea coasts
Longitudinal (Seif)	→		 Max. 300′ high, 60 miles long. Avg. 10′ high 200′ long	They form in areas with high, somewhat variable winds and small amount of sand available.	

Figure 9-14. Types of sand dunes.

Figure 9-15. Barchan dunes in the Columbia River Valley near Biggs, Oregon. Photo by G. K. Gilbert, U.S. Geological Survey.

Figure 9-16. Asymmetric wind ripples and the steep lee face of a complex
barchan dune. Photo by C. E. Erdman,
U.S. Geological Survey.

Figure 9-17. Aerial view of Imperial
Valley, California, showing a dune complex. Most areas of dunes are complex
and do not fall into the common classification of dunes. Photo by J. R. Balsley,
U.S. Geological Survey.

Figure 9-18. Edge of the dune complex of Imperial Valley, California. Photo by W. C. Mendenhall, U.S. Geological Survey.

QUESTIONS

9-1. What is the role of climate in erosion and the development of landforms?

9-2. Describe pediments.

9-3. Briefly describe wind erosion (two kinds).

9-4. Wind-deposited silt is called_____.

9-5. Deposits of wind-blown sand are called_____.

9-6. Describe the sorting of wind deposits.

9-7. How might wind-deposited sediments differ from water-deposited sediments?

9-8. Describe the development of desert pavement.

9-9. Why is wind not an important agent of erosion in humid windy climates?

SUPPLEMENTARY READING

Bagnold, R. A., *The Physics of Blown Sand and Desert Dunes.* London: Methuen & Co., 1941, 265 pp.

Sharp, R. P., "Wind-driven Sand in Coachella Valley, California," *Bulletin* Geological Society of America (September, 1964), Vol. 75, No. 9, pp. 785-804.

Thornbury, W. D., *Principles of Geomorphology* (2nd ed.), Chapter 12, pp. 288-302. New York: John Wiley & Sons, Inc., 1969, 594 pp.

Chapter 10

Mountain glaciers

Continental glaciers

Cause of glaciation

Geologic Work of Ice — Glaciers

Although glaciers are very much less important than rivers in overall erosion, they have shaped many of the landforms of northern North America and Eurasia. In addition, most of the high mountain ranges of the world have been greatly modified by mountain glaciers.

A glacier is a mass of moving ice. Glaciers form as a result of accumulation of snow in areas where more snow falls than melts in most years. This can occur at sea level at the poles to about 20,000 feet near the equator. The accumulation of snow must, however, become thick enough so that it recrystallizes to ice. The recrystallization process depends on the pressure of the overlying snow, which transforms the light, loosely packed snow into small ice crystals; with increased pressure, the small crystals become larger. (See Fig. 10-1.) The process is quite similar to the metamorphism of sandstone into quartzite and of limestone into marble. The tendency of snow to recrystallize and form larger crystals is well known to anyone who skis or who has lived in snowy areas. Falling snowflakes are generally small and light, but after a few sunny days, or in the spring, they are much larger at the surface; digging will reveal even larger crystals. When the amount of ice formed in this way becomes large enough that it flows under its own weight, a glacier is born. Ice, as we see it, is a brittle substance, so the fact that a glacier flows under its own weight, much like tar, may seem strange; however, a mass of ice a few hundred feet thick behaves as a very viscous liquid.

Glaciers are of two main types: the alpine or mountain glacier, and the much larger continental or icecap glacier. Present-day continental glaciers occur only near the poles, but mountain glaciers occur at all latitudes where high enough mountains exist. (See Table 10-1.)

225

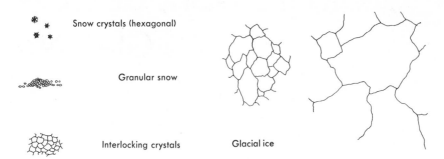

Snow crystals (hexagonal)

Granular snow

Interlocking crystals

Glacial ice

Figure 10-1. Some of the steps in the transformation from snow to glacial ice. Crystals in glacial ice can be several inches in diameter.

Table 10-1. Glaciers in the United States. Data from U.S. Geological Survey.

State	Approximate number of glaciers	Total glacier area in square miles
Alaska	?	About 17,000
Washington	800	160
Wyoming	80	18
Montana	106	10
Oregon	38	8
California	80	7
Colorado	10?	1
Idaho	11?	1
Nevada	1	.1
Utah	1?	.1?

Mountain Glaciers

Mountain glaciers (see Fig. 10-2) develop in previously formed stream valleys, which, because they are lower than the surrounding country, become accumulation sites for snow. When enough ice has formed, the glacier begins to move down the valley. How far down the valley it extends will depend on how much new ice is formed and how much melting occurs each year. As the glacier extends further down the valley, the amount of melting each year will increase until a point is reached at which it equals the amount of new ice added. Such a glacier is in equilibrium. If the amount of snowfall increases or the summers become cooler, it will advance further; and if the snowfall decreases or the summers become warmer, it will melt back. However, it generally takes a number of years for the more- or less-than-normal snowfall of a single year, or group

Figure 10-2. Mountain glaciers. Several mountain glaciers join in this valley. Note the crevasses on the surface of the ice and the moraines both on the valley sides and separating the ice tongues from the different glaciers. Photo from Swissair Photo Ltd.

of years, to become ice and to reach the snout of the glacier. This means that although a glacier is sensitive to climatic changes, it tends to average snowfall and temperature changes over a number of years. Hence, the advance or the retreat of glaciers gives some information on long-term climatic changes. Study of old photographs and maps shows, for instance, that the glaciers in Glacier National Park—and elsewhere—have retreated in the last few decades. Glaciers reached maximum advances about 1825, 1855, and 1895. In general they have been receding since the turn of the century although with some minor advances. One point needs emphasis: the glacial ice is always moving down-valley, even if the snout of the glacier is retreating because of melting.

The movement of glacial ice can be measured by driving a series of accurately located stakes into the glacier and surveying them periodically. As would be expected, because of friction, the sides of the glacier move more slowly than the center. The upper 100 to 200 feet of the glacial ice behaves brittlely so that the different rates of flow at the surface open large cracks, called *crevasses,* in this brittle zone. (See Figs. 10-3 and 10-4.) The movement of mountain glaciers varies from less than an inch a day to over 50 feet a day.

Erosion by mountain glaciers produces such spectacular mountain scenery as Yosemite Valley in the Sierra Nevada of California. Glaciers erode by plucking large blocks of bedrock and by abrasion. Plucking is accomplished by melt-water that flows

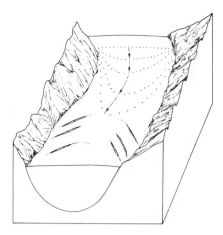

Figure 10-3. Movement of glaciers and the development of crevasses. The movement is shown by the successive locations of stakes driven into the ice in the upper part of the figure. The open-ing of crevasses in the brittle ice as a result of the greater movement of the center of the glacier is shown in the lower part of the figure.

Figure 10-4. Oblique aerial view of Mount Hood, Oregon, showing cre-vasses on the glaciers. Note the U-shaped valleys. Photo from U.S. Geological Survey.

into joints in the bedrock and later freezes to the main mass of ice which, on advancing, pulls out the block of loosened bedrock. These blocks arm the moving ice with rasp-like teeth that grind the sides and bottom of the glacial valley (Fig. 10-5). Glacial erosion

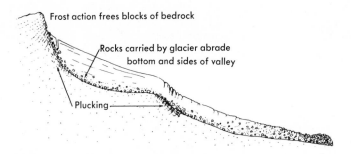

Figure 10-5. Longitudinal section through a mountain glacier showing development of a cirque by plucking, and reshaping of the valley by plucking and abrasion. This sketch shows an early stage in the development of a glacial valley.

greatly modifies the shape of the stream valley occupied by the glacier. Most of the erosion is probably done by plucking, and abrasion generally smooths and even polishes the resulting form. A glaciated valley differs from a stream valley in that it is deepened, especially near its head, and the sides are steepened so that its cross-section is changed from V-shaped to U-shaped (Figs. 10-6, 10-7, and 10-8). Glacial erosion is most active near the head of the glacier, and the deepening there flattens the gradient of the valley. This deepening, together with the steepening of the sides and the head of the valley, produces a large amphitheater-like form that closely resembles a teacup cut vertically in half. *Cirque* is the name applied to this form. (See Fig. 10-6.)

A glacially eroded valley is recognized by the U-shape, the cirque at the head, and the grooves, scratches, and polishing due to abrasion. (See Fig. 10-9.) In addition, any small hills or knobs in the valley have been overridden by the glacier, and they are rounded and smoothed by abrasion on the struck side and steepened by plucking on the lee side. (See Fig. 10-10.) Many glaciated valleys are nearly flat with rises or steps where more resistant rocks crop out. (See Figs. 10-11 and 10-12.) A glacier moving down a valley also tends to straighten the valley, because the glacier cannot turn as abruptly as the original river could. This has the effect of removing the spurs or ridges on the insides of curves of the stream valley. The amount of ice in a main-stream valley, whose source is near the crest of the mountain range, is generally much greater than

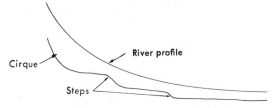

Figure 10-6. Glacial erosion of river valleys. Steps are commonly developed and may contain lakes.

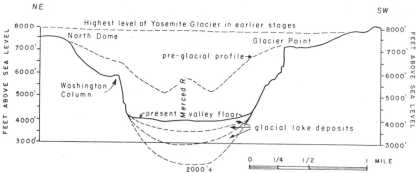

Figure 10-7. Glacial modification of Yosemite Valley, California. From Clyde Wahrhaftig, California Division of Mines and Geology, *Bulletin* 182, 1962.

Figure 10-8. Oblique aerial photograph looking south down the Kern River Canyon, Sierra Nevada, California. Note the U-shaped glacially eroded valley. Photo from U.S. Geological Survey.

the amount of ice in a tributary-stream valley; hence, the main valley is eroded deeper by the ice than is the tributary. After the ice melts, the tributary stream, in a valley called a *hanging valley,* is higher than the main stream into which it flows via a waterfall. (See Figs. 10-13 and 10-14.)

A.

B.

Figure 10-9. Evidence of glacial abrasion. A. Polished and grooved granite caused by the rock-studded base of a moving glacier. Photo by Mary Hill, California Division of Mines and Geology. B. Glacial grooves partly destroyed by mechanical weathering. Photo by C. W. Chesterman, courtesy California Division of Mines and Geology.

Figure 10-10. Small hills of resistant rock are smoothed by abrasion on the struck side and steepened by plucking on the lee side. The arrows show the direction of the main force of the glacier that moved from right to left. The jointing in the rock is indicated. From F. E. Matthes, U.S. Geological Survey, *Professional Paper* 160, 1930.

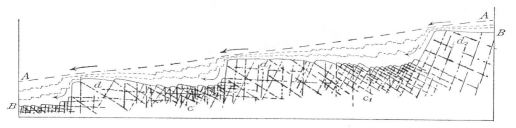

Figure 10-11.   Origin of steps in a glacial valley. *AA* is the profile of the pre-glacial valley and *BB* the post-glacial profile. Closely jointed rocks at *c* and c_1 are readily plucked by the glacier, but sparsely jointed rocks as at d, d_1, d_2 are reduced by abrasion. From F. E. Matthes, U.S. Geological Survey, *Professional Paper* 160, 1930.

Figure 10-12.   A glacial valley showing rock steps with lakes on the steps. Photo by S. A. Davis, courtesy California Division of Mines and Geology.

Even the high part of a mountain range that is above the level of the glaciers develops characteristic forms. Increased frost action here produces narrow, jagged ridges extending up to the peaks. The peaks develop pointed, pyramidal shapes, called *horns,* that are largely due to headward erosion by cirque development. (See Figs. 10-2 and 10-12.)

Glaciers also deposit the debris that they carry. *Drift* is the name used for all types of glacial deposits, and *till* is used for ice-deposited sediments. The material in the glacier is carried by the ice to the snout, where it is deposited to form the terminal moraine which marks the point of furthest advance of the glacier. (See Fig. 10-15.) Because the glacier carries a great variety of sizes of fragments, moraines are unsorted; huge boulders are mixed with all sizes of fragments down to silt and finer. (See Fig. 10-16.) The pebbles are generally faceted and striated by abrasion. Because the snout of a glacier is curved, the terminal moraines are characteristically curved. They continue along the sides of a glacier as lateral moraines. The lateral moraines develop, in

A.

B.

C.

D.

Figure 10-13. The development of Yosemite Valley illustrates both river and glacial erosion. A. The broad valley stage prior to the uplift of the Sierra Nevada. B. After uplift, the rivers down-cut about 700 feet. C. After the second uplift, the Merced River cut a canyon about 1300 feet below A. D. Yosemite Valley just after the glacier retreated. The lake was dammed by glacial moraine, and the filled lake is the present valley floor. B, Mount Broderick; BV, Bridalveil Creek; C, Clouds Rest; CC, Cascade Cliffs; CR, Cathedral Rocks; EC, El Capitan; EP, Eagle Peak; HD, Half Dome; LC, Liberty Cap; LY, Little Yosemite Valley; MR, Merced River; MW, Mount Watkins; ND, North Dome; R, Royal Arches; SD, Sentinel Dome; TC, Tenaya Creek; W, Washington Column; YC, Yosemite Creek. From F. E. Matthes, U.S. Geological Survey, *Professional Paper* 160, 1930.

Figure 10-14. Yosemite Valley. Figure 10-13 shows the steps in the development. Note that the present river is very small for the size of the valley. Photo by S. A. Davis, courtesy of California Division of Mines and Geology.

Figure 10-15. Idealized sketch of a glacier leaving successive moraines as it melts back. Both end moraines and lateral moraines are shown. From F. E. Matthes, U.S. Geological Survey, *Professional Paper* 160, 1930.

Figure 10-16. Glacial moraine containing many kinds of rock in all sizes from boulders to silt. Photo by Mary Hill, courtesy California Division of Mines and Geology.

much the same manner as terminal moraines, as a result of melting of the slow-moving ice at the valley sides. (See Figs. 10-2 and 10-15.) Many of the rocks deposited by glaciers are scratched or striated from abrasion during transport. The melt-water leaving the snout of a glacier also transports and deposits glacial material, forming outwash deposits that in some cases form thick sheets. The melt-water from an active glacier is colored white due to fine material, called *rock flour,* from the abrasive erosion of the glacier. This material resembles clay, but microscopic examination shows that it is composed of fine mineral fragments, not clay. It is the source of the varved "clays" to be discussed later. When a glacier melts, its transported load is dropped wherever it happens to be and forms ground moraine. Upon melting, blocks of ice buried in ground moraine leave depressions, which may become small ponds. (See Fig. 10-17.)

Figure 10-17. Pond created by the melting of a block of ice buried in moraine. Note the variety of sizes of rocks in the moraine. Baird Glacier, southeastern Alaska. Photo by A. F. Buddington, U.S. Geological Survey.

Continental Glaciers

The geologic work of continental glaciers is very similar to that of mountain glaciers. Their erosive effects are much less and are generally limited to rounding the topography and scraping off much of the soil and overburden. (See Figs. 10-18 and 10-19.) Their deposits are like those of mountain glaciers, but are more extensive. One type of deposit that helped lead to the recognition of continental glaciation is the *erratic boulder.* (See Fig. 10-20.) This term is applied to boulders, some of which weigh up to 18,000 tons, that are foreign to their present locations. In some areas, occurrences of distinctive erratic boulders can be traced back to their origin, thus indicating the path of the glacier.

Continental glaciers occupied parts of North America and Europe in relatively recent geologic time, and they occupy Greenland and Antarctica now. During the recent glacial period, there were four main times of glacier development in both Europe and America. In North America, there were several centers of ice accumulation in Canada. Glaciers moved out from these centers in all directions. Thus, although the ice moved generally southward in the United States, it moved northward in northern Canada. In no sense did the last glacial age result because ice from near the north pole moved south. Recognition of four advances depends largely on field observations. In some places fresh, unweathered moraines overlie weathered moraines which in turn overlie more deeply weathered moraines. Tracing such relationships led to the recognition of four main stages with the extents shown on the map in Fig. 10-21.

Moraines, however, are not the only types of deposits left by continental glaciers. In some places, blocks of ice were buried in the moraine and when the ice melted, depressions formed. These depressions are called *kettles,* and lakes form in many of them (Fig. 10-17).

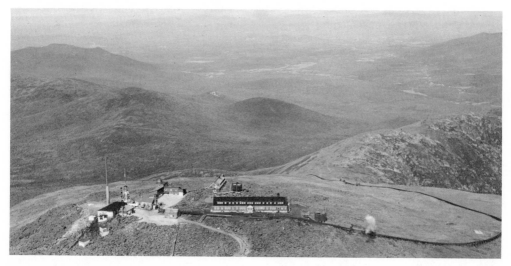

Figure 10-18. Aerial view of the 6288-foot summit of Mount Washington, New Hampshire, the highest point in the northeast. The continental gla- ciers overrode and smoothed all of the rounded mountains in the view. Photo by Dick Smith, New Hampshire Division of Parks.

Figure 10-19. Glacially eroded area near Hidden Lake, Northwest Territories, Canada. Continental glaciers have removed the soil and caused many small depressions occupied by lakes. The bedrock structures can be seen. The light-colored areas are granite bodies, and the white lines are pegmatite dikes; both features are discordant to the foliation of the bedrock. Oblique aerial photo from Department of Energy, Mines and Resources, Ottawa, Canadian Government Copyright (A5032-62.R).

Figure 10-20. Erratic boulders dropped by a glacier when it melted. Photo by Mary Hill, courtesy California Division of Mines and Geology.

Eskers are long, sinous ridges, generally 10—50 feet high, that run for fractions of a mile to many miles. They are believed to be the filling of melt-water channels that ran under the glacial ice (Fig. 10-22).

Drumlins are streamlined hills 25—200 feet high and several hundred feet long. Their origin is not known, but the streamlined shape suggests that they formed under the moving glacial ice. They are not symmetrical, and the blunt end is in the direction of the origin of the ice. (See Fig. 10-23.)

Figure 10-21. Maximum extent of Pleistocene ice sheets.

Figure 10-22. Esker. The long, somewhat curving ridge is an esker and probably originated as a meltwater channel under the continental glacier. Oblique aerial photo from Department of Energy, Mines and Resources, Ottawa, Canadian Government Copyright (A2711-94).

Figure 10-23. Drumlins. Note the many boulders on the one in the foreground. Saskatchewan, Canada. Photo by W. G. Pierce, U.S. Geological Survey.

The most recent continental glaciation of Pleistocene age began about three million years ago. This is estimated from the degree of weathering of deposits, from the amount of soil produced between glacial advances, from the study of fossils, and from other information obtained by a variety of methods. The last glacial advance began about 30,000 years ago, reached its maximum advance about 19,000 years ago, and retreated about 10,000 years ago. Another minor advance occurred between 1550 and 1850 and is sometimes called the Little Ice Age. All of these dates are fairly accurately known from carbon-isotope studies. This method is based on the fact that organic carbon has a fixed ratio of carbon isotopes when formed, but this ratio changes after the death of the organism due to radioactive decay of one of the carbon isotopes (carbon 14). The method is not suitable for material more than about 40,000 years old. Radioactive dating methods are discussed elsewhere.

Before radiocarbon was used, other methods, particularly the study of varved clays, were used. This method is much like the study of tree rings, as the varves are assumed to be yearly deposits in glacial lakes. The dark part of each varve is deposited during the summer, and the light part in the winter. (See Fig. 10-24.) Thus the thickness of varves records the climatic conditions, and the sequence of relative thicknesses is correlated from lake to lake until the life span of the glacier is covered; then the total number of varves is counted. Apparently due to error in correlation, this method gave too high an age estimate.

Cause of Glaciation

To discover the cause of glacial periods, we must first see under what conditions glaciers form. Average temperatures must be lower than at present, but not greatly lower; and precipitation must be high. These two conditions will cause the accumulation of snow, which will form glaciers. The temperature must be low enough to insure that precipitation will be in the form of snow, but if it is too low, it will inhibit precipitation. This latter point is illustrated by the lack of glacier formation, because of low precipitation, in the many very cold regions today. Heavy precipitation during at least parts of the glacial age is suggested by the huge lakes that formed in areas south of the glaciers, where it was too warm for snow to accumulate. On the other hand, a humid climate without much evaporation may also have been involved in the formation of these lakes. These lakes formed in the basin areas of western United States, and the present Great

Figure 10-24. Varves or yearly layers. Photo shows varves in the Green River Formation, Colorado. Diameter of area photographed is 2 inches. Photo by W. H. Bradley, U.S. Geological Survey.

Salt Lake is but a small remnant of Lake Bonneville, which covered much of northwestern Utah. Other lakes covered large areas in Nevada, and some of the present saline lakes, such as Carson Sink, are remnants. (See Fig. 10-25.). These lakes are saline because, without outlets to the sea, dissolved material builds up in them while evaporation keeps their levels fairly constant.

A number of theories of glacier origin have been proposed. A successful theory must explain the change in climate just discussed and must account for the four separate advances of the recent ice age as well as for other periods of glaciation at different places on the earth. The other periods of glaciation occurred about 275 and 600 million years ago. The younger of these two, and the better known, occurred in Africa, India, and Australia—far from the present poles.

The main theories are as follows:

1. Changes in the amount of energy received from the sun due either to changes in output by the sun, or in the amount of energy that penetrates our atmosphere. There is no evidence to suggest any changes in output by the sun of the type necessary to support this theory. Dust in the atmosphere, perhaps from volcanoes, could cool the earth, but whether this effect is enough to cause glacier development is debatable. Decrease in the amount of carbon dioxide in the atmosphere would cool the earth, as this gas allows radiant energy from the sun to pass through, but prevents radiation from the earth from leaving. Thus the atmosphere acts much like a greenhouse; however, it is doubtful if the necessary changes in the amount of carbon dioxide are possible.

2. Movement of the continents relative to each other (continental drift). This is discussed in Chapter 16.

Figure 10-25. Map showing the known lakes of glacial times in western United States. From J. H. Feth, U.S. Geological Survey, *Professional Paper* 425B, 1961.

3. Changes in the location of the earth's rotational poles. This could change the climate of some areas by moving them toward the cool poles. The regular periodic movements of the poles, the precession and other movements, which are mentioned later, are well known, but are not great enough to cause glaciers; moreover, the known glacial periods do not coincide with these movements. However, other, larger movements could have occurred in the past, although no reasonable cause for such movements has been suggested yet.

4. Changes in the circulation of the oceans. This is a promising, recently proposed hypothesis, that seems adequate for the recent glacial periods. The hypothesis begins with an ice-free Arctic Ocean. With the present pole positions and our present atmospheric circulation, the ice-free Arctic Ocean would cause the heavy precipitation

necessary to initiate a glacial advance. By the middle of the glacial advance the Arctic Ocean would freeze, cutting off much of the supply of moisture. Precipitation would continue until the North Atlantic became too cold to provide enough moisture. Melting would then begin and continue until the Arctic Ocean was again ice-free. The stage is now set for the cycle to begin again.

This hypothesis can account for glacial periods only when the poles are located as they are now. If both poles were over open ocean, then the atmospheric circulation could not develop glacial periods. Thus, to initiate a glacial period, the poles would have to shift. Thus, this theory accounts for the four glacial advances of the recent glacial period; it can also explain the irregular periods of older glaciation. Recent studies, however, suggest that the present Arctic ice may be much older than the last glacial advance, a finding which casts doubt on this theory.

It should be obvious from this discussion that the climatic conditions that cause glaciation also had far-reaching effects, even far from the areas of actual glacial accumulation. It is estimated that sea level was lowered about 500 feet during the glacial advances, and that if the present-day glaciers melt, sea level will rise between 100 and 150 feet.

QUESTIONS

10-1. What is a glacier?

10-2. The two main types of glaciers are _____ and _____.

10-3. What changes does a mountain glacier produce in its valley?

10-4. How does a present-day glacier show climatic changes?

10-5. Why do crevasses form on the surface of glaciers?

10-6. Describe the processes of glacial erosion.

10-7. What are the erosional effects of continental glaciers?

10-8. What is a glacial erratic?

10-9. How many glacial stages are recognized in the Pleistocene?

10-10. Which way did the ice move in northern Canada? What is the evidence?

10-11. What type of climate is most favorable for the development of glaciers?

10-12. What evidence do we have as to the climate during glacial times in southwestern United States?

10-13. List some suggested causes of continental glaciation.

10-14. Outline the new theory of glacier origin that depends on changes in ocean currents between the Arctic and the Atlantic oceans.

10-15. Which of the three agents of transportation—wind, water, and ice—is able to transport the largest boulders; which is most limited as to the largest size it can transport?

10-16. How do mechanical weathering and glacial erosion work together?

10-17. What is the evidence for continental glaciation?

10-18. How do glacially deposited rocks differ from those found in river deposits?

10-19. How does the fine material in a river with active glaciers at its head differ from the material in a river in a non-glacial area?

10-20. Describe several types of glacial deposits.

SUPPLEMENTARY READING

Davis, W. M., *Geographical Essays,* especially "The Sculpture of Mountains by Glaciers," "Glacial Erosion in France, Switzerland, and Norway," and "The Outline of Cape Cod." New York: Dover Publications, Inc., (1909), 1954, 777 pp. (paperback).

Denton, G. II. and S. C. Potter, "Neoglaciation," *Scientific American* (June, 1970), Vol. 222, No. 6, pp. 101-110.

Donn, W. L., "Causes of the Ice Ages," *Sky and Telescope,* (April, 1967), Vol. 33, No. 4, pp. 221-225.

Dyson, J. L., *The World of Ice.* New York: Alfred A. Knopf, Inc., 1962, 292 pp.

Flint, R. F., *Glacial and Pleistocene Geology.* New York: John Wiley & Sons, Inc., 1957, 553 pp.

Hough, J. L., *Geology of the Great Lakes.* Urbana: University of Illinois Press, 1958, 313 pp.

Thornbury, W. D., *Principles of Geomorphology* (2nd ed.), Chapters 14, 15, 16, pp. 345-444. New York: John Wiley & Sons, Inc., 1969, 594 pp.

Chapter 11

Shorelines

Waves breaking on the shore are also agents of erosion, transportation, and deposition. Breaking waves have a great deal of energy, and anyone who watches them cannot fail to be impressed by this. (See Fig. 11-1.) It is so obvious that they are agents of erosion that for many years they were considered more important than the much less obvious, less spectacular rivers. Before discussing shoreline processes, we will consider the origin of the various movements of ocean waters.

Circulation of the Oceans

The movement of the surface waters of the oceans is caused by wind and the earth's rotation. The wind blowing over the water exerts a drag and sets the water in motion. In this process, first ripples and then waves are generated. The circulation developed in this manner is shallow, at the most only a few hundred feet deep; and the water moves slowly, generally between a fraction of a mile per hour and a few miles per hour. The resulting circulation is rotational because the earth's rotation and the continents deflect the moving water. The earth's rotation causes a deflection of the water's forward motion to the right in the northern hemisphere, and to the left in the southern hemisphere. In the southern part of the southern hemisphere, where no continents intervene, the circulation is latitudinal. (See Fig. 11-2.)

Figure 11-1. Breaking waves are agents of erosion. Photo from California Division of Beaches and Parks.

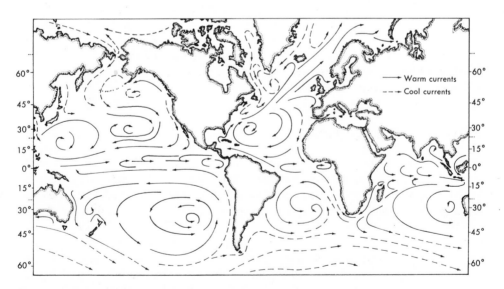

Figure 11-2. Circulation of surface waters of the oceans.

The deep circulation is caused largely by density differences in the ocean water. The cold, heavy Arctic, and especially the Antarctic, waters flow near the bottom toward the equator and in some cases to points beyond. In contrast, the waters of the warm Mediterranean, which have a high salinity because of evaporation, flow into the Atlantic at middle depths.

Much more study is needed before the circulation of the oceans can be described in detail, but the circulation is much less important geologically than waves. Oceanic circulation is, however, a very important means of heat transfer, greatly affecting the climates of the world and so affecting geologic processes.

Tides

Tides are caused in part by gravitational attraction, mainly due to the moon. Newton's law of gravity shows that there is a mutual attractive force between two bodies. The oceans are free to move and so are deformed by this force. Thus, the water on the side of the earth toward the moon moves a little closer to the moon, causing a high tide at that point. It is not quite so apparent that a high tide forms at the point on the opposite side of the earth at the same time. The reason for the two tidal bulges is that centrifugal forces caused by the earth's rotation are also involved in producing tides. In an equilibrium system such as the earth-moon system, the gravitational force between the two bodies is exactly balanced by the centrifugal force. This balance is exact at the centers of the two bodies, but at the earth's surface the two forces are not equal. On the side near the moon, the gravitational force toward the moon is greater; and on the opposite side, the centrifugal force acting in the opposite direction is greater. (See Fig. 11-3.) Because the moon orbits the earth, it rises about 51 minutes later each day. Thus, successive high tides occur about 12 hours and 25 minutes apart. The sun also affects tides because of its gravitational attraction, and all of these effects change somewhat with the season. This is the origin of tides, but this theoretical approach cannot be used to predict either the height or the time of actual tides at a coastline because tides are greatly affected by the shape of the coastline and the configuration of the ocean bottom.

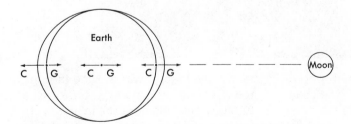

Figure 11-3. Origin of tides. On the side of the earth toward the moon, the gravitational attraction (G) between the moon and the earth is greater than the centrifugal (C) force. On the other side of the earth, the centrifugal force is greater. At the earth's center, the two forces balance. The tidal bulge is greatly exaggerated, and the earth-moon distance is greatly shortened.

Tidal currents may move sediments at some places; but except at these places, tides probably are not important agents of transportation or erosion. A notable exception occurs where tides move through narrow inlets, constantly scouring these entrances to many good harbors that would otherwise be blocked. (See Fig. 11-4.) The vertical movement of the tides is in the range of a few feet to a few tens of feet, and this constantly changing datum complicates the near-shore processes.

Figure 11-4. Vertical aerial photograph of inlet to Apalachicola Bay, Florida. The outgoing tide has carried out into the ocean the sediment brought into the bay by the Apalachicola River. At the time the photo was taken, clear water was moving through the inlet and the cloudy water forms a turbidity current in the ocean. In the upper right of the photo, clouds of fine sediment are being moved by wave action.

Do not confuse tidal currents with the misnamed *tidal waves* (seismic sea waves) that are caused by earthquakes in the oceans, discussed in Chapter 13.

The solid rock earth is, like the oceans, also deformed by the tidal forces. Although difficult to measure, the movement is three to six inches. Other aspects of tides are discussed in Chapter 18.

Waves

Waves can be described by their height and wavelength. These terms are defined in Fig. 11-5. The wavelength and period are related to the velocity at which the wave travels by the equation,

$$\text{Wavelength} = \text{period} \times \text{velocity}$$

The velocity, however, depends on the wavelength; and longer wavelength waves (longer period waves) travel at higher velocities. Period is much easier to measure than wavelength, so most relations are expressed in terms of period. In deep water, the velocity in miles per hour is about 3.5 times the period in seconds. The height of a wave has almost no effect on its period or velocity.

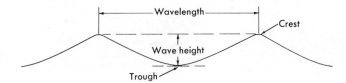

Figure   11-5. Characteristics   of   a wave.

Waves, like surface circulation, are caused by wind. Wind blowing over water causes ripples to form; and, once formed, the ripples provide surfaces on which the wind can act more efficiently. The buildup of ripples into waves is a complex process. Turbulence in the wind forms ripples of differing size and direction that combine to form other waves. The energy of the shorter wavelength waves is largely absorbed by longer waves, increasing the height of the longer waves. The height of the waves produced depends on the strength of the wind, how long the wind blows, and how large an area the wind blows over. Most large waves originate in storms; but, once formed, the waves can travel great distances with little loss of energy. Thus, a storm's energy may be transmitted by waves to a distant shore.

Typical waves in a storm area have periods near 10 seconds. As the waves move out of the storm area, their periods (and wavelengths) increase. At distances of a few thousand miles from the area of their formation, waves have periods between 15 and 20 seconds and may travel faster than the winds that generated them.

In deep water, waves are surface shapes; and though the shape moves forward with the wind, the water essentially does not. An individual water particle in a wave moves in a circular path and returns to almost its original position. (See Fig. 11-6.) Thus, in deep water a floating object bobs up and down. The wind does drag the water slowly forward, causing the slow surface circulation (see Fig. 11-7); but in the case of waves, it is the form, not the water, that moves. The motion is much like a wind wave through tall grass or wheat.

In shallow water the motion is very different. (See Fig. 11-8.) Because waves form at the surface, their energy decreases downward. At a depth of one-half the deep water wavelength, the water motion is negligible. As long as the water is this deep, a wave is unaffected by the depth of the water. If a wave travels into water shallower than this, however, it feels the bottom and changes. The wave travels slower, and some of its

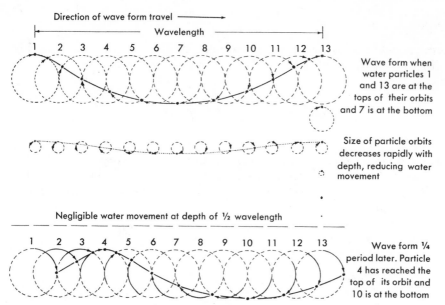

Direction of wave form travel ———————→

Wavelength

1 2 3 4 5 6 7 8 9 10 11 12 13

Wave form when water particles 1 and 13 are at the tops of their orbits and 7 is at the bottom

Size of particle orbits decreases rapidly with depth, reducing water movement

Negligible water movement at depth of ½ wavelength

1   2   3   4   5   6   7   8   9   10   11   12   13

Wave form ¼ period later. Particle 4 has reached the top of its orbit and 10 is at the bottom

Figure 11-6. Water particles move through a circular orbit as a wave passes. Almost no water motion occurs below a depth of one-half wavelength.

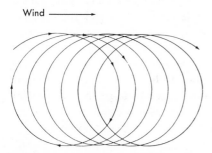

Wind ———————→

Figure 11-7. The wind moves the water slowly forward, causing the surface circulation of the oceans. The circles show the actual movement of water particles.

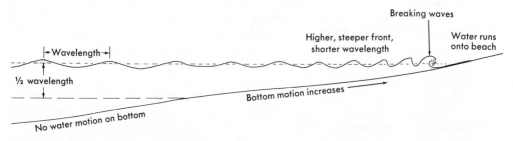

Breaking waves

Higher, steeper front, shorter wavelength

Water runs onto beach

|←Wavelength→|

½ wavelength

Bottom motion increases

No water motion on bottom

Figure 11-8. The changes that occur when a wave moves onto the shore.

energy is used in moving small grains on the bottom back and forth. The height of a wave increases as the water becomes shallower; and on a sloping shore, the front of a wave is slowed more than the rear. These effects cause waves to break, creating surf. In the surf zone, the breaking waves run up on the shore and are active agents of erosion and transportation. (See Fig. 11-9.) Their return to the ocean creates the "undertow"

Figure 11-9. Aerial view of breaking waves and surf zone. Photo from California Division of Beaches and Parks.

feared by many swimmers. On the landward side of breaking waves, the surface water moves on shore and the bottom water moves outward, thus tending to upset a bather. At times the returning water is concentrated at a few places, creating what are called *rips,* or *rip currents.* (See Fig. 11-10.)

Along many coasts, after breaking, the water moving onshore forms new waves that in turn break closer to the shore. This process may be repeated several times, especially if the waves are large and the bottom slopes gently.

Wave erosion, then, is concentrated in shallow water. Waves generally break at depths of about one and one-half times their height. Waves over 20 feet high are rare, although waves up to 100 feet high have been reported. At depths less than 30 feet, therefore, the majority of wave energy is expended and most of the geologic work of waves, or surf, is done. Because waves first begin to act on the bottom at about one-half their wavelengths, however, some wave erosion can occur as deep as 1000 feet. Waves with 10 second periods have wavelengths of about 500 feet, and 20 second waves have about 2000 foot wavelengths.

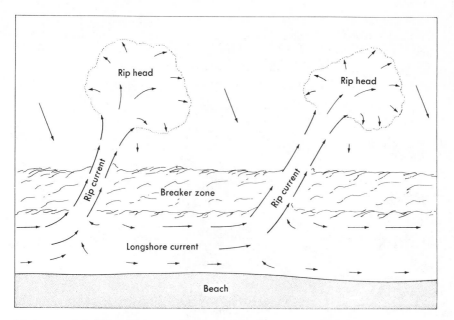

Figure 11-10. Rip currents are
formed by the return of the water that
is moved on shore by the waves. The
location of the rip currents is variable.
They create the "undertow" feared by
swimmers.

Another effect of shallow water on waves is to cause their bending, or refraction.
When a wave feels the bottom it slows. If parts of the wave do not feel the bottom, these
parts are not slowed. In this case, because different parts of the wave are traveling at
different speeds, the wave bends or refracts. The effect of refraction is to concentrate
the waves on headlands, as shown in Fig. 11-11.

Wave Erosion

Waves erode mainly by hydraulic action and abrasion, and to a much smaller extent
by solution.

Hydraulic action is the impact and pressure caused by waves. Storm waves, in
particular, are able to move huge blocks of bedrock. Breaking waves can cause great
pressure in cracks by the water's force alone or by compressing air in cracks. A
breakwater weighing 2,600 tons has been moved in this manner. Windows in light-
houses hundreds of feet above sea level have been broken by rocks thrown up by storm
waves. The effects of storm waves in moving large fragments may be likened to river
floods, during which the largest fragments are moved. Newspapers occasionally show
pictures of large and small boats thrown long distances inland in this way. Waves
breaking on a coastline very commonly develop cliffs by undercutting the bedrock. (See
Fig. 11-12.)

Figure 11-11. Wave refraction along the California coast. Note the changes in wavelength that occur in the bay on the right. Photo from U.S. Department of Agriculture.

Abrasion is the grinding done by material moved by the waves. It is especially effective in the surf zone, but can occur, though only to a much smaller extent, in the entire area where the waves feel bottom and where some of their energy is dissipated on the bottom. This action rounds and reduces the size of the pebbles. Waves running up on the shore carry pebbles and sand grains up with them and then back as they return to the sea. This rolling, churning action is very similar to the movement of the bed load of a river. Acting in the surf zone, such abrasion cuts a smooth surface, generally at a depth of 30 feet or less. The surface is called a surf-cut bench. (See Fig. 11-12.) The presence of such a bench may cause the waves to break far from the shore and so limit the effect of the waves on the actual shore.

Figure 11-12. Surf-cut terrace and undercut cliff. Photo by C. A. Kaye, U.S. Geological Survey.

Wave erosion is less effective than generally thought. It is estimated that only about one per cent of the sediments in the oceans came directly from wave erosion. Most beaches are formed of material moving onshore. At some places wave erosion is very active, and rates as high as 30 feet per year of recession are known. Most wave erosion occurs during large storms.

Wave Transportation

The movement of sediment by waves is a selective process that effectively sorts the sediment by size, especially in the near-shore surf zone. Only sand and coarser material are found in the surf zone, and the main movement of this material is along the shore. In this active zone the waves run up on the beach, carrying sand and coarser material as both bed load and suspended load. The slope of this area is such that the receding wave, aided by gravity, carries this material back with it, even though it loses some of its water by percolation into the beach. Any finer material in the receding wave is carried to deeper water by the "undertow." Thus the sand and coarser materials move back and forth in this zone. This material moves parallel to the beach because refraction of the waves generally causes them to move diagonally toward the shore. The waves moving onshore carry the sand forward in this diagonal direction. The receding water, however, moves the sand back down the slope under the influence of gravity. Thus, in each back and forth movement the sand is carried further down the beach. (See Fig. 11-13.) The magnitude of this longshore transportation has been measured with marked sand grains and at places amounts to a few thousand tons (dry weight) per day under normal conditions and several times this much during storms. Another measure of the longshore movement is the effect of barriers sometimes constructed to protect beaches or provide harbors for small boats, as shown in Fig. 11-14.

In the zone where the waves feel bottom, sand moves shoreward when the bottom is disturbed by the waves. The returning water flows offshore with a more constant, slower movement that can only move finer material. Thus, in this zone sand moves shoreward and finer materials may tend to move offshore. (See Figs. 11-4 and 11-15.)

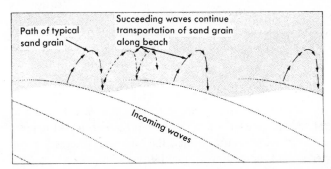

Figure 11-13. Movement of sand along a beach. The incoming waves are not parallel to the shoreline and move sand grains forward perpendicular to the wave. On retreat, the water flows perpendicular to the shore. Thus each wave moves material a little further along the beach.

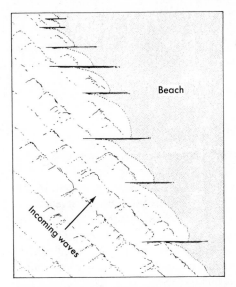

Figure 11-14. The effect of barriers on longshore movement of sand is shown in this oblique view of a beach.

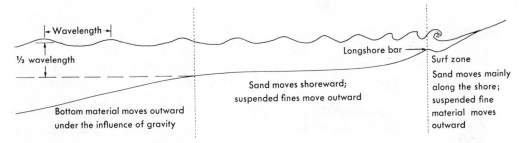

Figure 11-15. Transportation in the near-shore area.

In deeper water, however, only the long wavelength waves disturb the bottom materials, and the movement is symmetrically back and forth as the wave passes. The bottom generally slopes seaward so gravity may help to move some of the disturbed sediments seaward.

During storms beaches lose sand. This typically occurs during winters. The sand returns to the beach soon after the waves become normal again. On some shores the high storm waves move some of this sand inland beyond the beach. Wind may also move sand inland. In general, storm waves remove sand from the beach and deposit it as a bar at the edge of the surf zone, probably because the higher wave crests of the storm waves need a shallower slope to maintain the equilibrium slope on which they move the sand back and forth. (See Fig. 11-15.) On some coasts some of the sand is carried to deeper water, and it continues to move seaward down the gentle slope of the shelf.

Rip currents transport fine sand offshore, especially during times when onshore winds as well as waves move surface water onshore.

Sand also moves offshore in submarine canyons. This occurs where sand moving along the shore meets a canyon.

Storm waves or storm winds may stir up the bottom sediments, and the resulting turbid water may move offshore. Turbid water a few tens of miles offshore has been noted from satellites.

Wave Deposition

Wave deposits are of two main types: beaches; and sand bars, either offshore or attached to the shore.

Beaches are the more common wave-deposited features. Some of the material forming a beach comes from wave erosion, and some of it comes from other agents of erosion, especially rivers. In recent years many rivers have been dammed so they deliver little sand to the beaches. As a result many beaches are becoming smaller. The size of the sediment in a general way controls the slope of the beach, with pebble beaches being steeper than sand beaches. As noted before, beaches are in delicate balance and change drastically when wave and current conditions change. (See Fig. 11-16.)

The sediment carried along the shore is deposited as sand or gravel bars at headlands and across bays, as shown in Fig. 11-17. These bars and offshore bars are important in the development of shorelines.

Development of Shorelines

Understanding the development of shorelines is much like understanding the erosional development of rivers or land areas in that an inductive approach is necessary. The great variety of present-day coastlines suggests that shorelines are complex areas; many factors must be considered. Only a few of the better understood cases will be considered here. Any actual shoreline must be studied individually, taking into account such factors as rock type, rock structure, size of waves, direction of waves, number of storms, tidal range, and submarine profile. All coastlines, like all other parts of the crust, are constantly changing, and the type of changes depends partly on the recent changes in regimen. As with rivers, major changes are initiated mainly by changes in elevation.

A.

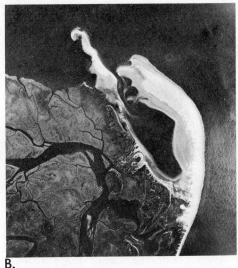

B.

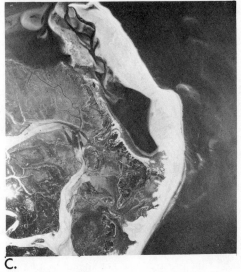

C.

Figure 11-16. Shoreline changes at Little Egg Inlet, New Jersey. A. 1940. B. 1957. C. 1963. These three photographs taken in 1940, 1957, and 1963 show changes in Island Beach south of the inlet but do not provide stereoscopic coverage. The tidal channels in the salt marsh serve as reference marks for comparison of shoreline changes from year to year. In 1940, a long narrow north-trending beach and spit separated the marsh from the ocean. By 1957, the north-trending beach and spit had been replaced by a northwest-trending beach and spit, shorter and slightly west of those of 1940. In addition a new northwest-curving spit about a quarter of a mile wide had been built east of the position of the 1940 spit and extending northward for about a mile. By 1963, the curving spit had grown northwestward an additional 1,000 feet and changed shape to some extent. Photos and caption from C. S. Denny and others, U.S. Geological Survey, *Professional Paper* 590, 1968.

Figure 11-17. Oblique aerial photo of Morro Bay, California, showing a sand bar enclosing a bay. Photo from California Division of Beaches and Parks.

Two types of sea level changes are possible. The water level itself may rise or fall, or the land may be uplifted or depressed. Sea level may change as a result of uplift or depression of the ocean basins. Evidence that such changes have taken place will be presented in later chapters. Another way that sea level is changed is by the formation and melting of glaciers. This has happened in the recent past, and it is very difficult to separate the changes caused by the recent withdrawal and return of the sea from the normal shoreline processes. Uplifts and depressions of the continents are also common and will also be described later. Changes in sea level from this second cause are more localized, but cannot always be distinguished from rise or fall of the ocean itself. Many coastlines show evidence of repeated relative changes of land and sea. The melting of the last glacial advance caused a rise in sea level, but the effects of this are not apparent on many coasts.

A suitable classification of coastlines has not yet been devised because of the difficulty of recognizing the many possibilities. Some possibilities can be discussed, but examples are hard to find. One reason for this may be that most coasts are unstable and the relation between land and sea is constantly changing, making classification difficult. The factors involved are probably whether the coast is being eroded or built-up, on the one hand, and whether emergent or submergent, on the other.

One way to classify coasts is to determine whether they are emergent or submergent. A submergent coastline would develop if either the land went down or the ocean rose. A submerged coastline should have deep embayments and estuaries caused by the filling of coastal valleys with sea water. (See Fig. 11-18.) This is, of course, not the only way that such a coast can develop. Few coastlines are of this type, although the recent melting of the glaciers should have made almost all coasts submergent.

Some types of emergent coastlines are easily recognized. Rising land or falling water level will bring beaches and surf-cut benches above sea level. In some cases, several benches are visible above a present beach. This is common at many places along the

A. Neutral coastline

B. Submergent coastline. Wave action may develop cliffs on the headlands, and the eroded material may form sand bars across the bays.

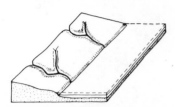

C. Emergent coastline

D. Profile of a coastline that has been up-lifted (or the sea level fallen) several times, producing a series of wave - cut terraces and cliffs.

Figure 11-18. Possible types of shorelines. A. A neutral coastline. B. The same coastline after the land has sunk or sea level has risen. Note the drowning of the river valleys, producing bays and estuaries. Such irregular coastlines are typical of submergence. C. The same coastline after the land has risen or sea level has sunk. Note the wave-cut cliff produced by breaking waves on the newly exposed land. The rivers form falls when they meet the ocean. Offshore sandbars may form. D. The profile of a coastline that has emerged in several stages similar to the coastline in C. The recent rise of sea level caused by the melting of glaciers has resulted in submergence of most coastlines; however examples similar to D are common on the West Coast.

California coast where recent faulting and uplift are responsible. (See Figs. 11-18, 11-19, and 11-20.) At some places warping of the wave-cut benches reveals differential uplift.

Assuming stability, it is possible to describe the development of some types of shorelines. An initially irregular coastline with headlands and embayments will be greatly modified by wave action. Refraction of the waves will concentrate the wave erosion on the headlands (see Fig. 11-21), and a wave-cut bench and sea cliffs will develop there. The waves will carry some of the eroded material into the embayments, which, together with river-eroded material, will tend to fill them. Sand bars of various types will also develop across the embayments. Ultimately, a smooth coast will develop. A coastline of this type tends to retreat landward.

Shallow, smooth coastlines, such as much of the Atlantic and Gulf coasts, have offshore bars or barrier islands with shallow lagoons behind them. The origin of these islands is a problem. Recent studies have suggested that they may be former beaches above which onshore winds had caused the development of sand dunes such as are common at many of our present beaches. These beaches are believed to have formed during glaciation when sea level was lower and winds may have been more active. Melting of the glaciers would cause sea level to rise and flood the area behind the beach-dune complex. The resulting lagoon may be partially filled by material carried into it by rivers. (See Fig. 11-22.)

Figure 11-19. Sea cliffs undergoing active erosion by waves and uplifted surf-cut benches on the skyline. Photo from Oregon State Highway Department.

Figure 11-20. Wave-cut beaches showing the many former levels of Great Bear Lake, Northwest Territories, Canada. In the right foreground, frost polygons and an old landslide can be seen. Photo from Department of Energy, Mines and Resources, Ottawa, Canadian Government Copyright (A4116.37).

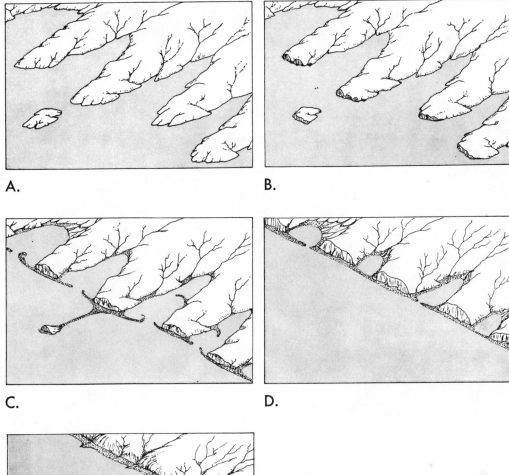

A.

B.

C.

D.

E.

Figure 11-21. Development of an initially irregular coastline.

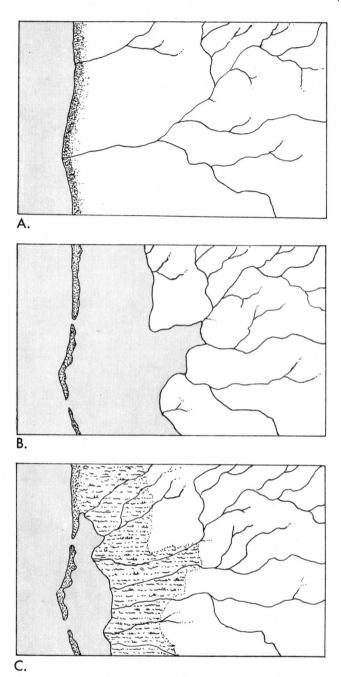

A.

B.

C.

Figure 11-22. Development of a gently sloping, smooth coastline. A. Beach-dune complex forms at a lower sea level during glaciation. B. Melting of the glaciers floods the area behind the beach-dune complex forming a lagoon. C. River deposition begins to fill the lagoon.

QUESTIONS

11-1. What causes the surface circulation of the oceans?

11-2. Explain why high tides occur simultaneously on both sides of the earth.

11-3. What are the relationships among wavelength, period, and velocity of water waves?

11-4. Describe the motion of water in waves in both deep and shallow water.

11-5. Describe the processes and the results of wave erosion.

11-6. Discuss the transportation of both sand and finer material by waves.

11-7. Do sea cliffs always mean uplift of land (or lowering of sea level)?

11-8. How might one distinguish between uplift of the continent and depression of sea level at a given place?

11-9. Discuss whether sea level is a horizon to which absolute movements of the continents can be referred.

SUPPLEMENTARY READING

Bascom, Willard, "Ocean Waves," *Scientific American* (August, 1959), Vol. 201, No. 2, pp. 74-84. Reprint 828, W. H. Freeman and Co., San Francisco.

_____, "Beaches," *Scientific American* (August, 1960), Vol. 203, No. 2, pp. 80-94.

_____, *Waves and Beaches.* Garden City, New York: Doubleday and Co., Inc., 1964, 267 pp. (paperback).

Carson, Rachel, *The Sea Around Us* (revised edition). New York: Oxford University Press, 1961, 230 pp. (Also available in paperback.)

Gross, M. G., *Oceanography.* 2nd ed. Columbus: Charles E. Merrill Publishing Co., Inc., 1971, 150 pp. (paperback).

Higgins, C. G., "Causes of Relative Sea-level Changes," *American Scientist* (December, 1965), Vol. 53, No. 4, pp. 464-476.

Hoyt, J. H., "Field Guide to Beaches," *ESCP Pamphlet Series,* PS-7. Boston: Houghton Mifflin Co., 1971, 45 pp.

Inman, D. L., "Beach and Nearshore Processes Along the Southern California Coast," California Division of Mines Bulletin, No. 170, Chapter 5, pp. 29-34, 1954.

King, C. A. M., *Beaches and Coasts.* London: Edward Arnold, 1959, 403 pp.

Kort, V. G., "The Antarctic Ocean," *Scientific American* (September, 1962), Vol. 207, No. 3, pp. 113-128. Reprint 860, W. H. Freeman and Co., San Francisco.

Munk, Walter, "The Circulation of the Oceans," *Scientific American* (September, 1955), Vol. 193, No. 3, pp. 96-104. Reprint 813, W. H. Freeman and Co., San Francisco.

Russell, R. J., "Instability of Sea Level," *American Scientist* (December, 1957), Vol. 65, No. 5, pp. 414-430.

Stommel, Henry, "The Anatomy of the Atlantic," *Scientific American* (January, 1955), Vol. 192, No. 1, pp. 30-36. Reprint 810, W. H. Freeman and Co., San Francisco.

Chapter 12

Ground Water and Water Resources

All the rivers run into the sea;
Yet the sea is not full:
Unto the place from whence the rivers come,
Thither they return again.

Ecclesiastes 1:7

Ground Water

It was not until the latter part of the seventeenth century that it was shown that the origin of river water is rainfall. When this measurement was undertaken, however, it was discovered that only a fraction of the amount of water that fell as rain on a drainage basin flowed out as the river. The fate of the rest of the precipitation is not hard to imagine. Some of it evaporates, some is used by plants, and some seeps into the ground. The latter is the origin of ground water.

Ground water is a very important part of the water resources of many areas. The scientific study of ground water in the United States did not begin until about 1875. Since World War II the problem of water resources has received much attention, and all aspects, including ground water, are being studied vigorously. Water studies have important legal, social, and economic aspects as well as the mainly scientific aspects considered here.

Movement of Ground Water

The surface below which rocks are saturated with water is called the *water table.* (See Fig. 12-1.) Some water is retained above the water table by the surface tension of water. (See Fig. 12-2.) In general, the water table is a reflection of the surface topography, but is more subdued, that is, has less relief, than the surface topography. (See Fig. 12-3.) Lakes and swamps are areas where the land surface is either below or at the water table.

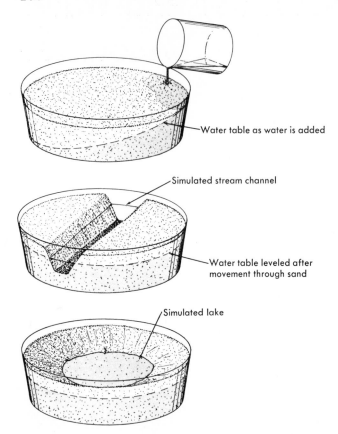

Figure 12-1. An illustration of ground water. The dish is filled with sand, and the water poured in percolates into the sand, filling the pore spaces. A hole in the sand can represent a well, or a slot a stream channel. This experiment works best in a transparent dish with water colored by a few drops of ink.

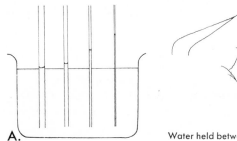 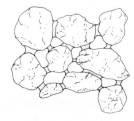

Water held between thumb and forefinger **B.** Water held between sand grains

Figure 12-2. Capillary action is caused by surface tension and can cause water to rise above the water table. A. In a very narrow tube, water rises well above the water level. B. In the ground, the narrow spaces between particles take the place of the tubes. For this reason, the top of the water table is a thin zone rather than a sharp line. This is also the way that soil retains its moisture.

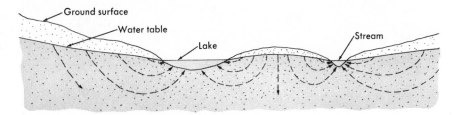

Figure 12-3. The water table has less relief than the topography. The arrows show the movement of the ground water.

Springs occur where the water table is exposed as on a valley side. The position of the water table changes seasonally, which explains why some springs are dry in summer. Springs may form where the water table is *perched.* This situation results where the downward percolation of rain water is stopped by a relatively impermeable rock, such as a shale. Springs of this type may flow only during the wet season. (See Fig. 12-4.)

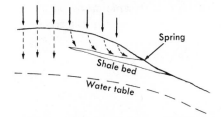

Figure 12-4. Spring formed by a perched water table created by impervious shale bed. Such springs tend to dry up in late summer.

The movement of ground water is controlled by the physical properties of the rocks. The amount of water that can be stored in the rocks is determined by the amount of pore, or open, space. The availability of the water is determined by the interconnections of the pore space, which is, of course, the permeability (Fig. 12-5). The most common

Well-sorted sandstone. Large amount of pore space gives high porosity and permeability.

Poorly sorted sandstone. Has much lower porosity and permeability.

Figure 12-5. The sorting of sandstone affects its porosity and permeability.

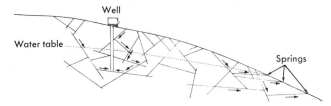

Figure 12-6. Springs and wells can also occur in fractured rocks such as granite.

reservoir rock is sandstone, although fractured granite or limestone, as well as many other rock types, can serve equally well. (See Fig. 12-6.)

A very important type of well is the artesian well. The reservoir of an artesian well is confined above and below by impermeable rocks. The reservoir is charged at a place where it is exposed, and the water is forced to move through the reservoir. A well drilled into an artesian reservoir may flow without pumping because the released water seeks its own level—that of the recharge area. However, the resistance to flowage of the water through the reservoir prevents the water from flowing to the same elevation as its recharging area; therefore, flowing wells must be lower in elevation than the recharge area. (See Fig. 12-7.)

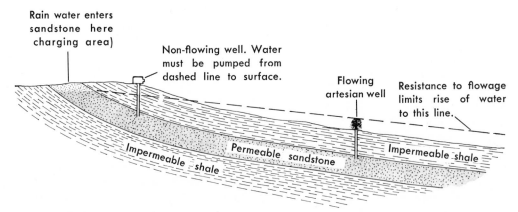

Figure 12-7. Artesian water results when the flow of ground water is restricted by impermeable layers.

Ground water is a very valuable economic commodity. It is recovered from wells for domestic, industrial, and agricultural use. (See Fig. 12-8.) In areas where ground water is used extensively, care must be taken that, on the average, no more water is withdrawn in a year than is replaced by natural processes or artificial infiltration. This can be determined by seeing that the water table is not lowered. The conservation of ground water is very important because it moves very slowly and many years may be required to replace hastily pumped water. The average rate of movement of ground water through rocks is only about 50 feet per year, although at some places the movement is much faster.

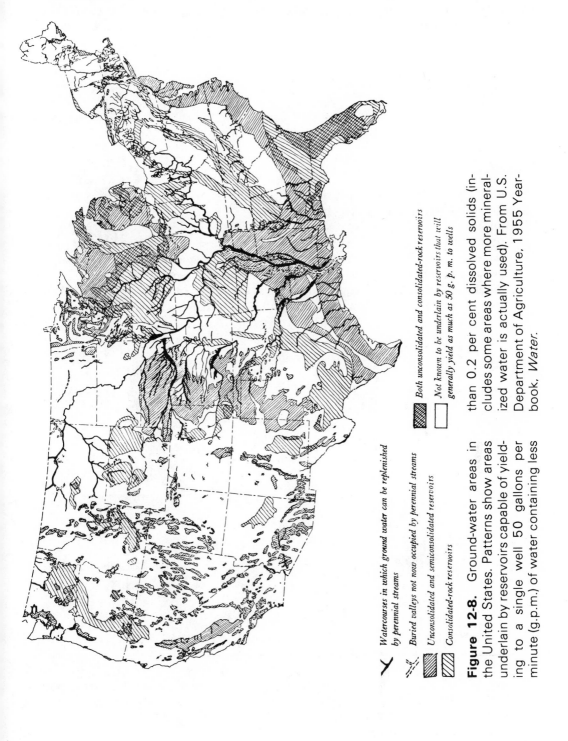

Y *Watercourses in which ground water can be replenished*
   by perennial streams

/. *Buried valleys not now occupied by perennial streams*

▨ *Unconsolidated and semiconsolidated reservoirs*

▧ *Consolidated-rock reservoirs*

▨ *Both unconsolidated and consolidated-rock reservoirs*

☐ *Not known to be underlain by reservoirs that will*
   generally yield as much as 50 g. p. m. to wells

Figure 12-8. Ground-water areas in the United States. Patterns show areas underlain by reservoirs capable of yielding to a single well 50 gallons per minute (g.p.m.) of water containing less than 0.2 per cent dissolved solids (includes some areas where more mineralized water is actually used). From U.S. Department of Agriculture, 1955 Yearbook, *Water*.

This slow movement of the water requires another caution. If water is withdrawn from a well by pumping too rapidly, the water is removed from the immediate vicinity of the well faster than it can be replaced. The result is called a *cone of depression* (Fig. 12-9). If cones of depression are developed around several closely spaced wells, the

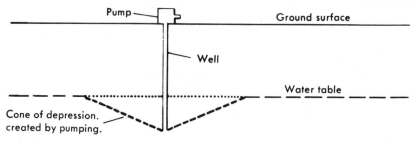

Figure 12-9. A cone of depression results when water is pumped too rapidly from a well.

water table can be drastically lowered. Thus, the spacing and the rate of pumping of wells need to be regulated in order to insure the most efficient use of the water. Figure 12-10 shows the changes that have occurred as ground water use has increased on Long Island. Note that as fresh water was removed faster than it was replaced, the heavier

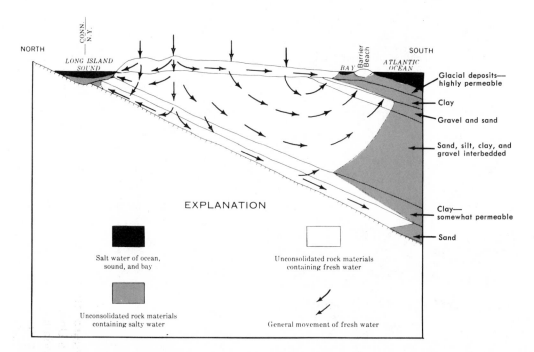

Figure 12-10. The encroachment of salt water from the ocean into groundwater reservoirs in Long Island, N.Y. This was caused by removal of ground water faster than it was recharged. From U.S. Geological Survey *Circular* 524, 1966.

salt water from the ocean encroached. At some places the water has been withdrawn so rapidly that hundreds of years may be required to replace it. Such usage is similar to mining, in the sense that a natural resource is used up.

Another effect of the withdrawal of ground water at a more rapid rate than it is replaced is the gradual subsidence of the surface of the ground. Subsidence has also occurred at some places because of the withdrawal of oil from the ground. (See Fig. 12-11.) Up to 20 feet of subsidence caused by withdrawal of ground water has occurred in parts of the Central Valley of California.

Pollution is a problem in some areas where well water is used. Drainage from cesspools can contaminate nearby wells. This danger is always present but is most acute where the ground water flows freely as in wide joints or cavernous limestones. If the polluted water seeps slowly through sand or gravel, especially above the water table, it is commonly purified in only a few tens of feet of travel. The purification is accomplished by filtering and oxidation. (See Fig. 12-12.) Advantage of this means of purifying water has been taken lately by spraying sewage and industrial waste on the ground surface, especially in wooded areas. In this way the ground water is recharged and the contaminants removed.

Geologic Work of Ground Water

Much of the geologic work of ground water was covered in the discussion of weathering in Chapter 4. Water is a very good solvent, especially if it contains carbon dioxide; and the role of ground water in solution and cementation have been described. It is not surprising that most ground water contains dissolved material and so is what is termed hard water. This is the origin of the taste of most spring and well water. In some cases

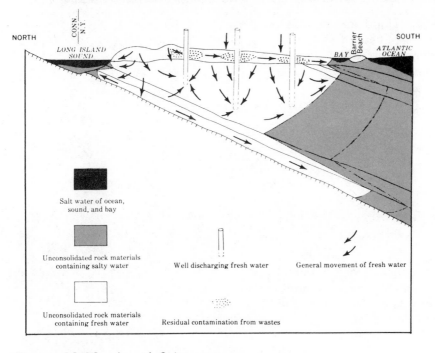

Figure 12-10. (cont.) Salt water encroachment, Long Island, N.Y.

Figure 12-11. Flooding at Long Beach, California, caused by subsidence resulting from withdrawal of oil. Photo courtesy of City of Long Beach.

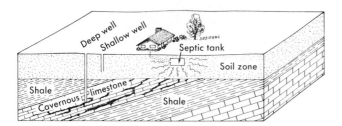

Figure 12-12. Pollution of wells. In this example, because of the cavernous limestone, the shallow well is more likely to produce pure water than is the deep well that is further away from the septic tank.

ground water contains so much dissolved material that it is not suitable for domestic or industrial use. Some ground water used for irrigation deposits dissolved salts in the soil and so can ruin soil unless enough water is used to flush the salts continually from the soil. Some of the water encountered in drilling is salty, and some of these waters were once sea water that was buried with the sediments when the bed was deposited in the ocean.

Ground water forms caves, but exactly how is not known. They are formed mainly in limestone, but can form in any soluble rock. (See Fig. 12-13.) Calcite is soluble in

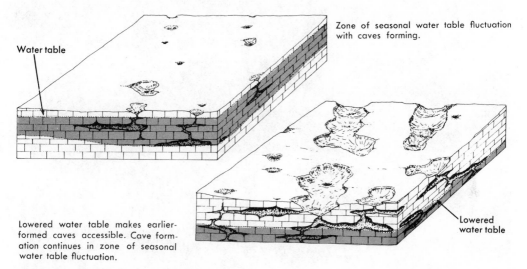

Water table

Zone of seasonal water table fluctuation with caves forming.

Lowered water table makes earlier-formed caves accessible. Cave form-ation continues in zone of seasonal water table fluctuation.

Lowered water table

Figure 12-13. Development of caves. The caves begin by solution of limestone in the zone of seasonal water table fluctuation. Lowering of the water table makes the caves accessible. An area of cavernous limestone generally has few, if any, surface streams.

water, especially in water that has carbon dioxide dissolved in it, as noted in the discussion of chemical weathering. Probably caves form in the zone of seasonal water table fluctuation by solution of limestone along joint planes. Lowering of the water table may make the cave accessible. In times of flood caves may be enlarged by erosion by running water. A process going on in caves (Fig. 12-14) is the deposition of calcite in the form of stalactites, stalagmites, and other features, all of which are caused by the evaporation of carbonate-charged water.

In limestone areas, caves or other channels may carry most of the water. The surface water may sink underground and flow through caves. In such an area there are only a few short surface streams, and they end in closed depressions called *sinks*. (See Fig. 12-15.) This, as might be expected, produces an uncommon type of surface topography. It is worth noting that only in the case where caves are developed is anything like an underground stream developed; many people have the erroneous impression that all

Figure 12-14. Stalactites and stalagmites in Carlsbad Caverns, New Mexico. The stalactites form by dripping
water that evaporates, depositing its
dissolved calcite. Stalagmites build up
when drops of water with dissolved calcite fall on the floor of the cave and
evaporate. Photo by National Park Service.

ground water flows in underground streams similar to surface streams. In the same
sense, there are no underground lakes, but only reservoir rocks whose pore space is
filled with water.

Hot Springs and Geysers

Geysers are periodic discharges of hot ground water, in contrast to hot springs which
flow more or less continuously. (See Fig. 12-17.) Studies of the isotopes in hot spring
and geyser waters have shown that most, if not all, of it is ordinary ground water that
has been heated. This heating is due either to deep circulation of the water or to contact
with hot igneous rock bodies. (See Fig. 12-16.) Geysers result when water accumulates
in vertical underground chambers where it is heated. The pressure of the overlying
water causes the boiling point near the bottom to rise. The heating also causes the
column of water to expand and spill over near the top. This reduces the pressure, and

Figure 12-15. A sink hole caused by solution of limestone by ground water. This sink probably formed by collapse into a cavern. Photo by N. H. Darton, U.S. Geological Survey.

the superheated water at the bottom flashes into steam and causes the geyser to erupt. (See Fig. 12-18.) This process is repeated more or less periodically in a geyser. All ground water contains dissolved material, and the hot waters contain, in general, more. Thus, hot springs and geysers commonly deposit calcite and other minerals. (See Fig. 12-19.)

Petroleum

Much of the foregoing information on ground water is applicable to the other natural fluids—petroleum and natural gas. These materials originate mainly from organic material in marine sedimentary rocks, although some do form in some lake-deposited sedimentary rocks. The beds that contain the oil-forming organic material are generally shale. It takes millions of years and increased temperature and pressure, due to burial, to form petroleum and natural gas. Petroleum and natural gas move very slowly in the shale source rocks; they are not economically recoverable by drilling until they have migrated to more permeable reservoir rocks. Thus, to have an oil field, it is necessary to have source rocks, reservoir rocks, and a trap, generally an impermeable cover. Because petroleum is lighter than water, it rises to the top of a reservoir. Several types of traps are shown in Fig. 12-20. The same cautions concerning well-spacing and rate of pumping that applied to ground water apply to oil fields.

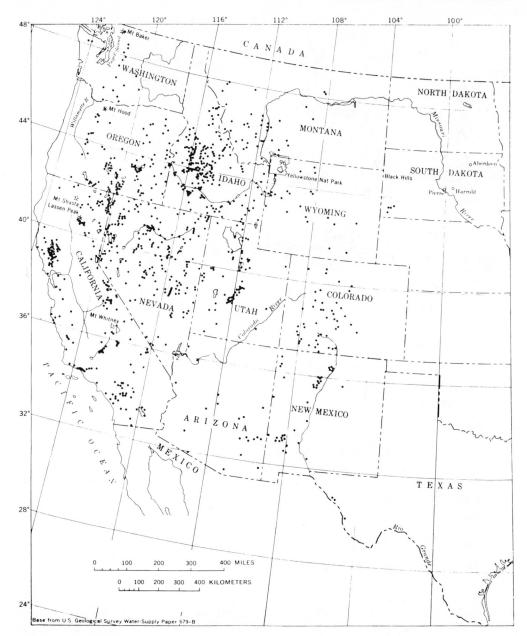

Figure 12-16. Locations of hot springs and geysers in the United States. They are mainly found in the geologically young, active areas. From G. A. Waring, U.S. Geological Survey, *Professional Paper* 492, 1965.

Figure 12-17. Old Faithful at Yellowstone National Park in eruption. Photo
by National Park Service.

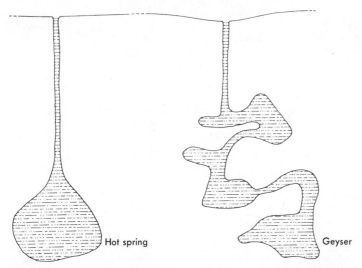

Figure 12-18. Idealized diagrams showing hot spring and geyser. A hot spring in which the water temperature no place exceeds the boiling point may have almost any configuration. If, however, high temperatures are involved, the configuration must be such that the heat is easily distributed by convection as in the figure shown. A geyser can form if the heat is not distributed by convection. In the figure above, the water near the bottom is heated to near its boiling point. The boiling point is higher there than at the surface because the weight of the water above increases the pressure. The water higher in the geyser system is also heated and so expands and flows out at the top. This reduces the weight of the water on the bottom and, at the reduced pressure on the bottom, boiling occurs. The bottom water flashes into steam, and the expanding steam causes an eruption.

Figure 12-19. Terraces built by hot spring deposits at Mammoth Hot Springs, Yellowstone National Park. Photo by National Park Service.

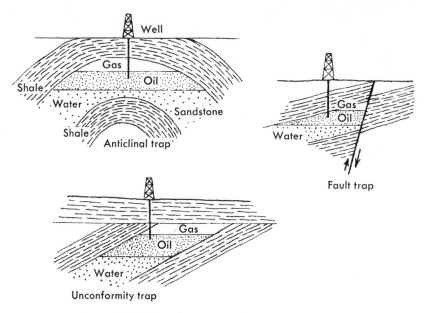

Figure 12-20. Several types of oil and gas traps. The terms anticline, unconformity, and fault are discussed in later sections.

Water Resources

Water is one of the cheapest and most useful and important commodities that we have. Most people take water for granted, but more and more often water shortages in different areas are reported in the newspapers. Are these shortages ominous signs that some regions have reached their limit of development, or are we mismanaging our water resources? Water is a world-wide problem, but only the United States, excluding Alaska and Hawaii, will be discussed here. The same natural problems occur everywhere, and they are aggravated to a greater or lesser extent by the stage of industrial and technological development of the country.

The United States receives an average of 30 inches of rain per year over the entire country. Of this 30 inches, about 21.5 are lost by evaporation or used by plants. The remaining 8.5 inches become the runoff of rivers and the ground water; this is the water available for use by man. (See Fig. 12-21.) This available water amounts to about 1,200 billion gallons per day, and in 1960 total withdrawals of water were about 270 billion gallons per day. It thus appears that we have more than four times as much available water as we need. The true situation is not this simple, unfortunately, and several other factors must be considered. They are the natural distribution of water, the need for runoff to carry away refuse, and the water used by plants that are the base of the food chain on which all life depends.

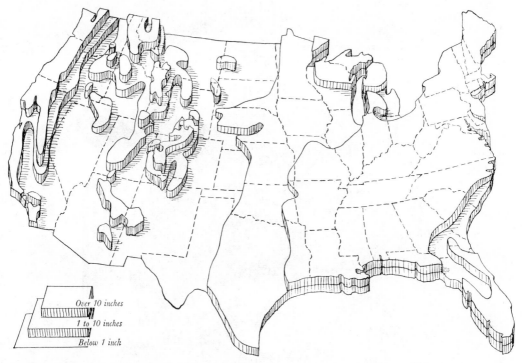

Figure 12-21. Average annual runoff
of rivers in the United States. From U. S.
Department of Agriculture, 1955 Yearbook, *Water.*

Natural Distribution of Water

Precipitation is not uniform, and neither is the distribution of population. Precipitation
ranges from about 85 inches per year in northwestern United States to about 4 inches
in the southwest. (See Fig. 12-22.) If individual stations are considered, the range is even
greater. These are average figures and may vary from year to year by large amounts,
especially in the drier regions. Climate, too, varies; and as shown in Fig. 12-23, the
potential or possible evaporation and consumption of water by plants exceeds the
rainfall in most of western United States. These are the arid regions where runoff occurs
only during infrequent heavy storms. The season at which rain occurs also affects the
use of the water. In most of eastern United States, much of the rain falls during the
growing season, but in the west, most of the precipitation comes during the winter
months. For all of these reasons, some places have sufficient water, a few places have
a surplus, and many areas have a deficiency, even though the overall available water
is more than four times the amount withdrawn.

In an effort to overcome these problems of natural distribution, dams and other
structures have been built to control and divert rivers. Many of these projects have
multiple purposes and may provide such benefits as municipal water supplies, flood

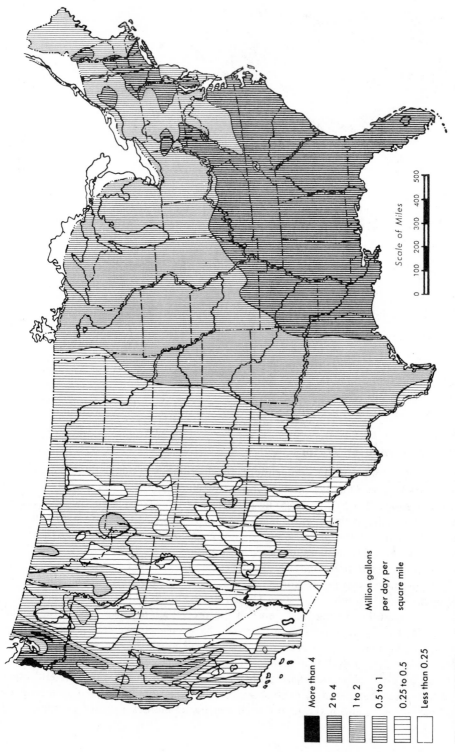

Figure 12-22. Average annual precipitation in the United States. Units are millions of gallons per day per square mile. From A. M. Piper, U. S. Geological Survey, *Water Supply Paper 1797*, 1965.

Scale of Miles

0 100 200 300 400 500

Million gallons
per day per
square mile

More than 4

2 to 4

1 to 2

0.5 to 1

0.25 to 0.5

Less than 0.25

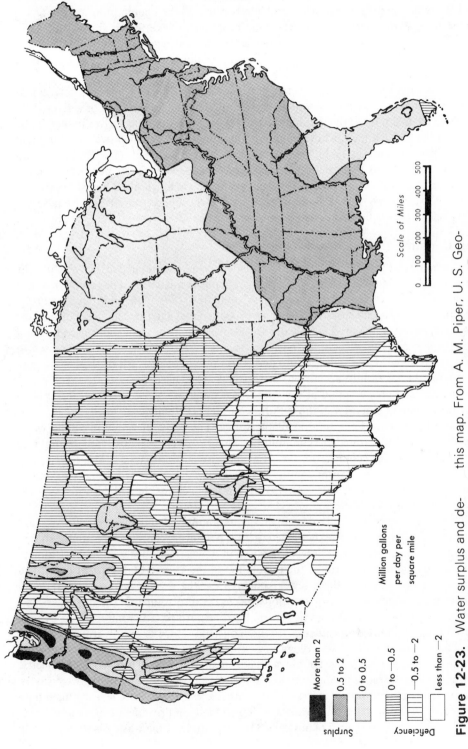

Figure 12-23. Water surplus and deficiency in the United States. Average precipitation less potential evaporation and consumption by plants is plotted on this map. From A. M. Piper. U. S. Geological Survey, *Water Supply Paper* 1797. 1965.

Million gallons
per day per
square mile

Surplus
- More than 2
- 0.5 to 2
- 0 to 0.5

Deficiency
- 0 to −0.5
- −0.5 to −2
- Less than −2

Scale of Miles
0 100 200 300 400 500

control, irrigation, electricity, and recreation areas. Deciding which rivers to control, and for which benefit, is always a problem because of the competition for use of the water and use of reservoir sites. Also, any reservoir or lake that is formed increases overall evaporation, and thus some water is lost. In arid regions this loss can be high, but recent studies suggest that much of it can be eliminated by placing a film of fatty alcohol on the water surface.

Like precipitation, river flow is seasonal; and during times of heaviest flow, floods can occur and much runoff is lost to the ocean. The maximum flow of a river is, in general, five to ten times the minimum flow. The most regular major river in the United States is the St. Lawrence, whose average maximum flow is less than twice its minimum flow. This is because its main flow comes from the Great Lakes. (See Table 12-1.) In the same way, dams that create lakes can make the flow of rivers more uniform and so increase the use of the water. At places such dams may eliminate most floods; but if, in an exceptional year, a flood does occur, much damage may result because the previously avoided flood area is now in use.

Use and Consumption of Water

Distinction must be made between the water *used* and the water *consumed.* Only the water lost to the atmosphere by evaporation or plants is consumed. The rest of the water withdrawn for use is returned either to the ground or to the runoff, and thus is available for other use. In some cases the water returned is unaffected by the use, but in most cases its quality is impaired. The water withdrawn is not the only water used by man; the runoff itself is used for some purposes in most rivers. Most of the water used to grow crops comes from soil moisture and so is not withdrawn. All of these complications must be taken into account to understand water needs.

The uses and consumption of the withdrawn water are shown in Table 12-2. Note that only about one-quarter of the withdrawn water is actually consumed, and that this is a small per cent of the total available water. The reuse of water in public water supplies is well illustrated by communities along a river. An upstream community withdraws its water supply from the river and discharges its wastes at a lower point on the river. The downstream community also draws its water from the river and so reuses some of the water. It is estimated that 60 per cent of the public water supplies withdraw some used water.

This reuse works well only if upstream communities adequately treat their sewage before discharging it into the river. Detergent types have recently been changed so that sewage treatment plants can prevent foaming in downstream water supplies. Unfortunately, only about two-thirds of the population of the United States have sewers, and only about two-thirds of these sewer systems treat the sewage before discharging it. Much of the water withdrawn for industrial use is also returned. In some cases this industrial return, like the domestic return, must be purified either naturally or by man before it can be reused. In other cases the industrial return is only warmer, but even this can drastically alter the ecology of a stream.

The runoff is not wasted water, but is used by man. It is used to produce hydroelectric power, for recreation, for transportation, and, as described above, to carry away wastes. The widespread pollution of rivers and streams suggests that we are nearing maximum usage for waste disposal in many areas, although many improvements in treating waste before discharge into rivers are possible. The magnitude of the waste disposal problem

Table 12-1. Large rivers. Data from U. S. Geological Survey, Water Resources Division, 1961.

A. World

Rank	River	Country	Drainage area (thousands of square miles)	Average discharge at mouth (thousands of cubic feet per second)
1	Amazon	Brazil	2,231	3,000 to 4,000
2	Congo	Zaire	1,550	1,400
3	Yangtze	China (mainland)	750	770
4	Bramaputra	Bangladesh	361	700
5	Ganges	India	409	660
6	Yenisei	USSR	1,000	614
7	Mississippi	USA	1,244	611
8	Lena	USSR	936	547
9	Parana	Argentina	890	526
10	St. Lawrence	USA & Canada	498	500
26	Danube	Romania	315	218 (Largest river in Europe)

B. United States

Rank	River	Length (miles)	Drainage area (thousands of square miles)	Average discharge at mouth (thousands of cubic feet per second)	Discharge at gaging station nearest mouth (thousands of cubic feet per second) Average	Maximum recorded
1	Mississippi	3,892	1,243.7	611[a]	553	2,080
2	Ohio	1,306	203.9	259	257	1,850
3	Columbia	1,214	258.2	256	183	1,240
4	St. Lawrence	—	302.0[b]	241[b]	236	314[c]
5	Missouri	2,714	529.4	69	69	892
6	Tennessee	900	40.6	64	61	460
7	Mobile	758	42.3	58	—	—
8	Red	1,300	91.4[d]	57	51[d]	233
9	Snake	1,038	109.0	49	49	409
10	Arkansas	1,450	160.5	42	41	536

a—About 25% of flow occurs in the Atchafalaya River; b—At international boundary, lat. 45°; c—Maximum monthly discharge; d—Flow of Ouachita River has been added.

is shown by the fact that the city of St. Louis, on the Mississippi River, produces 200,000 gallons of urine and 400 tons of solid body waste every day, in addition to industrial wastes.

The main consumption of withdrawn water is by irrigation. (See Table 12-2.) It was noted earlier that over 70 per cent of total precipitation either evaporates or is used by

Table 12-2. Use and consumption of withdrawn water. Amounts are in billion gallons per day. Data adapted from pp. 14-15, A. M. Piper, *Has the U.S. Enough Water?* U. S. Geological Survey, *Water Supply Paper* 1797, 1965, 27 pp.

	Withdrawn			Consumed					
				1960		1980		2000	
	1960	1980	2000		Per cent of with-drawal for this use		Per cent of with-drawal for this use		Per cent of with-drawal for this use
Use	Amount	Amount	Amount	Amount		Amount		Amount	
Municipal	21	29	42	3.5	16.7	3.5	12.1	5.5	13.1
Industry	140	363	662	3.2	2.3	11	3	24	3.6
Agriculture	109.6	167	184	61.3	56	104	62.3	126	68.5
"On site"[1]	—	—	—	—	—	71	—	97	—
Total	270.6[2]	559	888	68.0	24.6	189.5	34	252.5	28.4

[1] Water consumed by "on-site" uses comprises the effects of land treatment and structures, enlarged swamps and wetlands, and fish hatcheries. In large part, water consumed by such uses is intercepted before it enters a perennial stream; in other words, streamflow is depleted even though water may not be withdrawn in the usual sense.

Owing to past and present depletion of this kind, accepted values of streamflow as measured and published by the U.S. Geological Survey presumably are smaller than natural flows. Thus, present on-site consumption is not charged as an encumbrance against measured water supply. However, the estimates of on-site consumption as of 1980 and 2000 are for expected increases in such consumption; these must be charged against available supply as now measured.

[2] In 1965, 310 billion gallons per day were withdrawn.

plants. Plants, then, consume most of our water and must be considered in any study of water resources.

All life on the earth ultimately depends on vegetation. Plants are the base on which the food chain, or pyramid, is built. The amount of water needed to grow plants is very large. An acre of corn gives off about 4,000 gallons of water per day to the atmosphere. Five hundred pounds of water are required to grow one pound of wheat (dry weight). Most of this water is lost to the atmosphere through the leaves. About one-half of the wheat is discarded in milling, so about 1,000 pounds of water are required to produce one pound (dry weight) of bread. Thus the production of the 2½ pounds of bread that could be a daily diet requires 2,500 pounds of water, or 300 gallons. Meat requires even more water. About 30 pounds of alfalfa and 12 gallons of drinking water are required to produce one pound of beef. The 30 pounds of alfalfa require 2,300 gallons of water to grow. Thus the minimum water needed to maintain life, on bread alone, is 300 gallons per day, and if meat is included, is about 2,500 gallons per day. Water is also required to produce clothing, and in the same way, cotton requires much less water than wool.

Water Needs Present and Future

Water resources are two types: the available water composed of runoff and ground water, and the soil moisture that is either used by plants or evaporates. Both types of water are necessary to man, but most estimates of water needs are concerned with withdrawn water alone. The average rainfall on the United States is 30 inches per year or 4,200 billion gallons per day. Of this, 8.5 inches or 1,200 billion gallons per day is the available water and 21.5 inches or 3,000 billion gallons per day either evaporates or is used by plants. The water needs of plants on which all life ultimately depends will be considered first.

The minimum amounts of water necessary to produce food can best be determined by multiplying the number of people times the water consumed by plants to produce food. The 1960 population of the United States was about 180 million people. To feed these people on bread alone would require 300 gallons per day (2½ pounds of bread) times 180 million or 54 billion gallons of water per day. Our present diet is closer to a pound of meat plus some vegetables each day and consumes about 2,500 gallons per day per person, or about 450 billion gallons per day for the total population. Only about 51 billion gallons of this is withdrawn for irrigation (Table 12-2). Thus about 400 billion gallons per day of soil water is used to produce food. This soil water comes from the 3,000 billion gallons of water per day that either evaporate or are used by plants. It is estimated that about half of this 3,000 billion gallons per day evaporates and half is used by plants. However, it is also estimated that about one-quarter of the total rainfall falls in arid areas and so is not used productively for useful crops. Although these are very rough estimates, they suggest that at most 1,200 billion gallons per day are available for crops and that 400 billion gallons of this are currently being used to feed us. It would be overly optimistic to conclude that we can feed three times as many people as we do now, because this would mean replacing all our natural vegetation, including forests, with productive crops. How much more food can be grown and where the water will come from is a problem that must be faced in the relatively near future.

The withdrawn water that will be needed in the future has been estimated in Table 12-2. These estimates show that we will have to withdraw more than twice as much water in 1980 as we now do and over three times as much by the year 2000. The amounts of water actually consumed will increase even more, but all of these figures are less than the total available water. If these estimates are accurate, we will withdraw almost three-quarters of the total available water in 2000. To be able to do this, our water resources will have to be much better controlled and managed than they are today. This is an area where much thought, research, and money must be expended soon if we hope to have enough water for the future.

It may be possible to increase the available water for the future. Controlling the evaporation of water bodies and soil moisture is one possibility. Reducing the need of runoff by better waste disposal is another possibility. Regulating rivers so that more of the runoff can be used will be necessary, and this will entail moving surplus water from one drainage area to another. Desalinization of sea water, probably by atomic power, will help in some areas but may be expensive. More efficient use of ground water will help in some areas, but it must be recognized that ground water is really part of the runoff and is the source of the flowing water in most perennial rivers. To use ground water faster than it is replaced will only reduce the average runoff and will increase the available water only temporarily.

With an appreciation that he may be oversimplifying, the writer ventures that the United States can be assured of sufficient water of acceptable quality for essential needs within the early foreseeable future, provided that it (1) informs itself, much more searchingly than it has thus far, in preparation for the decisions that can lead to prudent and rational management of all its natural water supplies; (2) is not deluded into expecting a simple panacea for water supply stringencies that are emerging; (3) finds courage for compromise among potentially competitive uses for water; and (4) accepts and can absorb a considerable cost for new water-management works, of which a substantial part will need be bold in scale and novel in purpose.

U.S. Geological Survey Water Supply Paper No. 1797,
"Has The United States Enough Water?", *by A. M. Piper.*

QUESTIONS

12-1. What is the general relationship between the water table and the topography?

12-2. What is porosity?

12-3. What is permeability?

12-4. What happens near a well that is pumped too fast?

12-5. The average rate of movement of ground water is _____.

12-6. Sketch an artesian well. How long ago did the water in such a well fall as rain?

12-7. Describe capillary action as it applies to geology.

12-8. Sketch several types of springs.

12-9. Sketch a perched water table.

12-10. In what type of rock is pollution most likely to become a problem?

12-11. Describe one way a geyser eruption may occur.

12-12. How are caves formed?

12-13. Describe how the topography is different in an area underlain by soluble limestone compared to an area of sandstone. Assume humid climate.

12-14. Sketch several types of oil traps.

12-15. What is the nature of the United States' water problem?

12-16. What is the main use of runoff at present?

SUPPLEMENTARY READING

GENERAL

Bradley, C. C., "Human Water Needs and Water Use in America," *Science* (October 26, 1962), Vol. 138, No. 3539, pp. 489-491.

Hackett, O. M., *Ground-Water Research in the United States.* U.S. Geological Survey Circular 527, Washington, D.C.: U.S. Government Printing Office, 1966, 8 pp.

Kuenen, P. H., *Realms of Water.* New York: John Wiley & Sons, Inc., 1955, 327 pp. (Also in paperback edition, 1963.)

Leopold, L. B., and K. S. Davis, *Water.* Life Science Series, New York: Time, Inc., 1966, 200 pp.

Leopold, L. B., and W. B. Langbein, *A Primer on Water.* Washington, D.C.: U.S. Government Printing Office, 1960, 50 pp.

Meinzer, O. E., *Ground Water in the United States.* U.S. Geological Survey Water Supply Paper 836-D, Washington, D.C.: U.S. Government Printing Office, 1939, pp. 157-229.

Piper, A. M., *Has the United States Enough Water?* U.S. Geological Survey Water Supply Paper 1797, Washington, D.C.: U.S. Government Printing Office, 1965, 27 pp.

Sayre, A. N., "Ground Water," *Scientific American* (November, 1950), Vol. 183, No. 5, pp. 14-19.

Thomas, H. E., and L. B. Leopold, "Groundwater in North America," *Science* (March 5, 1964), Vol. 143, No. 3610, pp. 1001-1006.

U.S. Department of Agriculture, *Water.* U.S.D.A. Yearbook, Washington, D.C.: U.S. Government Printing Office, 1955, 218 pp.

Vogt, E. Z., and Ray Hyman, *Water Witching, U.S.A.* Chicago: University of Chicago Press, 1959.

CAVES

Mohr, C. E., and H. N. Sloane, *Celebrated American Caves.* New Brunswick, N.J.: Rutgers University Press, 1955, 339 pp.

Moore, G. W., "Origin of Limestone Caves: A Symposium," *Bulletin,* National Speleological Society, 1960, Vol. 22, Part 1, 84 pp.

Moore, G. W., and G. Nicholas, *Speleology: The Study of Caves.* Boston, Mass.: D.C. Heath and Co., 1964, 128 pp. (paperback).

PART 3

INTERNAL PROCESSES – STRUCTURAL GEOLOGY

In Part II the efficiency of erosion was seen. Part III explores the processes occurring within the earth that deform the surface rocks and so prevent erosion from leveling the earth. These processes cause the structures of the earth, both large and small. The term structure is used in its widest sense to include the whole scale of structures from the largest to the smallest.

The organization of Part III follows the scale of structures. In Chapter 13, the largest structures, the shape and internal layering of the earth, are considered. Chapter 14 covers the largest structures of the crust, and the continents. Chapter 15 describes the structures found on the oceans. The structures of the continents and the oceans are in separate chapters because of the differences in the types of data available for each. The last chapter of Part III explores hypotheses that have been suggested to explain crustal structures.

Chapter 13

Earthquakes and the Earth's Interior

From the scourge of the earthquake, O Lord, deliver us.
— from the Litany of the Roman Catholic Church

Knowledge of the earth's interior cannot be obtained by direct observation and, as a result, comes from very diverse kinds of data. Astronomical observations reveal the size and shape of the earth and indicate that its interior is much different from the surface. Study of earthquake vibrations reveals much of the internal structure, and meteorites provide clues to the possible internal composition. The use of data from such diverse fields shows why the geologist's interest and training must be so broad.

Shape, Size, and Weight of the Earth

Shape and Size of the Earth

The Greeks and other ancient peoples were excellent observers. They observed the earth's round shadow during eclipses of the moon. From this, and from noting that ships sailing away appear to sail downhill, they concluded that the earth is a sphere. They felt that this inference must be correct because they considered the sphere to be the most perfect of shapes.

About 235 B.C., Eratosthenes went a step further and measured the diameter of the earth. His method was simple (Figure 13-1) and involved only two assumptions:

1. The earth is a sphere.
2. The sun's rays are all parallel.

His method required the measurement, at the same time at two places, of the angle between the sun's rays and a plumb bob in the plane defined by the sun's rays and the plumb bob. The two places had to be widely separated to obtain a measurable angle, and he had to know the distance between the two cities. He chose noon, the time when the shadow is shortest and points due north (he had no other way to synchronize the time), and two cities that he believed to be on a north-south line so that the line between the two cities would be in the same plane as the sun's rays and the shadows. He chose to make the measurement on the day that the sun was directly overhead at noon at

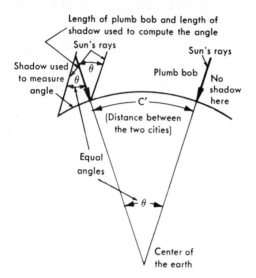

Figure 13-1. Measurement of the earth's diameter.

Syene (Assuan); he measured 7½ degrees at Alexandria. He then used simple geometry to compute the diameter.

He knew that for a circle,

$$C = \pi D$$

and for any arc of a circle,

$$C' = \frac{\theta}{360} \pi D \qquad \text{or} \qquad D = \frac{C'}{\pi} \frac{360}{\theta}$$

His measurement was about 15 per cent too high (we are not sure about the length of his linear unit), which was remarkably accurate for the time. Later, less accurate measurements indicated a much smaller earth and resulted in Columbus' thinking that he had reached Asia.

More precise repetition of this measurement at different places in the last few hundred years has shown that the diameter of the earth is less if the measurement is made nearer the poles. From this we reach the conclusion that the earth is ellipsoidal, not truly spherical. The equatorial bulge is apparently due to the rotation of the earth. The difference is not great:

Equatorial radius:	6,378,388 meters =	3,963.5 miles
Polar radius:	6,356,912 meters =	3,950.2 miles
Difference:	21,476 meters =	13.3 miles

The most precise knowledge we have on the shape of the earth comes from observations made on satellites. The earth's gravitational pull on a satellite determines its orbit; therefore, the shape of the earth's mass controls the exact orbit. The orbit can be measured by radar, and the shape of the earth can be computed. As a result, we now know that the cross-section at the equator is not an exact circle and that the southern hemisphere is slightly larger than the northern. These discrepancies are, however, very small.

Weight of the Earth

To weigh the earth is a more difficult problem than to determine its general shape. Work on this problem was not possible until after Newton, in 1687, deduced the behavior of bodies at rest and in motion. One of the results of the three laws of motion that bear his name was his law of universal gravitation. With this law, which states that the gravitational attraction between two bodies is proportional to their masses and inversely proportional to the square of their distance apart, he was able to explain the motion of the solar system. Stated mathematically,

$$F_{\text{gravitational}} = G\,\frac{m_1 m_2}{D^2}$$

where G is the universal gravitational constant.

The universal gravitational constant G should not be confused with the acceleration of gravity (generally called g and equal to about 32 ft/sec^2). One of Newton's important contributions was to differentiate between the *weight,* or force acting on an object, and its *mass,* an inherent, unvarying property. The background to this is that Aristotle stated that a heavy object falls faster than a light one; Galileo showed experimentally that both fall at the same speed; and Newton showed why this is true. Newton's second law states that the acceleration (rate at which speed changes) of an object is proportional to the force acting on it and inversely proportional to its mass:

$$a = \frac{F}{m} \quad \text{(more commonly written } F = ma)$$

The force F acting on a falling object is its weight. Because weight and mass are related, the quantity F/m is a constant for all falling bodies at any one place; hence, they all accelerate at the same rate. In the case of a falling body, a is the acceleration due to gravity, g. Note that because the earth is not a sphere, all points on the surface are not equidistant from the center of mass of the earth, and the weight F of an object, which is due to its gravitational attraction, is not a constant. Thus g, the acceleration of gravity, varies with latitude and elevation. Perhaps this should be taken into account in determining world's records in athletic events.

Note that although it is relatively easy to measure g at any point, this is not the universal constant of gravitation G, and there is no relationship between them that will enable calculation of G unless the mass of the earth is known.

In the case of any object on the earth, its weight is equal to its mass times the acceleration of gravity, or

$$F_{\text{weight}} = mg$$

This weight is also the gravitational force between the earth and the object:

$$F = G\,\frac{m\,m_{\text{earth}}}{R^2_{\text{earth}}}$$

Combining, we get

$$F = mg = G\frac{mm_e}{R^2_e}$$

Cancelling and rewriting,

$$m_e = \frac{gR^2_e}{G}$$

This equation cannot be solved until either m_e (mass of earth) or G (gravitational constant) is known.

The earth was accurately weighed in 1798 by directly measuring the force between two masses so that G could be calculated. (See Fig. 13-2.)

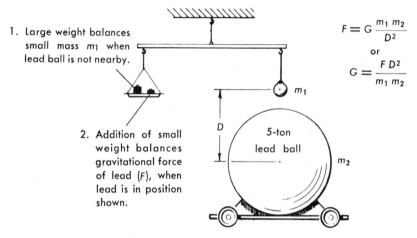

1. Large weight balances small mass m_1 when lead ball is not nearby.

2. Addition of small weight balances gravitational force of lead (F), when lead is in position shown.

$$F = G\frac{m_1\,m_2}{D^2}$$

or

$$G = \frac{F\,D^2}{m_1\,m_2}$$

Figure 13-2. Determination of the gravitational constant. The figure is diagrammatic, and the balance used was of a different, more sensitive design.

The mass of the earth is computed using the relation derived above:

$$m_e = \frac{gR^2_e}{G}$$

All quantities on the right side of the equation are known, and substitution leads to the result that the earth has a mass of about 7×10^{21} tons (or 7 followed by 21 zeros). Note that in English units the pound is considered by some a force, and by others a mass, unit. It is considered a unit of mass here.

Distribution of Mass in the Earth

A surprising result (predicted by Newton on the basis of planetary motion) is that this mass requires the average specific gravity of the earth to be about 5.5; that is, on the average, a given volume of it weighs 5.5 times as much as does the same volume of water. This is surprising because the rocks at the surface have average specific gravities less than three. Thus the rocks at depth must have much greater specific gravities than those at the surface. This suggests that the composition of the deeper parts of the earth is very different from that of the crust.

The distribution of mass in the earth determines its motion in space. One can imagine two spheres of equal size and weight; one has a uniform composition, but the second has its weight concentrated near the center with the rest hollow, except for spokes that separate the shell from the core. These two spheres will have different mechanical behavior. If we roll them down an inclined plane, they will travel at different rates, the uniform sphere moving more slowly.

The earth's distribution of mass is determined by study of the earth-moon system. The moon's gravitational attraction on the earth's equatorial bulge causes the earth's axis to wobble very slowly. This motion is called the *precession of the equinoxes* and is quite similar to the wobbling motion of a top as it slows down. Precession is a term used to describe this type of wobble, and equinox is a term that designates a relationship between the earth's position and the "fixed" stars on certain days, occurring twice a year when day and night are of equal duration. This wobble requires 25,735 years to return to the same point, and accounts for the fact that the Egyptian astronomers noted a different star in the position of Polaris, which is our present north star.

Because the amount of wobble (precession) caused by the moon's attraction depends on the mass distribution in the earth, it can be used to calculate the distribution of mass in the earth. This method also reveals that the earth is much denser at deep, than at shallow, levels.

Earthquakes and Seismology

Although earthquakes cause great damage and are greatly feared, they do provide much useful information about the earth's interior. Study of waves generated by earthquakes deep in the earth, for example, reveals that the earth has a layered structure. Combining this information with the distribution of mass deduced from astronomical observations leads to a model of the earth. First, the cause of earthquakes will be examined, and then the types of vibrations produced by earthquakes. The effects of these vibrations at the surface, which include the damage they cause, will be considered next; and finally, how their travel times through the earth reveal the earth's deep structures will be described.

Cause of Earthquakes

The ultimate cause of faulting, folding, volcanism, earthquakes, and other geologic processes is not known. The immediate cause of at least some shallow earthquakes is known, however. These earthquakes are caused by faults, breaks in crustal rocks along which movement has occurred. This is known from the formation of fault scarps at the time of earthquakes. The amount of displacement of surface features is not generally an accurate indicator of the amount of movement at depth, as shown in Fig. 13-3. (See also Figs. 13-4 and 13-5.) These movements cause vibrations; seismology is the study of these vibrations. The speed at which such vibrations, caused either by earthquakes or man-made blasts, travel through the earth provides much information about rock types and structure. Fault types will be described more completely in Chapter 14.

Faulting is one of the few geologic processes that is rapid or catastrophic. Earthquakes result from sudden movement by faulting; but apparently the actual movements within the earth are slow, and rupture only occurs when the forces have built up to

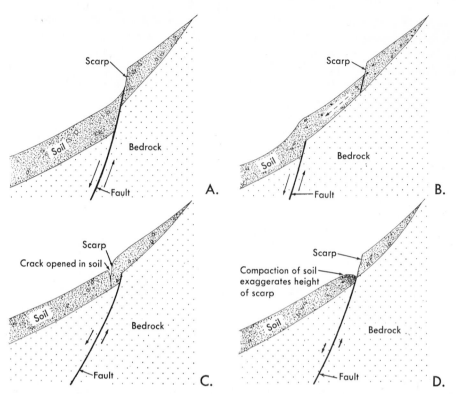

Figure 13-3. Types of fault scarps.
Parts A. and B. after I. J. Witkind, U. S.
Geological Survey, *Professional Paper*
435, 1964.

the point that they exceed the strength of the rocks. This is suggested by the slow movements that have been detected on some active faults. This was first noted after the 1906 earthquake in San Francisco on the San Andreas fault. Accurate surveys, which had been made on both sides of the fault before the earthquake, showed that a small amount of movement had occurred between the time of the surveys. Resurvey after the earthquake showed that about 20 feet of absolute horizontal movement occurred in the earthquake, in agreement with the 20 feet of relative movement measured by surface features. This lead to the theory that earthquakes occur when the energy stored by elastic deformation in the rocks on both sides of the fault is enough to rupture the rocks or to overcome the friction on an existing fault plane. This elastic rebound process is shown in Fig. 13-6. This theory explains the surface deformation of most earthquakes. Friction on the fault plane may cause sticking after some movement has occurred, and so the total strain may not be relieved in a single earthquake.

Slow creep has been detected along several faults in California in agreement with the elastic rebound theory. (See Fig. 13-8.) At several places buildings, roads, pipelines, and railroads cross active faults, and this slow movement has caused damage. Recording instruments have been installed at some of these places. Some examples of this movement are:

Figure 13-4. Scarp produced by the Hebgen Lake, Montana, earthquake in 1959. Note the subsidence below the scarp. Photo by J. R. Stacy, U. S. Geological Survey.

Warehouse at Niles on Hayward fault: 6" between 1921 and 1966 (Fig. 13-7)
Railroad tracks at Niles on Hayward fault: 8" between 1910 and 1966
University of California stadium at Berkeley on Hayward fault: 2¾" since 1948;
    1¼" between 1954 and 1966
Winery near Hollister on San Andreas fault: 3¼" in 7½ years

The movement appears to occur in spasms of a small fraction of an inch followed by weeks or months of quiet.

The elastic rebound theory is adequate for shallow earthquakes, that is, those that occur in the crust. Deeper earthquakes may have a different mechanism, however, because the pressure at depth would increase the friction on a fault plane to more than the strength of the rocks, making movement unlikely or impossible. The suggested mechanism to explain such deep faults is that the increase in pressure causes a recrystallization of the rocks to denser minerals that take up less volume. The earthquake is caused by the collapse of a volume of rock.

Wave Types

Earthquakes are recorded by seismographs. (See Figs. 13-9 and 13-10.) Three types of waves are sent out by an earthquake. (See Fig. 13-11.) They are

1. P waves—primary or push (compression) waves. "Primary" because they travel fastest and, therefore, arrive first at a seismic station. "Push" because they vibrate as compressions and rarefaction.

Figure 13-5. Tension crack in volcanic rocks on the eastern side of the Sierra Nevada. It is possible to descend 55 feet below the surface in this crack that is believed to have been formed by a prehistoric earthquake. Scarps over 50 feet high are found in this vicinity. Photo from U. S. Forest Service.

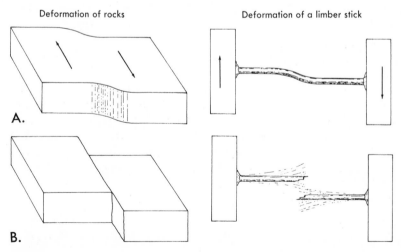

Figure 13-6. Origin of earthquakes. A portion of the earth's crust is shown on the left and a limber stick on the right. A. Slow deformation of the crust is caused by internal forces. B. When the strength of the rocks is exceeded, they rupture or fault, producing earthquake vibrations. Earthquakes on old faults result when the friction along the plane of the old break is exceeded.

Figure 13-7. Damage to warehouse built across the Hayward fault at Niles, California. The originally straight wall is bent several inches near the far end of the second window.

2. S waves—secondary or shake (shear) waves. "Secondary" because they are the second arrivals at a seismic station. "Shake" because they vibrate from side to side.

3. Surface waves—slow-moving waves with motion similar to the waves caused by a pebble tossed into a pond.

The propagation, or movement, of these waves will show more of the differences among them. The energy of an earthquake travels in all directions away from the source. We are interested only in the particular rays that are received at one point. The travel of the P waves is easily demonstrated. If a part of a spring is compressed and then quickly released, the resulting wave moves along the spring by successively compressing and spreading the spring. The S-wave movement is shown by displacing, and then releasing, one end of the spring. The wave moves along the spring by successive displacements to one side and the other. (See Fig. 13-12.) This same motion can be demonstrated by shaking a rope. (See Fig. 13-13.) Surface waves are ground motion caused by P and S waves reaching the surface. The transmission of energy by the surface wave is more complex, with two types of waves involved, and cannot be discussed here beyond the previous mention of the water-wave analogy. Typical seismic records show more than three waves because these main waves are commonly reflected by the layers within the earth.

Damage

Earthquakes cause damage in a number of ways. They set off landslides, they produce seismic sea waves, and they cause uplift or subsidence of large areas. However, the

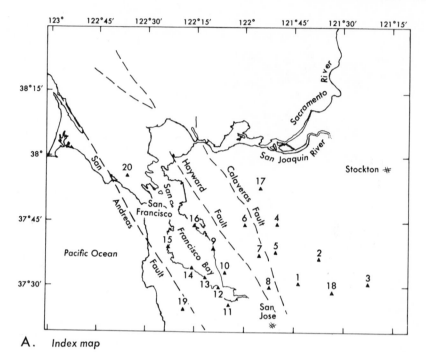

A.   Index map

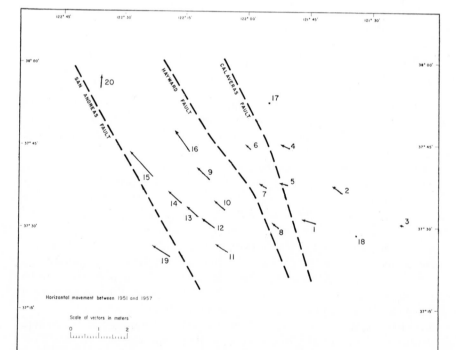

B.   *Horizontal displacement, 1951 to 1957*

Figure 13-8.   Recent horizontal movements in the San Francisco Bay area revealed by surveying. After A. J. Pope, J. L. Stearn, and C. A. Whitten, *Bulletin, Seismological Society of America,* Vol. 56, 1966.

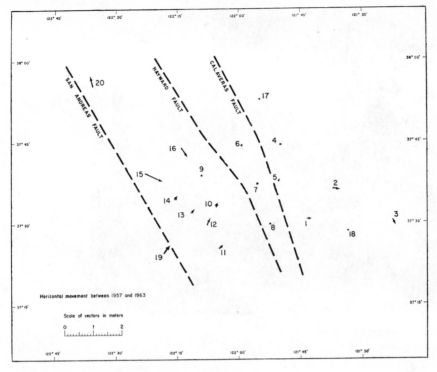

C. *Horizontal displacement, 1957 to 1963*

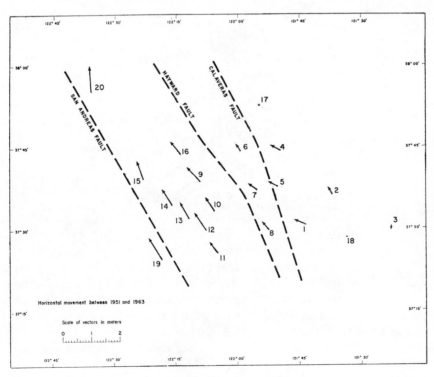

D.   *Total horizontal movement between 1951 and 1963*

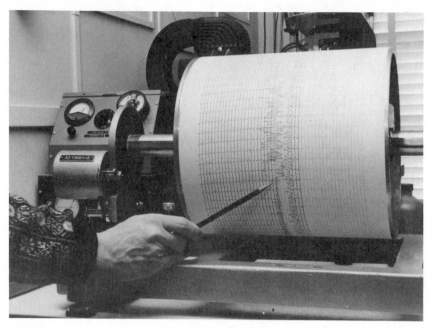

Figure 13-9. Long-period seismograph showing record of earthquake near Formosa on March 12, 1966. Photo from Environmental Science Services Administration, U. S. Coast and Geodetic Survey.

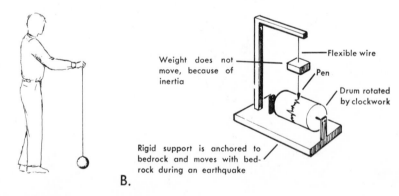

A.   B.

Figure 13-10. The operation of a seismograph. A. A weight on a string illustrates the principle of the seismograph. If the arm moves rapidly, the weight remains still. B. In this diagrammatic model of a seismograph, the support is attached to bedrock and moves when the earth quakes. The weight tends to remain still, so the pen records the relative movement between the chart that moves with the bedrock and the weight that does not move. Actual seismographs are designed to respond to movement in a single direction. Instead of using a pendulum, the weights are suspended like doors and so respond only to horizontal movements at a right angle to the plane of the weight.

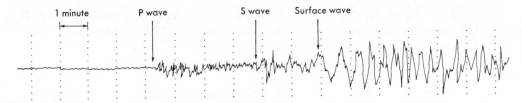

Figure 13-11. A seismograph record showing the three waves.

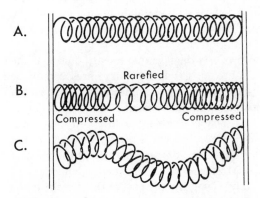

Figure 13-12. Types of waves. A. A spring stretched between two supports. B. A push wave showing compression and rarefaction. C. A shake wave.

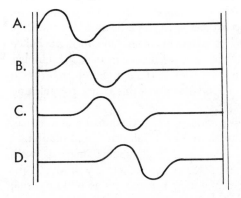

Figure 13-13. Propagation of an S wave shown by displacement of a rope. Part A. shows initial displacement; B., C., and D. show displacements at later times.

principal damage done by earthquakes is due to differential movements of buildings. These movements are caused mainly by the surface waves. The P and S waves vibrate fast but with small movement and, therefore, cause very little damage to structures. The surface waves, on the other hand, have much larger amplitude (movement) in lower frequencies and are the cause of most direct damage done by earthquakes. This is easy to understand, for if one part of a building is moved a few inches in one direction, while another part moves in another direction, the building will be damaged. Most earthquake-resistant buildings are designed to be flexible, so that such movements will not damage them. The building site will also affect the amount of movement that a structure experiences. Thus, in general, buildings on bedrock will be damaged less than those built on less consolidated, easily deformed material such as natural or artificial fills. These statements apply only to earthquake-wave damage and do not include the sometimes total destruction that occurs at places where there is an actual break or displacement at the surface.

In urban areas, another major cause of property damage due to earthquakes is the fires started by crossed electric wires and broken gas lines. These fires cannot be controlled because of broken water mains and disrupted communications. In general, private homes have not been too badly damaged in most recent earthquakes. This is because the usual wood-frame house is a reasonably flexible structure. In some cases the home may be too badly damaged to be salvageable, but it rarely collapses, killing the occupants. Particularly in many older homes the damage may be in the form of the building being pushed off its foundation. Most modern building codes require a tight bond between the foundation and the building. Typically the part of the average home which is damaged in an earthquake is the chimney. Because it is made of brittle bricks, a chimney tends to be cracked or fall down.

In certain parts of the world these statements are not true, particularly in the Near East where most dwellings are built of adobe brick. In an earthquake the brick, being a strong, rigid material, withstands the earthquake until its strength is overcome, at which time the building collapses, commonly killing all of the occupants. In the Near East, after such an experience, a city will be completely rebuilt right on top of the old ruins. This is one cause of the layered cities that archeologists excavate in these areas. In high-rise buildings the damage pattern is quite different. Most modern high-rise buildings are built of steel and concrete and are designed to be flexible. Much of the major structural damage to such buildings occurs where horizontal members are pulled away from vertical members. The tying of such joints is a critical part of such buildings. In general, an attempt is made in high-rise buildings to tie the foundation to bedrock. Generally this means a very deep foundation. In some areas this is not possible and other types of foundations are designed.

Some recent earthquakes, particularly the January, 1968, earthquake in Caracas, Venezuela, the earthquake of March, 1970, in Santa Rosa, California, and the February, 1971, earthquake in San Fernando, California, caused much damage to well-designed buildings. Engineers and engineering geologists are closely studying the damage that occurred at these places in hopes of finding ways to design buildings that are more earthquake-resistant. Part of the problem of this unexpected damage is apparently related to the vibration frequency that the building experiences. If the building is shaken in such a way that it begins to resonate at its natural frequency, much damage can result. The amount of movement is, in this way, magnified. The vibration frequencies of most high-rise buildings are quite low. If such buildings are built on relatively soft foundation material, even with well-designed foundations, apparently the high-frequency earthquake waves are filtered out by the soft foundation material, and the low frequencies move the foundation. Thus, in some cases, well-designed buildings on soft foundation materials are shaken at the natural frequency of the building. The length of time that the earthquake vibrations last is probably also very important in determining the amount of damage. The longer the earthquake, the more opportunity for the high-rise to begin moving at its natural frequency. Although eye-witness observers are not too reliable, earthquake vibrations seem to last anywhere from about a minute to three or four minutes. In general the amount of damage caused by a nearby earthquake is more related to the local geology, that is the underlying rock, than it is to distance from the fault or from the epicenter of the earthquake. This applies to those places within a few miles to perhaps a few tens of miles from the actual site of the earthquake.

Many people have a great fear of falling into a fissure created by an earthquake. Fig. 13-3 shows how small fissures can form in earthquakes, but this figure also shows that such cracks are generally shallow. The only recorded cases of loss of life in this way are a woman in a Japanese earthquake and a cow in the 1906 San Francisco earthquake. In both cases the walls of the cracks failed, and the victims were buried before they could climb out.

The likelihood of being injured or killed in a major earthquake in the United States is exceedingly small, but when a major earthquake does strike, the results can be severe, as witnessed by the 116 deaths in 1964 in Alaska; 28 in 1959 at Hebgen Lake, Montana; 115 in 1933 at Long Beach, California; and 64 in 1971 at San Fernando, California. If you feel the earth shaking under your feet and realize that you are experiencing an earthquake, you should dive under a table or desk, get in a doorway, or, in short, get into some position where you will be protected against material falling from the ceiling. The worst possible thing to do is to run out into the street, where you may be struck by falling cornices, chimneys and other flying objects. The elimination of buttresses and cornices and other masonry structures that may fall in an earthquake is one of the major considerations in designing earthquake-resistant buildings.

Most earthquakes are followed by aftershocks, some of which can be as severe as the original earthquake itself. For this reason, it is quite dangerous to reoccupy a building after an earthquake, even though the building has withstood the earthquake, because it may now, in its weakened condition, fail in the less severe aftershocks.

Seismic sea waves cause much earthquake damage. They are commonly called tidal waves, but this is a very poor term because tides have nothing to do with this type of wave. For this reason the Japanese word for these waves, *tsunami,* is in common use for them; but the translation of tsunami means tidal wave, so it, too, is not really suitable. *Seismic sea wave* is a much better, more descriptive term.

Seismic sea waves are generated by faults or earthquakes under the sea. These undersea movements can cause waves that travel across oceans as small waves imperceptible to ships at sea. In shallow water at shorelines, these waves become huge and cause great destruction. The initial movement of the water is commonly a recession followed in a few minutes by huge waves, some of which have been more than 200 feet high, although they are generally a few tens of feet high. Because they are so destructive to life and property, warnings are now sent by radio when seismographs detect an earthquake under the sea. These waves travel at velocities of 400 to 450 miles per hour and can cause damage on the opposite side of an ocean. (See Fig. 13-14.)

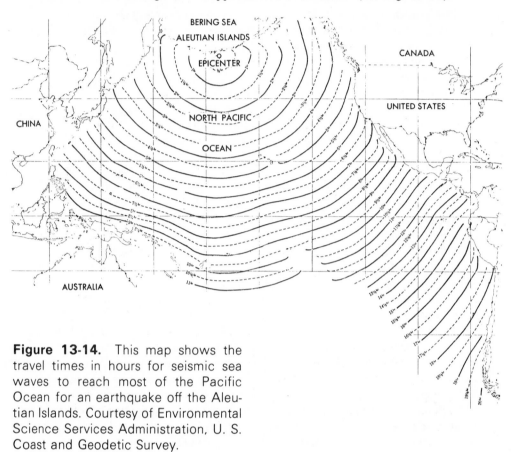

Figure 13-14. This map shows the travel times in hours for seismic sea waves to reach most of the Pacific Ocean for an earthquake off the Aleutian Islands. Courtesy of Environmental Science Services Administration, U. S. Coast and Geodetic Survey.

Size of Earthquakes

The size of an earthquake can be measured in many different ways. The amount of surface displacement is one measure; however, as shown in Fig. 13-3, the surface displacement may be different from the displacement at depth. Surveying shows the absolute displacement, which may differ from the relative displacement shown at the surface. Moreover, most earthquakes have no surface displacement, although in some it is appreciable. The maximum surface displacement so far recorded is 47 feet 4 inches at Yakutat Bay, Alaska, in 1899, and in the Good Friday earthquake of 1964 parts of the sea floor off Alaska were raised over 50 feet.

Another measure of the size of an earthquake is the size of the area in which the displacement occurs. In the Good Friday earthquake in Alaska in 1964, the observable crustal deformation covered an area of between 66,000 and 77,000 square miles. (See Fig. 13-15.) This is the largest of any one historical earthquake, although the 1906 San Francisco earthquake had the longest length of surface movement in a single earthquake. A 205 mile stretch of the nearly 600 mile long San Andreas fault moved in that earthquake.

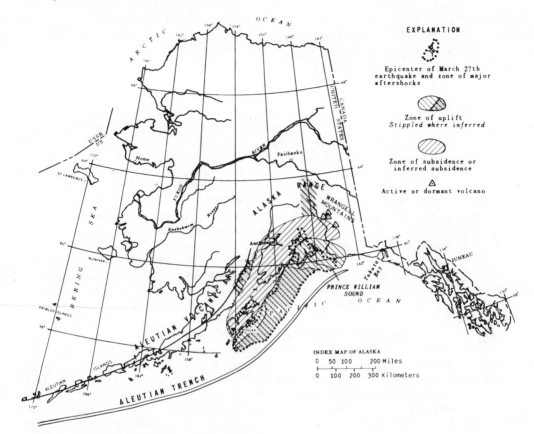

Figure 13-15. Map of Alaska showing the area where elevations changed in the Good Friday earthquake of 1964. From George Plafker, *Science,* Vol. 148, p. 1676, June 25, 1965.

The duration of shaking is another measure of an earthquake.

None of these ways of measuring an earthquake enables direct comparison of earthquakes, and none is useful for the many earthquakes without surface displacement. Seismologists have devised *intensity* scales to overcome these difficulties. Intensity scales rate the damage that occurs to substantial buildings. (See Table 13-1.) The estimation of intensity on the basis of damage and eyewitness reports is somewhat subjective, but works reasonably well. The main problem is that the damage depends somewhat on the local surface conditions, and so the intensity grades of many earthquakes follow the surface geology. (See Fig. 13-16.) Intensity maps are made by sending postcards to numerous people near the earthquake, asking them to describe the surface effects. In this way the epicenter of an earthquake can be located reasonably well. Notice that intensity enables direct comparison of surface effects only and that a deep earthquake will, in general, have a lower intensity than a shallow earthquake, even though both release the same amount of energy.

The *magnitude* of an earthquake is a measure of the amount of energy released. This figure is what is reported in newspapers and on the radio soon after an earthquake

Table 13-1. Earthquake intensity
scale. This scale indicates the amount
of surface damage caused by an earthquake. (Modified Mercalli Intensity
Scale of 1931)

    I. Not felt except by a very few under especially favorable circumstances.
   II. Felt only by a few persons at rest, especially on upper floors of buildings. Deli-
cately suspended objects may swing.
  III. Felt quite noticeably indoors, especially on upper floors of buildings, but many
people do not recognize it as an earthquake. Standing motorcars may rock slightly.
Vibration like passing of truck. Duration estimated.
  IV. During the day, felt indoors by many, outdoors by few. At night, some awakened.
Dishes, windows, doors disturbed; walls make creaking sound. Sensation like
heavy truck striking building. Standing motorcars rocked noticeably.
   V. Felt by nearly everyone, many awakened. Some dishes, windows, etc., broken; a
few instances of cracked plaster; unstable objects overturned. Disturbances of
trees, poles, and other tall objects sometimes noticed. Pendulum clocks may stop.
  VI. Felt by all, many frightened and run outdoors. Some heavy furniture moved; a
few instances of fallen plaster or damaged chimneys. Damage slight.
 VII. Everybody runs outdoors. Damage negligible in buildings of good design and
construction; slight to moderate in well-built ordinary structures; considerable in
poorly built or badly designed structures; some chimneys broken. Noticed by
persons driving motorcars.
VIII. Damage slight in specially designed structures; considerable in ordinary substantial buildings with partial collapse; great in poorly built structures. Panel
walls thrown out of frame structures. Fall of chimneys, factory stacks, columns,
monuments, walls. Heavy furniture overturned. Sand and mud ejected in small
amounts. Changes in well water. Persons driving motorcars disturbed.
  IX. Damage considerable in specially designed structures; well-designed frame structures thrown out of plumb; great in substantial buildings, with partial collapse.
Buildings shifted off foundations. Ground cracked conspicuously.
   X. Some well-built wooden structures destroyed; most masonry and frame structures
destroyed with foundations; ground badly cracked. Rails bent. Landslides considerable from riverbanks and steep slopes. Shifted sand and mud. Water splashed
(slopped) over banks.
  XI. Few, if any, (masonry) structures remain standing. Bridges destroyed. Broad
fissures in ground. Underground pipelines completely out of service. Earth slumps
and land slips in soft ground. Rails bent greatly.
 XII. Damage total. Waves seen on ground surfaces. Lines of sight and level distorted.
Objects thrown upward into air.

occurs. The magnitude of an earthquake is determined from the amount of movement
on a standard seismometer. This amount of movement is corrected for the distance
between the instrument and the earthquake. The early reports of an earthquake commonly give slightly different magnitudes from different seismic stations. This occurs
because the movement of the standard seismometers is somewhat affected by the
foundations on which they rest. Another reason for this slight uncertainty is that
magnitude determinations are based on the wave with the largest amplitude. For most
earthquakes this is the surface wave, but the very deep earthquakes have no surface

waves. Therefore, a magnitude scale based on P and S waves had to be devised for deep earthquakes. The relationship between these two scales has slight discrepancies.

The magnitude scale in common use is the Richter Scale. The largest magnitude yet recorded on this scale is 8.9, and the smallest felt by humans is 2. The San Francisco earthquake of 1906 was 8.3. For each increase of one unit on the scale, for example from 6.5 to 7.5, the energy released increases about 31 times and the recorded vibrations at a distant station increase about 10 times.

Many more earthquakes occur than cause damage great enough to be reported in the newspapers. Although the average annual number of earthquakes strong enough to be felt at least locally is estimated at about a million, most of the earth's seismic energy is released by the few large earthquakes.

Examples

Alaska's Good Friday earthquake occurred at 5:36 P.M. local time on March 27, 1964. It was centered in south-central Alaska and had a magnitude of 8.4 to 8.6. It thus released about twice as much energy as the San Francisco earthquake of 1906 that had a magnitude of 8.3. Figure 13-15 shows the large area in which the land level changed. This figure shows the epicenter of the main shock and also the zone of aftershocks that presumably outlines the area where fault movement occurred at depth. Aftershocks are common in most big earthquakes and are believed to be caused by readjustments made necessary by the main movement. In many earthquakes, the aftershocks cause almost as much damage as the main earthquake. The seismic sea waves caused by this earthquake caused much damage in Alaska and as far away as California, and were recorded in Japan. The vibrations caused many landslides, which were especially damaging to coastal installations built on unconsolidated rocks. Examples are shown in Figs. 7-15, 7-16, and 7-17. In this case, a previous geologic report had suggested that the particular clay of the area was susceptible to landsliding.

The Hebgen Lake, Montana, earthquake of August 17, 1959, was a shallow earthquake of magnitude 7.1. It produced two more or less parallel fault scarps seven miles long and up to 20 feet high. (See Fig. 13-4.) Tilting movements are clearly shown by the displaced shorelines of Hebgen Lake. This area is close to Yellowstone National Park, and the shocks affected some of the geysers there, mainly by increasing their activity at least temporarily. The main damage in this remote area was the triggering of a huge rockslide. Unfortunately, the slide fell on a campground, burying a number of people. (See Fig. 7-12.) It also dammed the Madison River, creating a problem. It was feared that when the lake created by the dam overtopped the dam, the flowing water might rapidly erode the rockslide debris and cause a flood. A channel was later cut through the rockslide to control the water flow.

Earthquake Prediction and Control

The immediate goal of seismologists is the prediction of the time, place, and magnitude of earthquakes so that property damage and casualties can be reduced. This goal may be reached in the next ten years. Recently and inadvertently man has triggered some earthquakes, suggesting the possibility that earthquakes can be controlled by man. Such control would enable man to cause several small, harmless quakes instead of one natural destructive quake.

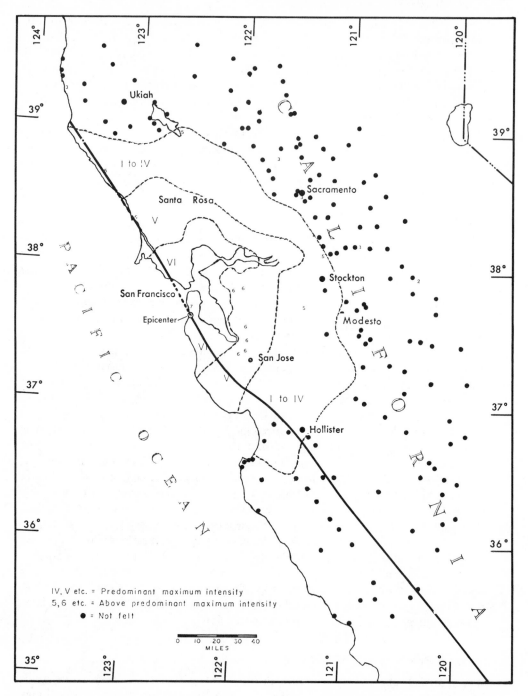

Figure 13-16. Map of the reported intensity of the March 22, 1957 earthquake. Movement apparently occurred on the San Andreas fault that is also shown. After W. K. Cloud, California Division of Mines and Geology, *Special Report* 57, 1959.

"Earthquake weather" is believed by many to be a precursor of earthquakes. Studies of many destructive earthquakes over the last few hundred years have revealed no correlation between weather and earthquakes, but the folklore persists, probably because it is so longstanding. Many large earthquakes have occurred at new or full moon when the highest tides occur because the earth, moon and sun are in line.

The places where earthquakes are most apt to occur are easily predicted on the basis of past experience. (See Fig. 13-17.) In Chapter 16 the reasons why earthquakes occur at these places will be described. As the maps show (Figs. 13-18 A and B), most earthquakes in the United States occur in California and in a belt through part of the Rocky Mountains. One obvious way to reduce earthquake damage would be to avoid these areas. Such a solution is not practical because it would mean avoiding most of California, the most populous state; however, anyone moving into an earthquake area should be aware of the potential danger.

Within these broad earthquake areas, the epicenters are located mainly along known faults so it is possible to avoid these high-danger areas. The faults to avoid are those that are active. Such faults are those that are creeping or are the sites of previous earthquakes. Some other faults, generally much older, are no longer active and so are much less dangerous. Just because no historical earthquake has occurred on a fault and there is no evidence of recent movement, the fault is not proven inactive. Faults are generally weak areas, and movement on one could be caused by a nearby large earthquake on an apparently unrelated fault. Also, as noted earlier, the amount of damage is controlled by the nature of the building and the rocks on which it is built.

The San Andreas fault in California is one of the longest faults; and because it has been the site of several large earthquakes, it has been studied closely. It extends about 700 miles through California from the northern Coast Range to the Mexican border. It is a fracture zone a few hundred yards to over a mile wide, composed of many more or less parallel faults. The total amount of movement is difficult to measure, and estimates vary from a few miles to 350 miles. The kinds of evidence used to determine the amount of movement are shown in Fig. 14-22. The behavior of the fault varies along its length. Fig. 13-19 shows that there are three active areas characterized by creep and/or many small, frequent earthquakes. In marked contrast are the areas where large movements (up to 20 feet) occurred in recent large earthquakes. These two parts of the fault are apparently locked, and few earthquakes occur at those places. These large segments of the earth's crust are capable of storing large amounts of strain and may sometime release their strain in a large earthquake. Thus these are the areas where large destructive earthquakes may occur. Most quakes on the San Andreas fault are shallow and so are capable of causing much damage.

The detailed prediction of earthquakes is based on observation of events that occur just before an earthquake, such as creep, tilting, increase in micro-earthquakes, changes in strain, and changes in such physical properties of rocks as electrical conductivity and magnetic susceptibility. Instruments set up near faults are necessary to make such observations, and such data gathering is one of the main types of current research into earthquake prediction. These observations have only been made on a few earthquake faults to date, and, so far, each area seems to behave differently. The Japanese have been issuing general earthquake predictions for some areas since 1966. Their predictions are based on slight tilting and swarms of micro-earthquakes that precede potentially damaging earthquakes by a number of weeks. Such events do not occur near the San

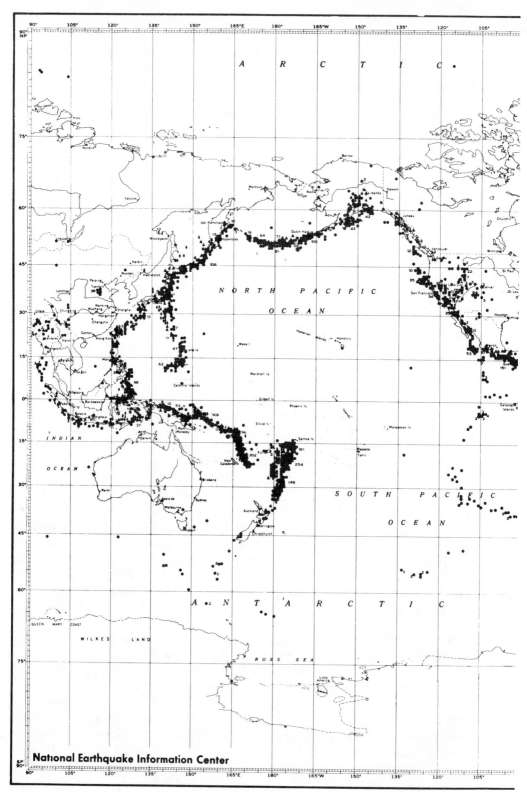

Figure 13-17. Earthquake zones of the earth shown by the locations of 1966 earthquakes. Courtesy of Environmental Science Services Administration, U.S. Coast and Geodetic Survey.

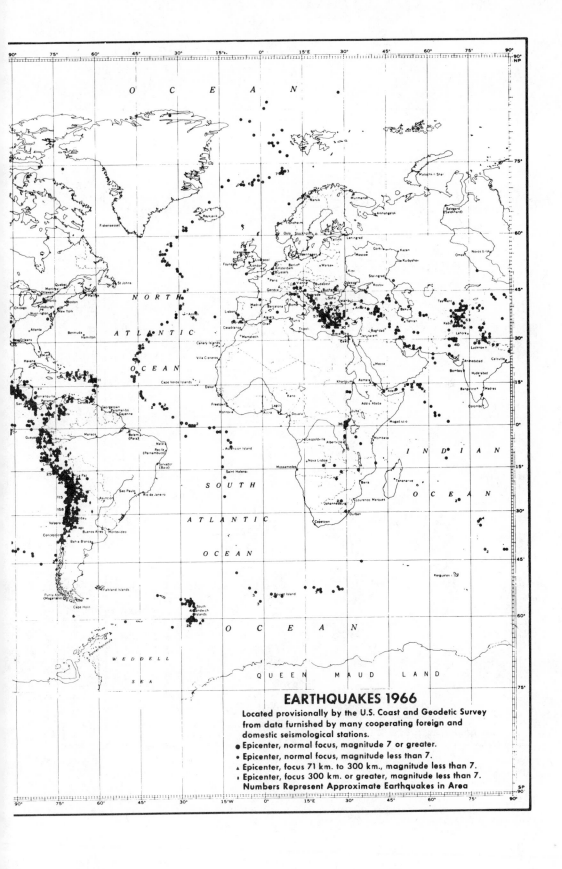

EARTHQUAKES 1966

Located provisionally by the U.S. Coast and Geodetic Survey from data furnished by many cooperating foreign and domestic seismological stations.

• Epicenter, normal focus, magnitude 7 or greater.
• Epicenter, normal focus, magnitude less than 7.
▲ Epicenter, focus 71 km. to 300 km., magnitude less than 7.
• Epicenter, focus 300 km. or greater, magnitude less than 7.
Numbers Represent Approximate Earthquakes in Area

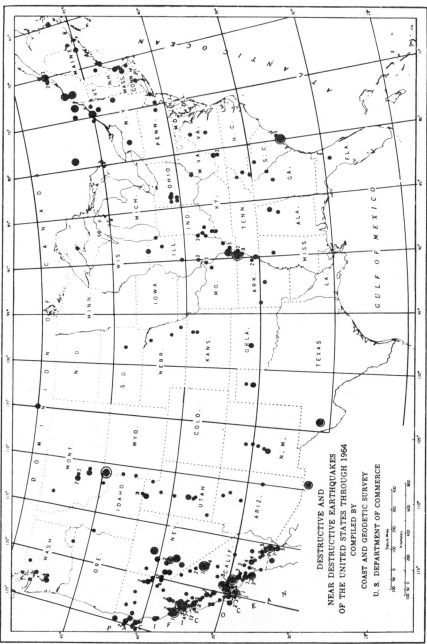

Figure 13-18. A. Destructive and near-destructive earthquakes in the United States through 1964. The smallest circles are intensity VII to VIII, the large solid circles are intensity VIII to IX, the smaller ringed circles are intensity IX to X and the large ringed circles are intensity X to XII. The numbers are the number of earthquakes. Courtesy of Environmental Science Services Administration, U. S. Coast and Geodetic Survey.

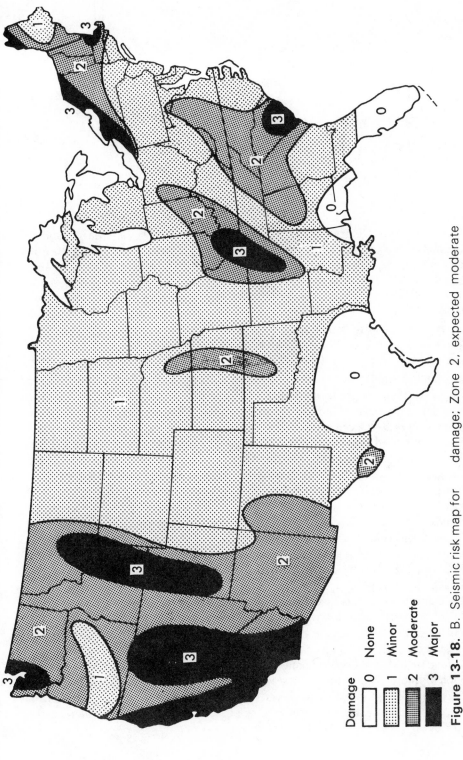

Figure 13-18. B. Seismic risk map for conterminous U.S. The map divides the U.S. into four zones: Zone 0, areas with no reasonable expectancy of earthquake damage; Zone 1, expected minor damage; Zone 2, expected moderate damage; and Zone 3, where major destructive earthquakes may occur. Both A. and B. courtesy of Environmental Science Services Administration, 1969.

Damage

0 None

1 Minor

2 Moderate

3 Major

Figure 13-19. Earthquake activity on the San Andreas fault in California. The three stippled regions are areas of creep and many small to severe earthquakes. Few earthquakes occur in the areas of the 1857 and 1906 breaks, but because the strain energy may build up to high levels in these "locked" areas, a few very severe earthquakes may be expected. In the area south of San Francisco, the San Andreas fault may be locked because the activity there occurs along other faults. After C.R. Allen, 1968, and L.C. Pakiser and others, 1969.

Andreas fault in California, our best instrumented fault. There, deformation of deep rocks changes their magnetic properties and so causes slight changes in the local magnetic field a few tens of hours before creep of a few millimeters occurs. So far these magnetic changes have occurred only before such movements. In a few cases earthquakes have followed these movements by a few days. In other places along the San Andreas, changes in rate of creep or even reversals in direction of creep have preceded earthquakes. Creep also occurs periodically after some earthquakes on the San Andreas fault, with the period between successive creep episodes increasing. Interestingly, at such places creep does not seem to occur at locations where aftershocks do occur.

The most promising approach to earthquake safety may be control of quakes. Our knowledge of this came from accidentally triggering earthquakes. The best publicized of these events were the many small earthquakes near Denver that were caused by pumping waste fluids into a disposal well at the Rocky Mountain Arsenal. These earthquakes, some of which had magnitudes up to 5.5 and caused some damage, began in 1962. Until that time there had been few earthquakes in the Denver area, and it was noted that the earthquakes followed periods of pumping waste fluids into the well. A series of seismic stations were set up to monitor the well, and they revealed that the earthquakes occurred in a zone about five miles long and up to a mile deeper than the 12,000 foot well. Analysis of the data suggested that the movements resulted from a reduction in friction caused by the increased pore pressure due to pumping. Soon after this discovery similar micro-earthquakes were observed at the Rangely oil field in northwestern Colorado. Here water was being pumped into the ground to increase the recovery of oil.

These discoveries stimulated the search for other man-made earthquakes. In 1945 a paper had appeared that documented hundreds of small earthquakes in the area of Lake Mead since the building of Hoover Dam. One of these quakes had a magnitude of almost 5, and two had magnitudes of about 4. It was suggested that the weight of the water had caused movements on old faults. More recently at Koyna, India, a lake created in 1962 and 1963 by construction of a dam apparently caused a number of earthquakes that culminated in 1967 with a magnitude 6.5 earthquake that caused about 200 deaths. The geologic environment at Koyna is similar to that at Grand Coulee, Washington, where no earthquakes associated with the lake have been observed. Other occurrences of earthquakes caused by dam-impounded lakes are a magnitude 5.8 earthquake at Lake Kariba on the Zambezi River, Zambia, and a 6.3 magnitude quake at Kremasta Dam, Greece.

The exact cause of any of these earthquakes is not clear. At the Denver well, they could be due to reduced friction as suggested above, or to temperature changes caused by the injection of cool fluids. Withdrawal of fluids at Denver suggests that very little mixing of waste fluids and ground water occurred. At the dams, increased weight could be the trigger, or the lakes could cause inseeping of fluids and so the situation could be much like that at the injection wells. No earthquakes are associated with many high dams. Much more study is necessary, and hopefully such studies will be included in the construction costs of new large dams. The subsurface geology of oil fields is generally well studied, making them good places for earthquake study if fluid injection is used for secondary recovery late in the production history of the field.

The third way that earthquakes have been triggered is with underground nuclear blasts. The earthquakes produced in this way have all been within about 15 miles of the actual blast and of lower magnitude than the nuclear blast. At the Benham test of 1968, the nuclear blast had a magnitude of 6.3, and the largest associated aftershock had a magnitude of 5.

Thus, at the moment, we are on the verge of being able to predict earthquakes, and we can trigger them. This combination may enable us to reduce the dangers of earthquakes, but there are many serious problems that must be overcome before this goal is reached. Prediction alone will reduce the likelihood of casualties but not all property damage. Control by triggering many small earthquakes instead of allowing a large natural one could reduce both casualties and property damage, especially if earthquake-resistant construction is practiced more extensively. Fluid injection wells currently appear best for such control. For instance, a series of such wells along the San Andreas fault could be used to dissipate the strain in a number of small harmless quakes or in creeping movement. The problem is that the San Andreas fault, which is our best studied fault because it causes so many earthquakes, crosses some densely populated areas. Any experimental program triggering earthquakes along it is likely to be blamed for any earthquake damage that occurs. The resulting lawsuits would become a serious problem because the results of such a program are difficult to predict and it is almost impossible to distinguish between natural earthquakes unassociated with fluid injection and those caused by man's interference. In spite of such hypothetical problems, the benefits appear so great that a program of earthquake control should be started as soon as practical. Until such time, the protection of people and structures must depend on zoning laws, building codes, and foundation and structure designs.

Location of Epicenters

The different speeds of the P and S waves can be used to locate the epicenter of an earthquake. A P wave can travel across the earth in about 20 minutes, an S wave travels at about one-half this speed and the surface wave would require several hours to make the trip. Because the S wave travels slower than the P wave, the difference in arrival time between the P and S waves is proportional to the distance away from the station that the earthquake occurred. As shown in Fig. 13-20, reports from three stations are needed to locate an epicenter.

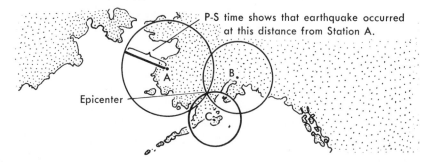

Figure 13-20. A, B, and C are seismic stations. The P-S times at each station indicate that the earthquake occurred somewhere on the circle drawn around each station. The point where all three circles intersect is the epicenter.

To use this simple principle, it was first necessary to determine the velocities of the P and S waves. This was done by finding the approximate location of an earthquake from intensity reports. Then the P and S times at many seismic stations were compared with the distance between the stations and the earthquake. The simplest way to do this is to plot a graph of distance *versus* arrival time for each wave. Generally, it was necessary to refine such a determination by adjusting the location of the earthquake, which was known only approximately from the intensity, until the best velocity determination was found. This was repeated for many earthquakes until the travel-time curves for the whole earth were determined. In this process the internal structure of the earth was determined.

Internal Structure of the Earth

Travel times of earthquake waves have revealed a layered structure of the earth. The details of the analysis are best seen in Fig. 13-21. Three main layers have been distinguished: the crust, a very thin shell between 3 and 35 miles thick; the mantle, the next layer, about 1800 miles thick; and the core, the central part, with a radius of about 2150 miles. The velocity of earthquake waves changes abruptly at the contacts between these layers, and the S waves cannot penetrate the core. (See Fig. 13-21.) This last observation shows that the core is liquid because shake or shear waves cannot pass through a liquid, although push waves can. Actually, the structure in each layer is probably more complex than suggested here. For example, P waves speed up as they travel through the central part of the core, thus suggesting that this part of the core may be solid.

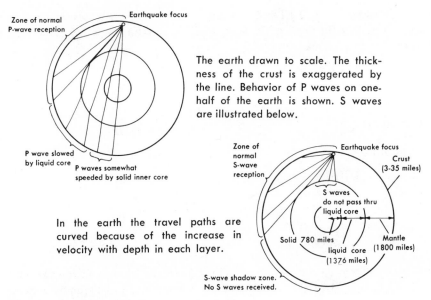

The earth drawn to scale. The thickness of the crust is exaggerated by the line. Behavior of P waves on one-half of the earth is shown. S waves are illustrated below.

In the earth the travel paths are curved because of the increase in velocity with depth in each layer.

Figure 13-21. Behavior of P and S waves in the mantle and the core reveals the layered structure of the earth.

The crust is the thin shell that overlies the mantle. A very marked change in seismic velocity separates the crust from the mantle. This change is the Mohorovičić discontinuity, commonly called M-discontinuity or Moho. It was named for the seismologist who discovered it in 1909. The thickness of the crust (or distance to the Moho) varies between about 3 miles and 35 miles. (Fig. 13-22.) It is thicker under the continents (20 to 35 miles) than under the oceans, and is thickest under some young mountain ranges.

The Mohorovičić discontinuity is recognized from seismic evidence. Shallow earthquakes reveal a layered structure by producing two sets of P and S waves at seismic stations less than 500 miles away from the epicenter. Because one set of P and S waves travels through the crust and the other mainly through the mantle, the sets travel at

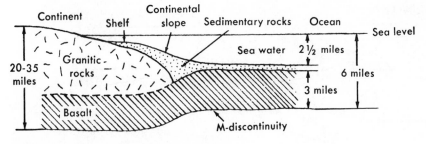

Figure 13-22. Generalized cross-section of continent-ocean boundary at a geologically stable area. The sedimentary rocks on the continental shelf and slope generally have more complex structures than shown. From seismic data.

different speeds. (See Fig. 13-23.) Through study of the travel times at several stations, it is possible to calculate the depth of the Mohorovičić discontinuity, which is the name applied to this velocity change.

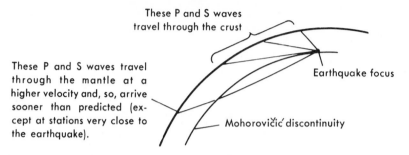

Figure 13-23. Determination of the depth of the Mohorovičić discontinuity.

Other smaller differences in travel times (velocity) reveal a layering within the crust itself, as shown in Fig. 13-24. The continental crust is believed to be granitic because intrusive and metamorphic rocks of this composition underlie the sedimentary rocks that cover much of the surface, and because granitic rocks have the observed seismic velocities. In the same way, the oceans are believed to be underlain by basalt, and the lower layer under the continents is believed to be basalt. Strengthening this inference is the fact that basalt volcanoes occur in both oceanic and continental areas. Although this layer has the composition of basalt, seismic data suggest that it is in the form of gabbro or amphibolite, both of which have the same chemical composition as basalt.

The nature of the Mohorovičić discontinuity is not known, except that there is a marked change in seismic velocity with a thickness of 0.1 to 2 miles. For many years this was assumed to be caused by a change in composition from peridotite or dunite in the mantle to basalt in the lower layer of the crust. Recently, however, it has been suggested that the change may not be due to a difference in chemical composition, as between dunite and basalt, but due to a change in density or in mineralogy in a rock

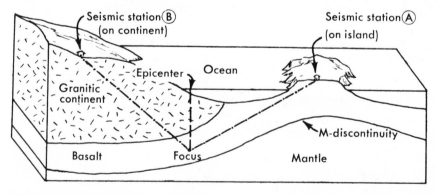

Figure 13-24. Block diagram showing paths of seismic waves in oceanic basalt and continental granite. Both stations Ⓐ and Ⓑ are equidistant from the focus of the earthquake, but station Ⓐ receives waves first because they travel faster in basalt than in granitic rock.

of the same chemical composition. Such changes are known to take place as a result of increased pressure and temperature, and might occur in a zone narrow enough to account for the rapid change in velocity. A change in mineralogy occurs when basalt is subjected to high temperature and pressure. Under these conditions, the augite and plagioclase of basalt become unstable, and the elements recombine to form a type of garnet and a green pyroxene, omphacite. Such rocks are called *eclogite,* and their rare appearance at the surface is at places where uplift and erosion have exposed deep-seated rocks or where intrusive bodies have carried them to the surface. The weight of the overlying rocks supplies the pressure needed to change the mineralogy. Therefore, if the Moho is in fact caused by changes in mineralogy rather than in composition, it should follow the surface topography; however, it does not. Figure 13-25 shows the depth of the Moho from seismic data and the calculated depth of the transition zone in part of the United States. Figure 13-26 shows other lateral variations in the structure of the United States.

Although the Moho is seismically the most obvious discontinuity in the upper part of the earth, it is probably not the most important zone. A low seismic velocity zone in the upper mantle is probably more involved in the formation of surface structures. This zone will be described in the next chapter, and its role in the formation of surface structures is discussed in later chapters.

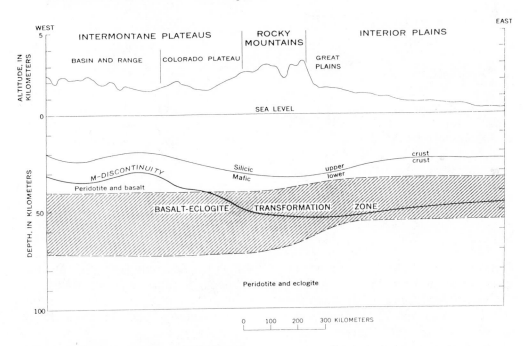

Figure 13-25. Section showing the crustal and upper-mantle structure and the inferred basalt-eclogite transformation zone from the Basin and Range area to the Great Plains. From L. C. Pakiser, U. S. Geological Survey, *Professional Paper* 525B, 1965.

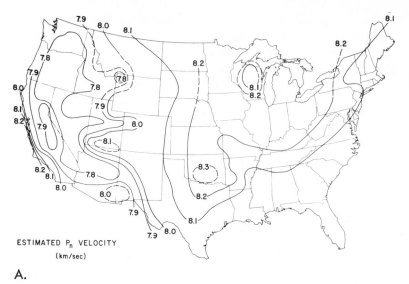

ESTIMATED P_n VELOCITY
(km/sec)

A.

B.

Figure 13-26. Lateral variation in the structure of the United States. A. Seismic velocity in the upper mantle. The velocity is less than 8 kilometers per second in much of western United States. After Eugene Herrin and James Taggart, Seismological Society of America *Bulletin,* Vol. 52, 1962. B. Crustal thickness. The contours are in kilometers. After L. C. Pakiser and J. S. Steinhart, 1964.

It is possible to determine the depth of an earthquake from the arrival times of vibrations at nearby stations, as shown in Fig. 13-27. The point at which the vibrations originate is called the *focus,* and the point on the surface directly above is called the *epicenter.* Most earthquakes are shallow, less than 35 miles, but some are as deep as 435 miles. Deep earthquakes were first suspected from the lack of surface waves from

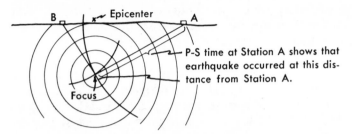

Figure 13-27. Determination of the focus (depth) of an earthquake.

some earthquakes. Deep earthquakes suggest that the mantle acts as a solid to rapid deformations at least to this depth. These deep-focus earthquakes occur only at the continent-ocean boundary. We will discuss them later.

The velocity of waves is related to the physical properties of the medium through which they pass; hence, it is possible to make some inferences about the material that composes the deep parts of the earth. Remember also the density distribution of the deeper parts of the earth, which was learned from the earlier discussion of the weight of the earth. Both methods of investigation lead to the specific gravity of the mantle, varying from about 3 near the top to about 6 at its base. Similarly, the core has a specific gravity of about 9 at its top and about 12 at the center.

Composition of the Layers of the Earth

Because the earth's interior is not accessible to direct sampling, all knowledge of its composition is based on speculation. Our only evidence concerning the composition comes from meteorites. In Chapter 18 it will be shown that the earth and the meteorites are believed to have formed at the same time and from the same material. Meteorites are the objects that fall into the earth from space. Many of them are burned by frictional heat in the atmosphere to form what are called "shooting stars." Some meteorites are quite large and form impact craters, such as Meteor Crater in Arizona. The craters on the moon, which is not protected by an atmosphere, are believed to have formed in this way.

Meteorites are of three types: stony meteorites, composed mainly of olivine and pyroxene (peridotite is the rock name for this composition); iron meteorites, composed mainly of iron and nickel; and stony-iron meteorites, which are mixtures.

The specific gravity and other physical properties of peridotite or dunite fit those inferred for the upper mantle from seismic and other indirect methods of study. The lower mantle (below the low velocity zone described in the next chapter) is probably composed of simple oxides of iron, magnesium, and silicon. In the same way, the core is generally believed to be mainly molten iron with some nickel. Recently on both seismic and chemical evidence, it has been suggested that iron and sulfur are the main elements in the core. In reaching these assumptions, allowance must be made for the increased temperature and pressure deep within the earth, for these will greatly modify the properties of rocks within the earth. Because these temperatures and pressures are too high to reach in the laboratory, some uncertainties exist. The conditions in the mantle, particularly in the upper mantle, are of importance in our study of the crust, and so the uncertainties will lead to several possible hypotheses concerning processes in the crust.

The pressure deep in the earth is reasonably well known, but the temperature is less well known. The pressures are due to the weight of the overlying material, and, with our knowledge of specific gravity gained from seismic studies, can be calculated from the law of gravity. (See Fig. 13-28.) These pressures are hydrostatic; that is, they are

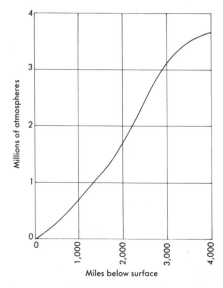

Figure 13-28. Pressures within the earth. Adapted from John Verhoogen, *American Scientist,* Vol. 48, 1960.

equal in all directions because the rocks in the mantle and crust are not, as we will see, completely rigid, even though they are solid. A simple analogy will illustrate hydrostatic pressure. Bricks piled one on the other exert a pressure only on their base, because they are rigid. A pile of sand, however, unless confined by a wall, will assume a cone shape because the sand is not rigid; hence, the pressure due to its weight is exerted in all directions.

Measurements of the rate of heat flow to the surface of the earth were made in the nineteenth century and led Lord Kelvin to the conclusion that the total age of the earth was less than 100 million years—probably in the range 20-40 million years. His calculation was based on the assumption that the earth was molten at its time of origin and has been cooling ever since. This explanation of the origin was commonly accepted in his day, and because radioactivity was unknown, his conclusion appeared to be reasonable. It was not accepted by geologists, however, because they realized that several times that length of time is required to deposit the thickness of sedimentary rocks found at the surface. We now know that the amount of radioactive material in the crust alone is enough to account for all of the heat flow from the earth. We shall return to this heat source during our study of the crust.

The temperatures in the earth are estimated from the physical properties of the rocks because the origin of the heat is not known. The earth's heat may be a remnant of its formation, or may be due to radioactive sources in the earth. It clearly is not due to heating by the sun, because heat from the sun can be accounted for and maintains only the surface temperature. Data from drill holes and deep mines reveal that the temperature rises with depth in the earth at the approximate rate of 1°F per 50 feet. If this rise continues to the center of the earth, the temperature there would be 400,000°F, an unreasonably high temperature—hotter than the sun.

The upper limit of temperatures in the earth is estimated from knowledge of the melting points of the materials assumed to form each layer, taking into account the rise in melting point at high pressures. The lower limit is estimated from calculations of the lowest temperature which will permit heat transfer by convection. Convection (see Fig. 16-23) is believed to be an important process in the mantle and is discussed later. Although convection requires a fluid, the material in the mantle could flow slowly (and still transmit S waves as a solid) in much the same way that tar behaves at room temperature. Under such conditions tar acts as a solid to rapid deformation, such as the passage of S waves, but over a long period it will flow under its own weight. From this reasoning, the temperature distribution shown in Fig. 13-29 is derived.

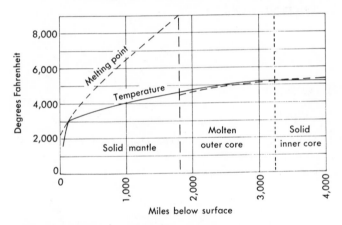

Figure 13-29. Estimated temperatures in the earth. Adapted from John Verhoogen, *American Scientist,* Vol. 48, 1960.

QUESTIONS

13-1. How was the size of the earth first measured? (Use a diagram.)

13-2. How was the weight of the earth determined?

13-3. What is the average specific gravity of the earth, and what problem does this pose?

13-4. What is the immediate cause of an earthquake? What is the ultimate cause of an earthquake?

13-5. How does a seismograph work?

13-6. Briefly describe each of the three seismic waves.

13-7. Which of the earthquake waves causes the most damage? How is most of the property damage due to earthquakes caused? How might some of this damage be avoided?

13-8. What does the magnitude of an earthquake mean?

13-9. How is the epicenter of an earthquake determined? (Diagram might help.)

13-10. Describe the behavior of S waves passing through the earth.

13-11. What effect does the moon have on earthquakes?

13-12. How can earthquakes be predicted?

13-13. How can man trigger earthquakes?

13-14. Describe the behavior of P waves passing through the earth.

13-15. How is the focus of an earthquake determined?

13-16. The deepest earthquakes are _____ miles.

13-17. The core has a radius of _____ miles and is believed to be composed of _____. What is the evidence for this composition?

13-18. The mantle of the earth is _____ miles in thickness and is believed to be composed of _____. What is the evidence for this composition?

13-19. The crust is separated from the mantle by the _____.

SUPPLEMENTARY READING

EARTH'S SIZE, SHAPE, and WEIGHT

Bullen, K. E., "The Deep Interior," in *The Earth and Its Atmosphere* (D. R. Bates, ed.), pp. 31-47 (also covers much seismology). New York: Science Editions, Inc., 1961, 324 pp. (paperback). This book was also published under the title *The Planet Earth*. London: Pergamon Press, 1957.

King-Hele, Desmond, "The Shape of the Earth," *Scientific American* (October, 1967), Vol. 217, No. 4, pp. 67-77.

SEISMOLOGY

Anderson, D. L., "The Plastic Layer of the Earth," *Scientific American* (July, 1962), Vol. 207, No. 1, pp. 52-59.

Anderson, D. L., "The San Andreas Fault," *Scientific American* (November, 1971), Vol. 225, No. 5, pp. 52-68.

Benioff, Hugo, "Earthquake Source Mechanisms," *Science* (March 28, 1964), Vol. 143, No. 3613, pp. 1399-1406.

Bullen, K. E., "The Interior of the Earth," *Scientific American* (September, 1955), Vol. 193, No. 3, pp. 56-61. Reprint 804, W. H. Freeman and Co., San Francisco.

Christensen, M. N., and G. A. Bolt, "Earth Movements: Alaskan Earthquake, 1964," *Science* (September 11, 1964), Vol. 145, No. 3627, pp. 1207-1216.

Grantz, Arthur, and others, "Alaska's Good Friday Earthquake, March 27, 1964," U.S. Geological Survey Circular 491, 1964, 35 pp.

Hansen, W. R., and others, "The Alaska earthquake, March 27, 1964: Field investigations and reconstruction effort," U.S. Geological Survey, Professional Paper 541, 1966, 111 pp.

Hodgson, J. H., *Earthquakes and Earth Structure.* Englewood Cliffs, N.J.: Prentice-Hall, Inc., 1964, 166 pp. (paperback).

Iacopi, Robert, *Earthquake Country.* Menlo Park, Calif.: Lane Book Co., 1964, 192 pp. (paperback).

Oliver, Jack, "Long Earthquake Waves," *Scientific American* (March, 1959), Vol. 200, No. 3, pp. 131-143. Reprint 827, W. H. Freeman and Co., San Francisco.

Pakiser, L. C., and others, "Earthquake Prediction and Control," *Science* (December 19, 1969), Vol. 166, No. 3912, pp. 1467-1474.

Plafker, George, "Tectonic Deformation Associated with the 1964 Alaska Earthquake," *Science* (June 25, 1965), Vol. 148, No. 3678, pp. 1675-1687.

Press, Frank, and W. F. Brace, "Earthquake Prediction," *Science* (June 17, 1966), Vol. 152, No. 3729, pp. 1575-1584.

Smylie, D. E., and L. Mansinha, "The Rotation of the Earth," *Scientific American* (December, 1971), Vol. 225, No. 6, pp. 80-88.

Tocher, Don, "Earthquakes and Rating Scales," *Geotimes* (May–June, 1964), Vol. 8, No. 8, pp. 15-18.

Wallace, R. E., "Earthquake Recurrence Intervals on the San Andreas Fault," *Bulletin,* Geological Society of America (October, 1970), Vol. 81, No. 10, pp. 2875-2890.

Witkind, I. J., "The Hebgen Lake Earthquake," *Geotimes* (October, 1959), Vol. 4, No. 3, pp. 13-14.

Witkind, I. J., and others, "The Hebgen Lake, Montana, Earthquake of August 17, 1959," U.S. Geological Survey, Professional Paper 435, 1964, 242 pp.

COMPOSITION OF LAYERS (See also SEISMOLOGY)

Hales, A. L., "A Look at the Mantle," *Geotimes* (July-August, 1964), Vol. 9, No. 1, pp. 9-13.

Takahashi, Taro, and W. A. Basset, "The Composition of the Earth's Interior," *Scientific American* (June, 1965), Vol. 212, No. 6, pp. 100-108.

Verhoogen, John, "Temperatures within the Earth," *American Scientist* (June, 1960), Vol. 48, No. 2, pp. 134-159.

Chapter 14

The Continents

Topographic Features of the Earth's Surface

The relief features of the earth are summarized in Fig. 14-1. This diagram shows that the very high and very low portions of the earth are very small in area, and that two levels, the continental and deep ocean platforms, make up most of the surface. The reason for the two distinct levels is one of the major problems of geology. Note that the difference between the highest and the lowest points is about 12 miles, which is very small compared with the radius of the earth ($12/3957 = 0.3\%$), so that the earth is smoother than a billiard ball. This comparison shows that the features we observe on the surface are very slight compared with the earth as a whole.

The topographic features shown by the diagram are not randomly distributed. In general, the rock types and structures of these features are also distinct. Analysis leads to the following main features of the earth's surface:

Continental features
Continental platforms (including continental shelves which are submerged parts of continental platforms)
Mountain ranges
Oceanic features
Deep ocean platforms

Submarine mountains*
Mid-ocean ridges*
Continent-ocean border features
Continental slopes
Volcanic island arcs*
Ocean deeps

One interesting aspect of the earth's surface is that, for the most part, ocean occurs on the opposite side of the earth (antipodal) from continents. Only six per cent of the earth is not this way, or, said another way, 82.6 per cent of the continental areas have ocean antipodal. The reason for this is not known, but the odds against this being a chance relationship have been calculated at 14 to 1.

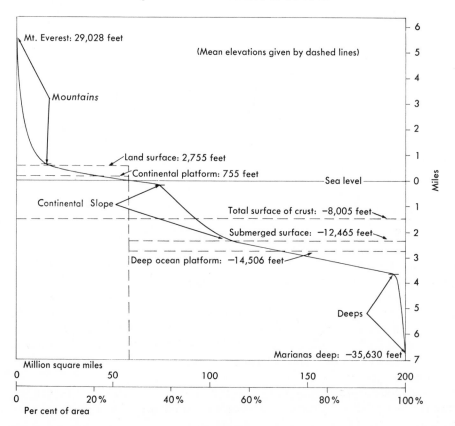

Figure 14-1. Elevations of the earth's crust.

Differences Between the Oceans and the Continents

Certainly the differences between ocean and continent are the most pronounced contrast on the earth's surface. One important contrast is in rock type—the ocean basins

*These features do not show on the diagram and are described later.

are composed mainly of basalt; the continents, of igneous and metamorphic rocks of granitic composition. To understand how these compositional differences might account for the two main levels in the elevation diagram (Fig. 14-1), we must turn to a different type of study.

Surprisingly, an important clue came from a surveying error made about the middle of the nineteenth century when the English were mapping India. Their methods of surveying were much the same as our modern ways; only the instruments were different. Because it is easier and more accurate to measure angles than distances, most surveys use triangulation. To use this method, a base line is accurately measured, generally by actually taping it. From each end of this measured base line, the angle is measured to a third station that must be visible from both stations on the base line. The third station can now be located, either by calculating its position by trigonometry or by making a scale plot. The triangle is checked by measuring the angle at the third station. Now, two of these stations can be used to locate another station, and so forth until all stations are located. (See Fig. 14-2.)

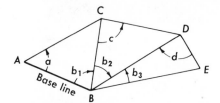

Figure 14-2. Location of points by triangulation. *AB* is base line. Angles *a* and b_1 are measured so that *C* is located relative to *A* and *B*. Angles *c* and b_2 are now used to locate *D*. Angles b_3 and *d* will now locate *E*, and the process can be continued.

The location of each station in the Indian survey was determined also by astronomical observations in the same way that a navigator at sea locates his position. It was discovered that the northern stations near the Himalaya Mountains were located too far south by astronomical methods, according to the locations obtained by the method of triangulation. Although the differences in location by these two methods were small, they were greater than could be due to errors in measurements, especially as corrections for the curvature of the earth had been applied. Apparently, the great mass of the Himalaya Range attracted the plumb bob that was used to determine the zenith in the astronomical determination of location. A level bubble and a plumb bob are basically the same, and one or the other must be used to establish the zenith from which astronomical measurements are made. Knowing the size of the mountain range and using Newton's law of gravitation, it is easy to calculate the expected amount of attraction of the plumb bob by the range; in this case, the observed deflection was much less than the expected deflection (Fig. 14-3). This implies either that the mountains are composed of rock lighter than the plain, or, if both mountains and plains are composed of the same rock type, that the rock under the mountains is lighter. The latter case is supported by seismic data and geologic mapping and is called the *roots-of-mountains hypothesis.* (See Fig. 14-4.)

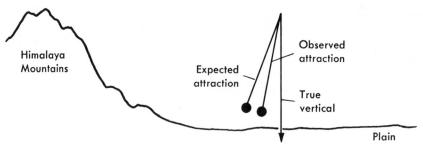

Figure 14-3. The gravitational attraction between the Himalaya Mountains and the plumb bob causes the plumb bob to be deflected from the true vertical. The deflection is less than expected, suggesting that mountains are underlain by light rocks.

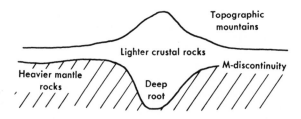

Figure 14-4. The root of a mountain range.

This suggests that the granitic continents, like the mountains, may stand higher because they are less dense and thicker than the basaltic oceans, granite being less dense than basalt. A possible layer in which the crustal rocks may float is now sought, and we will begin by considering the strength of crustal rocks.

Laboratory tests show that granite, the main rock composing the continents, has a compressive strength of about 22,000 pounds per square inch. The specific gravity of granite is about 2.7; that is, it weighs 2.7 times as much as an equal volume of water. From this we can calculate how high a column of granite can support itself; that is, at what height the granite will weigh more than its compressive strength.

1 cubic foot of water weighs 62.4 pounds

1 cubic inch of water weighs $\dfrac{1}{1728} \times 62.4 = 0.036$ pounds

∴ 1 cubic inch of granite weighs $2.7 \times .036 = 0.097$ pounds

1 mile $= 5280$ feet $= 63,360$ inches

∴ A mile-high column of granite 1 inch in cross-section weighs

$63,360 \times .097 = 6,145.92$ pounds

Therefore, the weight of a column of granite a mile high produces a pressure of 6,146 pounds per square inch at its base. Thus, if the granite approaches four miles in thickness, its compressive strength is exceeded and the base will crumble under its own weight. The continental crust exceeds this thickness, and some mountain ranges are higher. How is this possible? A factor we have so far neglected, but which will not basically alter the problem, is the change in strength of granite due to the increase in temperature and pressure at depth. The specific gravity of rock will increase due to pressure, the pressure will tend to strengthen the rock, and the higher temperature will weaken it. The total of these effects is to increase greatly the height of the column of rock that is self-supporting, but clearly a limit will be reached. Because the conditions deep in the crust and below it cannot be duplicated in the laboratory, and because the exact composition of the rocks is not known, an accurate height cannot be calculated. It is clear, however, that the continents cannot be supported by their rigidity or strength.

Then what does support the continents? Mountain ranges stand high because they have light roots, much as a big iceberg stands higher above the water than a small iceberg because its root is deeper. In a similar way, the crust of the earth may float on the mantle, and the lighter rock of the continents floats higher than the denser rocks underlying the oceans.

Apparently, then, the higher temperature and pressure at some depth make the rocks plastic, and the crust floats on these plastic rocks. Because the rocks in this zone are weak, probably because they are near the melting point for the pressure there, this could be the zone where magmas are generated. The theory that the crust floats on this viscous substratum is called *isostasy*. The depth at which the rocks become plastic is not easy to determine. It is not at the Mohorovičić discontinuity, which marks the base of the crust, but is between 60 and 125 miles below the surface, and so is in the mantle. Apparently below this depth the increased pressure increases the strength of the rocks and they behave as solids, at least to rapid deformations such as seismic waves generated by deep earthquakes. A low-velocity layer in the upper mantle was, on the basis of sparse seismic data, thought to exist and was proved in two ways. Study of the seismic waves generated by underground nuclear blasts revealed shadow zones due to refraction of waves by the low-velocity zone (Fig. 14-5). The other proof was the observation that the intense Chilian earthquake of 1960 set the zone above this layer vibrating, much like the ringing of a bell. Proof of the long-suspected weak layer adds much credence to the isostasy hypothesis. Seismic evidence shows that this zone is closer to the surface under the oceans, suggesting that the differences between continent and ocean may persist below the crust. This low-velocity zone is apparently important in the other processes that form the earth's surface structures that are described in later chapters.

Further evidence for isostasy can be found in areas recently covered by continental glaciers, around some recent volcanoes, and by measurements of the gravitational attraction near most mountain ranges. Parts of the earth that were covered by thick accumulations of ice during the recent glacial period were apparently depressed by the weight, and now that the ice has melted, they are rising. This rising may be due in part to elastic or plastic compression of the crust, but calculation suggests that these effects are not enough to account for the post-glacial uplifts and that isostasy was involved. The Hawaiian Islands are a group of young volcanoes, and their weight has apparently

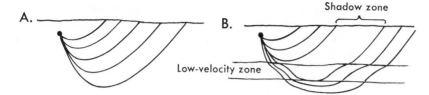

Figure 14-5. The recognition of the low-velocity zone in the upper mantle. Part A. shows the paths of seismic waves with no low-velocity zone. Part B. shows diagrammatically how refraction in the low-velocity zone produces a shadow zone at the surface. In both diagrams, for simplicity, the refraction that occurs at the M-discontinuity is omitted.

depressed the crust, creating a moat of deeper water surrounding them, as shown in Fig. 14-6. On the continent, erosional debris would mask a similar occurrence where a rapidly emplaced volcano depresses the crust. Measurements of the gravity force—which is, of course, affected by the mass distribution under the point—when corrected for such effects as elevation and nearby mountain mass, show that most of the topographic features of the earth are compensated at depth, and thus tend to prove the hypothesis of isostasy. This suggests that movements in the plastic zone that tend to restore isostatic balance are fairly rapid in geologic terms.

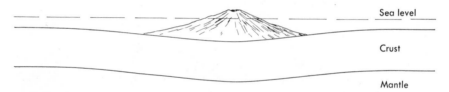

Figure 14-6. Depression of the crust into the viscous mantle by addition of a weight such as a volcano. As erosion removes the weight, the crust would return to its original position. The Hawaiian Islands and some other volcanic piles in the Pacific Ocean have apparently depressed the crust in this manner, because they are surrounded by a ring of deeper water.

Deformation of Rocks

The continents display structures on many scales. Bent, broken, and otherwise deformed rocks are seen in small and large exposures. Study of these deformed rocks reveals much information on the kinds of forces that caused the deformation. The occurrence of these deformed rocks in distinct belts shows larger, more fundamental structures that occupy many thousands of square miles. These larger structures are best developed in mountain belts. Further study reveals many other unusual features of these large belts of deformed rocks and suggests that they may show how continents form. Thus, this chapter progresses from the small, easily studied features of deformed rocks to much larger, more conjectural structures.

When rocks are subjected to forces, they may either break or be deformed. Abundant evidence of both kinds of response is common as faults and folds respectively. Whether a rock breaks or is folded depends on its strength, which depends on the type of rock and the prevailing temperature and pressure. Thus weak rocks tend to be folded and brittle rocks tend to be faulted. Folding is favored at depth, and faulting is more common near the surface.

Folds

Some general facts about the nature of the deforming forces can be learned from the study of folded and faulted rocks. Folded rocks generally record compressive forces, and folded rocks occupy less length than they did before they were folded (Figs. 14-7 and 14-8).

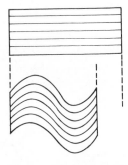

Figure 14-7. Folded rocks occupy less length than unfolded rocks.

Figure 14-8. Folding shown by a resistant sandstone in West Virginia. Photo by G. W. Stose, U. S. Geological Survey.

Folds may be upright or overturned, open or tight. The basin-like folds are called *synclines* and the hill-like folds are called *anticlines* (Fig. 14-9). A better definition, especially if the folds are overturned (as shown in Fig. 14-9) is that in anticlines the oldest rocks are in the center and in synclines the youngest rocks are in the center. Remember that the oldest rocks are at the bottom of horizontal sedimentary rocks. A bed is overturned (upside-down) if it has been bent more than 90 degrees.

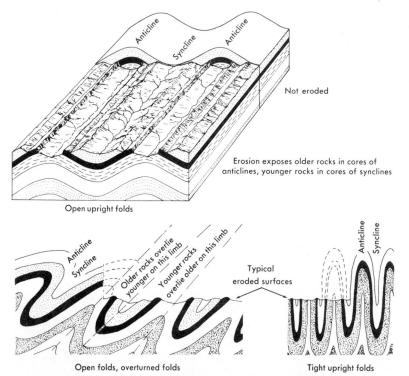

Figure 14-9. Types of folds.

Folds are not all horizontal but may plunge, as shown in Figs. 14-10 and 14-11. Notice the difference in the pattern of the ridges between the horizontal folds in Fig. 14-9 and the plunging folds shown in Fig. 14-11. A short anticline that plunges in both directions can be called a dome.

All rocks do not respond in the same way to the same forces. (See Fig. 14-12.) For example, if a series of sandstone and shale beds are folded, the weaker shales will flow into the crests of the anticlines and the bottoms of the synclines.

A special type of fold, caused in most cases by a vertical movement, is called a monocline. (See Fig. 14-13.) A fault is generally found below most monoclines. In the usual case, brittle basement rocks respond to the forces by faulting, and the more pliable overlying sediments respond by bending.

Faults

A fault is a break in rocks along which movement has taken place. Four types of faults can be distinguished, but there is much gradation among these types. Two of these fault types have predominantly vertical movement, and two have mainly horizontal movement.

Figure 14-10. Vertical air photograph of an anticline plunging toward the lower left. Virgin Valley, Utah. Photo from U. S. Department of Agriculture.

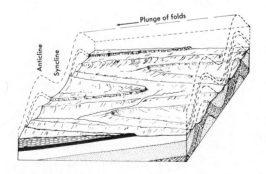

Figure 14-11. Plunging folds. In synclines erosion exposes younger rocks in the core and in anticlines, older rocks.

Figure 14-12. Folded rocks in the Wallowa Mountains, Oregon. Irregular folding of this type is common. The dark discontinuous bodies are dikes that were apparently emplaced before the folding. Photo by M. Guymon, Oregon Game Commission.

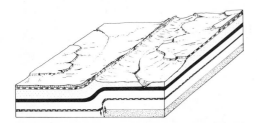

Figure 14-13. Monocline. This type of fold is commonly caused by faulting at depth.

The two types of faults with mainly vertical movement are shown in Fig. 14-14. Note that *reverse* faults (see Fig. 14-15) are caused by compression and that the crust is shortened as a result of this type of faulting. *Normal* faults (see Fig. 14-16) lengthen the crust and so are tensional features. The term normal is poor because normal faults are not normal in the sense of being more common than others. At the time the term

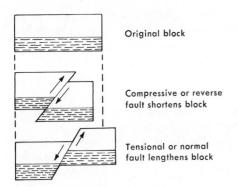

Original block

Compressive or reverse
fault shortens block

Tensional or normal
fault lengthens block

Figure 14-14. Normal and reverse
faults.

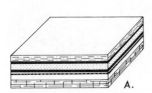 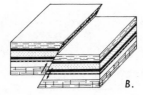 

A.   B.   C.

Figure 14-15. Development of a reverse fault. Step B. is never found in
nature because erosion would remove
the overhang. In C. the bending of beds
in the fault zone is shown.

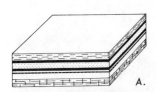 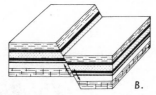 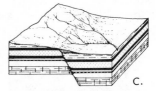

A.   B.   C.

Figure 14-16. Development of a normal fault. In C. the bending of beds in
the fault zone is shown.

was first applied they were thought to be more common, so the term is an historical
mistake. Compressive features such as folds and reverse faults are much more common,
suggesting that most orogenic forces are compressional. Normal faults are generally
associated with the stretching that accompanies vertical forces, such as uplifts, rather
than horizontal tension in the crust. Very few measurements of stress in rocks have been
made; but where the vertical stress has been measured, it is generally more than the
theoretical value, which is the weight of the overlying rocks. This suggests that vertical
forces do exist in the crust.

    Thrust faults are reverse faults that are nearly flat and so have mainly horizontal
movement. Although this definition may lead one to think that they are merely a special

case of reverse fault, in practice they are very different. This is not to imply that there are no borderline cases where either the term *reverse* or *thrust* can be used. Many thrust faults have relatively thin upper plates that have moved long distances. Such instances pose real problems because it can be calculated that the friction on the fault plane is so great that the force necessary to overcome this friction and move the upper plate would crumple the upper plate. The problem is discussed below. At other places thrust faults involve basement rocks and clearly reveal crustal compression and shortening. Some thrust faults are formed by extreme folding where the lower limb of the fold is pushed out as a thrust fault. In some cases this has occurred at such depth that the rocks were plastic, and metamorphic rocks are in the cores of the folds. The most extreme example of thrust structure is in the Alps of Europe. Here, in the various ranges of the Alps, thrust sheets and great overfolds have moved many miles and have piled up, one on the other, as shown in Figs. 14-17 and 14-18.

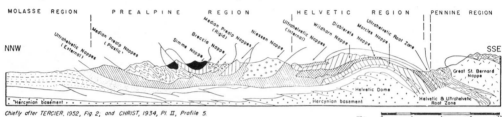

Figure 14-17. Cross-section through part of the Alps. Nappe is the term applied in the Alps to the overthrust rocks. Note the large distances that the overthrust rocks have moved. From J. C. Crowell, *Bulletin,* Geological Society of America, Vol. 66, 1955.

Figure 14-18. Mount Pilatus in the Alps south of Lucerne, Switzerland. A syncline apparently modified by faulting forms the summit of the mountain. The discordant beds on the left are probably thrust faulted against the syncline. Photo from Swissair Photo Ltd.

The second type of horizontal-moving faults will be called *lateral faults* (see Fig. 14-19), although many other terms, such as strike-slip, wrench, and rift, have been

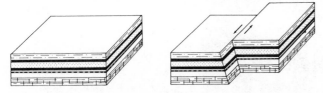

Figure 14-19. Lateral fault.

applied to these faults. Lateral faults are generally nearly vertical and may have considerable displacement. The San Andreas fault in California is of this type although it may also be a transform fault, a type of fault considered in Chapter 16. (See Figs. 14-20 and 14-21.) This fault is over 600 miles long, and the total lateral movement has been estimated at 350 miles, although many geologists who have studied it believe the lateral movement to be much less, and some believe the main movement has been vertical. (See Fig. 14-22.) Interestingly, the earthquake foci on the San Andreas are at depths of five or six miles, perhaps indicating that this long fault is rather shallow. Most lateral faults are much smaller, and it is very difficult to prove long movements on any fault. In the Sangre de Cristo Mountains of New Mexico, a vertical fault of this type is believed to have a 23 mile horizontal displacement, based on the separation of a distinctive body of metamorphic rock. Lateral faults that apparently offset oceanic features will be discussed in Chapter 16.

Most faults have neither all horizontal nor all vertical movement, but have diagonal movement. (See Figs. 14-23 and 14-24.) This is known from study of the grooves and scratches on fault planes that reveal the direction of the last movement. It should be

Figure 14-20. San Andreas and related faults in California show locations of historic surface movement. From G. B. Oakeshott, California Division of Mines and Geology, *Bulletin* 190, 1966.

Figure 14-21. Aerial photo of the San Andreas fault, California. The crushed fault zone forms the diagonal line on the photo. The fault is crossed by a highway in the foreground. Photo by J. R. Balsley, U. S. Geological Survey.

clear from Chapter 13 that most faults have recurring movements; these scratches show only the last movement. It is not generally possible to determine the exact amount of movement on a fault. To determine the exact movement, a point that was once joined must be found on both sides of the fault. (See Fig. 14-25.) Generally, only planar features, such as dikes or sedimentary beds, are available to determine the movement, and such planar features do not always permit unique solutions to the amount of actual movement. Also many fault planes are curved. The movement problem is further complicated on most faults by the deformation of the rocks near the fault. On the fault plane the rocks are ground and milled by the movement, creating what is called *fault gouge.* This is a type of metamorphism described in Chapter 5; on deepseated faults some recrystallization may occur. The rocks near the fault are generally bent or dragged by the movement, making reconstruction of the pre-faulting geometry very difficult.

Joints are cracks or breaks in rocks along which no movement has occurred. It is possible in some cases to show that the jointing was caused by the same forces that caused faulting or folding. Most rocks are jointed, and the origin of the joints is not always clear. Deep-seated rocks such as batholiths commonly have joints more or less parallel to the surface, suggesting that such joints form when the rock expands when the overlying rocks are eroded away. (See Fig. 4-3.)

Inferences from Deformed Rocks

Some inferences about crustal forces can be made solely from study of deformed rocks. One of these is the importance of compression in production of folded and thrust-faulted rocks. This conclusion seems clear, especially in areas where rigid metamorphic basement rocks have been folded or thrust along with the more plastic sedimentary rocks. At other areas, only the near-surface rocks have been compressed; the deeper rocks are not deformed. This is the case of some thrust faults where the upper plate

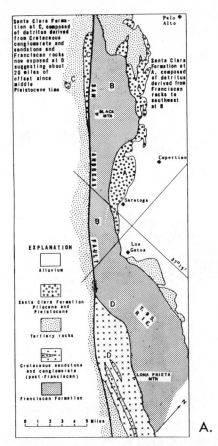

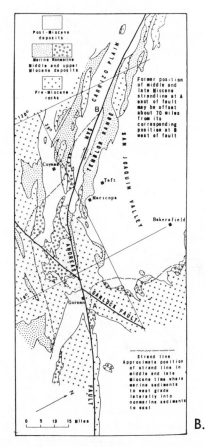

Figure 14-22. Evidence for movement on the San Andreas fault. Map A. shows an area east of Pilarcitos fault in Figure 14-20, and Map B. is located at the junction with the Garlock fault. The geologic time terms are defined in Figure 1-1. Middle Pleistocene was about one-half million years ago and middle Miocene about 20 million years ago. Do you feel that this evidence proves conclusively that large movements have occurred on the San Andreas fault? From T. W. Dibblee, Jr., California Division of Mines and Geology, *Bulletin* 190, 1966.

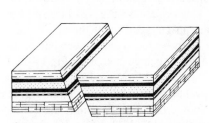

A. Normal fault with diagonal movement

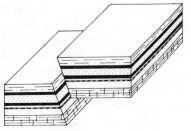

B. Reverse fault with diagonal movement

Figure 14-23. Faults with oblique movements.

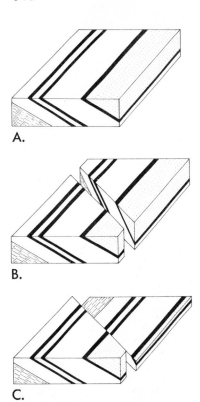

A.

B.

C.

Figure 14-24. Map patterns in faulted rocks. Part A. shows the original block. Part B. shows the oblique-vertical movement on the fault. Part C. shows the same area after erosion has leveled the uplifted block. Can the actual movement on the fault be determined from this map pattern alone?

is not strong enough to withstand the pressure necessary to push it. In such cases, perhaps uplift produces a slope down which the block slides under the influence of its own weight. Such a mechanism is mechanically possible on rather gentle slopes. Folding can also be caused by plastic beds slumping or sliding downslope. This can be seen, on a small scale, in many sedimentary beds. (See Figs. 14-26 and 14-27.) Another possible way to explain thrust faults is that the rocks may in some way be buoyed up or floated, at least partly, by water in the rocks. This process seems possible, and the high water pressures encountered in some wells are also suggestive. If the upper plate were unweighted in this way, the friction force would be reduced and movement could occur with a very small force. This hypothesis is currently being tested. (See Fig. 14-28.)

Vertical movements on the continents are easy to prove. Marine sedimentary rocks high on mountain ranges certainly prove such movements. Igneous and metamorphic rocks that formed deep in the earth and are now exposed in mountain ranges also show uplift. Other evidences of vertical movements are the uplifted, wave-cut beaches on many seacoasts, as discussed in Chapter 11. Areas below sea level, such as Death Valley, imply downward movement. The only agent that could have eroded Death Valley below sea level is the wind, and the wind is much too feeble an eroder to have done this.

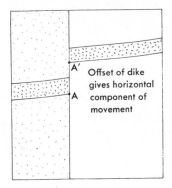

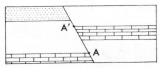

Offset of dike gives horizontal component of movement

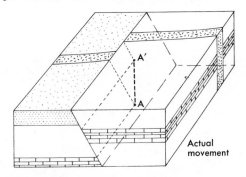

A and A' locate the points of intersection of the near margin of the dike with the top of bed at the fault plane

Actual movement

Displacement of horizontal bed gives vertical component of movement

Figure 14-25. Determination of the actual movement on a fault. In most cases only one planar feature such as the displacement of the dike or of one of the sedimentary beds is available, and such features only reveal one component of the movement. In this figure the actual movement can be determined by finding the displacement of the intersection of the dike and one of the beds. Practically, in this case, the fault movement can be determined by finding the horizontal component by measuring the offset of the dike, and the vertical movement from the displacement of the horizontal sedimentary beds. This figure shows a highly idealized case. In most instances the rocks are more deformed, especially near the fault plane.

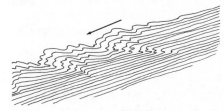

Figure 14-26. Gravity thrust fault. The block broke away and slid down the slope.

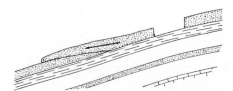

Figure 14-27. Gravity folding. Plastic sediments tend to flow downslope, producing folds.

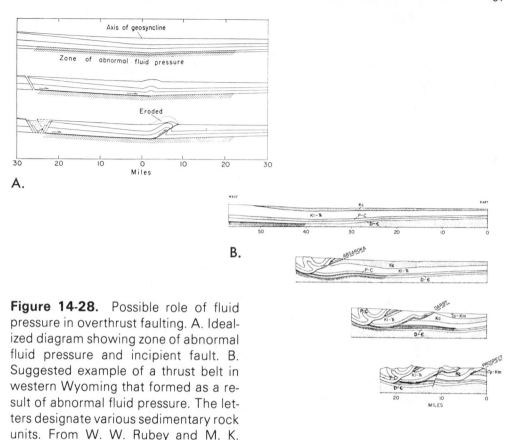

Figure 14-28. Possible role of fluid pressure in overthrust faulting. A. Idealized diagram showing zone of abnormal fluid pressure and incipient fault. B. Suggested example of a thrust belt in western Wyoming that formed as a result of abnormal fluid pressure. The letters designate various sedimentary rock units. From W. W. Rubey and M. K. Hubbert, *Bulletin,* Geological Society of America, Vol. 70, 1959.

Although most of our evidence of forces in the crust comes from study of ancient rocks, some movements can be seen directly. Recent earthquake faults show rapid movements, both horizontal and vertical. Slow creep measured by surveying also is currently occurring on some faults, as described in Chapter 13.

Dating folding and faulting is discussed in Chapter 17, but the principles can be seen in question 28 at the end of this chapter.

Other types of deformation are the igneous and metamorphic processes considered earlier. These processes are commonly part of the mountain-building processes that create folds and faults. Some of the relationships among these processes will be seen in the next section.

Larger Features of the Continents

The most obvious features of the continents are relatively flat continental platforms and linear belts of mountains. (See endpapers.) The earliest idea on the origin of the continents was that they formed during the initial solidification of the earth. Study of the older parts of the continents disclosed that the oldest rocks are metamorphosed sedimentary rocks that are themselves the products of erosion of a still earlier terrane.

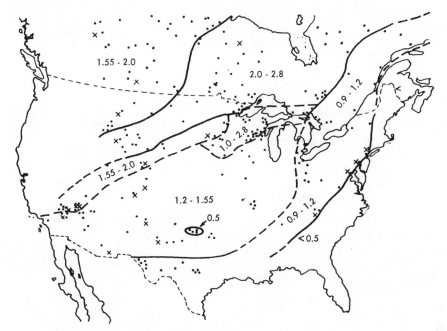

Figure 14-29. Distribution of ages in crystalline rocks from the central part of North America. Circles: ages that are within the limits specified for a given zone on the map; crosses: ages that are outside the limits. Age limits are given in billions of years. After G. R. Tilton and S. R. Hart, *Science* (April 26, 1963), Vol. 140, p. 364.

Radioactive dates have revealed that, in a general way, the oldest rocks are in long belts. Because each of these belts has structures similar to parts of some of our present mountain ranges, they may be the eroded stumps of older mountain ranges. (See Fig. 14-29.) Thus, mountains may be the key to understanding continental structure.

This discovery suggests that the present processes going on at the surface have been going on since the formation of the crust, and that the continents may have formed by accretion. The oldest rocks so far discovered are about 3.98 billion years old, which suggests a total age of the earth between 4 and 6 billion years. Thus, the process of continent formation may still be going on. We can look for evidence of this in the mountains where much of the present volcanic and earthquake activity occurs.

Mountains

A mountain is a topographic feature that rises above the level of the surrounding area, more or less abruptly. Various types of mountains are shown in Fig. 14-30. The definition and the illustration do not emphasize the most important geologic aspect of mountains, which is the deformation of the rocks that form the mountains. Most mountains occur in ranges of the complex type, and although single isolated mountains of the other types shown in Fig. 14-30 do occur, they are relatively rare.

Thus, typical mountains have complex internal structures formed by folding, faulting, volcanic activity, igneous intrusions, and metamorphism. In a sense, complex mountains are residual, too. The origin of the internal mountain structure is different from the origin of the topographic expression of mountains. Topographic mountains

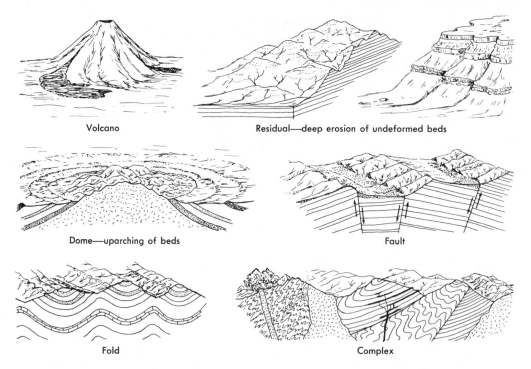

Volcano   Residual—deep erosion of undeformed beds

Dome—uparching of beds   Fault

Fold   Complex

Figure 14-30. Types of mountains.

are formed by uplift and erosion, and so are residual. Thus most mountain ranges have complex internal structure formed at an earlier time, and the topographic expression of mountains is the result of later uplift and erosion—a two-step history. When the work of erosion is completed, only deformed rocks remain as evidence of an earlier mountain range.

Dating mountain building is difficult because three dates are involved:

The age of formation of the rocks,
The age of the internal structure, and
The date of uplift when erosion formed the topographic mountains.

Commonly the internal structure is formed in several stages, all of which should be dated. Determining each of these three dates in actual cases involves many difficulties.

Geosyncline Theory—Origin of Mountains

In 1859, James Hall, an American geologist who made important studies of fossils, made an important observation. He discovered that rocks in the northern part of the Appalachian Mountains were much thicker and more clastic than were rocks of the same age located further in the interior of the continent. In other words, the present mountains are on the site of a much deeper, earlier basin than is the area to the west. (See Fig. 14-31.)

This observation was extended by James Dwight Dana, who recognized that it was true of the whole Appalachian Mountains. His studies culminated in 1873 when he applied the name *geosyncline* to the great, elongate basin of deposition that was later

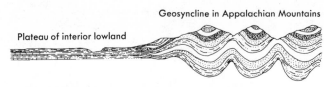

Geosyncline in Appalachian Mountains

Plateau of interior lowland

Figure 14-31. Sedimentary rocks of the same age are thicker and more clastic in mountain ranges than in the interior of continents.

folded, faulted, and uplifted to form a mountain range. Similar observations have been made all over the world, namely that many of the present mountain ranges were once elongate basins, or geosynclines.

It has also been noted that parts of some of the geosynclinal mountain ranges are composed of metamorphic rocks and large intrusive bodies of granitic rock termed batholiths. Such rocks form the oldest parts of the continents, and are sometimes called basement rocks because they generally underlie the thin veneer of sediments on the stable parts of the continent.

These observations have been combined to form the geosyncline theory. The following description is of a very idealized geosyncline. The geosyncline begins as a broad, elongate downwarp, generally at the edge of a continent, that is flooded by sea water. Because the geosyncline acts as a trap for sediments, the sedimentation stage begins with this downwarp. In general, the part of the geosyncline near the ocean is more active than the rest, is depressed deeper, and is the site of volcanic activity. These geosynclines are large, the order of many hundred miles long and up to several hundred miles across.

In spite of the great thicknesses of rocks in geosynclines, there is abundant evidence that the water was at no time very deep. The cross-bedding, ripple-marks, mudcracks, and types of fossil life displayed in these rocks are all features that form today in relatively shallow water. Thus, the reason for the great thicknesses of sedimentary rocks must be that the geosynclines continued to sink as they were filled with sediment. In some places the features in the volcanic, or at least more active, part of the geosyncline suggest that the water may have been deep there at times. In addition to the evidence of deep-water fossils, such features as graded bedding and the channeling due to mud and muddy water sliding down slopes suggest some deep water. Such slides are known to occur today on the slope between the continental shelf and the deep ocean. It is interesting that almost no deep-water deposits of the type now forming in the deep oceans far away from the continents have been recognized in old rocks, perhaps because such deposits are thin and difficult to recognize. The currently forming deep-water deposits consist mainly of the fallen shells of tiny animals that live near the surface and of volcanic dust that has settled on the ocean. Sediments formed in this way that are composed of more than 30 per cent biologic material are called *oozes.*

Differences in environment are shown by differences in sedimentation. As shown in Fig. 14-32, the geosyncline gradually tapers off to form the shelf that borders the continent. In most cases the continent has little relief; this is shown by the sediments (shale, limestone, and fairly pure quartz sandstone) that form on the shelf. Such sediments reveal rather complete weathering. This same suite of sedimentary rocks also develops nearby in the geosyncline, but here the rocks are much thicker because of the

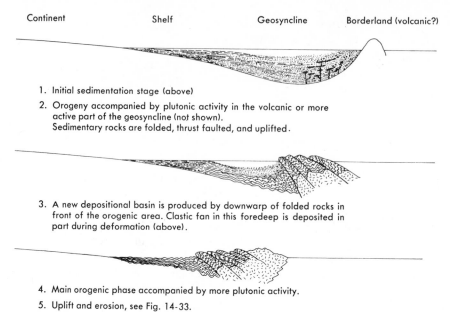

1. Initial sedimentation stage (above)
2. Orogeny accompanied by plutonic activity in the volcanic or more active part of the geosyncline (not shown).
   Sedimentary rocks are folded, thrust faulted, and uplifted.

3. A new depositional basin is produced by downwarp of folded rocks in front of the orogenic area. Clastic fan in this foredeep is deposited in part during deformation (above).

4. Main orogenic phase accompanied by more plutonic activity.
5. Uplift and erosion, see Fig. 14-33.

Figure 14-32. Stages in development of a geosyncline.

greater downsinking. The volcanic part of the geosyncline, on the other hand, is the scene of more rapid sedimentation as a result of deeper and more rapid downwarping. In addition to volcanic rocks, the sediments here are rapidly deposited, poorly sorted, quickly buried clastic rocks such as impure shales, dark sandstones with clay matrix (graywacke), and coarser clastic rocks.

The source of these "dumped-in" clastics is a problem. Some of the material may come from the continent, but the volume generally requires a source on the other side of the geosyncline. The problem is further complicated, as we shall see, by the later igneous and metamorphic activity, which obscures the early situation in this part of the geosyncline. Because a geosyncline commonly forms at the edge of a continent with the volcanic part on the ocean side, some geologists have suggested that the source of these clastics was a continent now below sea level; but recent studies show that this is not indicated by the structures of the continental margins. Possible present-day geosynclines will be studied in Chapters 15 and 16.

The sedimentation phase of a geosyncline ends when it stops sinking and stops receiving sediments, generally after 30,000 to 50,000 feet of sedimentary rocks have been deposited. The next phase is the orogenic (mountain-making) phase during which (1) the sediments are deformed to make mountain structure and (2) the volcanic part of the geosyncline may be changed by igneous and metamorphic activity into a rigid mass of granitic composition, thus becoming continent-like. The history of actual geosynclines is generally more complex, with overlap and repetition of these three phases. Some geologists believe that this is the way that the continents are formed— but not all geosynclines are at the edges of continents, and other theories have been suggested. The basement on which geosynclines form is exposed only in the less active parts of geosynclines, and here the basement is continental. No pre-geosyncline base-ment is exposed in the more active volcanic part of geosynclines, and here the geosyn-cline could have formed over oceanic crust.

The orogenic phase of the geosyncline is probably the most interesting, most impor-tant, most studied, and least understood phenomenon in geology. In this phase the geosyncline is folded, apparently by compressive forces. In the volcanic part of the geosyncline, the folding may be accompanied by metamorphism and emplacement of granitic rocks. In this way part of the geosyncline may be transformed into a rigid mass. It should be noted that granitic rocks and metamorphism are not always confined to only the volcanic part of the geosyncline, and, at places, parts of the volcanic geosyn-cline may not be affected at all. At a few places, batholiths, generally small ones, occur in non-geosynclinal areas. At this stage a new depositional basin generally develops over the border between the old geosyncline and the shelf or continent, and a new sedimentation cycle may ensue. This may be followed by other episodes of folding, faulting, and thrust faulting in this or other parts of the geosyncline. Metamorphism and emplacement of batholiths may or may not accompany these orogenic events. These events may be spread over a few hundred million years. Geosynclinal folding and faulting occur, probably episodically, over long lengths of time, but emplacement of batholiths appears to be much more rapid. There is an unfortunate tendency to date mountain building from the easily obtained radioactive dates of batholith crystalliza-tion, even though many batholiths are clearly earlier or later than the deformation.

The final pulse may be the emplacement of later, small granitic batholiths. These late batholiths are emplaced either before or during the post-orogenic volcanism, if any, and are similar in chemical composition to those volcanic rocks.

After the deformation the geosynclinal areas generally are uplifted to form mountain ranges. The geosynclinal deformation produces mountain structures, but topographic mountains do not exist until this uplift occurs. The uplift is probably caused by the development of a light granitic root by igneous and metamorphic processes. Then the light rocks tend to float up under the influence of isostasy. The uplift of some mountain ranges occurs much later than the deformation. The development of the root and the uplift are apparently late features and do not occur until after the forces that caused the initial downwarp and those that caused the deformation cease to operate.

Once a mountain range exists, isostasy may control its future to a large extent. As soon as it rises, a mountain range is subjected to erosion. As erosion removes weight from the mountains, the buoyancy of the root probably causes the range to rise, as shown in Fig. 14-33. This process should go on until the root has risen to the level of the rest of the continent and the topographic mountains have been reduced by erosion to the level of the rest of the continent. In this manner a new section of the continental platform may be formed. The rocks of this section would display mountain structures, and this process may produce an age distribution similar to that shown in Fig. 14-29.

Why geosynclines form is not known, but some possible hypotheses will be consid-ered in Chapter 16. The formation of continents is an important problem. Are they original features of the crust merely modified by geosynclines, or have they grown during geologic time? The most common theory of origin of continents is that they have been distilled somehow from the upper mantle. It is clear from earlier chapters that unless the continents had been rejuvenated, they would have been eroded to base level soon after their formation. This rejuvenation could have occurred by either building

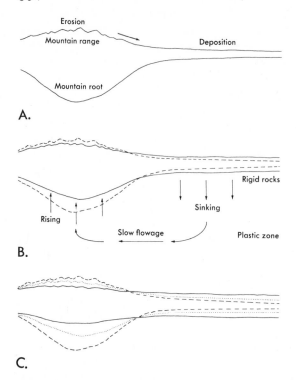

A.

B.

C.

Figure. 14-33. Removal of a mountain root by erosion and isostasy. Erosion removes material from the mountains, thereby lightening them. This can result in isostatic uplift that reduces the depth of the root. The place where the eroded material is deposited may also be depressed by the additional weight. Several stages in the process are shown in the diagram.

up the continents from below or by lateral spreading. The radioactive dates of the basement rocks and the geosyncline theory suggest that the continents have grown through geologic time.

Examples of Continental Structure

The main structural provinces of the United States will show that the distribution of deformed rocks is not random. The structural provinces considered are shown in Fig. 14-34, which should be compared with the landforms map. (See endpapers.)

The *Continental Interior* is the stable part of the continent between the geosynclines. For the most part this province is low lying and is covered by a thin veneer of sedimentary rocks. To the northeast the sedimentary rocks are thin and the ancient metamorphic rocks of the Canadian Shield are exposed. To the west the elevation increases on the Great Plains. The Continental Interior is, for the most part, only slightly deformed and contains a number of gentle basins and domes. (See Fig. 14-35.) Faulting is minor, folds are very gentle, and igneous rocks are few and far between.

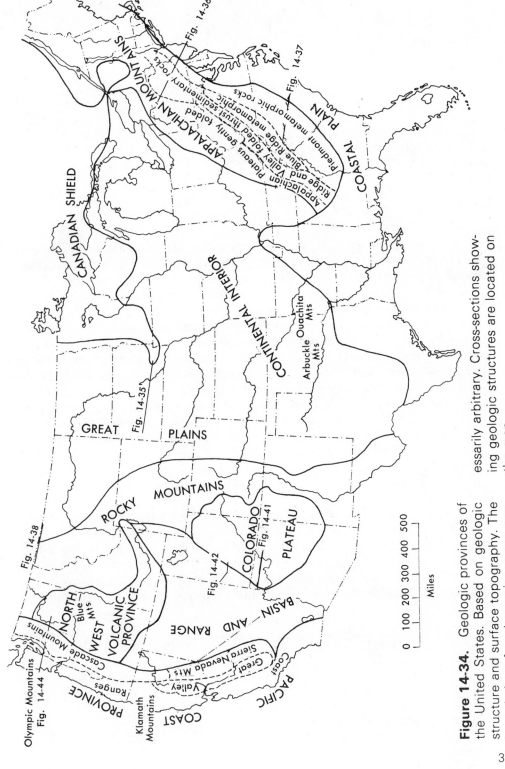

Figure 14-34. Geologic provinces of the United States. Based on geologic structure and surface topography. The boundaries of such provinces are necessarily arbitrary. Cross-sections showing geologic structures are located on the map.

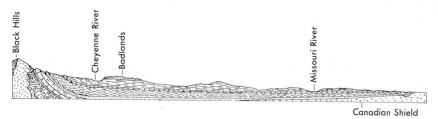

Figure 14-35. Cross-section across the Great Plains from the Black Hills of South Dakota to the Canadian Shield in eastern South Dakota. After David White and others, U. S. Geological Survey, *Bulletin* 691, 1919.

As the mountains in either direction are approached, the sedimentary rocks become thicker, folds and faults can be easily seen in the outcrops, and intrusive and volcanic rocks become more common. In addition, the Continental Interior is interrupted by the strongly deformed rocks of the Arbuckle and Ouachita Mountains in southern Oklahoma and adjacent Arkansas.

The *Appalachian Mountains* are typical geosynclinal mountains. The eastern side is dominated by igneous and metamorphic rocks, and the west is mainly sedimentary rocks. In the southern Appalachian Mountains, metamorphic rocks at low elevations form the Piedmont, and similar rocks at higher elevations form the Blue Ridge. In this area the sedimentary rocks are mainly thrust faulted in the Ridge and Valley area. (See Fig. 14-37.) Further north, in Pennsylvania, the sedimentary rocks are mainly folded. (See Fig. 14-36.) West of the Ridge and Valley area is the Appalachian Plateau where the sedimentary rocks are only mildly deformed. New England has mainly igneous and metamorphic rocks, and the folded and thrust belt narrows.

The *Atlantic Coastal Plain* is a thin cover of undeformed sedimentary rocks that presumably overlies eroded metamorphic and batholithic rocks similar to those of the Piedmont.

The *Rocky Mountains* are also geosynclinal mountains but are more complex than the Appalachian Mountains. The northern Rocky Mountains are folded and thrust faulted like the western part of the Appalachians (see Figs. 14-38 and 14-39), and the many thrust faults indicate considerable shortening. The western part of the northern Rocky Mountains is the Idaho batholith and the batholiths of northeastern Washington. Again this is similar to the Appalachians. The rest of the Rocky Mountains is

Appalachian Plateau :   Ridge and Valley   Piedmont : Coastal Plain

Figure 14-36. Cross-section showing the structures in the Appalachian Mountains in Pennsylvania. Here the sedimentary rocks are folded rather than thrust as in Figure 14-37. After Douglas Johnson.

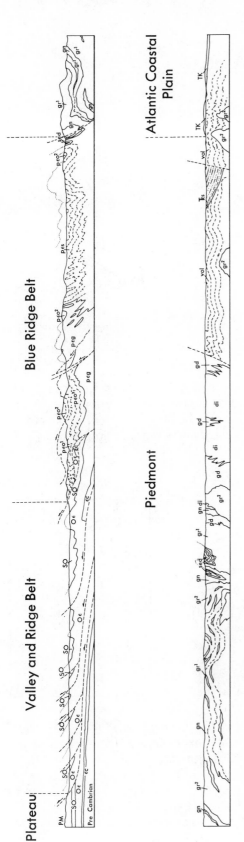

Figure 14-37. Cross-section through the southern Appalachian Mountains. The Piedmont and Blue Ridge areas contain mainly metamorphic rocks, and the Valley and Ridge Belt is largely thrust-faulted sedimentary rocks. From P. B. King, *Guides to Southeastern Geology*, 1955.

357

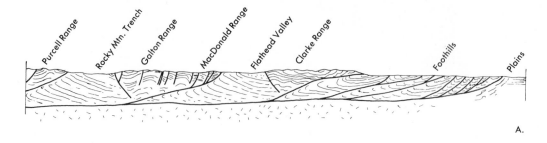

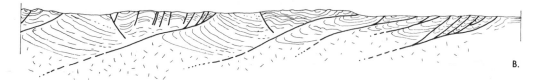

Sedimentary rocks

Continental crust

Figure 14-38. Cross-section through the northern Rocky Mountains near the International Boundary. The sedimentary rocks here are thrust faulted, and the total movement may be many miles. Two interpretations of the structure are shown. A. Assumes that the thrusts join, and the deformation is mainly in the sedimentary rocks. Recent seismic data support this interpretation. B. Assumes that the basement rocks are involved in the deformation.

A.

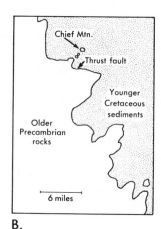

B.

Figure 14-39. A. Chief Mountain, Glacier National Park, Montana. Chief Mountain is an erosional remnant of a thrust plate of older rocks thrust over much younger rocks. B. Map showing thrust fault near Chief Mountain. The minimum displacement on the thrust is the distance between the furthest east and the furthest west that the thrust fault is exposed, a distance of about 12 miles (not shown here). The movement was slightly north of east as shown by gouges on the fault plane. Part A. from National Park Service. Part B. from C. P. Ross and others, *Geologic Map of Montana,* U. S. Geological Survey.

different. In Wyoming the mountains are mainly uplifts with some faulting on the margins, and there are large sedimentary basins between the mountain ranges. These structures are something like those of the Basin and Range Province but differ in that the ranges and the basins are much larger in Wyoming. The age of faulting, and the rock types and the internal structures exposed in the mountains, are also very different. In the central Rocky Mountains ancient rocks like those of the Canadian Shield are exposed. In Colorado, unlike Wyoming, there are more ranges than basins. (See Fig. 14-40.) In New Mexico these structures merge with those of the Basin and Range Province.

Figure 14-40. Looking north at the Rocky Mountain front south of Denver, Colorado. The ranges on the left are composed of metamorphic and igneous rocks. In the foreground and middleground the eroded remnants of sedimentary rocks are prominent. Thus the basic structure shown is one flank of a large anticline. In the distance on the right is Table Mountain, formed by a resistant lava flow. Photo by T. S. Lovering, U. S. Geological Survey.

The *Colorado Plateau* may be the cause of the narrowing of the Rocky Mountains. This is an unusual structural feature in that it is an area at high elevation with a fairly thick accumulation of sedimentary rocks that is relatively undeformed. Why this province is undeformed is unknown because the rocks surrounding it are deformed. The sedimentary rocks are flat-lying and are broken by a few nearly vertical faults and monoclines. (See Fig. 14-41.)

The *Basin and Range Province* is an area of many relatively small block-faulted mountain ranges. (See Figs. 14-42 and 14-43.) The rocks that make up these ranges are greatly deformed by folding, faulting, and igneous and metamorphic processes. Thrust faults with many miles of displacement can be traced from range to range. The farthest west of these ranges is the Sierra Nevada in California that is the batholithic part of the geosyncline in this area.

The *Pacific Coast Province* is a very complex, geologically young area. It is geologically active with much seismic activity and with the only active volcano in conterminous United States. The San Andreas fault, described earlier, is in this area. The Coast Ranges, consisting of complexly folded and faulted rocks, form the western part of this

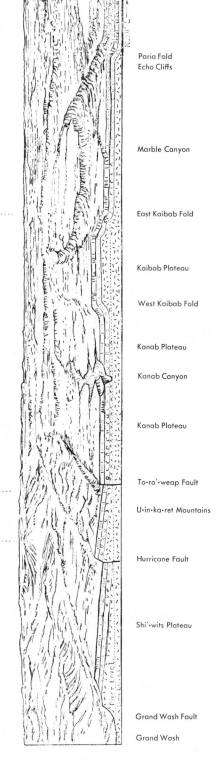

Paria Fold
Echo Cliffs

Marble Canyon

Paria Plateau

East Kaibab Fold

Kaibab Plateau

West Kaibab Fold

Kanab Plateau

Kanab Canyon

Kanab Plateau

Virgin Valley

To·ro'·weap Fault

U·in·ka·ret Mountains

Pine Valley Mountain

Hurricane Fault

Shi'·wits Plateau

Grand Wash Fault

Grand Wash

Figure 14-41. Cross-section of part of the Colorado Plateau in Arizona. From J. W. Powell, U. S. Geological and Geographical Survey of the Territories, 1876.

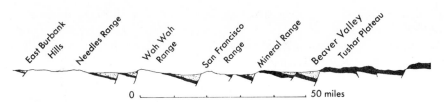

Figure 14-42. Diagrammatic cross-section of part of the Basin and Range Province. From J. H. Mackin, *American Journal of Science,* Vol. 258, 1960.

Figure 14-43. Oblique aerial view of a range in the westernmost part of the Basin and Range Province in southern California. Erosion has removed much of the range front, and recent movement on the fault has revealed its loca-tion far in front of the range. Note the difference in the dissection on both sides of the fault scarp. Compare with Figure 9-3. Photo from U. S. Geological Survey.

province. The central part of the province is a lowland except in northern and southern California. The eastern part of the province is, in California, the Sierra Nevada Mountains that are related to the Basin and Range Province. North of the Sierra Nevada is the Cascade Range. This range is composed of volcanic and continental sedimentary rocks capped by very young, high volcanoes. The volcanic rocks of the Cascade Mountains have more in common with the Northwest Volcanic Province to the east than with the Pacific Coast Province, but other structural and sedimentary features are related to the rest of the province.

The *Northwest Volcanic Province* consists of volcanic rocks of several ages. (See Fig. 14-44.) The Columbia River Basalt covers southeastern Washington and adjacent

Figure 14-44. Cross-section through central Washington. The Cascade Mountains are composed of volcanic and continental sedimentary rocks that grade into marine rocks to the west. Small batholiths invade the Cascade Mountain rocks, and the large volcanoes such as Mount Rainier were built on top of the existing range. The Columbia River Basalt covers the area to the east.

Oregon with several thousand feet of basalt flows. Much of Oregon and the Snake River Valley in Idaho are covered by younger volcanic rocks. In this area are exposed many recent volcanic features not yet old enough to be affected by weathering. The volcanic rocks extend into the northwest corner of the Basin and Range Province and are here considered part of that province because they have been deformed by basin and range faulting. The Blue Mountains in Oregon are composed of much older rocks that give our only indication of the structures of the geosynclinal rocks in this region.

QUESTIONS

14-1. What are the main topographic features of the earth's surface, and what is the geologic significance of the two most important features?

14-2. The continents are composed of _____.

14-3. The oceans are underlain by _____.

14-4. What is the evidence for mountain roots?

14-5. How is the crust of the earth supported?

14-6. Why do the continents stand higher than the ocean floors?

14-7. Is it the strength of rocks that enables mountains to stand above the general level of the continents? Prove it.

14-8. At what depth does isostatic compensation occur? Why?

14-9. Can isostasy account for the formation of sedimentary basins? To answer this question, a simple calculation is necessary. Assume an initial depth of 100 feet of water and that this is completely filled with sediment. Calculate how much the mantle is depressed by equating the net gain in weight in the basin to the weight of the depressed mantle. (Ans. $h = 240$ feet)

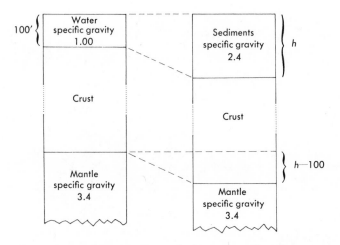

14-10. What are the ages of the basement rocks forming North America?

14-11. How are these rocks distributed?

14-12. Why have geologists concentrated much of their attention on mountain ranges?

14-13. What steps are involved in the formation of mountains?

14-14. How are mountains dated?

14-15. Outline briefly the history of a geosyncline.

14-16. Discuss the origin of continents.

14-17. What type of force produces folding?

14-18. What type of force produces normal faults?

14-19. What type of force produces reverse faults?

14-20. What is an anticline? Discuss several ways one can be distinguished from a syncline.

14-21. What is a thrust fault?

14-22. What evidence would you look for in the field to tell whether a crack in the rocks is a fault or a joint?

14-23. What determines whether a rock folds or faults if it is subjected to a compression? List all factors.

14-24. How could you tell in the field whether a fault is still active?

14-25. How does one determine the amount of movement on a fault?

14-26. Prove to your girlfriend (boyfriend), who is an art major and has taken no science courses, that crustal movements have occurred.

14-27. Compare the continental structure in the Mississippi Valley with the Rocky Mountains.

14-28. The following diagram is a cross-section through an area. Starting with the earliest event, list the sequence of events that have occurred in this area. You may wish to review your answers after reading Chapter 17.

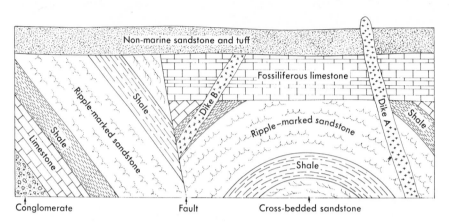

SUPPLEMENTARY READING

EARTH'S SIZE, SHAPE, and WEIGHT

Daly, R. A., *Strength and Structure of the Earth.* New York: Prentice-Hall, Inc., 1940, 434 pp.

Ewing, Maurice, and Leonard Engel, "Seismic Shooting at Sea," *Scientific American* (May, 1962), Vol. 206, No. 5, pp. 116-126.

Heiskanen, W. A., "The Earth's Gravity," *Scientific American* (September, 1955), Vol. 193, No. 3, pp. 164-174. Reprint 812, W. H. Freeman and Co., San Francisco.

Oliver, Jack, "Long Earthquake Waves," *Scientific American* (March, 1959), Vol. 200, No. 3, pp. 131-143. Reprint 827, W. H. Freeman and Co., San Francisco.

CONTINENTS

Croneis, Carey, and W. C. Krumbein, *Down to Earth,* Chapter 16, "What Price Continents?" pp. 129-136; Chapter 17, "Folds and Faults," pp. 137-144. Chicago: University of Chicago Press (1936), Phoenix edition, 1961 (paperback).

Engel, A. E. J., "Geologic Evolution of North America," *Science* (April 12, 1963), Vol. 140, No. 3563, pp. 143-152.

Kay, Marshall, "The Origin of Continents," *Scientific American* (September, 1955), Vol. 193, No. 3, pp. 62-66. Reprint 816, W. H. Freeman and Co., San Francisco.

King, P. B., *The Evolution of North America,* Chapter 4, pp. 41-75. Princeton, N.J.: Princeton University Press, 1959, 190 pp.

Milne, L. J., and others, *The Mountains.* New York: Life Nature Library, Time, Inc., 1962, 192 pp.

Wilson, J. T., "The Crust," in *The Earth and Its Atmosphere* (D. R. Bates, ed.), pp. 48-73. New York: Science Editions, Inc., 1961, 324 pp. (paperback). This book was also published under the title *The Planet Earth.* London: Pergamon Press, 1957. This paper is reprinted in White, J. F. (ed.), *Study of the Earth,* pp. 89-105. Englewood Cliffs, N.J.: Prentice-Hall, Inc., 1962 (paperback).

Wilson, J. T., "Geophysics and Continental Growth," *American Scientist* (March, 1959), Vol. 47, No. 1, pp. 1-24.

Chapter 15

The Oceans and the Continent– Ocean Borders

The birth of the oceans is a matter of conjecture, the subsequent history is obscure, and the present structure is just beginning to be understood. Fascinating speculation on these subjects has been plentiful, but not much of it predating the last decade holds water. . . .

H. H. Hess, "History of Ocean Basins," p. 599, in *Petrologic Studies, A Volume to Honor A. F. Buddington,* Geological Society of America, November, 1962, pp. 599-620.

Oceans

Until quite recently very little was known about the more than seventy per cent of the earth's crust that is covered by the oceans. The continents, except for the continental shelves that are parts of the continents covered by ocean, can be studied directly. The oceanic crust must be studied by such indirect means as depth soundings and seismic studies. This difference in the kinds of data available results in knowing different things about these two major parts of the earth's crust. One very obvious aspect of this is that, with a few exceptions, only large structures are known in the oceans. Hopefully, future, more detailed, studies and new methods will provide as much information about the oceanic crust as is known of the continental crust.

Oceanic Platforms

The deep ocean platforms cover about forty per cent of the earth's surface. (See Fig. 15-1.) The reason for this vast platform was explored in an earlier section; it was found to result from the level at which the heavier basaltic crust under the ocean floats isostatically.

Seismic studies have revealed a remarkably uniform structure of the oceanic platforms. (See Fig. 15-2.) Under a depth of two and three-quarter miles of water lies a thin layer of unconsolidated sediments averaging about 1,000 feet thick. This first layer is composed of red clay and oozes. Below this is the second layer of irregular thickness,

367

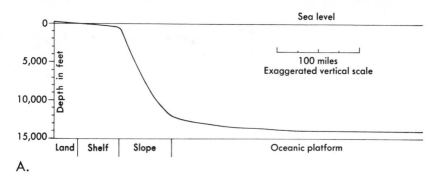

A.

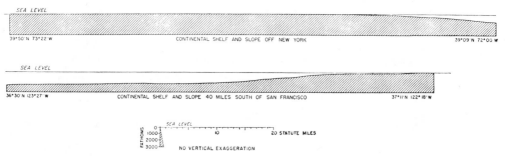

B.

Figure 15-1. A. Continental margin showing typical continental shelf, continental slope and oceanic platform. B. Actual profiles of continental margins with no vertical exaggeration. Part B. from K. O. Emery and others, U. S. Geological Survey, *Professional Paper* 260-A, 1954.

averaging about a mile and a quarter thick. This layer is believed to be composed of basalt and consolidated sediments. The third layer extends to the Moho and is everywhere very close to three miles thick. The composition of this layer is not known but is traditionally thought to be basalt.

The oceanic platforms are generally fairly smooth surfaces. They are interrupted at places by low hills, submarine mountains, mid-ocean ridges, and fracture zones, all of which will be described below. The submarine mountains indicate some volcanic activity. The low hills are probably also volcanic, but may in part be caused by faulting or other processes. Erosion is very slow on the oceanic platforms, so minor features tend to be preserved. Sedimentation, too, is slow for the most part and in areas far from land is limited to dust settling on the ocean and biologic material, especially shells, sinking from the near surface. Perhaps the smoothest surfaces on the earth are parts of the oceanic platforms near the continental slopes. The smoothness is probably the result of burial of the surface irregularities by sediments moving down the continental slopes, perhaps as density currents. (See Fig. 15-3.)

The age of the ocean floors is remarkably young in geologic terms—less than 160 million years. The youngest parts are near the mid-ocean ridges, and the oldest near continents. The distribution of ages will be discussed further in the next chapter.

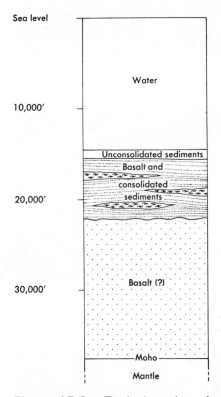

Figure 15-2. Typical section of rocks on the oceanic platforms.

Submarine Mountains

Mapping the bottoms of the oceans really began in the 1920's with the invention of the echo sounder. (Fig. 15-4.) Prior to that time soundings had to be taken with a weight on a line. The many soundings made during World War II resulted in the discovery of flat-topped seamounts. During the late 1950's, in connection with the International Geophysical Year, enough soundings were made to discover the unity of the mid-ocean ridges. These discoveries were made possible by the development of the continuous echo sounder, which can operate while a ship is under way. Using sound waves, it is possible to obtain a topographic profile of the bottom. By using other energy sources, such as dynamite or electric sparks, it is possible to penetrate to various depths into the bottom rocks and so to learn more about the structures. These techniques have been used mainly on the continental shelves in exploration for potential oil-bearing structures. (See Fig. 15-5.)

The submarine mountains are mainly in the Pacific Ocean. These mountains, rising thousands of feet above their bases, are apparently basaltic volcanoes that tend to form in lines. Some of these, such as the Hawaiian Islands, extend above sea level.

In some areas these submarine mountains are flat topped. (See Fig. 15-6.) These flat tops are now at depths up to a mile below the surface of the ocean. They are especially

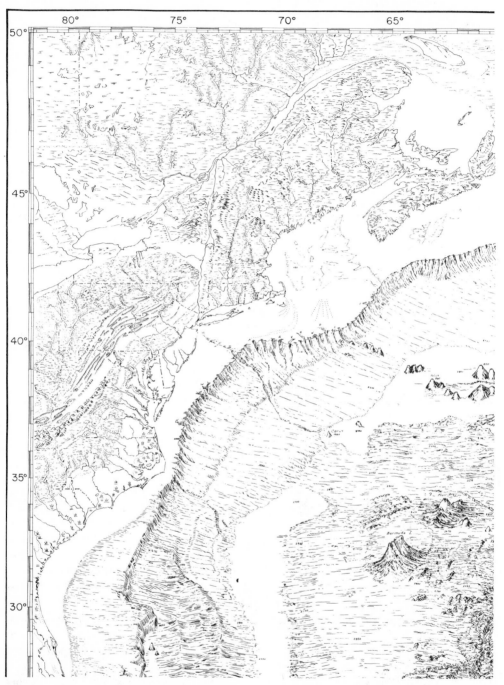

Figure 15-3. The continental margin and ocean floor off eastern United States. The continental shelf and slope are cut by submarine canyons. The smooth surfaces in the deep ocean below the continental slope are probably depositional. The rough topography to the southeast and the scattered hills are from volcanic activity. This is a portion of the Physiographic Diagram of the North Atlantic Ocean published by the Geological Society of America © Copyright by Bruce C. Heezen and Marie Tharp. Reproduced by permission.

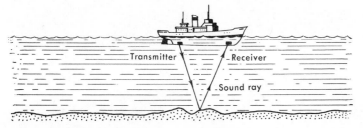

Figure 15-4. An echo sounder. Depth = 1/2 speed of sound X echo time.

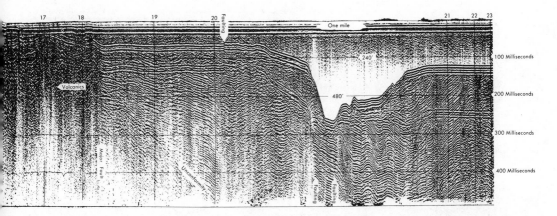

Figure 15-5. Sparker profile near Santa Cruz Island, California. An electric arc created waves that penetrated the bottom sediments and were reflected back to the receiver. The sea floor is shown by the strong reflection at a depth of about 300 feet on the right side. The sedimentary rocks that form most of the cross-section are in fault contact with the volcanic rocks on the left (major fault). Three other faults in the sedimentary rocks are also labeled. Note that the sea floor is displaced about 8 feet by the fault labeled on the top. Courtesy of R. F. Herron who made the profile using EG&G International, Inc., equipment.

interesting because their flat tops were probably produced by wave erosion at sea level, although some types of submarine volcanism may produce initially flat-topped mountains. Wave-cut benches are being formed now on the present islands, and if this process continues, such islands will become flat topped. Lowering sea level or uplifting an island can preserve such benches, and this can be seen on present-day islands and coastlines. (See Fig. 11-18). Similar benches are revealed in profiles of some of the flat-topped seamounts (Fig. 15-6). Samples dredged from flat-topped seamounts contain basalt fragments and Cretaceous and Early Cenozoic shallow-water fossils, suggesting the age of the wave erosion.

The coral atolls of the Pacific may also reveal subsidence. The corals live in the near-surface zone and form an offshore reef. (See Fig. 15-7.) If the island sinks slowly enough, the corals will build up the reef so that they remain near the surface where their food supply is. This process may continue until the island is completely submerged, perhaps quite deeply, and only a coral reef remains. At Eniwetok the coral is 4,500 feet deep.

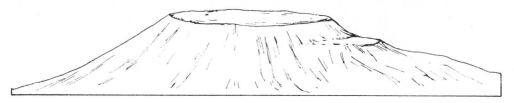

Figure 15-6. Flat-topped seamount.
Note the small wave-cut bench.

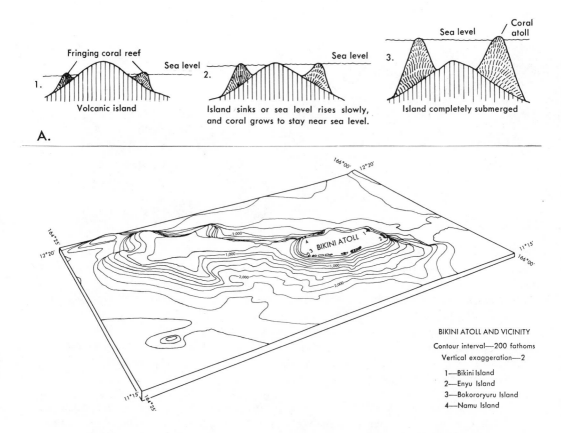

Figure 15-7. Coral atolls. A. Development of a coral atoll. B. Perspective diagram of Bikini Atoll and Sylvania flat-topped seamount adjoining it on the west. Bikini Atoll formed on a flat top, unlike the atoll shown in A. Most of Bikini Atoll is submerged and only the small black areas in the diagram are islands. Part B. from K. O. Emery and others, U. S. Geological Survey, *Professional Paper*, 260-A, 1954.

The submarine mountains reveal much of the history of the oceans. Because they are basalt volcanoes, they record the importance of volcanic activity in the oceans. Only about 40,000 submarine mountains have been discovered. This is a small number compared to the number of continental volcanoes, which suggests that either volcanoes form at a slow rate on the oceans, or that the ocean floors are young features. The latter point will be discussed in the next chapter. The flat-topped seamounts and the atolls reveal large changes in elevation of the deep ocean floors. Figure 15-8 shows that the distribution of flat-topped seamounts and atolls is not random, but rather that a large

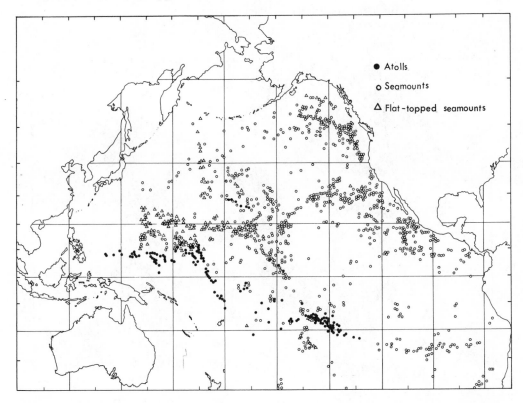

Figure 15-8. Map of the Pacific Ocean showing atolls, seamounts, and flat-topped seamounts. These volcanic features tend to form lines, suggesting eruption along faults. The coral atolls are almost entirely confined to the warm waters within 30 degrees of the equator. The flat-topped seamounts also form distinct clusters. The data on atolls, seamounts, and flat-topped seamounts is from H. W. Menard, *Marine Geology of the Pacific,* McGraw-Hill, 1964.

elongate area in the southwest Pacific Ocean has subsided, probably indicating a former position of the mid-ocean ridge or rise. The Cretaceous and Early Cenozoic shallow-water fossils dredged from the flat tops suggest the time when this subsidence occurred. Note that the atolls are mainly on the south flank of this area, suggesting that climate may have controlled the growth of coral. Present areas of coral growth are restricted to about 30 degrees on each side of the equator. The present depths below sea level of the flat-topped seamounts are not regular; therefore, as suggested above, differing rates of subsidence may also have influenced coral growth.

Mid-ocean Ridges

The mid-ocean ridges form the longest continuous mountain range on the earth. Their extent can be seen on the map in Fig. 15-9. Their length is about 40,000 miles; their width is 300 to 3,000 miles; they rise up to about 10,000 feet above their base; and at places, the highest points form islands. Two parts of this ridge are fairly well known: the Mid-Atlantic Ridge and the East Pacific Rise.

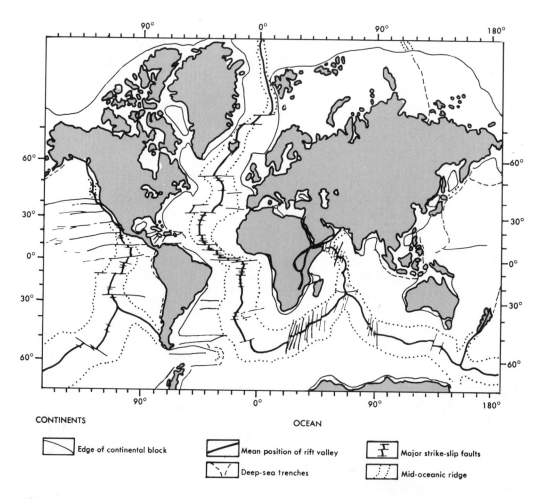

Figure 15-9. The mid-ocean ridge and associated features. After B. C. Heezen, in *Continental Drift,* Academic Press, 1962.

The Mid-Atlantic Ridge has been known for many years because islands such as Iceland, the Azores, St. Paul Rocks, Ascension, St. Helena, and Tristan da Cunha are parts of it. That it is part of a much larger feature was not known until the 1950's; at that time, many soundings greatly increased our knowledge of its nature. It consists of numerous ridges and lines of peaks paralleling the main ridge that extend outward for hundreds of miles on the flanks (Fig. 15-10). The central crestal zone has one or more discontinuous rift valleys that are up to thirty miles wide. The seismic activity of the ridge is shown by the numerous earthquake epicenters located on it. The bedrock, revealed in dredged samples, is basalt, some of which is pyroclastic; and some of the peaks that form in lines are apparently basalt volcanoes that formed along the faults. (See Fig. 15-11.) At many places these features are offset by what appear to be cross

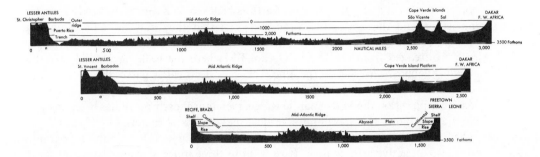

Figure 15-10. Transatlantic topographic profiles showing the Mid-Atlantic Ridge. Vertical exaggeration 40:1. From B. C. Heezen, Geological Society of America, *Special Paper* 65, 1959.

Figure 15-11. The ocean floor on the Mid-Atlantic Ridge. Basalt rocks and some bottom-dwelling organisms can be seen. Courtesy R. M. Pratt.

faults that, as can be seen in Fig. 15-9, apparently have large horizontal displacements. The unusual nature of these faults is described in the next chapter.

The Mid-Atlantic Ridge has been studied directly on the few islands that are part of the ridge. The islands are mainly basaltic, and recent volcanic activity on Tristan da Cunha and Iceland has also been basaltic. St. Peter and St. Paul Rocks are tiny islands on the ridge, and sheared peridotite (mylonite) is found there. Radioactive dates on these rocks suggest that they may be as old as 4.5 billion years. As will be seen in Chapter 18, this is about the time the earth is believed to have crystallized. Thus the composition and the age of these rocks suggest that they may be mantle material somehow brought to the surface by the process that formed the ridge. Peridotite has also been dredged from the Indian Ocean. Elsewhere the rocks are mainly basaltic (including gabbro) with some serpentine. The central rift valley passes through Iceland; and though the evidence is not clear, dilation of 1 to 240 miles may have occurred.

At other places on the Mid-Atlantic Ridge, metamorphosed basalt has been dredged. Like the peridotite, this too implies vertical movements of material.

The East Pacific Rise is similar to the Mid-Atlantic Ridge. It has similar topography and is also an active seismic area. It rises up to three miles above its base and forms the Galápagos and Easter Islands. It meets North America just south of Baja California, leaves in northern California, and is last seen off British Columbia. Measurements on the East Pacific Rise show that the heat flow near the crest is up to five times greater than average, and on the flanks is much lower than average (Fig. 15-12). This high heat flow extends into the Gulf of California. Seismic studies using underwater explosions have revealed that the crust under the East Pacific Rise is only about three-quarters of its thickness elsewhere in the Pacific.

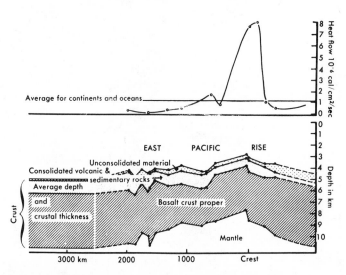

Figure 15-12. Cross-section and heat flow across the East Pacific Rise. The heat flow is very high on the crest of the rise but lower than normal on the flanks. The crust is thin on the crest of the rise. The section on the left is average for the ocean. From H. W. Menard, *Science* (December 9, 1960), Vol. 132, p. 1741.

The mid-ocean ridge goes inland in Africa and joins the African rift valleys that, up until the discovery of the mid-ocean ridge, were thought to be unique. These African areas are now being studied closely in the hope of learning more about these ridges. The African rift valleys are unsymmetrical but similar to the rifts on the mid-ocean ridges, and are associated with high plateaus, indicating uplift like the mid-ocean ridges. The African rift valleys are not all the same age, and most show evidence of recent geologic activity, such as volcanoes and changes in elevation shown by drainage features. The volcanoes are not ordinary basalt, but are alkali-rich rocks, some of which are extremely rare types. (See Fig. 15-13.)

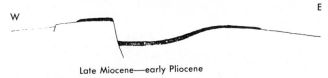

Late Miocene—early Pliocene

1. Volcanics erupted on crest of uplift
2. Faulting on west side of rift, flexuring on east side

Late Pliocene

3. Faulting of floor of rift; renewal of movement on main fractures; new fractures on rift shoulders
4. Volcanic activity in rift floor

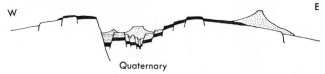

Quaternary

5. Further uplift of rift shoulders; renewal of movement on faults in rift floor; new closely spaced fractures develop in median zone
6. Small plugs and larger volcanoes built in rift floor; Some volcanoes on the rift shoulders

50 miles

Figure 15-13. Development of the Central Rift in Kenya. From B. H. Baker in *East African Rift System,* University College, Nairobi, 1965.

Fracture Zones

Fracture zones and faults, both of which are associated with mid-ocean ridges, are prominent oceanic features. Some of the faults are indicated by the offsets of the mid-ocean ridges shown in Fig. 15-9. The total displacement of the southern part of the Mid-Atlantic Ridge may be quite large, and recently a probable 300-mile offset northeast of Greenland was discovered. In the northeastern Pacific Ocean, similar features associated with the East Pacific Rise are found. In this area great fracture zones and the San Andreas fault are also found. These faults are different from those described in Chapter 14, and they are the *transform faults* discussed in the next chapter.

The fracture zones of the Pacific are shown in Fig. 15-14. Most follow great circle routes, and the longer ones are from 2,500 to 6,000 miles long. The Mendocino fracture

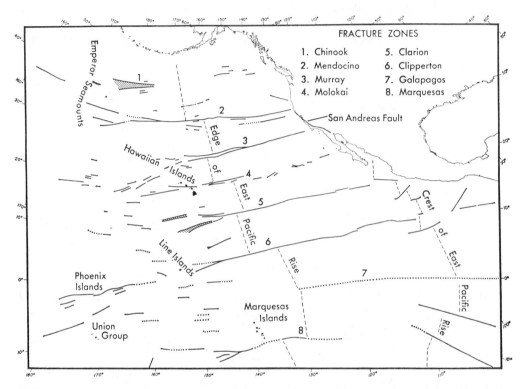

Figure 15-14. Principal fracture zones of the Pacific plotted on a great-circle projection. A straight line on this projection is part of a true great circle. Cross hatching indicates areas of submarine volcanoes. Possible displace-ments on the fracture zones are indicated by the displacements of the edge of the East Pacific Rise. From H. W. Menard, *Science*, January 6, 1967, Vol. 155, p. 73.

zone is the most northerly of the long fracture zones and departs most from a great circle route. It is over 3,500 miles long and has a south-facing scarp up to 8,600 feet high. The Murray fracture zone is about 2,500 miles in length and is a north-facing scarp up to 7,000 feet high. The Molokai fracture zone is less regular than the others. The Clarion fracture zone is a trough up to a mile deep for much of its length. The Clipperton fracture zone is over 6,000 miles long, about one-quarter of the earth's circumference. It is a zone about 50 miles wide of irregular relief of 1,000 to 2,000 feet, and is less pronounced than the other fracture zones. Seismic studies show that the crustal thickness changes abruptly at some of these fracture zones. These zones are believed to be faults. Possible displacements of features are indicated in Fig. 15-14.

Magnetic data, which will be discussed further in Chapter 16, also suggest faulting. The fracture zones do not extend inland. Some of the volcanic islands, such as the Hawaiian Islands, cross the fracture zones without displacement, suggesting that the islands are younger than the fractures. Debris from such islands apparently masks the fracture zones in some areas.

Continent-Ocean Borders

The change in elevation between the continents and the oceans is one of the main features on the elevation diagram, Fig. 14-1. Two types of continent-ocean borders can be easily distinguished. They are the continental slopes that are geologically relatively stable and the active volcanic island arcs. Although these two border types seem distinct, studies suggest that they are related. It also seems likely that both types of border are related to geosynclines. This relationship will be explored in the following chapter. Sedimentation on the shelf and slope was described in Chapter 4.

Continental Slope

The continental slope appears to be the simpler of the border types, but this may be only apparent because of the lack of information on crustal structure. The continental slope begins abruptly at the edge of the continental shelf. The shelves of the world average about 45 miles wide, but the range is from near zero to 800 miles. Their average depth is near 200 feet, and their edge is generally at 400 to 500 feet, although the range is from near sea level to 2,000 feet. The slope angle changes from about 10 feet per mile on the shelf to between two to six degrees on the slope in a distance of under 20 miles. Some continental slopes are as steep as 45 degrees. Compared to the average slope of most mountain ranges, five degrees is very steep. The average drop on the continental slope is about 12,000 feet, and the slope angle lessens with depth, making the distinction between the ocean platforms and the slope difficult. Thus, the slopes are a few tens of miles on the average. The region between the slope and the ocean platform is called the *continental rise*. The rise is apparently composed of sediments.

The east coast of the United States is typical of continental slopes (Fig. 15-1). Most areas of the west coast are similar, but the East Pacific Rise and the fracture zones are also present.

Continental slopes are composed of sediments. One question is whether the slope is an area of erosion or deposition. Submarine canyons are cut into the continental slope at many places, as described in Chapter 4. This implies that, at least at these places, erosion is at work to lower the angle of the continental slopes. However, some deposition is taking place also and the sediments were described in Chapter 4. The deltas built by large rivers modify the continental slopes. The average angle of deltas is less than one and one-half degrees, much less than the continental slopes, which implies that the continental slopes are not entirely constructional. Some of the erosion may be caused by deep currents parallel to the slopes as well as by turbidity currents moving down the slopes. At places, the slope is bare of unconsolidated sediments. Thus, both erosion and deposition occur on the continental slopes. (See Fig. 15-15.)

Some continental slopes are in part formed by reefs, as shown in Fig. 15-15. The barrier reef of Australia is an example.

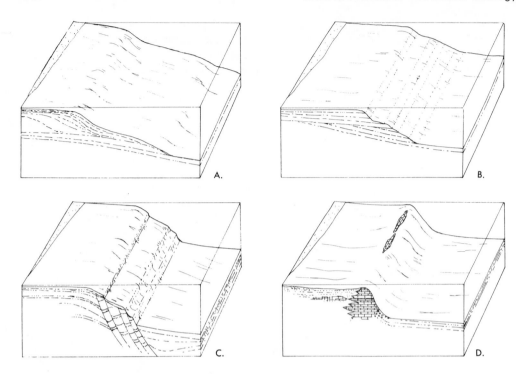

Figure 15-15. Continental slopes. A.
Depositional. B. Erosional. C. Faulted.
D. Reef.

Seismic studies show that the continental crust ends at the continental slope, but the
details of the structure in this area are not clear. Faulting has been suggested by some
as the origin of continental slopes because many are seismically active. These more
complex continental slopes may grade into the active volcanic island arcs. In the next
chapter the origin of geosynclines will be considered, and one possibility is that the
continental shelf may be one part of a geosyncline and the continental slope may be
the more active part.

Island Arcs

At other places, particularly on the rim of the Pacific Ocean, the border is more active
with chains of volcanic islands forming arcs. (See Fig. 15-16.) The island arcs were
noted early in the history of geologic oceanography, but recognition of many of their
features came much later. Although oceanography began in the 1870's with the voyage
of the *Challenger,* little was learned in detail until the 1920's when echo sounders were
first employed. It was soon discovered that deep trenches exist on the ocean side of the
volcanic island chains. During the 1920's and early 1930's, the Dutch geologist F. A.
Vening-Meinesz devised a method of measuring the value of gravitational attraction at
sea. Using a method that measured the period of a pendulum, he had to work in Dutch
and U.S. Navy submarines to avoid the effects of ocean waves. He made the extremely
important discovery that the acceleration of gravity is low (negative gravity anomaly)
over these deep trenches. This discovery shows that the formation of these deep
trenches was an anti-isostatic process that upset isostasy locally. (See Fig. 15-17.) The

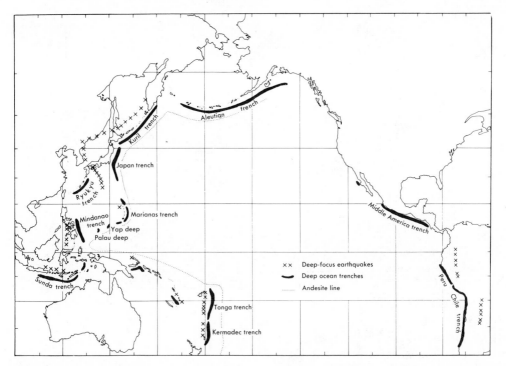

Figure 15-16. Map of the Pacific Ocean showing the relationships among island arcs, deep-focus earthquakes, and deep trenches. The volcanic rocks on the continent side of the andesite line are andesite, and basalt occurs on the ocean side. The epicenters of shallower earthquakes lie between the deep earthquakes and the island arcs.

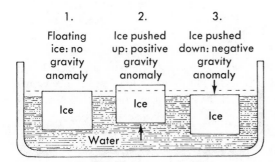

Figure 15-17. Gravity anomalies: Station 1 gravity is normal. At Station 2 the attraction of gravity is higher because more mass is present below the station. At Station 3 there is less mass below the station because the lighter ice displaces the heavier water, so the gravity force is less.

island arcs are also active seismically, and about the same time it was discovered that some of the earthquakes come from very deep in the mantle. (See Figs. 15-16 and 15-18.) Such a concentration of features on the narrow border between the continent and the ocean must have important geologic implications.

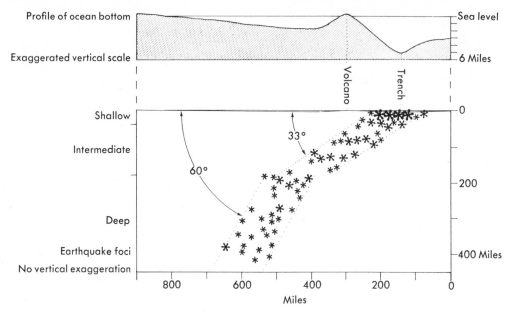

Figure 15-18. Locations of earthquake foci associated with a volcanic island arc (Kamchatka-Kurile area). The foci may locate a fault plane. After Hugo Benioff.

The volcanic island arcs pose many problems, but also give much insight into earth processes. The volcanoes of the island arcs are andesitic in composition. Small batholiths of similar chemical composition are found on some, such as the Aleutian Islands. This is also the composition of the volcanoes in the Andes Mountains of South America and the Cascade Range in North America that, together with the island arcs, form what has been somewhat poetically called the "rim of fire" of the Pacific Ocean. The deep oceanic trough off South America and the deep-focus earthquakes inland suggest that the structure there may be similar to that of the island arcs of the western Pacific. On the oceanward side of this rim of andesite volcanoes, all of the volcanic islands are basaltic (see andesite line in Fig. 15-16). Shallow (35 miles or less) and intermediate (35 to 180 miles) depth earthquake foci parallel the island arcs and lie on planes that dip about 35 degrees under the continents.

Deep-focus earthquakes (180 to 400 miles) occur only on the Pacific rim. (See Figs. 15-16 and 15-18.) Their foci lie on a plane continuous with the shallow and intermediate earthquakes, but dipping about 60 degrees. The change in slope occurs at about 180 miles.

QUESTIONS

15-1. What kinds of sediments are found on the oceanic platforms?

15-2. Describe the submarine mountains of the Pacific (composition, arrangement, etc.).

15-3. How do flat-topped seamounts form? What does this suggest about the history of the oceans?

15-4. How long are the mid-ocean ridges?

15-5. Describe a cross-section of the mid-ocean ridges. (A sketch might help.)

15-6. What is the evidence of faulting on the mid-ocean ridges?

15-7. Describe the fracture zones of the Pacific Ocean.

15-8. How is the distinction between continent and ocean made?

15-9. Sketch the ocean-continent boundary at a stable place, and label the parts.

15-10. What does a negative gravity anomaly mean?

15-11. Describe a typical island arc and sketch a cross-section.

15-12. It has been suggested that the East Pacific Rise runs inland near the west coast of North America. How do the structures in that area compare with the structures found on the mid-ocean ridges?

15-13. Describe the similarities between island arcs and geosynclines.

15-14. What are the similarities and differences between geosynclines and continental slopes?

SUPPLEMENTARY READING

Dietz, R. S., "Origin of Continental Slopes," *American Scientist* (March, 1964), Vol. 52, No. 1, pp. 50-69.

Emery, K. O., "The Continental Shelves," *Scientific American* (September, 1969), Vol., 221, No. 3, pp. 107-122.

Ewing, Maurice, and Leonard Engel, "Seismic Shooting at Sea," *Scientific American* (May, 1962), Vol. 206, No. 5, pp. 116-126.

Fisher, R. L., and Roger Revelle, "The Trenches of the Pacific," *Scientific American* (November, 1955), Vol. 193, No. 5, pp. 36-41. Reprint 814, W. H. Freeman and Co., San Francisco.

Heezen, B. C., "The Rift in the Ocean Floor," *Scientific American* (October, 1960), Vol. 203, No. 4, pp. 98-110.

Menard, H. W., "Fractures in the Pacific Floor," *Scientific American* (July, 1955), Vol. 193, No. 1, pp. 36-41.

Menard, H. W., "The East Pacific Rise," *Scientific American* (December, 1961), Vol. 205, No. 6, pp. 52-61.

Menard, H. W., *Marine Geology of the Pacific.* New York: McGraw-Hill Book Co., 1964, 271 pp.

Raff, A. D., "The Magnetism of the Ocean Floor," *Scientific American* (October, 1961), Vol. 205, No. 4, pp. 146-156.

Stetson, H. C., "The Continental Shelf," *Scientific American* (March, 1955), Vol. 192, No. 3, pp. 82-86. Reprint 808, W. H. Freeman and Co., San Francisco.

Tazieff, Haroun, "The Afar Triangle," *Scientific American* (February, 1970), Vol. 222, No. 2, pp. 32-40.

Chapter 16

Origins of Crustal Structures – Sea-Floor Spreading and Continental Drift

"Having gathered these facts, Watson, I smoked several pipes over them, trying to separate those which were crucial from others which were merely incidental."

Memoirs of Sherlock Holmes
Sir Arthur Conan Doyle

A World System of Crustal Structure

Seismic and magnetic studies of recently discovered oceanic features, together with modern ideas of continental drift, based on geologic and magnetic studies, have resulted in a new unified concept of crustal structures, and, although many problems remain, it promises to revolutionize geology. In brief, this theory envisions the continuous formation of new sea floor at the mid-ocean ridges, movement of the resulting crustal plates, carrying with them the continents, and finally the descent of the plates at the deep-sea trenches.

Sea-floor Spreading

Age

The evidence that new sea floor is created at the mid-ocean ridges is abundant. Radioactive dating of basalt samples shows a progressively older age away from the ridges. The sediments on the sea floor also thicken away from the ridges, and examination of drill cores reveals that the bottom sediments are older further away from the ridges. The sea floors are young geologically, and only a few samples older than 130 million years have been found. The oldest sediments so far found are latest Jurassic (about 160 million years old) sandstone off South America and in the western Pacific.

385

Magnetic Studies

The earth's magnetic field is impressed on a rock at the time the rock is formed. Basalt, because it contains small amounts of magnetite, an especially magnetic mineral, faithfully records the earth's magnetic field at the time the basalt cools through a certain temperature (Curie point) that is well below its melting point.

Magnetic surveys near the mid-ocean ridge show bands of alternating high and low magnetic intensity that parallel the ridge as shown in Fig. 16-1. The meaning of these magnetic bands was not clear until magnetic studies were made on the continents. It was soon discovered that many rocks recorded a magnetic field opposite or reversed to the present magnetic field. Magnetic studies of each flow in areas of thick accumulations of basalt flows showed that these reversed magnetic field rocks did not occur in

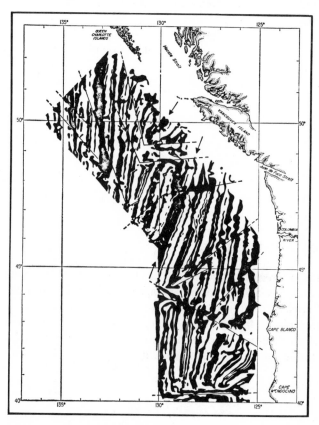

Figure 16-1. Magnetic anomalies in part of the ocean along the west coast of North America. Positive magnetic anomalies are black. They are believed to be formed where the remanent magnetism of the underlying rocks reinforces the earth's field. The intervening white areas have lower than normal magnetism and may be places where the underlying rocks have reversed magnetism. The straight lines are faults that displace the pattern. The arrows indicate short lengths of oceanic ridge, and the pattern is parallel and symmetrical to the ridges. From F. J. Vine, *Science,* Vol. 154, p. 1406, December 16, 1966.

a random way but, rather, that if radioactive dates of the rocks were also obtained, the magnetic reversals occurred simultaneously all over the earth. Starting with the youngest rocks, it has been possible to determine the times of reversals accurately enough to use the magnetic direction to date some young rocks. (See Fig. 16-11.) These reversals suggest that the bands of alternating high and low magnetic intensities paralleling the mid-ocean ridge could be caused by bands of ocean floor basalt that have normal and reversed magnetic fields. The normally magnetized basalt would reinforce the present magnetic field, giving a high magnetic intensity; and the reverse magnetized rocks would partially cancel the present field, giving a lower than normal magnetic intensity. Careful comparison of the magnetic intensity on each side of the mid-ocean ridge shows that the magnetic intensity on one side is nearly the mirror image of that on the other side. (See Fig. 16-2.) The final argument is to construct a model of the ocean floor by calculating the positions of bands of normal and reversed magnetic basalt and then to calculate the theoretical magnetic field produced by such a model. As shown in Fig. 16-3, this model produces a field almost identical to that found in nature. Other uses of magnetic data will be described when continental drift is considered.

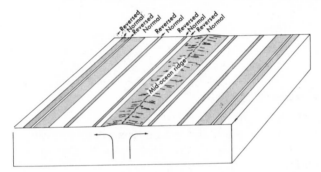

Figure 16-2. Formation of magnetic anomalies at mid-ocean ridges. In this hypothesis the lavas that form the ocean floor cool at the ridge and are magnetized by the earth's field at the time they cool. Convection currents carry the lavas away from the ridges and so create magnetic anomaly patterns parallel and symmetrical to the ridges.

Movement of the Sea Floors

Crustal Plates

Much of the evidence for sea-floor spreading just described indicates that, in a general way, the newly created sea floor moves away from the mid-ocean ridge at rates between ½ and 2 inches per year. Seismic activity shows where this motion occurs. In the earlier discussion of earthquakes, it was noted that earthquakes are believed to result from the sudden movement of one segment of crust past another. Figure 16-4 shows where earthquakes occur, and their concentration is at the deep-sea trench—island arc areas, to a lesser extent on the mid-ocean ridge, and at many places on the faults or fracture zones that offset the ridge. These earthquakes suggest that the earth can be thought of

Profile Reversed

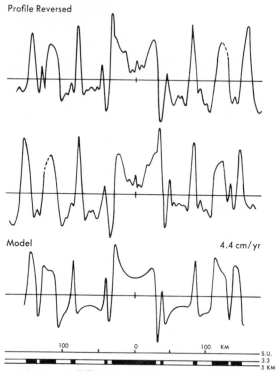

Model   4.4 cm/yr

Figure 16-3. Magnetic profiles across the East Pacific Rise. The two upper curves are an actual magnetic profile, shown both as recorded and reversed. The lower curve was computed from a model, assuming a spreading rate of 4.4 cm per year. After F. J. Vine, *Science,* Vol. 154, p. 1409, December 16, 1966.

as consisting of seven main plates and a few smaller plates, each one moving from the mid-ocean ridge to a trench or island-arc. These plates are indicated in Fig. 16-4. Basaltic oceanic crust is created at the ridge and absorbed at the trenches.

Transform Faults

More detailed seismic studies reveal the nature of the movements on the faults that offset the mid-ocean ridge. It is possible to determine the relative direction of the movement of the plates on each side of the fault from the direction of the first movement on seismograms. Such studies show that the movement on each block is away from the mid-ocean ridge, as expected if new crust is created there. Note, however, as shown in Fig. 16-5, that the relative movement is opposite to that expected by ordinary fault movement. This special kind of fault is termed a *transform fault,* and, so far, it is only known to occur in association with mid-ocean ridges. The San Andreas fault and the oceanic fracture zones are believed to be transform faults.

Continental Drift

The idea of continental drift has been with us since the first accurate maps showed that Africa and South America fit together. Recently, much new data have resulted in a resurgence of this theory. Continental drift has had many adherents in the past, and they have proposed several different schemes of drift with differing timetables and routes. Because the evidence used by the proponents of these schemes could, in many details, be disproved, and thus negate the scheme, the theory of continental drift fell into disrepute, especially with geologists in the northern hemisphere. Most of the proponents of drift have been southern hemisphere geologists. A brief survey of the geology of the southern hemisphere will show why this is so. It will be necessary, then, to digress into a discussion of the earth's magnetic field. Finally, the pros and cons of continental drift will be discussed.

During the last part of the nineteenth century, European geologists were deciphering the thrust structures in the Alps and American geologists were mapping the West, which was just opening up. As a result, they took little notice of some startling discoveries made in India, Australia, and South Africa. At each of these places a peculiar flora, called the *Glossopteris* flora after its most abundant genera, was found in beds overlying glacial deposits. Since then the same sequence has been found in South America and Antarctica. The glacial deposits are from continental, not mountain, glaciers and rest on polished and striated basement rocks. (See Fig. 16-6.) The striations suggest that in many cases the ice moved from areas that are now occupied by oceans. The moraines also suggest continental drift. They contain many boulders of rock types not found on that continent, and in some cases the foreign boulders match well with outcrops on the adjoining continent. The questions immediately arise: How could extensive glaciers form in these tropical areas, and how did the *Glossopteris* flora spread across the oceans now separating these areas? (See Fig. 16-7.) These areas are so widespread that shifting the rotational axis of the earth does not help because even in the most favorable location, some of the glacial areas are still within ten degrees of the equator. If all of the continents of the southern hemisphere had been in contact at one time, these occurrences would be explained; but the continent would have to break up and drift to explain the present geography. (See Fig. 16-8.) Thus began the idea of continental drift. Today the evidence is even more compelling because of the discovery of *Glossopteris* and other fossils, indicating a humid, temperate climate at one time in Antarctica. There is much more evidence, but before evaluating it, we will discuss the earth's magnetic field because magnetic data are used extensively in reconstructing continental drift.

The Earth's Magnetic Field

It is the earth's magnetic field that aligns a compass needle and makes it point toward magnetic north. The earth behaves much as if it had a giant bar magnet in its core. (See Fig. 16-9.) However, even though the core is made of iron and nickel, no such magnet can exist at the temperatures there. Iron loses its magnetism when heated above a certain temperature—well below its melting point. Also the earth's magnetic field is not that of a simple dipole, which is the field of a bar magnet. (See Fig. 16-10A.) The earth's field consists of a dipole field and a non-dipole field. One effect of the non-dipole field is that the south magnetic pole, at the edge of Antarctica, is 1,300 miles from the point that is antipodal to the north magnetic pole, which lies north of Canada. However, for the most part, the earth's field can be considered nearly a dipole field.

The earth's magnetic field causes a compass to point toward the magnetic pole, but if the compass is pivoted to move in the vertical direction also, it will indicate inclination as well. (See Fig. 16-9.) At the magnetic equator, the field is horizontal; at the poles,

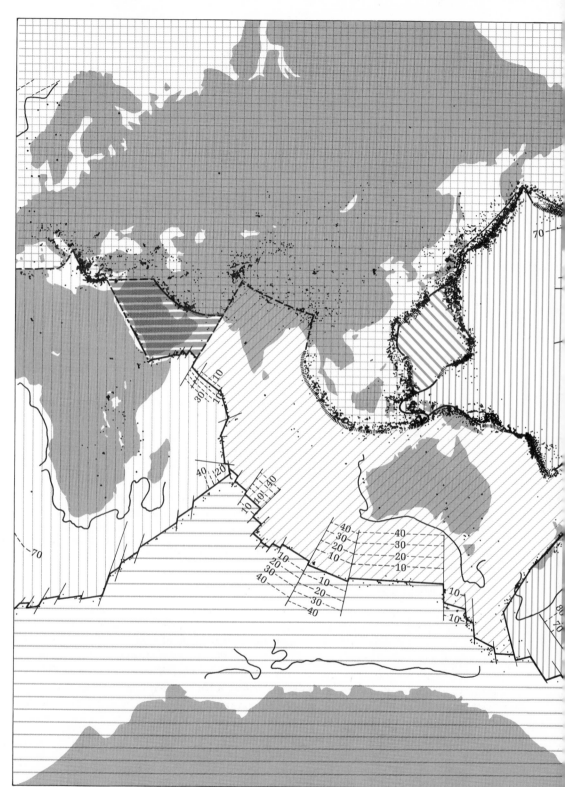

Figure 16-4. The crustal plates. In general, each plate begins at a ridge (heavy line) and moves toward a trench (hatching). Earthquake epicenters (dots) tend to outline the plates. Transform faults are shown by thin lines. The

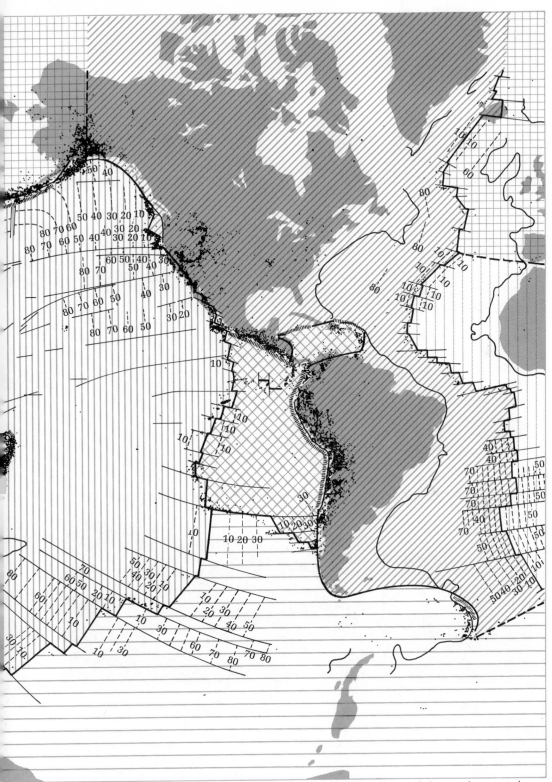

dashed lines show the ages (in millions of years), based on magnetic data. The information and interpretation are from many sources, mainly workers at Lamont-Doherty Geological Observatory.

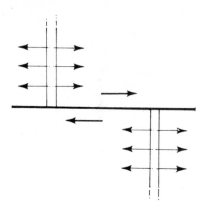

A.

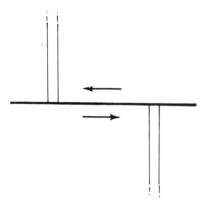

B.

Figure 16-5. Transform fault compared with simple fault. A. Transform fault. The relative movement on the transform fault is caused by the new sea floor created at the two segments of the ridge. The transform fault is shown by the single line, and the ridge

by the double line. B. Two segments of once-continuous ridge separated by a simple fault. Note that the relative movement, shown by the arrows, is opposite to that of the transform fault shown in A.

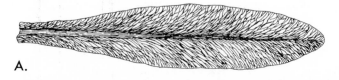

A.

B.

Figure 16-6. Evidence for continental drift. A. *Glossopteris.* This leaf has been found in beds overlying ancient glacial deposits in India, Australia, South Africa, and Antarctica. B. Glacial polish in Australia giving clear evidence of ancient glaciers there. Ice moved to-

ward upper right. C. Striated boulder from ancient glacial moraine in Australia. D. Glacial polish from glaciers of the same age in Africa. Etching by Bushmen on left. Parts B., C., and D., photos by Warren Hamilton.

vertical; and in between, it varies. The intensity of the field at the poles is about twice that at the equator. At some places records of the direction and inclination of the earth's field have been kept for hundreds of years, and they reveal changes in direction and inclination. (See Fig. 16-10.) More recent studies also show variation in intensity. Study of the remanent magnetism of rocks, to be discussed below, indicates that several times in the geologic past the direction of the earth's field has reversed, i.e., the north pole of a compass needle would point toward Antarctica. The timing of these reversals in the last few million years is well enough known that they can be used to date the rocks. (See Fig. 16-11.) Any theory for the origin of the earth's magnetic field must explain these observations.

The earth's magnetic field is believed to originate from electric currents in the core. It is a well-known law in physics that a moving current creates a magnetic field, and conversely, a current is induced into a conductor moving in a magnetic field. Electrical currents in the core could start in a number of ways, such as through a weak battery action caused by compositional differences. Once started, these currents could be amplified by a dynamo action in the earth's core. A dynamo produces electric current by moving a conductor in a magnetic field. In the earth the core is the conductor, and because it is fluid, it can move. The mechanism envisioned is similar to a dynamo which produces an electric current, and this current is fed to an electric motor that drives the dynamo. Friction and electrical resistance would prevent such a perpetual motion machine from continuing to operate unless more energy is added. In the earth this

C.

D.

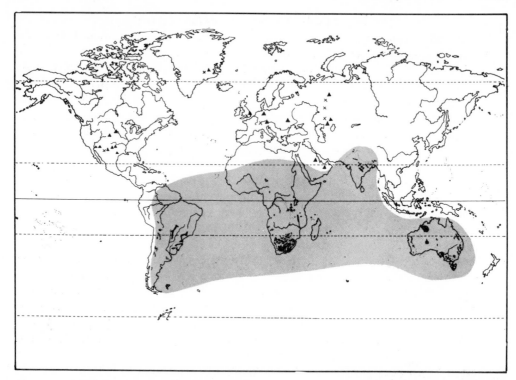

Figure 16-7. Map showing the present distribution of the ancient glaciation of the southern hemisphere. The area involved in the hypothetical southern continent existing before drifting is stippled. The areas of glacial deposits are indicated in black and extend from north of the equator to near the south pole. The arrows show the direction of glacial movement. Reefs (crosses) and evaporites (triangles) far to the north indicate much warmer climates. Reefs and evaporites from Opdyke, 1962.

Figure 16-8. Reconstruction of the ancient continent of the southern hemisphere. The glaciers and their directions of movement are indicated.

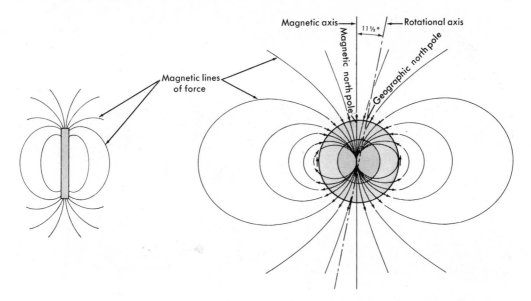

Figure 16-9. The earth's magnetic field is much like that of a bar magnet. The magnetic poles do not coincide with the geographic poles that are de-termined by the rotation axis. The short, heavy arrows show the vertical direction in which a compass free to move in the vertical plane will point.

additional energy can come from movement of the fluid core. Calculation shows that very little additional energy needs to be added in the case of the earth; and the necessary energy for core motion, probably in the form of convection, could come from many sources.

Thus, a current is somehow started in the earth's core, and this current produces a magnetic field. The conducting core moves through the magnetic field, and a current is induced in this part of the core, starting the whole process over. (See Fig. 16-12.) In a stationary earth, the movements of the core would probably be more or less random and the magnetic fields produced would probably cancel each other. The rotation of the earth tends to orient the motions of the core so that the motions and, therefore, the currents are in planes perpendicular to the rotational axis. Viewed from outside the earth, the overall current produced would be parallel to latitudes. Such currents produce a dipole magnetic field centered on the rotational axis.

A field produced in this way would have all of the characteristics of the earth's magnetic field. The main field would be a dipole field; superimposed on this would be other fields caused by variations in the core's motion. This can account for the variations in intensity and location of the earth's field. The earth's dipole field does not coincide with its rotational axis, however, as this theory predicts, although if averaged over a long enough period, the dipole field would perhaps appear centered on the rotational axis. The positions of the magnetic pole deduced from remanent magnetism of Pleistocene and younger rocks cluster around the rotational axis, suggesting that this is true. The present dipole field is 11.5 degrees from the rotational axis and lies northwest of Greenland, some distance from the magnetic north pole north of Canada.

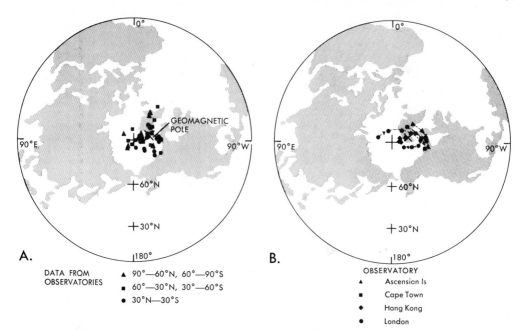

A.

DATA FROM   ▲ 90°—60°N, 60°—90°S
OBSERVATORIES   ■ 60°—30°N, 30°—60°S
                   ● 30°N—30°S

B.

OBSERVATORY
        ▲   Ascension Is
        ■   Cape Town
        ◆   Hong Kong
        ●   London

Figure 16-10. Variations in the earth's magnetic field. A. Variations in position of magnetic north pole computed in 1945 from direction and inclination of the earth's magnetic field at various places. These calculations were made assuming a dipole field and show the effect of the non-dipole part of the earth's field. B. Changes in the position of the magnetic north pole with time. Time between points is 40–50 years. In both figures the cross shows the position of the north geographic pole, and the X, the average 1945 position of the north geomagnetic pole. From Allan Cox and R. R. Doell, *Bulletin,* Geological Society of America, Vol. 71, 1960.

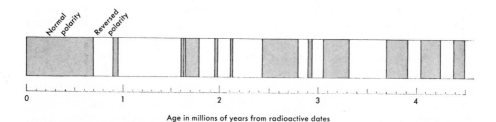

Age in millions of years from radioactive dates

Figure 16-11. A few of the many known reversals of the earth's magnetic field. From data by Cox.

Remanent Magnetism

Much of the recent data on continental drift have come from measuring the remanent magnetism of rocks. Remanent magnetism is the magnetism frozen into a rock at the time the rock forms. For instance, in the cooling of a basalt, when the temperature

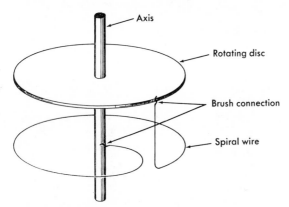

Figure 16-12. The disc dynamo illustrates in a general way how the earth's magnetic field is believed to originate. The disc is attached to the axis and rotates with the axis. The spiral wire is in electrical contact with the axis and the disc by means of brushes. To start this self-sustaining dynamo, suppose that a current is flowing in the spiral wire. This current produces a magnetic field; the disc is a conductor moving through the field, and so a current is induced in the disc. This current flows to the spiral wire by means of the brush and so continues the processes.

reaches a certain point, which is well below the crystallization temperature, the magnetic minerals, mainly magnetite, are magnetized by the earth's field. On cooling below this temperature, their magnetism is not greatly affected except by very strong fields, many times stronger than any in the earth. Thus it is possible to measure the direction and inclination of the earth's field at the time the basalt cooled. The inclination can be used to calculate the distance to the magnetic pole.

Remanent magnetic data for rocks of different ages have been collected on each of the continents. The study of magnetic data from old rocks is called *paleomagnetism* (old magnetism). The magnetic poles for each age (geologic period) on each continent group in a fairly small area. However, in general, for any given continent, each geologic period has a different location for the magnetic pole, and the poles for any period fall in different locations for each continent. (See Fig. 16-13.) Here, then, is evidence of continental drift. If the magnetic pole moved, it should be at the same place during each period for each continent; but instead, each continent gives a different location for each period. Thus, if the interpretation of the paleomagnetic data is correct, the continents must have moved relative to each other. The magnetic pole may also have moved. Note in Fig. 16-13 that the magnetic data indicate no drift of North America relative to Europe since Eocene.

Paleomagnetic data have certain limitations. They tell the direction and distance to the magnetic pole. Thus they show the rotation of a continent and its latitude. (See Fig. 16-14.) They do not indicate the longitude, and the continent may be anywhere on the indicated latitude. This gives the imagination rather free reign in proposing reconstructions, particularly since many of the rocks are incompletely mapped and dated. (See Figs. 16-15 and 16-16.) In spite of the close grouping of poles for a given period, the possibility remains that the magnetic pole may not have been coincident with the geographic pole. At present, they differ by 11.5 degrees.

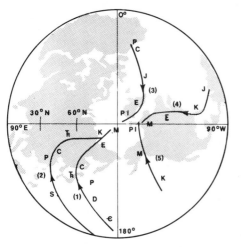

(1) Europe
(2) North America
(3) Australia
(4) India
(5) Japan

Pl	Pliocene	About	6 million years ago.
M	Miocene	"	18 " " "
E	Eocene	"	50 " " "
K	Cretaceous	"	100 " " "
J	Jurassic	"	160 " " "
Ŧ	Triassic	"	200 " " "
P	Permian	"	250 " " "
C	Carboniferous	"	310 " " "
D	Devonian	"	375 " " "
S	Silurian	"	420 " " "
€	Cambrian	"	550 " " "

Figure 16-13. On this diagram, for each continent the position of the magnetic pole relative to the present position of each continent for each geologic period is indicated by the letters. The lines drawn through these locations show the apparent drift of the magnetic pole for each continent. The diagram indicates that movement of the continents relative to each other must have occurred but does not show the route of drifting because of the limitations of paleomagnetic data shown in Figure 16-14. From Allan Cox and R. R. Doell, *Bulletin*, Geological Society of America, Vol. 71, 1960.

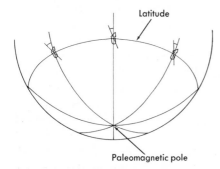

Figure 16-14. Paleomagnetic data can only show the latitude and orientation of a continent. The horizontal component of the paleomagnetism points the direction toward the pole, and the vertical inclination gives the latitude. The continent could have been in that orientation anywhere on that latitude.

Pro and Con

We will now examine the evidence for and against continental drift and discuss a few of the proposed schemes of drift.

Pro

The evidence for continental drift includes the following:

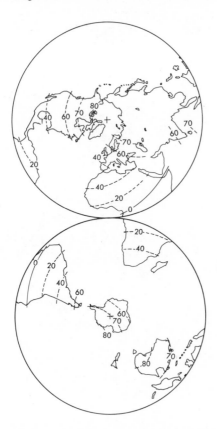

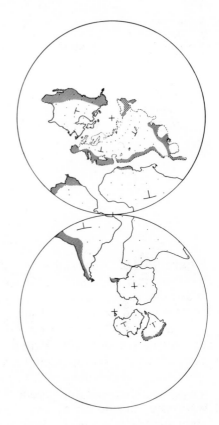

Figure 16-15. Reconstruction of the positions of continents in the geologic past from paleomagnetic data. A single geologic period (Jurassic) is shown. On the left the continents are shown in their present locations with the Jurassic latitudes computed from paleomagnetic data indicated. Obviously many locations of each continent are possible on the basis of these data. On the right the probable locations are shown. They are based on paleomagnetic and other data such as reasonable movements from the position during the last and to the next geologic period. Figure 16-16 shows the positions of the continents during other geologic periods. From D. Van Hilten, *Tectonophysics,* Vol. 1, 1964.

Paleomagnetism is discussed above.

Fit of the Continents. It has been said that it is too good to be accidental, and recent matches made by computers on submarine contours to include the continental shelves seem to bear this out. Folding, faulting, and intrusion since drifting apart (about 250 million years) would have modified the shapes of the continents.

Climate. Glaciation was discussed above.

In Antarctica, before that glaciation, the climate was much different from the present. Rocks bearing shallow marine fossils have been found, and at another place a thousand miles away dune sands suggesting deserts are widespread.

R   Recent
T   Tertiary
E   Eocene   About 50 million years ago.
K   Cretaceous   "   100   "   "   "
J   Jurassic   "   160   "   "   "
ƫ   Triassic   "   200   "   "   "
P   Permian   "   250   "   "   "

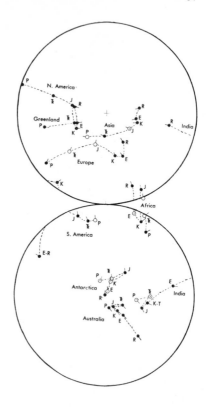

Figure 16-16. Reconstruction of the movements of the continents since Permian time. The circles are located at the centers of the continents. The tails attached to the circles give the orientation of the present north direction on each continent. The open circles show locations not in agreement with paleomagnetic data, and the open circles with dots indicate no available paleomagnetic data. From D. Van Hilten, *Tectonophysics,* Vol. 1, 1964.

It is hard to imagine that the *Glossopteris* flora recently found 210 miles from the South Pole in Antarctica would grow there in that region of cold, months-long nights as well as in tropical India. Other temperate- and warm-climate-indicating fossils, including coal beds, have been found in Antarctica. Some plants do grow inside the Arctic circle at the present time.

The distribution of evaporites through geologic time also suggests drift and appears to be in harmony with the magnetic data. Evaporites suggest hot arid conditions, and deserts occur today between 30 degrees north and south of the equator. Extensive salt deposits are not forming at present, however. Ancient salt deposits of approximately the same age as the glaciation are found in northeast Greenland, arctic Canada, Siberia, Germany, Kansas, and New Mexico, as well as other places. (See Fig. 16-7.)

Coral reefs are also currently restricted to about 30 degrees north and south of the equator, but ancient reef deposits are found far to the north.

One thing to keep in mind in evaluating the climatic evidence is that the last glaciation, the Pleistocene, occurred in what is now the temperate zone. Thus in the last million years the climate has changed; and so the climate could have changed in the past although it is difficult to imagine continental glaciation at the equator, even though mountain glaciers occur there now.

Apparent Rifting of the Continents is the probable origin of the Red Sea. (See Figs. 16-17 and 16-18.) The Gulf of California and the African Rift valleys may also have originated in this way. If a small sea can form in this way, it seems possible that a larger ocean, such as the Atlantic, could also.

Rock Layers and their Fossils. Piecing together the continents is like putting together a jig-saw puzzle whose pieces do not quite match. Fitting together the various layers of sedimentary rocks is like putting together a multi-layer jig-saw puzzle with many of the pieces missing. The generalized stratigraphy of the southern continents is shown in Fig. 16-19. The similarity among these continents in the Carboniferous Period is striking. This similarity persists through the Jurassic in all but Australia. This stratigraphy and fossils in Australia suggest that it became isolated from the other continents after the glaciation and that it has remained isolated to the present. The Australian fauna and flora of the present appear to have migrated from time to time very selectively from mainland Asia. The present flora and the marsupial fauna are very different from those of the other continents. The placental mammals that populate the other continents did not reach Australia.

Figure 16-17. View northeast across the Sinai Peninsula taken from Gemini XI. The southeast end of the Mediterranean Sea is in the upper left, and the north end of the Red Sea is in the lower right. The Sinai Peninsula is bordered by the Gulf of Suez and the Gulf of Aqaba. The Gulf of Aqaba is part of a linear depression on which the Dead Sea can be seen. The Red Sea may be a rift that originated when Africa and Arabia separated, and the two Gulfs shown here may be related to that fracture. Photo from NASA.

Figure 16-18. Looking northeast at the south end of the Red Sea and the Gulf of Aden, taken from Gemini XI at an elevation of 470 nautical miles above Africa that forms the foreground. The Indian Ocean is to the right, and the southwest end of the Arabian Peninsula forms the top of the picture. The Red Sea and the Gulf of Aden may be fractures that formed when Africa and Arabia separated. The part of Africa where the Red Sea narrows (Afar Triangle) is recently uplifted ocean floor. Photo from NASA.

The *Glossopteris* flora found just above the glacial deposits was cited earlier as evidence of previous connection among the southern continents. It is possible that seeds (spores) could have floated across the oceans to distribute these plants, but it seems rather unlikely. Today, a newly formed volcanic island soon develops plants, but the seeds are carried by birds. The first flying animals did not appear until tens of millions of years after the supposed separation and drift. Land bridges and island chains have been suggested in the past, but the known oceanic structure precludes this possibility.

Glossopteris leaves a foot long, tree trunks two feet in diameter, and coal seams 25 feet thick in Antarctica, within three degrees of the south pole, indicate, at the least, a marked change in climate. The similarity of the faunas in the underlying (Devonian) rocks strongly suggests that Antarctica, Africa, and South America were once a single continent. These faunas are so similar that the species cannot be distinguished.

The swimming reptile *Mesosaurus,* found in both South America and south Africa, is also strong evidence favoring connection of these two continents. Although a swimmer, this aquatic reptile lived in fresh water, and it is very unlikely that it could swim across an ocean. Other reptiles that lived during the time of the alleged connection have been found on the southern continents. The distribution of some of these reptiles requires a single connected continent.

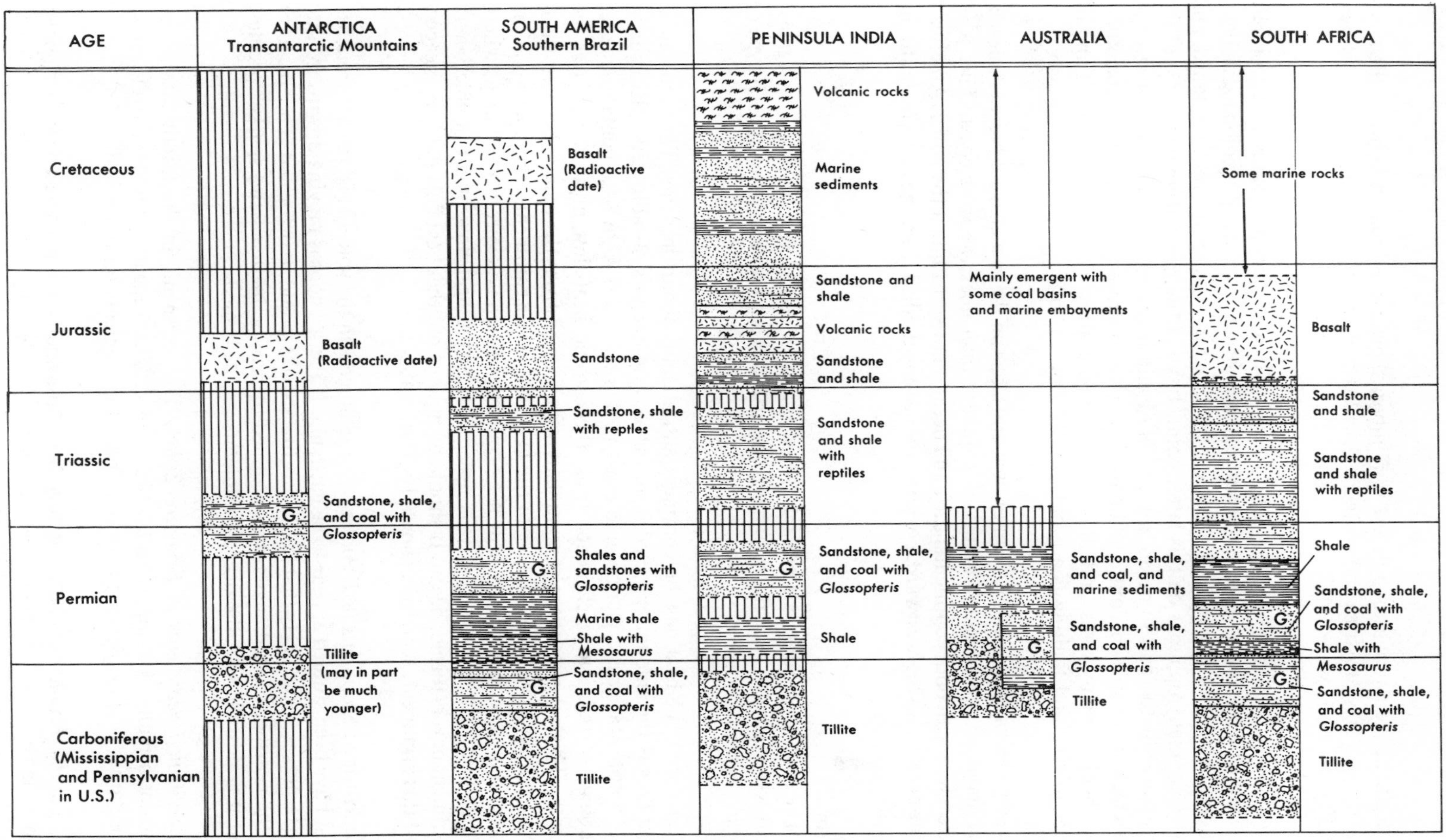

Figure 16-19. Simplified stratigraphic columns in the southern hemisphere. The age terms are described in Chapter 17. Vertical lines indicate no rocks of that age.

Radioactive dates, although few in number, show a good match between the basement rocks of west Africa and northern Brazil. The dates obtained on the basalts shown in Fig. 16-19, however, do not suggest a close correspondence.

Structures. Another test is to see if the folds, faults, and other structures of appropriate age match when the continents are put together. Much diversity of opinion exists here, in part because results depend to a great extent on how the continents are joined, and there are many opinions on this latter point. In all of the reconstructions, some of the structures match and some do not, so here again it becomes somewhat subjective as to which structures are considered most important. Dating the formation of structures is difficult in many cases, and allowance for modification by the proposed drift and by later folding and faulting is a problem. (See Fig. 16-20.)

Oceanic Sediments. The continental shelf and slope off the east coast of North America has about six times more volume of sediment than does the west coast. This is in agreement with both continental drift and ocean-floor spreading if the North American continent is moving westward, covering the sediments shed from the continent into the Pacific.

Con

The evidence against continental drift will now be considered. Some points against drift were included with the arguments for drift to keep the discussion of each type of evidence in one place.

Heat Flow. The heat flow on continent and ocean is about the same. The continental heat flow is the amount expected from the known amounts of radioactive elements in the continental crust. The oceanic crust is thinner and is known to contain smaller amounts of radioactive elements; thus, the oceanic heat flow must come from the mantle. This suggests that the continents were formed from the mantle beneath them because if the mantle contributes to the heat flow under continents, the flow would be higher than it is. Secondly, if drift has occurred, one would expect to find areas where the uniform pattern of heat flow is upset; that is, the heat flow might double where continent covers former ocean. It should be pointed out that heat flow data are meager, and this argument, therefore, should be held in abeyance until more data are collected.

Undeformed Sediments in Trenches. The sediments in the volcanic island arc trenches that have been studied appear to be flat-lying. If the sea floor is moving downward into the trench, it seems likely that the sedimentary rocks would be deformed. The structure of the near-bottom sediments is determined by seismic methods and sparker profiles, and it is possible that the deeper sediments are deformed.

Undeformed Sediments in Fracture Zones. The sediments in the fracture zones or transform faults also appear to be undeformed. This suggests that the movement in those areas is slow. Seismic data does reveal movements on these faults.

Geographic Problems. The fitting together of the main continents was described above and is fairly good. Problems arise for some smaller but significant areas. In most reconstructions of the pre-drift continents, there is no room for Central America

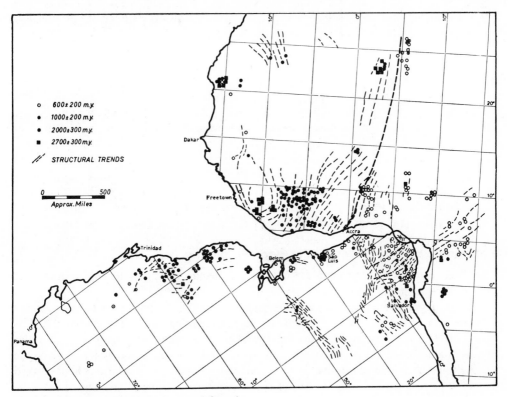

Figure 16-20. West Africa and South America fitted together, showing the coincidence in both structural trends and radioactive ages of the rocks (m. y. is million years). From P. M. Hurley and others, *Science* (August 4, 1967), Vol. 157, p. 496.

because North and South America are too close together. Spain also overlaps, and so in reconstructions it must be bent into the area now occupied by the Bay of Biscay.

Structures of Continental Interiors. One of the important failings of the theory of sea-floor spreading is that it does not explain structures not on the margins of continents. Sea-floor spreading can explain geosynclines on continental margins, but similar structures also occur in the interior of continents. An example is the basin that occupied the area of the Rocky Mountains in part of the Mesozoic Era. This feature had many similarities to geosynclines, and parts of it were later deformed. Thus not even all geosynclines or geosyncline-like areas can be explained by sea-floor spreading, but most true geosynclines do fit into the theory.

The evidence for continental drift is strong, as is the evidence for sea-floor spreading. Together they form a world system of crustal structure. It is now possible to relate the structures of the oceans and the continents.

The movements of the continents through geologic time are outlined in a general way in Part IV.

Origin of Crustal Structures

In this section an attempt will be made to find a mechanism that will account for the structures of the crust. Folded rocks suggest shortening, and one of the first theories suggested that this was caused by shrinking of the earth due to cooling. This contraction hypothesis had to be abandoned when calculation showed that thermal contraction could not be great enough, and when very old glacial deposits were discovered, showing that the climate has not changed radically since then. The features that need to be explained, in addition to the compressive structures of the continents, include geosynclines and the features of ocean-floor spreading.

Relationship of Structures to Sea-floor Spreading

Geosynclines are the main features of continental geology and appear to be the mechanism of continent formation. Sea-floor spreading is the main process in the oceans and accounts for the formation of oceanic crust. Geosynclines form on ocean-continent boundaries and appear to be related to sea-floor spreading as shown in Fig. 16-21. The sedimentation phase of the geosyncline is the development of the familiar continental shelf, continental slope, and continental rise. The sedimentary rocks on the shelf and the slope are similar in type and thickness to those found on the landward sides of geosynclines—that is, beds of shale, sandstone, and limestone. The rocks of the continental rises are very different and are largely emplaced by turbidity currents that flow down the continental slope. These rocks are poorly sorted, have graded bedding and other features mentioned in Chapter 4, as well as some features of deep-ocean sediments. The rocks of continental rises resemble the rocks on the oceanward side of geosynclines in all features except their lack of volcanic rocks.

The development of a trench and a volcanic island arc can be the answer to that problem. To form a trench, the oceanic crust must move downward. It seems likely that the oceanic crust could not bend down sharply near the trench but would move downward under the continent in the zone of the deep-focus earthquakes (Fig. 16-21B). The higher temperature in the mantle would melt the crustal material and so produce the magma that rises to form the volcanic island arc. It seems likely that some deformation of the sedimentary rocks would occur in the formation of a volcanic island arc and by subsequent movement of the descending oceanic plate.

The collision of a second continental plate with a geosyncline is another way in which a geosyncline can be deformed. In this case, the compression will result in thickening of the crust, and isostasy will cause a high mountain range to form. The Himalaya Mountains are believed to have formed in this way. The heat and friction of such a collision could also cause metamorphism and generation of magma. In this way new continental crust could be created. The oceanic side of geosynclines is generally most deformed and the site of most metamorphism and igneous activity. The sedimentary rocks on the continental side tend to be folded, and in some cases thrust faulted, but they lack the extreme deformation of the oceanic side. This may be because they were

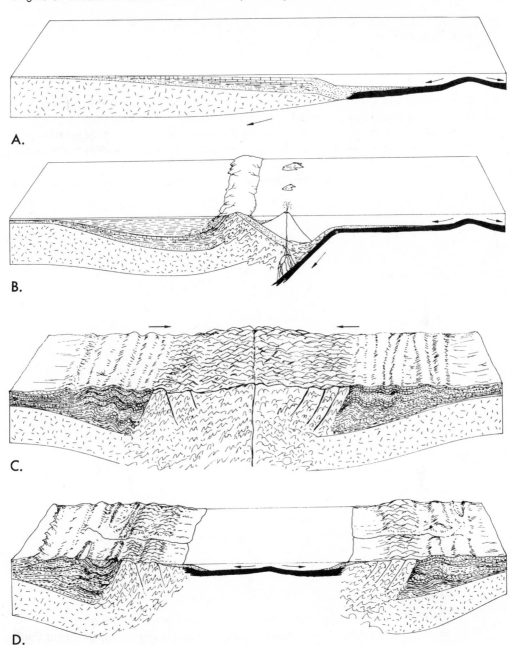

A.

B.

C.

D.

Figure 16-21. Geosynclines and sea-floor spreading. A. Continental shelf, slope, and rise. B. Development of volcanic island arc. C. Collision of continents. D. Separation of continents by the development of a new oceanic rise.

The events shown in these diagrams are an interpretation of the history of the Appalachian Mountains, and these events are described in Chapters 21 and 22.

deposited on continental shelves and so were above continental crust that protected
them from extreme deformation.

Thus crustal deformation may occur where two rigid plates, moving in opposite
directions, meet. If the collision is between two plates of oceanic crust, one will move
downward, forming a trench. If one of the plates is continental crust, it will not move
down because of its buoyancy; the oceanic plate will move downward, forming a trench.
If two continental plates meet, a large mountain range, such as the Himalaya Mountains, will form because of the compressive thickening of the crust. Where an oceanic
plate sinks, it and its associated oceanic sediments will be partially melted by the higher
temperatures in the mantle and by the frictional heat generated. The result will be the
andesitic volcanic rocks that form the volcanic islands associated with trenches. (See
Fig. 16-22.)

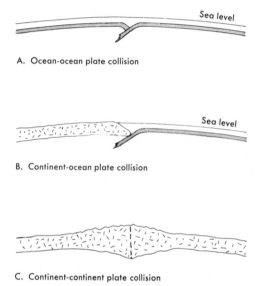

A. Ocean-ocean plate collision

B. Continent-ocean plate collision

C. Continent-continent plate collision

Figure 16-22. Types of collisions of
plates. A. ocean-ocean. B. continent-
ocean. C. continent-continent.

Convection Currents

Convection currents have been proposed as the cause of most geological features.
Convection cells develop when a fluid is heated at the bottom. As the warmed fluid
expands, it becomes less dense and tends to rise; denser, cooler fluid at the top moves
down to take the place of the rising, warm fluid, as shown in Fig. 16-23. In this process
heat is transported upward by the movement of the warmed fluid.

Although there is much disagreement on the viscosity of the mantle, convection has
been explored theoretically by a number of investigators who have established that it
could exist in the mantle, even though the mantle is viscous enough to behave as a solid
in transmission of earthquake waves. A substance may respond elastically to a rapid
stress but may be deformed viscously by a stress applied over a long period of time.

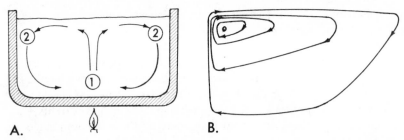

Figure 16-23. A. Two simple convection cells produced by heating at the bottom. Warm fluid at ① rises and cool fluid from ② falls to take the place of the rising warm fluid. B. Convection in a fluid whose viscosity depends on temperature and pressure. Heating is on the bottom.

The latter case could occur in the mantle, especially in view of the temperatures and pressures there. Thus it is possible for a solid to flow at temperatures well below the melting point. A possible example of this behavior is the slow movement of glacial ice under the influence of gravity and its very different behavior when subjected to a rapid stress. Convection currents are believed to move at the rate of one to a few centimeters per year.

Convection currents in the mantle could deform the crust or carry crustal plates. The convection could involve all or part of the mantle or just the low seismic velocity zone. The liquid core may be the heat source for convection, if convection does exist in the mantle. The development of the core is part of the early history of the earth described in Chapter 18.

The earliest model of the convection currents that might cause a geosyncline or deep sea trench is the meeting of two downward flowing currents. In the case of the deep trench in front of an island arc, such a downward buckle would also account for the negative gravity anomaly because light crustal material would replace heavier mantle rock. Notice that such an arrangement might also account for compressive folding of the material collected in the depression. This could be accomplished if, at a late stage, the horizontal parts of the convection cells pushed the sides of the depression together. (See Fig. 16-24.)

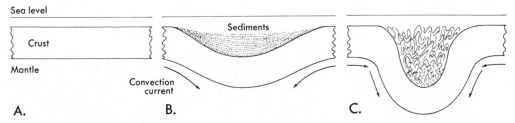

Figure 16-24. Geosyncline caused by convection currents. A. Initial condition. B. Downwarp caused by descending convection currents. C. At a later time, after geosyncline has filled with sediment, the horizontal component of the convection current squeezes the geosyncline. Thinning of the crust by downwarping may help to make this possible. The deep part of the geosyncline may be the site of metamorphism due to the increased pressure and temperature.

Many features of the mid-ocean ridges can also be explained by convection currents. The high heat flow suggests upwelling convection currents in the mantle. The convection currents could cause the rise and the consequent crustal thinning, as well as the faulting and earthquake epicenters. The upwelling mantle material could melt partially as a result of the reduced pressure near the crust-mantle boundary, forming basalt. The basalt would become new ocean floor. (See Fig. 16-25.) Horizontal currents could cause transform faults, continental drift, and sea-floor spreading. Descending currents could cause trenches and volcanic island arcs. That many parts of the mid-ocean ridges do not have a high heat flow would be explained if different parts of the ridge formed at different times. This seems likely in a feature of this size, and differences in topography may indicate differences in age.

Recently, however, some studies have suggested that convection might not be possible in the deep mantle. It also appears that thin convection cells, as extensive as the plates involved in sea-floor spreading confined to the low seismic velocity zone, would not be stable but would break up into much smaller cells. For these reasons it has been suggested that the plates, extending down to the weak zone in the upper mantle, move as units, and the return flow is in the weak zone of the upper mantle. Such movement resembles convection (Fig. 16-26). The movement of the plates may be caused by the addition of material to the crust at the mid-ocean ridges and/or by the weight of the down-sinking leading edge of the plate at the island arc pulling the plate (Fig. 16-25).

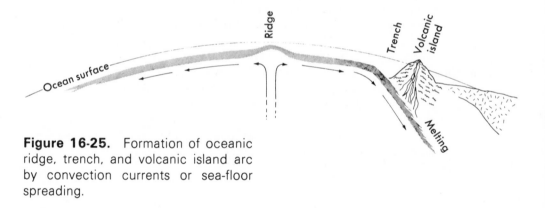

Figure 16-25. Formation of oceanic ridge, trench, and volcanic island arc by convection currents or sea-floor spreading.

Energy

Large amounts of energy are necessary to deform rocks, generate magma, and cause metamorphism. The heat flow from the earth indicates the presence of sources of heat energy large enough to account for geologic processes. The total heat flow from the earth is about 1,000 times more than the energy of earthquakes and 100 times more than volcanism, and is enough to cause geologic processes. Heat flow is calculated from temperature measurements in drill holes and on the bottom of the ocean. Reliable heat flow measurements were first made on the continents in the mid 1930's and at sea in 1950. An obvious source of the heat flow is the heat generated by the decay of radioactive elements. The known radioactive content of crustal rocks of the continents is enough to account for the continental heat flow. The average heat flow from the oceans is about the same as the average from the continents. However, the thin layer of oceanic basalt does not contain enough radioactive elements to account for the oceanic heat

Table 16-1. Summary of heat flow measurements. The units are micro-calories per square centimenter per second (10^{-6} cal/cm²/sec). From W. H. K. Lee and Seiya Uyeda, "Review of Heat Flow Data," in *Terrestrial Heat Flow*, American Geophysical Union, Geophysical Monograph 8, 1965.

Average: 1.5 ± 0.15 (Same on continents and oceans)	
Continents	
Undeformed areas such as interior lowlands	1.54 ± 0.38
Areas of metamorphic rocks deformed more than 600 million years ago	0.92 ± 0.17
Areas deformed between 600 and 225 million years ago	1.23 ± 0.40
Areas deformed less than 225 million years ago	1.92 ± 0.49
Oceans	
Trenches	0.99 ± 0.61
Floors	1.28 ± 0.53
Ridges	1.82 ± 1.56

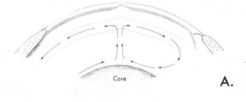

A.

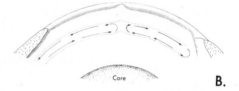

B.

C.

Figure 16-26. Convection currents. A. Deep convection involving much or all of the mantle. B. Thin convection confined to the low-velocity zone of the mantle. C. A convection-like flow in which the rocks above the low-velocity zone move and the return flow is in the low-velocity zone.

flow. Therefore, some of the heat flow from the oceans must come from the mantle, and this is especially true of the mid-ocean ridges such as the East Pacific Rise, where the heat flow is very high.

Radioactive elements are fairly evenly distributed in the continental crust; and geologically active areas, such as volcanoes and mountain belts, are no more radioactive than other areas. Radioactive energy is enough to cause all geologic processes, but the distribution of radioactivity suggests that it is not the energy source, so other sources must be considered.

Some of the heat reaching the surface seems to come from deep in the earth. The deep source of heat is believed to be the liquid core. The heat could be released by crystallization of the core, and the inner core, believed to be solid, may be slowly crystallizing. The internal heat could also come from recrystallization in the mantle, but we know very little about either the composition or the physical conditions there.

QUESTIONS

16-1. What is the evidence for sea-floor spreading?
16-2. Describe transform faults.
16-3. What are the components of the earth's magnetic field?
16-4. What changes have occurred in the earth's magnetic field?
16-5. How is the earth's magnetic field believed to be caused?
16-6. How is remanent magnetism impressed on rocks?
16-7. What data on the previous position of a continent is obtained from remanent magnetism?
16-8. How does remanent magnetism favor sea-floor spreading?
16-9. What evidence suggests that the southern continents were once joined?
16-10. What evidence suggests that drift alone may have occurred (as distinct from showing that the continents were once joined)?
16-11. What evidence do you feel is strongest in suggesting continental drift? Which against?
16-12. Discuss whether all of the lines of evidence that point toward continental drift should be considered as proof, or whether each piece of evidence must be evaluated individually and must stand alone or be dropped.
16-13. Where do magmas originate? What is the evidence?
16-14. Where does the energy come from that causes the deformation of rocks?
16-15. What are convection currents? What is the evidence for them in the mantle?
16-16. How could geosynclines be formed by tension? What kind of forces cause the deformation of geosynclines?

SUPPLEMENTARY READING

Bullard, Sir Edward, "The Origin of the Oceans," *Scientific American* (September, 1969), Vol. 221, No. 3, pp. 66-75.
Cox, Allan, G. B. Dalrymple, and R. R. Doell, "Reversals of the Earth's Magnetic Field," *Scientific American* (February, 1967), Vol. 216, No. 2, pp. 44-54.

Cox, Allan, and R. R. Doell, "Review of Paleomagnetism," *Bulletin,* Geological Society of America (June, 1960), Vol. 71, No. 6, pp. 645-768.

Dickinson, W. R., "Plate Tectonics in Geologic History," *Science* (October 8, 1971), Vol. 174, No. 4005, pp. 107-113.

Dietz, R. S., "Geosynclines, Mountains and Continent-Building," *Scientific American* (March, 1972), Vol. 226, No. 3, pp. 30-38.

Doumani, G. A., and W. E. Long, "The Ancient Life of the Antarctic," *Scientific American* (September, 1962), Vol. 207, No. 3, pp. 169-184. Reprint 863, W. H. Freeman and Co., San Francisco.

Eardley, A. J., "The Cause of Mountain Building—An Enigma," *American Scientist* (June, 1957), Vol. 45, No. 3, pp. 189-217.

Elsasser, W. M., "The Earth as a Dynamo," *Scientific American* (May, 1958), Vol. 198, No. 5, pp. 44-48. Reprint 825, W. H. Freeman and Co., San Francisco.

Heirtzler, J. R., "Sea-Floor Spreading," *Scientific American* (December, 1968), Vol. 219, No. 6, pp. 60-70.

Hurley, P. M., and J. R. Rand, "Pre-drift Continental Nuclei," *Science* (June 13, 1969), Vol. 164, No. 3885, pp. 1229-1242.

Jacobs, J. A., *Earth's Core and Geomagnetism.* New York: Pergamon Press, Inc., 1963, 201 pp. (paperback).

Kummel, Bernhard, *History of the Earth,* Chapter 11, "Gondwana Formations," pp. 328-347. San Francisco: W. H. Freeman and Co., 1961, 610 pp.

Kurtén, Bjorn, "Continental Drift and Evolution," *Scientific American* (March, 1969), Vol. 220, No. 3, pp. 54-64.

Maxwell, J. C., "Continental Drift and a Dynamic Earth," *American Scientist* (Spring, 1968), Vol. 56, No. 1, pp. 35-51.

Menard, H. W., "The Deep Ocean Floor," *Scientific American* (September, 1969), Vol. 221, No. 3, pp. 126-142.

Orowan, Egon, "The Origin of the Oceanic Ridges," *Scientific American* (November, 1969), Vol. 221, No. 5, pp. 102-119.

Takeuchi, Hitoshi, Seiya Uyeda, and Hiroo Kanamori, *Debate About the Earth.* San Francisco: Freeman Cooper & Co., 1967, 253 pp. (paperback).

Tazieff, Haroun, "The Afar Triangle," *Scientific American* (February, 1970), Vol. 222, No. 2, pp. 32-40.

Vine, F. J., "Sea-floor Spreading—New Evidence," *Journal of Geological Education* (February, 1969), Vol. 17, No. 1, pp. 6-16.

Wilson, J. T., "Continental Drift," *Scientific American* (April, 1963), Vol. 208, No. 4, pp. 86-95. Reprint 868, W. H. Freeman and Co., San Francisco.

_____, "Some Aspects of the Current Revolution in the Earth Sciences," *Journal of Geological Education* (October, 1969), Vol. 17, No. 4, pp. 145-150.

PART 4

EARTH HISTORY

Geology is an historical science, and in Part IV this facet is explored. The concept of geologic time and its measurement, one of the most important aspects of geology, is described in Chapter 17. The origin and early history of the earth are astronomical topics and are discussed in Chapter 18. Some biologic aspects of geology are the topic of Chapter 19. The following five chapters are the interpretation of geologic history. The final chapter is on the geologic future and environment.

Chapter 17

Principles of Historical Geology

The result, therefore, of our present enquiry is, that we find no vestige
of a beginning,—no prospect of an end.

James Hutton, 1788

Historical geology, the history of the earth, is perhaps the most important goal of geology. In this chapter the methods used to determine geologic history will be described. Some of these methods have already been discussed, such as interpretation of the formation of a rock based on study of the rock itself and its relationship to other rocks. Simple examples are the interpretation of a sandstone composed of well-rounded quartz grains as probably the product of deep weathering and long transport, and recognition that an intrusive rock is younger than the rock it intrudes. Geologic history is determined from interpretation of the way rocks were formed and the dating of these events. This chapter is concerned mainly with methods of dating so that rocks and events can be placed on a time scale.

Geologic Dating

Rocks can be dated both relatively and absolutely. Relative dates are of two types. The simpler of these is the determination of the local sequence based on relationships such as the fact that the oldest beds are at the bottom of undisturbed sedimentary rocks. Question 28 at the end of Chapter 14 illustrates the method. Determining these relationships by mapping the rocks—that is, plotting on an accurate map the positions of rock bodies—is the main method of geologic study. The second way of dating rocks is by referring them to a geologic time scale based on the study of the fossils in the rocks.

Absolute dating is a more recent development and is based on radioactive decay of elements in rocks; this method has been used to calibrate the relative geologic time scale developed from the study of fossils.

Geologic Time

The recognition of the immensity of geologic time and the development of methods for subdividing geologic time are among the great intellectual accomplishments of man. It is difficult to discuss the development of ideas in generalities, but a brief outline is presented here. Discoveries made near the end of the eighteenth century led to rapid progress, although there had been earlier studies. The culmination of this early work was the theory now known as *uniformitarianism,* published by the Englishman James Hutton in 1785 and 1795. Uniformitarianism simply means that the present is the key to the past; that is, the history of old rocks can be interpreted by noting how similar rocks are being formed today. Hutton's book was poorly written, and the theory did not gain wide acceptance until it was popularized by his colleague Playfair in 1802. Uniformitarianism is a simple idea, but it is the keystone of modern geology. A simple example will show the power of this theory. Many varieties of shell-bearing animals can be seen in the mud along the seashore at low tide. A similar mudstone with similar shells encountered in a canyon wall is interpreted, using uniformitarianism, as a former sea bottom that has been lithified and uplifted. If this seems too elementary, we need only remember that just a few years before Hutton's time, only a few people recognized that fossil shells were evidence of once-living organisms.

Recently uniformitarianism, as presented above, has been criticized as being oversimplified. The objection is that geological processes have modified the earth; therefore, at least in some cases, these processes do not operate now in the same ways that they did in the past. Thus, the present may not be the key to the past. For example, in later chapters it will be shown that the origin of life changed the composition of the atmosphere, making further creation of life impossible. Thus, though uniformitarianism in its simplified version can explain many geologic phenomena, the doctrine has limitations that must be taken into account.

Geologic history is based largely on the study of sedimentary rocks and their contained fossils. Such studies have led to the development of a geologic time scale showing the relative ages of rocks. In recent years it has been possible, using measurements of the radioactive decay of certain elements in rocks, to determine the absolute ages of some rocks. This ability to date igneous and metamorphic rocks has expanded the range of historical geology.

After several false starts, historical geology developed in the last part of the eighteenth century and the first half of the nineteenth. In 1782 the great French chemist Lavoisier demonstrated that near Paris the quarries dug for pottery and porcelain clay all exposed the same sequence of sedimentary rocks. Georges Cuvier and Alexandre Brongniart published maps in 1810 and 1822 showing the distribution of the various rock types around Paris, and in 1815 William Smith published a geologic map of England. Cuvier and Brongniart studied the fossils in each of the sedimentary layers that they mapped and discovered, as did Smith, that each layer contained a different group of fossils. They discovered that a sedimentary bed could be identified by its fossils. That work set the scene for the next 20 years, for during this period the geologic time scale was developed.

The geologic time scale, or column, was the result of individual work by a number of people in western Europe. These men studied the rocks and their fossils at well-exposed places, generally near their homes, and described these rocks and their fossils in books and papers. They called these sequences of rocks *systems* and named each of the systems. As the studies proceeded in this haphazard manner, the general sequence of older to younger was recognized and the gaps were filled in. It was also shown that rocks in similar stratigraphic position, although far removed from the type area, contained the same or very similar fossils. Thus it took work by many people spread over a wide area to demonstrate that fossils can be used to date rocks, not only in local areas, but all over Europe. In this way a geologic column, or time scale, was developed as a standard of reference; fossils from other areas could be correlated or dated by comparison with fossils on which the time scale was based. The geologic time scale in current use is shown in Fig. 17-1. Almost all of the systems or periods were originally defined in the first half of the nineteenth century.

The units used in historical geology are of several types. The *systems* just described are the rocks deposited during a time interval called a *period*. Thus the Cambrian System was deposited during the Cambrian Period. The systems and periods are arbitrarily defined, and their recognition away from the type area where they were defined is based on interpretation of fossils. A geologist mapping an area locates on his map the occurrence of easily recognized, objective units called *formations*. Formations are rock units named for a location, a type area, where they are well exposed, such as the Austin Chalk and the Tensleep Sandstone. All parts of a formation are not necessarily the same age. The sandstone bed in part B of Fig. 17-8, for example, might be defined as a formation. Thus there are three kinds of units, time units such as periods, time-rock units such as systems, and rock units such as formations. Table 17-1 summarizes the terms used.

The changes in marine invertebrates during geologic history are obvious to even a casual observer; for this reason such fossils can be used to date the enclosing rocks. (See Fig. 17-2.) The changes between periods are not as great but are just as obvious to a trained observer. The most marked changes between adjacent periods are between

Table 17-1. Stratigraphic units.

A. Time and Time-Rock Units

Time Units	Time-Rock Units
Eon	—
Era	Erathem
Period	System
Epoch	Series
Age	Stage
(Phase)	Chronozone

B. Rock Units

Group
Formation
Member

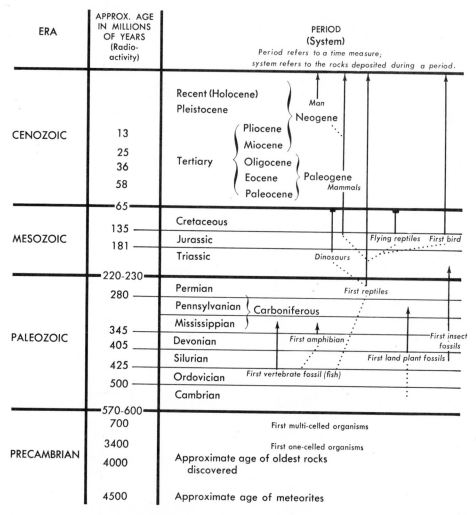

ERA	APPROX. AGE IN MILLIONS OF YEARS (Radio-activity)	PERIOD (System) Period refers to a time measure; system refers to the rocks deposited during a period.

Figure 17-1. The geologic time scale. Shown to the right is a very simplified diagram showing the development of life. Not included on the diagram are many types of invertebrate fossils such as clams, brachiopods, corals, sponges, snails, etc., which first appeared in the Cambrian or Ordovician and have continued to the present.

Cambrian and Ordovician, Permian and Triassic, and Cretaceous and Cenozoic. Marked changes also occur at the end of the Ordovician, Devonian, and Triassic. Not all of these marked changes come at the end of the cycles described in Chapters 21, 22, 23, and 24.

The geologic column or time scale provided only a relative time scale to which rocks could be referred. It also showed the general development of life and so set the stage for Darwin, who published his theory of evolution in 1859.

Although the idea of evolution was not new to Darwin, his book started a controversy that is still continuing in non-scientific circles. Until about this time no contro-

A.

B.

Figure 17-2. Brief comparison of these two fossiliferous slabs of rock show the great changes in life that have occurred during geologic time. A. Pennsylvanian fossils from Kansas. Bryozoa and brachiopods are most abundant. Photo from Ward's Natural Science Establishment, Inc., Rochester, N.Y. B. Miocene fossils from Virginia. Snails and clams are most abundant. Photo by W. T. Lee, U.S. Geological Survey.

versy existed between geology and religion because even the geologists interpreted geologic history in biblical terms. Biblical scholars thought that the earth was about 6000 years old, based on their interpretation of the Old Testament. However, in the hundred years before Darwin, some geologists began to recognize that a much longer time was necessary for the deposition of the sedimentary rocks and other geologic events than students of the Bible would allow. Before evolution was accepted, the changes in fossil life throughout the geologic column were explained by a catastrophic dying out of all life and a new creation of the animals found as fossils in the succeeding bed. This idea had its roots in the biblical Noachian Flood and the widely held theory of Werner, a late eighteenth century German, who believed that all surface rocks, including igneous rocks, were precipitated in a universal sea. Darwin had a pronounced influence on the development of geologic thought, but his theory also needed the long duration of geologic time envisioned by the uniformitarianists.

Thus the central point became the age of the earth. The geologists and the evolutionists needed a very old earth. Joly in 1899 calculated the age at 90 million years by dividing the total amount of salt in the ocean by the amount of salt annually brought into the ocean by the rivers. Estimates of the age of the earth based on the time necessary to deposit the rocks of the geologic time scale were made by many geologists in the latter part of the 19th century; and although they varied between 3 and 1584 million years, about 100 million years was an average figure. The lack of precision and the subjective nature of the assumptions made these estimates suspect. Lord Kelvin, one of the most influential physicists of the day, entered the fray and attempted to calculate the age of the earth from thermodynamics. He assumed that the earth began as a melted body and has been cooling ever since. He calculated that 20 to 40 million years had passed from the time when the earth had cooled enough for life to exist to the present. About the turn of the century, when Kelvin was an old man but still refining his calculations, radioactivity was discovered. As noted earlier, radioactivity is a source of heat within the earth, and so Kelvin's heat flow calculations were based on an incorrect assumption.

Radioactivity was discovered by Becquerel in 1896. In 1902 Rutherford and Soddy showed that by radioactive decay an element is changed in part to another element. This sounded much like the quest of the alchemists, who tried to change lead into gold, and was not immediately accepted by many scientists. However, in 1905 Boltwood noted the association of lead and uranium in uranium minerals and that the ratio of lead to uranium was higher in the older minerals. Boltwood used this information to calculate the age of these minerals, and one of them had a computed age of over two billion years. This greatly extended the possible age of the earth and was quickly accepted by a few geologists. Present methods of radioactive dating show the earth to be about five billion years old. These methods will be described later in this chapter. Thus, in a single stroke, radioactivity provided the energy source not taken into account by Kelvin and a means to find the age of the earth.

Unconformities

Once the time scale was set up, geologists traced the systems into previously unmapped areas. One problem that soon appeared was that the systems were originally defined, in many cases, as the rocks between two important breaks in the geologic record. In other areas there were no breaks, but continuous deposition. Thus, between the original systems were intervals not included in either system; and arguments about the assignment of the neglected intervals are, in some cases, still in progress. The breaks are of several types, such as change from marine to continental deposition, influx of volcanic rocks, and unconformities. (See Fig. 17-3.) The latter is the most important and requires discussion.

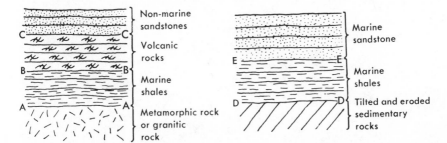

Figure 17-3. Natural breaks in the stratigraphic record. A-A and D-D are unconformities because the time necessary for erosion is not represented by sedimentary rocks. The other breaks may or may not be unconformities, depending on whether there is a time interval between the different rock types or continuous deposition. D-D is an angular unconformity.

All breaks in the geologic record that indicate a time interval for which there is no local record are called *unconformities* (see Figs. 17-4 and 17-5); they may take several different forms. If sedimentary rocks overlie metamorphic rocks, clearly the sediments were deposited after the metamorphism; and because no sedimentary rocks representing the time of metamorphism are present, this is a type of unconformity. In this case, if the metamorphic rocks were formed at considerable depth in the crust, as is generally the case, then the period of erosion which uncovered the metamorphic rocks prior to deposition of the sediments is also missing from the sedimentary record. Another type of unconformity may consist of a change in fossils, representing a short or a long period of time, that occurs between two beds in the sedimentary section. Such an unconformity may be due to erosion of previously deposited beds or to nondeposition, and may not be at all conspicuous until fossils are studied. The third type of unconformity is more obvious and consists of folded and eroded sedimentary rocks that are overlain by more sedimentary rocks. Such an unconformity is called an *angular unconformity* because the bedding of the two groups of sedimentary rocks is not parallel. In this case, the time of folding and erosion is not represented by sedimentary rocks.

Correlation and Dating of Rocks

So far we have considered how the geologic systems were defined, and it is obvious that they are recognized in previously unstudied areas by comparing the contained fossils with the fossils from originally defined type areas. In this manner the systems defined in Europe were recognized in North America and the other continents. This type of correlation by fossils is also most important in establishing the time equivalence of nearby beds.

A.

B.

Figure 17-4. Unconformities. A. An angular unconformity in the Grand Canyon. The rocks beneath the unconformity are Precambrian in age and those just above it are Cambrian. Photo by L. F. Noble, U.S. Geological Survey. B. An erosional unconformity in Silurian rocks in Niagara Gorge, N.Y. Photo by G. K. Gilbert, U.S. Geological Survey.

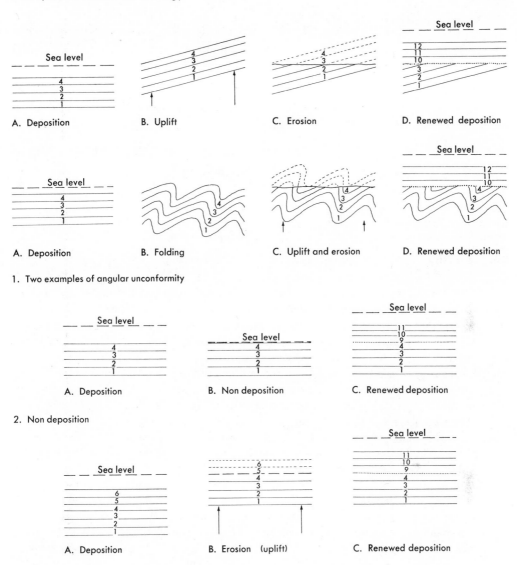

A. Deposition   B. Uplift   C. Erosion   D. Renewed deposition

A. Deposition   B. Folding   C. Uplift and erosion   D. Renewed deposition

1. Two examples of angular unconformity

A. Deposition   B. Non deposition   C. Renewed deposition

2. Non deposition

A. Deposition   B. Erosion (uplift)   C. Renewed deposition

3. Erosion

Figure 17-5. Development of unconformities. In each case a time interval is not recorded in the bedded rocks. It is commonly difficult to distinguish between cases 2 and 3. In each case, for simplicity, sea level is assumed to be the base level that controls whether erosion or deposition takes place.

Another type of correlation is to establish that two or more outcrops are parts of a once continuous rock body. The simplest case of such rock correlation is actually to trace the beds from one area to the other by walking out the beds. In other cases, distinctive materials in a bed, such as fragments of an unusual rock type or a distinctive or uncommon mineral or assemblage of minerals, may be used to identify a bed. Another method uses the sequence of beds. (See Fig. 17-6.)

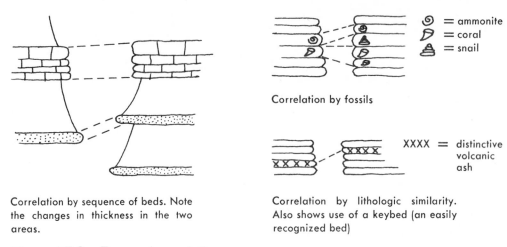

Correlation by fossils

⊚ = ammonite
ᗡ = coral
ᗍ = snail

XXXX = distinctive volcanic ash

Correlation by sequence of beds. Note the changes in thickness in the two areas.

Correlation by lithologic similarity. Also shows use of a keybed (an easily recognized bed)

Figure 17-6. Types of correlation.

By correlating from one area to another, the total thickness of sedimentary rocks in a region can be determined as shown in Fig. 17-7.

Note that the methods of correlation, with the exception of the use of fossils, establish only the continuity of a sedimentary bed. Fossils establish the fact that both beds are the same age. The difference between these two types of correlation can be seen by considering an expanding sea, a not uncommon occurrence in geologic history. As the sea advances, the beach sands, for example, form a continuous bed of progressively younger age. This is shown diagrammatically in Fig. 17-8.

This diagram also illustrates another difficulty in the use of fossils. Note that limestone, shale, and sandstone are all being deposited at the same time in different parts of the basin. Each of these different environments will attract and be the home of different types of organisms. Thus, the fossils found in each of these environments will be different, even though they are of the same age. This means that environment must be taken into account in establishing the value of a fossil or group of fossils in age determination. For this reason, free-swimming animals make the best fossils because their remains are found in all environments. Fossils of this type are very useful and have been called *index fossils* if they establish clearly the age of the enclosing rocks. Ideally, an index fossil should be widespread in all environments, be abundant, and have a short time span. Very few fossils meet all of these requirements, and generally, groups of fossils are used to establish the age of a bed. (See Figs. 17-9 and 17-10.)

Time-lines, or surfaces, can be established in sedimentary rocks by physical means alone in some rare instances. For example, a thin ash bed from a volcano may be spread over a very large area by wind. The deposition of such an ash bed may occur over a number of days or even weeks, but in terms of geologic time it is instantaneous. These rare occurrences help in local correlation problems and show the validity of the methods of dating and correlation described here.

The only rocks that have reasonably abundant fossils are marine sedimentary rocks, and not even all of these have enough fossils to establish their age clearly. On the previous pages, we have discussed dating of fossiliferous marine rocks. Similar techniques can be used to date continental sediments, but here the number of fossils is generally much smaller. This fact is easy to understand if one compares the number

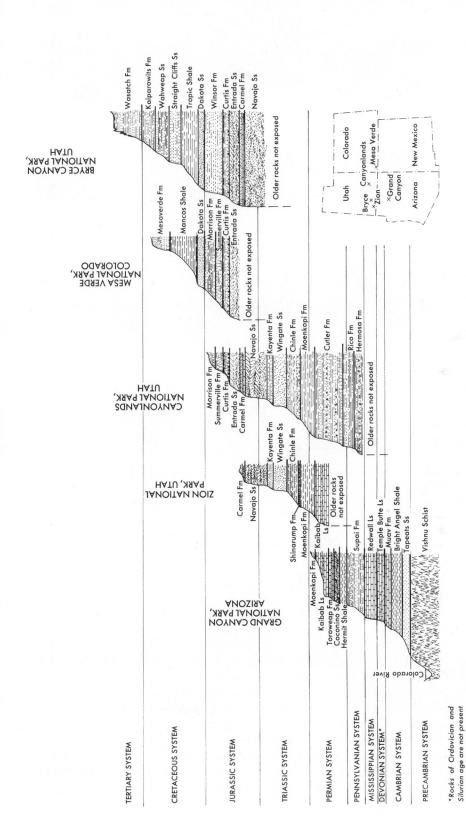

Figure 17-7. Correlation of the strata at several places on the Colorado plateau reveals the total extent of the sedimentary rocks. After U.S. Geological Survey.

427

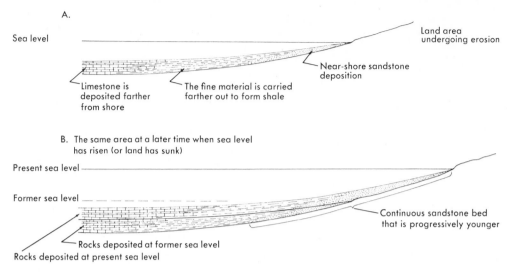

A.

Sea level

Land area
undergoing erosion

Near-shore sandstone
deposition

Limestone is
deposited farther
from shore

The fine material is carried
farther out to form shale

B. The same area at a later time when sea level
has risen (or land has sunk)

Present sea level

Former sea level

Continuous sandstone bed
that is progressively younger

Rocks deposited at former sea level

Rocks deposited at present sea level

Figure 17-8. Deposition in a rising or expanding sea. A. The deposition at the start. B. Deposition at a later time after sea level has risen.

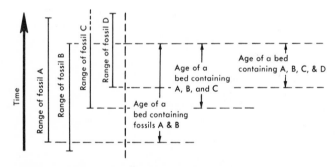

Time

Range of fossil A

Range of fossil B

Range of fossil C

Range of fossil D

Age of a bed
containing A, B, C, & D

Age of a
bed containing
A, B, and C

Age of a
bed containing
fossils A & B

Figure 17-9. The use of overlapping ranges of fossils to date rocks more precisely than can be accomplished by the use of a single fossil.

of easily fossilized animals exposed at low tide at the seashore with the much smaller number of animals living in a similar-sized area of forest or grass land, and if one considers how easy it is for a clam shell to be buried and so preserved, especially when compared to the chance that a plant leaf or a rabbit skeleton will be buried and so preserved as a fossil. A land organism, even if buried, may not be preserved, as oxidizing conditions that destroy organic material exist even at some depth. Also, a dead rabbit is apt to be eaten and dismembered by scavengers. Another problem in the use of continental fossils is that they are controlled even more by climate than are marine organisms. The progression of types of plant and animal life seen during a mountain climb will illustrate this point.

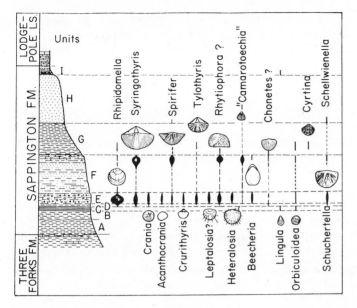

Figure 17-10. An example of the distribution of brachiopods in the Sappington Formation in western Montana. The relative number of each fossil found in the beds is indicated by the thickness of the black lines. The occurrence of these animals was at least in part controlled by the sedimentary environment. From R. C. Gutschick and Joaquin Rodriquez, *Bulletin,* American Association of Petroleum Geologists, Vol. 51, 1967.

Dating igneous and metamorphic rocks presents even more problems because such rocks almost never contain fossils. They must be dated by their relationships to fossil-bearing rocks. Volcanic rocks are dated by the fossiliferous rocks above and below and, in some cases, by inter-bedded sedimentary rocks. Intrusives must be younger than the rocks that they intrude; they cannot be dated closer unless the intrusive has been uncovered by erosion and, then, unconformably covered by younger rocks. (See Fig. 17-11.)

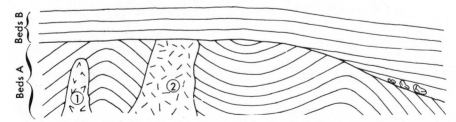

Figure 17-11. Dating of intrusive rocks. Intrusive body ① is younger than beds A; no upper limit can be determined. Intrusive ② is younger than beds A and older than beds B. Closer dating may be possible if the conglomerate composed of pebbles of ② in the basin at the right contains fossils. The contact between beds A and B is an angular unconformity.

Metamorphic rocks are even more difficult to date because both the age of the original rock and the date of the metamorphism should be obtained. The age of the parent rock can be found by tracing the metamorphic rock to a place where fossils are preserved. Finding the date of metamorphism may be more difficult because, in general, an unconformable cover, if present, gives only the upper limit.

Radioactive Dating

Some of these problems in dating igneous and metamorphic rocks can be overcome by modern radioactive methods. Such methods produce an absolute age in years and so have been used, in addition, to date the geologic time scale in years. (See Fig. 17-12.) In spite of the seeming precision of radioactive dating, it is not yet possible to date a rock as accurately this way as with fossils. This is because of inherent inaccuracies in the measurements of the amounts of the elements produced by radioactive decay. These inaccuracies produce an uncertainty in the radioactive date that is more than the span of zones based on fossils, especially in those parts of the geologic column that contain abundant short-lived fossils. As methods of analysis are improved, this limitation may be overcome and the present high cost of dating may also be reduced.

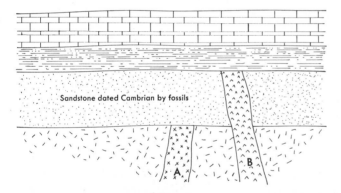

Figure 17-12. Determination of absolute age of one of the geologic periods. The sandstone contains Cambrian fossils. Dikes A and B have been dated by radioactive methods. The Cambrian Period is younger than dike A and older than dike B. Can the absolute age of the Cambrian be determined from a single occurrence such as this?

Radioactive dates are obtained by studying the daughter elements produced by a radioactive element in a mineral. The principles of radioactivity were covered in Chapter 2. Sometimes more than one element is studied in an effort to provide a check on accuracy. A radioactive date tells the time of formation of that mineral and so tells the time of crystallization of an igneous rock, or the time of metamorphic recrystallization in the case of a metamorphic rock. In some favorable cases, in which microscopic studies show that different minerals formed at different times in a rock that has undergone more than one period of metamorphism, it is possible to date the periods of metamorphism by dating elements in minerals formed during each period of metamorphism. In many radioactive decays a gas is one of the products, and heating during metamorphism may drive off the gas, thus resetting the radioactive clock to give the time of metamorphism.

A radioactive element changes to another element by spontaneously emitting energy. The rate of this decay is unaffected by temperature or pressure; hence, if we know the rate of formation of the daughter products, all we need to find is the ratio of original element to daughter product to be able to calculate the time of crystallization of the mineral containing the original radioactive element. The main assumptions, then, in using radiodates are that the decay rate is known and constant, and that no daughter or parent elements are lost. The latter assumption is the least sound because of weathering and metamorphism.

Several radioactive elements are used to date rocks, but only the most important are reviewed here. The methods in current use are tabulated in Table 17-2 (see also Chapter 2). A few minerals contain uranium, and these generally also contain thorium, which is about the same atomic size. These minerals are rare and with a few exceptions are almost completely confined to pegmatite veins in batholiths. Because such veins are probably the last to crystallize, these occurrences give only the upper limit of the age of the batholith. Each of the three radioactive isotopes that these minerals contain— the two of uranium as well as thorium—produces a series of radioactive daughter products; each member of each series decays to the next daughter until, finally, a stable isotope of lead is produced. (See Fig. 2-10.) These many steps constitute a possible source of inaccuracy, as it is possible for any one of the daughters to be removed by leaching or some other process. The elements and the final products are:

$$\text{Uranium } 238 \longrightarrow \text{Lead } 206$$
$$\text{Uranium } 235 \longrightarrow \text{Lead } 207$$
$$\text{Thorium } 232 \longrightarrow \text{Lead } 208$$

Thus, in this case, it is possible to measure three ratios which should all agree in age.

A problem with age determinations involving lead is that many rocks contain some lead of nonradioactive origin. The amount of nonradiogenic lead is quite small but may require an appreciable correction in calculating the age of very old rocks. This ordinary lead contains the isotope lead 204, so its presence can be easily recognized. The problem is that along with the minor amounts of lead 204 are the three isotopes, lead 206, lead 207, and lead 208, which also occur radiogenically, and the ratio of all four isotopes varies. Thus to correct for the pre-existing ordinary lead, one must analyze nonradioactive rocks in the area to find the ratio between lead 204 and each of the other isotopes. In this way the nonradiogenic lead 206, lead 207, and lead 208 can be excluded from that due to radioactive decay. Meteorites contain lead 204, and the correction for the original lead is a problem. As discussed later, meteorites are used to date the formation of the earth, so this problem results in some uncertainty in this important date.

Rubidium is a radioactive element that is generally present in small amounts in any mineral that contains potassium, because both elements are about the same size. Rubidium decays in a single step to strontium (rubidium 87 → strontium 87). Because potassium minerals that contain rubidium are so common (feldspar and mica) and only a single radioactive decay is involved, this process may become the most useful method of radioactive dating. However, at present the half-life of this decay cannot be measured accurately because the low energy of the beta particles emitted makes them difficult to detect.

Table 17-2. Physical methods of age determination. Some of the methods are described in Chapter 2. After E. J. Zeller, "Modern Methods for Measure- ment of Geologic Time," *Mineral Information Service,* California Division of Mines and Geology, January, 1965, Vol. 18, No. 1, pp. 9-15.

Method	Kind of sample used	Parameter measured	Range and accuracy	Remarks
		Isotopic or radioactive yield methods		
Uranium–Lead	Zircon, xenotime, and other igneous and metamorphic minerals with original uranium, but with no original lead.	Uranium and lead by analyses and α-counts per unit of mass per unit of time.	10 million to 3-4 thousand million years. Not as accurate as with isotopic analysis.	There should be no leaching of uranium during life of mineral and no original lead in mineral at time of crystallization.
Uranium 238– Lead 206	Any igneous or metamorphic mineral or whole rock with original uranium 238 (especially uraninite, pitchblende, samarskite, cyrtolite, thorianite, zircon, and thorite.)	Quantities of uranium 238, lead 206, and lead 204 by mass spectrometer analysis. Lead 204 gives indication of amount of original lead, of all isotopes, that is present.	10 million to approximately 10,000 million years. More accurate than the chemical uranium—lead method.	Uranium 238 is more abundant than uranium 235 so less chance of analytical error. Radon 222 half-life is long enough (3.82 days) to suggest that there may be error through radon escape.
Uranium 235– Lead 207	Any igneous or metamorphic mineral or whole rock with original uranium 235. (Same minerals as above.)	Quantities of uranium 235, lead 207, and lead 204 by mass spectrometer analysis. Lead 204 gives indication of amount of original lead, of all isotopes, that is present.	10 million to several thousand million years. More accurate than the chemical uranium—lead method.	Radon 219 in series has such short half-life (3.92 sec.) that escape is much less likely than in uranium 238 — lead 206 series.
Thorium 232– Lead 208	Accessory minerals in igneous rocks.	Thorium 232, lead 208 by mass spectrometer analysis.	10 million to several thousand million years. Accuracy 10 per cent.	Lead 208 may be removed selectively from thorium minerals resulting in ages which appear too low.
Lead–Lead	Galena or uranium minerals.	Lead 206, lead 207, lead 208, lead 204 by mass spectrometer analysis.	10 million to several thousand million years. 1 per cent over 100 million years. (Younger ages may be calculated with accuracy of 1 million years.)	Has the advantage that it involves an isotopic ratio instead of absolute measurement.

432

Method	Material	Measurement	Age range	Remarks
Lead 210	Fresh and marine water and glacial ice. Recently deposited sediments.	Count rate of bismuth 210 derived by radioactive decay of separated lead 210.	Up to about 100 years. Sequence of analyses gives fairly accurate estimate of rates of accumulation.	Part of chain of uranium 238 —lead 206. Radon 222, to atmosphere from rocks, yields lead 210. Lead 210 washed from atmosphere by rain. This disintegrates to bismuth 210—polonium 210—lead 206. Lead 210 is rapidly removed from water by precipitation or absorption on sedimentary particles. Rate of removal from water and original amount of lead 210 in sediments are necessary for data for age determination.
Potassium—Argon	Micas and feldspars.	Potassium 40, argon 40. Potassium measured by flame photometer, argon by mass spectrometer.	30,000 to several thousand million years. Error in micas less than 10 per cent.	Feldspars lose argon (up to 60 per cent). Metamorphism affects the results if rocks are heated.
Rubidium—Strontium	Mica, muscovite, biotite. Feldspar. Whole rock. Glauconite (in sediments).	Amount of strontium 86. Amount of strontium 87. Amount of rubidium 87. Ratios determined by mass spectrometry.	Not suited for ages below 50 million years due to long rubidium 87 half-life. Accuracy of method dependent on rubidium 87 half-life which is not accurately determined.	Decay constant of rubidium 87 uncertain due to low energy of β-particles emitted. Metamorphism may cause rehomogenation of rubidium and strontium isotopes so method may be used to date metamorphic events.
Thorium 230	Oceanic sediments, cores.	Count rate of thorium 230	Up to 300,000 years.	Part of chain of uranium 238—lead 206. Used to date marine sediments. Uranium 238 in sea water decays to thorium 230 which is precipitated and deposited on sea floor.
Thorium 230— Protactinium 231	Oceanic sediments, cores.	Count rate of thorium 230 and protactinium 231.	Up to 300,000 years.	Protactinium 231 is part of chain of uranium 235—lead 207. Accumulates similar to thorium 230 described above.

Table 17-2. Physical methods of age determination (cont.).

Method	Kind of sample used	Parameter measured	Range and accuracy	Remarks
Isotopic or radioactive yield methods (continued)				
Helium	Uranium and thorium bearing minerals. (The helium is the α particles emitted.)	Abundance of helium. Uranium 238, uranium 235, and thorium 232 abundance by their α activity. Uranium/thorium ratio.	Several million to several thousand million years. Accuracy in dispute.	Method is not extensively used at present due to discovery of the ease of diffusion by helium.
Carbon 14	Charcoal, wood, shell.	β-activity of carbon in sample by β-counting techniques.	0 to 50,000 years. Accuracy good for last 20,000 years.	Assumes a steady state for the production of carbon 14 in the atmosphere by cosmic radiation. Assumes uniform distribution of carbon 14 throughout the earth's atmosphere.
Hydrogen 3 (Tritium)	Water samples, 2-6 liters.	β-activity of tritium.	Last 30 years.	Recent bomb testing has greatly affected the usefulness of this method.
Radiation damage methods				
Metamict	Zircon (best); euxenite; sphene; uraninite; davidite.	Amount of radiation damage (i.e. metamict condition) in mineral usually by X-ray diffraction. Amount of radioactive material by α-count.	Now considered a comparatively inaccurate method.	Heating will cause crystal to become crystalline again; therefore, method may be used to date metamorphic or igneous events.
Pleochroic halo (Radiation halo)	Biotite with an inclusion of zircon, apatite, or monazite.	Intensity of the coloration of the biotite. Width of the coloration. α-activity of the inclusion.	Late Precambrian to approximately Eocene. Useful for relative age determination.	Coloration is temperature sensitive and may reverse at a high level of radiation. Assumes identical irradiation sensitivity of all biotites and that the coloration, once formed, does not fade through time.

Method	Materials	Process measured	Range / Accuracy	Remarks
Thermoluminescence	Carbonate rocks, fluorite, quartz, feldspars.	Luminescence of sample when heated to temperatures up to 400°C. Glow-curve is measured by a recording microphotometer.	One thousand million year maximum. Accuracy ± 15 per cent.	May be used to determine periods of metamorphism. Has specific uses for paleoclimatological measurements.
Radiation pitting (Fission particle tracks)	Mica. Other minerals may also be usable.	Particle tracks counted by optical microscopy. Uranium determined by α activity.	One thousand million years. Accuracy has not yet been tested fully.	Very promising new method. May be sensitive to heating in metamorphic processes.

Chemical methods

Method	Materials	Process measured	Range / Accuracy	Remarks
Devitrification of glass	Glass, either natural obsidian or man made.	Crystallinity of the glass, especially on the surface.	Useful for relative age determinations in past 5 million years for obsidian.	Rate of crystallization dependent upon chemical composition, temperature, and water present in proximity of the glass.
Degradation of amino acids	Shell or bone fragments.	Amino acids and peptide bonding.	Accuracy best in last 1 million years but amino acids are detectable in rocks over 500 million years old.	Involves a chemical process and rate is temperature dependent.
Protein-nitrogen	Shells of mature forms of mollusks.	Organic nitrogen content of water-soluble residues (by micro Kjeldahl method).	Probably limited to Pleistocene and Late Tertiary.	Only relative ages may be determined.

435

Potassium also has a radioactive isotope that is useful in dating, especially since it is such a common element. The small amount of radioactive potassium decays to form two daughters:

$$\text{Potassium } 40 \Big\langle \begin{matrix} \text{Calcium } 40 \\ \text{Argon } 40 \end{matrix}$$

Calcium 40 is ordinary calcium and so is present in most minerals. Therefore, the method used is to measure the amounts of potassium 40 and argon 40. The main problem is finding minerals that retain the argon gas. It appears that micas, in spite of their cleavage, retain most of the argon but feldspar loses about one-fourth of the argon.

The natural radioactive isotope of carbon is effective in dating carbonaceous material, such as wood or coal, less than 40–50,000 years old. This method depends upon the fact that all living matter contains a fixed ratio of ordinary carbon 12 and radioactive carbon 14. Cosmic rays react with nitrogen in the atmosphere to form the carbon 14, which ultimately gets into all living matter via the food cycle. The carbon 14 decays back to nitrogen. Therefore, the ratio of carbon 12 to carbon 14 can be used to calculate how long an organism has been dead, because a dead organism takes in no more carbon 14. Extensive use of this method is made in archaeology and in glacial studies, although carbon 14 dates do not always agree with dates obtained by other means.

Another type of radioactive dating is to measure the damage done to a crystal lattice by the particles released by radioactivity. These methods, mentioned in Chapter 2, are relatively inaccurate.

Ages of rocks can also be determined by chemical means. At present these methods yield mainly relative ages, but new advances may result in absolute ages. An obvious example is the relative dating of glacial moraines on the basis of weathering. Rates of weathering are dependent on so many factors that absolute ages are not possible in this way and great care must be exercised in application. Another example is the crystallization of volcanic glass. Obsidian, or any other glass, is not stable but tends to crystallize; however, the rate of crystallization depends on the composition and amount of water in the glass, as well as on other factors. A promising chemical method of dating is the breakdown of amino acids in fossils. The breakdown is dependent on many factors, such as temperature and composition of formation water, but gives fairly accurate results in young rocks.

Age of the Earth

The age of the earth itself is a geological problem to which radioactive dating has been applied. So far, the oldest rocks found are about 4,000 million years old. This, then, is the youngest that the earth can possibly be, and is probably about the time the crust formed; the earth itself is probably older. Meteorites were discussed earlier and are thought to be the material from which the solar system was formed. Radioactive dates from meteorites give ages of around 4,700 million years. This suggests that the earth may have formed 4,500 to 5,000 million years ago.

The first abundant fossils occur in rocks of the Cambrian system although a few fossils have been found in much older rocks. The base of the Cambrian has been dated radioactively as about 600 million years. Why abundant, relatively advanced animals appeared suddenly just prior to this time is a problem, but these fossils may record only the development of easily preserved hard parts such as shells. See Fig. 17-1. Because fossils are so much easier to use than radioactive methods in dating and correlating rocks, we know many more details of geologic history from Cambrian time onward than Precambrian time. We know only the broadest outlines of the history of the Precambrian, even though it includes almost nine-tenths of the earth's past.

Interpreting Geologic History

Deciphering the history of the earth or any part of it is a difficult reiterative process. It has many steps, and each affects every other step. Multiple working hypotheses must be used to test each step. The process begins with study of the rocks themselves and their relationships to other rocks. This leads to some ideas about their mode of origin. The rocks must be dated and correlated, using the methods described in this chapter. If the rocks have been deformed by folding and faulting, their original relationships and positions must be determined. Each layer or rock unit can now be reconstructed to show the geography at its time of formation. Remember that the rock record is incomplete because of erosion and because many rock layers are exposed only at a few places. The final step is to reconstruct the events that caused the changes between each of the steps.

   This brief outline shows how the incomplete rock record is interpreted. Note the various levels of abstraction that are involved. It is no wonder that the history of the earth is not completely known. Interpreting the history of the earth or a small area is difficult, and revision is constantly necessary as more data are obtained. Thus geology is an ongoing process that offers great challenge; many fundamental problems remain to be solved.

Interpreting Ancient Environments

The environment of the geologic past can be determined at least partially from study of rocks. Such features as coral reefs, evaporites, desert deposits, and glacial deposits are a few obvious examples. Some of the features that indicate environment were described in earlier chapters. Fig. 17-8A shows in a generalized way how the rock type changes from sandstone near shore to shale and limestone farther from shore. Fig. 4-23 is an example of a reconstruction of an ancient sedimentary basin from the study of scattered outcrops. Many of the features used in such studies are also described in Chapter 4. The grains that compose a sedimentary rock, such as a sandstone, also reveal much environmental information. The composition of the grains in a sandstone is influenced by both the nature of the source rocks and the amount of weathering. The topography and the climate greatly influence the weathering. The size and shape of the grains are influenced by the topography, and the distance and method of transportation. Obviously, only some of the factors involved in interpretation of ancient environments have been mentioned here, and a little thought will reveal many more aspects.

   Isotopes can be used to determine ancient temperatures as well as age. Oxygen isotopes are used in temperature determination. Oxygen has three isotopes: oxygen 16, oxygen 17, and oxygen 18. When water evaporates, the lightest isotope—common oxygen 16—evaporates at a slightly higher rate than the heavier isotopes. This difference in evaporation rates is proportional to temperature. The ratio of oxygen 16 to oxygen 18 can be measured and used to determine the temperature. In practice, this ratio is measured in fossil shells made of calcium carbonate ($CaCO_3$). There are many pitfalls in this method, but so far the results have agreed fairly well with other geological evidence.

Other means of estimating temperatures are coral reefs, evaporites, desert deposits, and glacial deposits. Fossils, both plant and animal, have also been used very successfully to reconstruct climates and ecology.

QUESTIONS

17-1. How does one interpret the depositional environment of a layer of ancient rocks?

17-2. Describe in general terms how the geologic column was established and discuss whether it consists of natural subdivisions in North America.

17-3. What are unconformities? Describe the various types.

17-4. How was the value of fossils in determining geologic age established?

17-5. What is meant by correlation?

17-6. List and briefly describe the methods of correlating rocks.

17-7. Why is it generally easier to date marine rocks than continental sediments?

17-8. How are metamorphic rocks dated? What dates are required?

17-9. Which method of radioactive dating is best and why?

17-10. What are index fossils?

17-11. How can metamorphic events be dated radioactively?

17-12. What effect would a small amount of modern root material mixed with a sample to be dated by the carbon 14 method have?

17-13. How can an angular unconformity be distinguished from a low angle thrust fault?

17-14. Fossils aid in determining the age of a rock unit. How else do they aid in geologic interpretation?

17-15. Prove to your girlfriend (boyfriend), who is an art major and has not taken any science, that the earth is very old.

SUPPLEMENTARY READING

Adams, F. D., *The Birth and Development of the Geological Sciences.* New York: Dover Publications, Inc., 1954, 506 pp. (paperback). (Originally published in 1938 by Williams & Wilkins Co., Baltimore.)

Deevey, E. S., Jr., "Radiocarbon Dating," *Scientific American* (February, 1952), Vol. 186, No. 2, pp. 24-33. Reprint 811, W. H. Freeman and Co., San Francisco.

Eicher, D. L., *Geologic Time.* Englewood Cliffs, New Jersey: Prentice-Hall, Inc., 1968, 150 pp. (paperback).

Eiseley, L. C., "Charles Lyell," *Scientific American* (August, 1959), Vol. 201, No. 2, pp. 98-106. Reprint 846, W. H. Freeman and Co., San Francisco.

Emiliani, Cesare, "Ancient Temperatures," *Scientific American* (February, 1958), Vol. 198, No. 2, pp. 54-63. Reprint 815, W. H. Freeman and Co., San Francisco.

Engel, A. E. J., "Geologic Evolution of North America," *Science* (April 12, 1963), Vol. 140, No. 3563, pp. 143-152.

Faul, Henry, *Ages of Rocks, Planets, and Stars.* New York: McGraw-Hill Book Co., 1966, 109 pp. (paperback).

Fenton, C. L., and M. A. Fenton, *Giants of Geology.* Garden City, New York: Doubleday & Co., 1952, 333 pp. (Also published in paperback, New York: Dolphin Books and Dolphin Masters, 1952, 318 pp.)

Geikie, Sir Archibald, *The Founders of Geology.* New York: Dover Publications, Inc., (1905, 2nd edition) 1962, 486 pp. (paperback).

Gillispie C. C., *Genesis and Geology.* Cambridge, Mass.: Harvard University Press, 1951. (Also published in paperback, New York: Harper & Row, Publishers, 1959.)

Hay, E. A., "Uniformitarianism Reconsidered," *Journal of Geological Education* (February, 1967), Vol. 15, No. 1, pp. 11-12.

Hurley, P. M., "Radioactivity and Time," *Scientific American* (August, 1949), Vol. 181, No. 2, pp. 49-51. Reprint 220, W. H. Freeman and Co., San Francisco.

Knopf, Adolph, "Measuring Geologic Time," *Scientific Monthly* (November, 1957), Vol. 85, No. 5, pp. 225-236. Reprinted in J. F. White (ed.), *Study of the Earth,* pp. 41-62. Englewood Cliffs, N. J.: Prentice-Hall, Inc., 1962 (paperback).

Kulp, J. L., "Geologic Time Scale," *Science* (April 14, 1961), Vol. 133, No. 3459, pp. 1105-1114.

Kummel, Bernhard, *History of the Earth.* San Francisco: W. H. Freeman and Co., 1961, 610 pp.

McAlester, A. L., *The History of Life.* Englewood Cliffs, New Jersey: Prentice-Hall, Inc., 1968, 152 pp. (paperback).

Mintz, L. W., *Historical Geology.* Columbus, Ohio: Charles E. Merrill Pub. Co., 1972, 786 pp.

Chapter 18

Early History of the Earth – Astronomical Aspects of Geology

The kindest words regarding these entertainments are those of H. H. Read, quoted by L. R. Wager (1958) in his address at the 1958 meeting of the British Association for the Advancement of Science: '. . . just as things too absurd to be said can yet with perfect propriety be sung, so views too tenuous, unsubstantial and generalized for ordinary scientific papers can yet appear with some measure of dignity in presidential addresses.' This is then an appropriate occasion to review some speculations, by no means original, on the large-scale evolution of the Earth. Views on the early history of the Earth and on happenings in the deep interior can hardly fail to be tenuous and generalized, so I think they may qualify under Read's dictum. Nevertheless, we cannot hope to account for the development of the surface film which naturally engages most of our attention without considering the enormously larger interior from which this film was derived and which continues to determine its evolution . . .

> Introduction to Francis Birch's presidential address to The Geological Society of America, "Speculations on the Earth's Thermal History," *Bulletin,* Geological Society of America, Vol. 76, pp. 133-154, 1965.

The Earth[1] is part of the solar system, so we must look to astronomy for clues to the Earth's origin. Our main interest is not astronomy, so this chapter will be descriptive and much information will be presented without proof. Details will be discussed only where they have important application to geology.

Our sun is not the only star thought to have planets. Although the other stars are too far away to see the other planets, even with our largest telescopes, their existence is inferred from the wavering motion of three stars and the as yet unconfirmed wavering motions of five more stars.

Solar System

The solar system is composed of the sun and the planets that orbit around the sun, the satellites of the planets, and comets, meteors, and asteroids. All of the orbital motions are explained by the law of gravity.

---

[1]In this section the astronomical convention of capitalizing the names of all planets will be followed.

The sun is an ordinary, average-size star, composed mainly of hydrogen and helium, but containing most of the elements found on Earth. The sun's energy, which makes life possible on Earth and ultimately causes most geologic processes, comes from nuclear reactions discussed in Chapter 2. The sun has 99.86 per cent of the mass of the solar system, and the Earth has about 1 per cent of the remaining 0.14 per cent.

The nine planets all revolve in the same direction around the sun. A tenth planet revolving in the opposite direction was reported in 1972, but little is known of this, the

Table 18-1. Summary of solar system data. These statistics are approximations and vary somewhat in the different sources from which they were gathered.

Body	Distance from sun (average) (millions of miles)	Radius[1] (miles)	Mass (Earth = 1)	Specific gravity	Eccentricity of orbit	Inclination of spin axis	Period of revolution around sun	Period of rotation
SUN	—	432,000	332,000	1.4	—	—	—	25 days at equator 34 days at poles
Mercury	36	1,500	0.05	5.1	0.206	10° (?)	88 days	58.6 days
Venus	67	3,800	0.81	5.0	0.007	6°	225 days	242.9 days[2]
Earth	93	3,963	1.00	5.5	0.017	23.5°	365.25 days	23 hrs. 56 min.
Mars	141	2,100	0.11	4.0	0.093	25.2°	687 days	24.5 hours
Jupiter	483	43,500	318	1.3	0.048	3.1°	11.9 years	9.9 hours
Saturn	886	37,000	95	0.7	0.056	26.7°	29.5 years	10.2 hours
Uranus	1,783	15,750	14.6	1.6	0.047	98°	84 years	10.8 hours[2]
Neptune	2,793	16,500	17.3	2.3	0.009	29°	165 years	15.7 hours
Pluto	3,666	1,800 (?)	0.1 (?)	?	0.249	?	248 years	6.4 days (?)

[1] Data from various sources varies up to ten per cent.
[2] Venus rotates clockwise, opposite to other planets.
   (Uranus does also because of its 98° tilt of its spin axis.)

most distant planet yet detected. The orbits are all elliptical, very close to circular, and lie close to a single plane, except Pluto and planet 10 whose orbits are inclined. (See Fig. 18-1.) Seen from the Earth, the sun and the planets follow the same path across the sky. The planetary data are summarized in Table 18-1. The inner planets, Mercury, Venus, Earth, and Mars, form a group in which the individual members are more or less similar in size. The next four planets, Jupiter, Saturn, Uranus, and Neptune, form a similar grouping of much larger planets. Pluto is much smaller than the other outer planets and has a tilted, eccentric orbit, all of which suggest that it may not be a true planet. It may once have been a satellite of Neptune and escaped into its own orbit. Between Mars and Jupiter are several thousand small bodies, the asteroids, that are probably the origin of meteorites.

Most of the information that we have about the planets is summarized in Table 18-1. Only a few important points will be described in the text.

Solar radiation received (Earth = 1)	Solar radiation reflected (Albedo) %	Temperature (°C)			Atmospheric pressure at surface (Earth = 1)	Composition of atmosphere
		Theo-retical[3]	Actual			
			Top of atmosphere	Surface		
--	--	--	--	5,500°	--	--
7	6	Similar to actual	--	425° sun; 17° dark	0.003	Carbon dioxide
2	73	$-35°$	$-38°$	425°	100	Mainly carbon dioxide, minor water
1	34	$-20°$	--	15°	1	78% nitrogen 21% oxygen
0.4	30	$-65°$	$-50°$	$-42°$[4]	.006	Mainly carbon dioxide, possibly minor water
0.04	45	$-168°$	$-123°$	2,000° est.	200,000	60% hydrogen, 36% helium, 3% neon, 1% methane and ammonia
0.01	50	$-195°$	$-167°$	2,000° est.	200,000	Same as Jupiter
0.003	66	$-185°$	$-185°$ approx.	?	8	Similar to Jupiter, with more methane, no ammonia
0.001	62	$-185°$	$-185°$ approx.	?	?	Same as Uranus
0.0006	15 (?)	$-212°$	Large variation because of eccentricity		?	?

[3] Theoretical temperatures are calculated from the energy input from the sun.
  Actual temperatures are generally higher because of absorption in the atmospheres.
[4] Average varies from 20° to $-95°$ depending on latitude.
  Actual measurements range from $-123°$ to 17°.

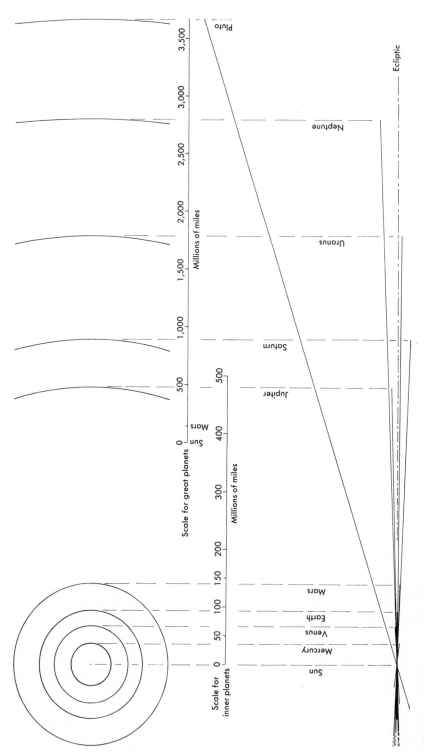

Figure 18-1. The orbits of the planets. Because Pluto is almost 40 times further away from the sun than is the Earth, it is not possible to draw this diagram to scale in a page-size figure. The inner planets, the circles, are drawn to scale, and the distances among the great planets are drawn to a different scale. Eccentricity of the orbits is neglected. The edge view shows that, with the exception of Pluto, the orbits lie nearly in the same plane (ecliptic).

444

Inner Planets

The inner planets are Mercury, Venus, Earth, and Mars. Mercury is small and too close to the sun for detailed telescopic study. Earth is the main subject of this book. Venus and Mars, our closest neighbors, are of great interest to us and have been studied by telescope and with space probes. Because they are our closest neighbors, they are the planets on which life might occur. For life, such as we know it, to exist, liquid water is probably necessary, and only Mars has temperatures in this range.

Mars has a reddish color and so was named for the god of war. It is farther from the sun than is Earth, and so has lower temperatures. Its rotational axis is tilted with respect to its orbital plane, and so it has seasons like the Earth. Telescopic observation reveals that polar ice caps form and disappear each year, and the color of the planet changes from bluish-green in spring to brown in the fall, suggesting seasonal growth to some observers. (See Fig. 18-2.) Mariner IX photographs reveal dust clouds, so the observed color changes may be caused by dust storms and clouds. Mars' atmosphere is composed mainly of carbon dioxide and a little water vapor. Mariner spacecraft revealed that this inhospitable atmosphere is less than one per cent as dense as the Earth's. The average temperature at the equator at noon is probably 40°F, and the temperature may fall to –200°F at night. Thus the conditions are not very favorable for life on Mars, but some may exist. The pictures returned by Mariner space probes show evidence of both internal and surface activity.

Nix Olympia (see Fig. 18-3) is the largest volcano ever seen by man. It is more than four miles high and 310 miles across. It has a broad complex crater and what appears to be flow material.

Faulting on Mars is suggested by Fig. 18-4. The valleys shown in Fig. 18-4A are part of a parallel system that extends more than 1100 miles. They are interpreted as tensional features, perhaps associated with uplift. Many years ago, some astronomers reported a network of faint lines that were incorrectly called canals. The report of these "canals" caused speculation as to the existence of intelligent life on Mars. Perhaps these valleys are the "canals" seen by some astronomers. Fig. 18-4B shows another valley that also may be of fault origin. Note the lines of craters that parallel the main valley.

The surface processes shown in the pictures are of many types. Craters are common in many areas, and, although some are of volcanic origin, many are caused by meteorite impact. This is not unexpected because Mars is near the asteroid belt and has a thin atmosphere. The effects of meteorite impacts are clearly shown on Mars' moon Phobos. (See Fig. 18-5.) Some areas on Mars are featureless, and because it seems unlikely that meteorites would not hit all parts, some process apparently obscures the craters. (See Fig. 18-6.)

Dust storms have been observed on Mars. As the storm subsides, first the higher elevations come into view and finally, the lowlands. In spite of the thin atmosphere, winds up to 300 miles per hour are possible and 170-mile-per-hour winds are not uncommon. The dust may result from meteorite impacts or other processes. The wind may move the dust to lower elevations, and in this way, create the featureless plains of Mars. The pits and basins of Fig. 18-7 could also be caused by wind erosion.

Although it seems unlikely, water may be present, at least at times, and cause some of the surface features of Mars. Long gullies that resemble river courses have been noted. (See Fig. 18-8.) The south polar cap appears to be a dual ice cap. Much of it is composed of frozen carbon dioxide, and this is the part that disappears and reforms

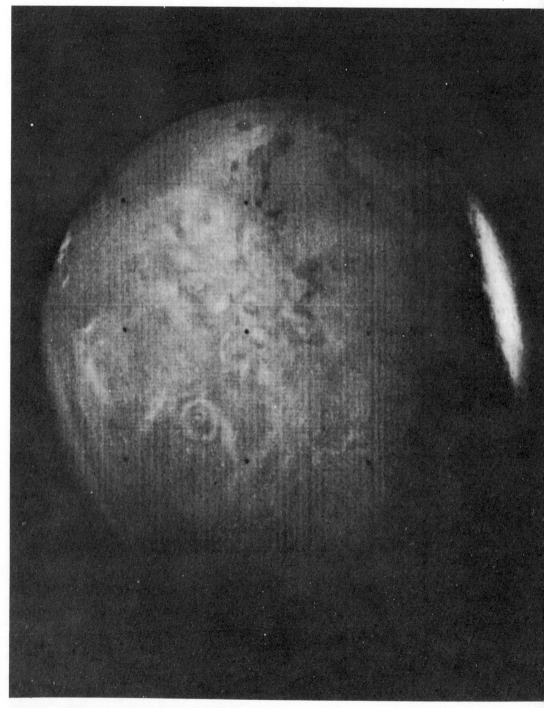

Figure 18-2. Mars from 293,000 miles. The circular crater is about 300 miles in diameter. White area is south polar cap. Mariner 7 photo from NASA.

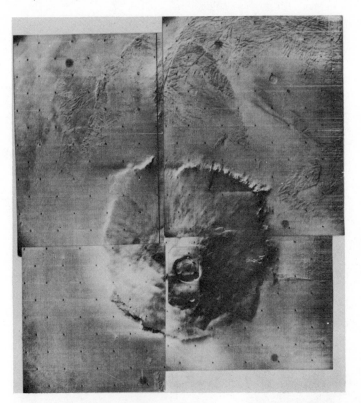

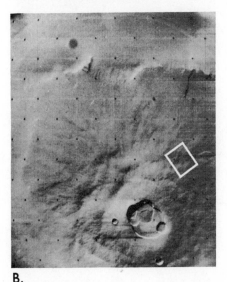

B.

C.

Figure 18-3. Nix Olympia, a gigantic volcanic mountain on Mars. A. The mountain is 310 miles in diameter across the base. The complex crater is 40 miles in diameter. B. Closer view of the summit. C. Telephoto view of the flank showing features that resemble flow material on earth. Part C shows an area 27 x 34½ miles. Photos from NASA.

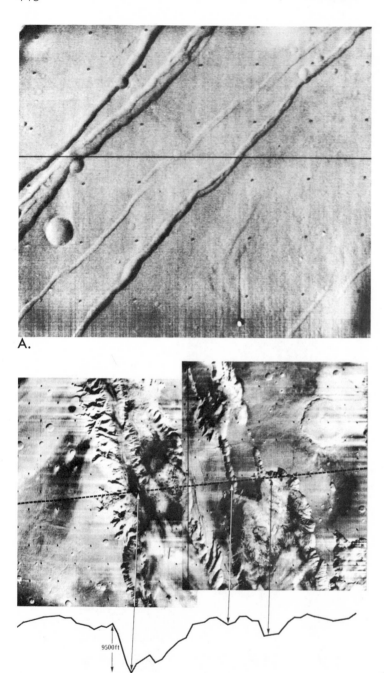

Figure 18-4. Valleys of probable fault origin on Mars. A. These structures are part of a system of parallel fissures extending more than 1100 miles. The widest valley is about one mile across and has a smaller valley inside it. The photo covers an area 21 by 26 miles. B. A valley on Mars much deeper than the Grand Canyon. This valley is 75 miles across. Photos from NASA.

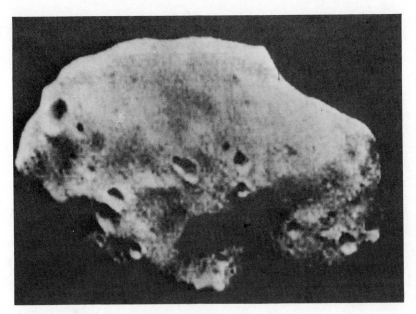

Figure 18-5. Mariner IX photograph of Mars' moon Phobos. Phobos is about 13 by 16 miles. The age and strength of Mars' innermost moon is shown by the impact crater. Photo from NASA.

seasonally. The other cap may be made of ordinary ice. If the water in this second ice cap melted at times, it might cause some of Mars' surface features. The pits and basins shown in Fig. 18-7 are near the south pole, and so could have formed by the melting of subsurface ice. Other areas have a chaotic terrain of short ridges and small valleys that resemble landslides or collapsed areas on Earth. (See Fig. 18-9.)

Venus is named for the goddess of love because it is visible only near sunrise and sunset. It is closer to the sun than is the Earth and thus hotter. It has a thick atmosphere, composed almost entirely of carbon dioxide and, as discussed in Chapter 10, the carbon dioxide causes a greenhouse effect, which would account for Venus' high temperature. Also present in minor amounts are hydrogen, water vapor, carbon monoxide, and hydrofluoric and hydrochloric acids. The surface temperature is near 536°C, making life a very remote possibility. The atmosphere has a solid cloud cover 45 to 60 miles thick so the surface is not visible. In spite of its slow rotation, the surface temperature does not change as much as expected from day to night, suggesting a circulation in the atmosphere that transfers the heat. Venus is the only planet that rotates on its axis in a clockwise direction (see discussion of Uranus below). No magnetic field has been detected, suggesting that it, too, lacks a liquid core, although its rate of rotation may be too slow to develop a magnetic field. Most of this recent information came from space probes that passed near Venus and from the Russian soft-lander in 1967. Radar observations suggest that Venus may have mountain ranges, but little can be said about their type or origin.

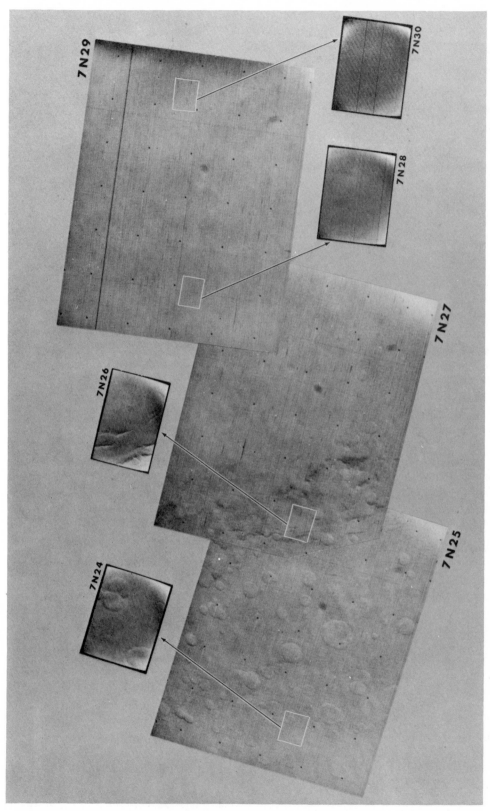

Figure 18-6. Boundary between cratered terrain and smooth or featureless terrain on Mars. The photos show the ridged boundary with the smooth terrain to the right. Mariner 7 photo from NASA.

7N29

7N30

7N28

7N27

7N26

7N25

7N24

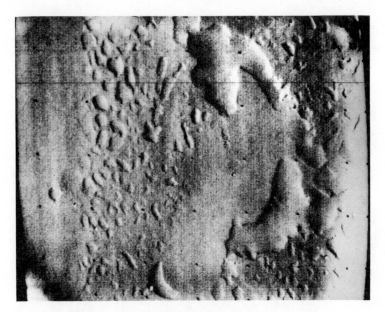

Figure 18-7. Pits and hollows on Mars. The small pits are one to two miles across. These features may have been formed by wind erosion, or they may be collapse features. Photo from NASA.

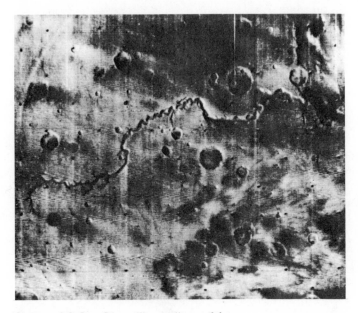

Figure 18-8. River-like gully on Mars. The origin of this 250-mile-long feature is unknown. Photo from NASA.

Figure 18-9. Chaotic terrain on Mars
is shown on the right. The chaotic terrain is a jumble of ridges, each of which
is one half to two miles wide and one to
five miles long. Mariner 6 photo from
NASA.

Great Planets

The great planets, Jupiter, Saturn, Uranus, and Neptune, are so called because they
are much larger than the other planets. Although larger in diameter and mass, they
are much less dense than the other planets. They all have thick atmospheres, which
are, of course, all that we can see. Their rapid rotation causes pronounced bulges of
the gaseous exteriors. The upper surfaces of their atmospheres are very cold because
of their distances from the sun. They all have methane in their atmospheres, with
ammonia, also, in Jupiter and Saturn; but most of their atmospheres are probably
hydrogen and helium. Their low overall densities suggest that they, like the sun, are
composed mainly of hydrogen and helium. Their cores are denser than the Earth's;
Uranus and Neptune may have rock cores, and the others may be liquid. The temperatures of the surfaces of the great planets are not known, and they may be cold;
gravitational compression of the thick atmospheres may have caused the surface temperatures of Jupiter and perhaps Saturn to be as high as 2,000°C, making their structure

more like that of the sun than a planet. Saturn's rings are unique and probably are dust similar in composition to a moon. (See Fig. 18-10.) This material is too close to the planet to consolidate and form a moon. The rings may be left over from the formation of the planet, or they might be satellites that approached the planet too closely and were broken up when the gravitational attraction overcame their strength. Uranus is unique in having its rotational axis tilted 98 degrees to its orbital plane; alternately, it could be considered tilted 82 degrees and revolving in the opposite direction to the majority of the planets. Its moon's orbit is in the same direction as its rotation.

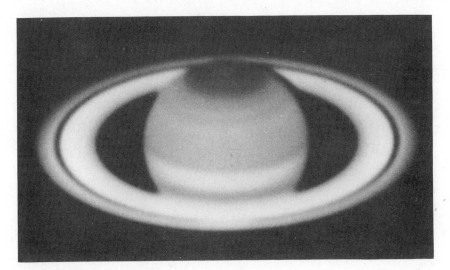

Figure 18-10. Saturn and its ring system. Photo from Mount Wilson and Palomar Observatories.

Moon

Just as the planets orbit the sun, most of the planets have moons, or satellites, that revolve around them. (See Table 18-2.) Most of the moons revolve in the same direction as the planets, but some move in the opposite direction (retrograde). Most of the moons are much smaller than their planets. Ours is much closer in size to the Earth so that the Earth-moon system should probably be considered a double planet. The moons that orbit in the opposite direction and have very inclined orbits may have been captured by the planets instead of having been formed with the planets. Origin of satellites is a problem discussed later.

The moon orbits the Earth and rotates on its axis at the same rate and so only one side of the moon is visible from the earth. (See Fig. 18-11.) Our only knowledge of the far side of the moon comes from space probes that have photographed the far side. They have revealed a cratered surface with almost no smooth areas. (See Fig. 18-12.)

The visible face of the moon has large, dark, smooth areas and lighter colored, more rugged areas. The early astronomers called the dark areas seas [maria (pl.), mare (sing.)] and the lighter areas, lands, in a fanciful analogy to the earth. The craters range from hundreds of miles across down to a few inches in diameter. Most of these craters

Table 18-2. Satellites of the planets. The planets that are not listed do not have known moons. Data are mainly from Werner Sandner, *Satellites of the* *Solar System,* American Elsevier Publishing Co., Inc., New York, 1965, 151 pp.

Satellite of	Distance from planet (thousands of miles)	Diameter (miles)	Orbital eccentricity	Inclination of satellite's orbit to orbit of planet (degrees)	Remarks
Earth	238.9	2,160	0.055	5	
Mars	5.8	9-12	0.017	2	
	14.6	6-9	0.003	2	
Jupiter	113	100	0.003	3	
	262	2,450	0.000	3	
	417	2,040	0.000	3	
	666	3,560	0.001	3	
	1,170	3,340	0.007	3	
	7,120	100	0.152	28	
	7,290	40	0.207	28	
	7,300	16	0.132	28	
	13,000	16	0.200	160	Retrograde
	14,000	19	0.207	163	Retrograde
	14,600	40	0.378	147	Retrograde
	14,700	20	0.222	155	Retrograde
Saturn	52	100-200	Unknown	Unknown	Discovered Dec. 1966
	115	400	0.019	27	
	148	500	0.000	27	
	183	800	0.000	27	
	234	750	0.002	27	
	327	1,100	0.001	27	
	760	2,600	0.029	27	
	920	310	0.104	26	
	2,200	990	0.028	16	
	8,030	200	0.166	175	Retrograde
Uranus	80.8	Unknown	Unknown	98	⎫ Rotation axis at almost right angle to orbit, so can also be considered to be inclined at 82° and have retrograde movement.
	119	370	0.007	98	
	166	250	0.008	98	
	272.6	620	0.002	98	
	364.5	500	0.001	98	⎭
Neptune	220	2,500	0.000	139	Retrograde
	3,700	180	0.760	Small	

probably were formed by meteorite impacts. The moon, unlike the Earth, is not protected by an atmosphere, so more meteorites hit the moon than the Earth. The relative ages of various parts of the moon's surface can be determined by the density of the craters, the older surfaces having more craters. The present estimated rate of meteorite impact is not enough to have formed all of the moon's craters.

The craters, however, are of several types and may have more than one origin. The rayed craters are the most spectacular. (See Fig. 18-13.) The rays are believed to be material splashed from the crater by impact. They are visible from Earth at full moon and appear lighter colored in such illumination. The rays cover older craters, and the rays have more small, sharp, young craters than their surroundings. At places these young craters form in rows radial to the main crater, suggesting that their origin is from impact of material thrown out of the main crater.

Evidence of volcanic activity on the moon is abundant. Many of the craters may be volcanic in origin, and volcanic activity probably modifies many of the craters. Figures 18-13 and 18-14 show domes; and these domes could be volcanoes, laccoliths, or hills of other origins, partially buried by volcanic flows or other surface layers. Probably all of these features exist on the moon, but Figs. 18-13 and 18-14 show a number of domes with pits or craters at the top, strongly suggesting volcanoes.

Other probable volcanic features are the ridges shown in Fig. 18-14. Ridges like these are several hundred feet high and extend for hundreds of miles. They are almost entirely confined to the maria. Their origin is unknown, and it has been suggested that they may be caused by downsinking of the maria. More likely, they may be the location of dikes that fed the surface flows.

The nature of the moon's surface layer and surface processes is very different from the Earth's because of the lack of water and atmosphere. Downslope movements occur on steep crater walls. (See Fig. 18-15.) Some very slow erosion does occur because the older craters are more rounded than younger ones. This rounding is believed to be caused by the impact of micrometeorites and atomic particles from the sun.

Many other features of the moon have been studied and described. A history of the formation of the visible features has been developed from study of photographs. Discussion of most of these features is beyond the scope of the present work. One important point is that the moon has a rugged surface with many mountains; however, these mountains have apparently been formed by impacts and volcanic activity. Thus, they are very different from the compressive folded and faulted mountains of the Earth.

The rocks returned from the moon resemble earth rocks but with some important differences. In composition most of the rocks resemble earth basalt, but have much more titanium and much less of the more volatile elements such as sodium and potassium. Gas holes are common. The main minerals are plagioclase feldspar and titanium-rich pyroxene, so that the mineralogy and texture are much like earth rocks. (See Fig. 18-16.) The plagioclase feldspar, unlike most earth rocks, is almost pure calcium plagioclase. One unusual aspect is the presence of small blebs of glass of two different compositions inside some minerals. This suggests that at some stage in crystallization, the melt may have separated into two immiscible liquids like oil and water. Basalt, gabbro, anorthosite, and peridotite have all been returned from the moon. In texture, the surface rocks of the moon are more varied and include basalt and coarser-grained gabbro, breccia, glass, and a few fragments of plagioclase rock (anorthosite). Much of the material is dust size. (See Fig. 18-17.) The dust is apparently mainly made by meteorite impacts, but some appears to be volcanic ash. Melting and fracturing from impact shocks are visible in most samples, and spherical glass particles and iron-nickel

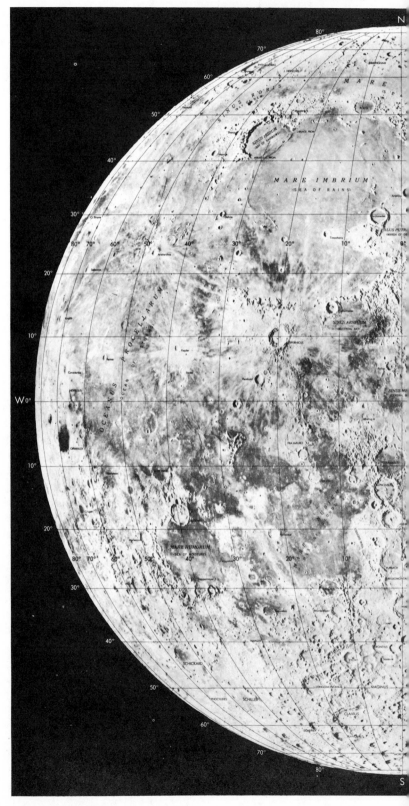

Figure 18-11. Map of the visible side
of the moon. U.S. Air Force, 1962.

Figure 18-12. Hidden side of the moon photographed by Lunar Orbiter III from an altitude of about 900 miles, looking south from a point about 250 miles south of the lunar equator. The prominent crater is about 150 miles in diameter, and its rim shows evidence of inward slumping like many of moon's craters. Photo from NASA.

Figure 18-13. Oblique view of the crater Kepler taken by Orbiter III from a height of 32.7 miles. Kepler has rays extending radially in all directions. The rays are believed to be the material splashed out by the impact and are visible from earth only near full moon. The light color streaks in the lower right may be ray material, especially as the associated smaller craters form lines radial from Kepler. Many domes are visible; and one in the foreground has a crater at its summit, suggesting that the domes may be volcanic. The line in the foreground may be a fault. Photo from NASA.

spheres of impact origin are common. (See Figs. 18-18 and 18-19.) The breccia contains fragments of all of these materials and is probably also formed by impact. (See Fig. 18-20.) Layering suggests that the surface formed in stages. (See Fig. 18-21.)

Most of the samples have been from the maria and so are basaltic in composition. The feldspar-rich rocks (anorthosite) are believed to come from the highlands. Other

Figure 18-14. An array of domes near the older impact crater Marius. The domes are from two to 10 miles in diameter and from 1,000 to 1,500 feet high. Many have pits on the summit and so resemble volcanoes. The prominent low ridges are also believed to be volcanic features, probably feeder dikes. Photo from NASA.

surveys made by satellite show that the composition of the highlands is different from the maria and that the highlands probably are made of rocks with larger amounts of feldspar than the maria.

Carbon compounds have been found in the moon rocks, but no trace of life has been found.

The moon rocks have specific gravities between 3.4 and 3.5. The entire moon is 3.36, suggesting that the whole moon cannot be of the same composition as these samples. Compaction of these samples at depth would form a much denser moon. Study of the orbits of spacecraft around the moon has shown that some parts of the moon are much denser than others. The areas of high mass concentrations (called mascons) are the large circular maria. The origin of these mascons is not known. They could be caused by large iron-nickel meteorites that were buried by impact, or they may be buried concentrations of mare basalt.

The seismographs left on the moon reveal that most of the moon's "earthquakes" are caused by tidal attraction and occur when the moon is closest to the earth. In comparison to the earth, the moon is an almost dead planet. Most of the moonquakes have magnitudes less than two on the Richter Scale and most appear to originate from a single active zone. Meteorite falls have also been recorded by the seismographs. Moonquakes build up and decay much slower than earthquakes, probably because of the thick surface layer of breccia on the moon. Seismic studies reveal that the outer part of the moon is layered. The surface to about 12 miles is apparently rubble. Between

Figure 18-15. Closer view of Copernicus from Orbiter II. Hummocky terrain in foreground is ejecta from the crater. Details of the sliding of the rim can be seen. The central peaks within the crater are about 1,000 feet high. The low area on the left foreground is apparently a fault trough filled by younger volcanic rocks. Photo by NASA.

Figure 18-16. Thin section of basalt from the moon. The texture closely resembles terrestrial basalt. White is plagioclase, gray is pyroxene, and black is ilmenite, a titanium mineral. Photo from NASA.

Figure 18-17. Astronaut Aldrin deploying the seismometer. Note the footprints in the fine surface material and the many pits and rocks on the moon's surface. To the left of the astronaut is the reflector that was used with a laser beam from earth to measure the earth-moon distance. Photo from NASA.

Figure 18-18. Glass spheres from the moon's surface. The largest is .5mm in diameter. Colors are clear, yellow, red, and black. Their origin is probably from melting during meteorite impact. Photo from NASA.

about 12 and 40 miles, the seismic velocity resembles that of mare basalt. At about 40 miles the velocity again increases.

The history of the moon deduced from the data collected in the space program begins with its formation about 4600 million years ago. This is the same date as the origin of the earth and the solar system. The moon probably was formed by accumulation of material in the same manner as the rest of the solar system—a topic that will be covered later. Although it is not confirmed, the moon probably melted at this time. The infall of material and the resulting compaction would cause heating. This initial melting of the moon would allow lighter material to float to the surface and heavier material to sink toward the center. To this point, the history of the earth, which will be detailed later, and the moon are similar, although these events did not occur simultaneously on both bodies. The oldest rocks so far found on the moon are the feldspar-rich rocks of the highlands that are about 4150 million years old. These are the rocks that are believed to have formed in the initial melting of the moon. Another result of the melting may have been a liquid iron core that produced a magnetic field. This would account for the residual magnetism found in some moon rocks. If such a liquid core did exist, it has now solidified so that the moon now has a very weak magnetic field.

The next major event in the moon's history was the outpouring of the basalt of the maria. This occurred between 3700 and 3200 million years ago. The mare basins must have formed at this time or earlier, and the basalt layers fill the basins. The mare basins may have been formed by the impacts of large meteorites. The formation of the mare basalt was apparently the last moon-wide event. Since that time, volcanism has probably occurred locally, and some volcanic activity may still be occurring. The source of the heat energy that melted the mare basalt was probably radioactivity. Local areas of high radioactivity have been measured, suggesting that this may also be the cause of any later volcanism.

Meteorites

Meteorites are objects from space that hit the Earth. Their composition and some other features were discussed earlier in relation to the layered structure of the Earth. The iron-nickel meteorites are believed to be similar in composition to the core of the Earth, and the stony meteorites similar to the mantle. One interesting aspect of meteorite recognition is that over 90 per cent of those seen to fall are stony meteorites, but only

A.

B.

Figure 18-19. Moon rocks. A. Basalt sample returned by Apollo 12. The cavity on the lower right was probably caused by a gas bubble. Such features are common in volcanic rocks. The specimen is 9 cm in diameter. B. This rock, also returned by Apollo 12, was collected from the bottom of a one-meter-deep crater. The bottom part is covered by a thin layer of glass that was probably produced by melting caused by impact. The rest of the rock also appears to be shattered by impact.

a third of the finds not seen to fall are stony. Only a few aspects of meteorites will be discussed here; much more information can be found in the references at the end of this chapter.

Meteorite specimens range from dust to 60 tons. Those that have produced the large impact craters on the Earth were much larger, weighing perhaps hundreds or thousands of tons. (See Figs. 18-22 and 18-23.) The shock wave caused by such impacts can cause metamorphism of the rocks as well as shattering. A few very high pressure minerals form in this way and are found only at suspected impact craters. Meteorite craters on the Earth are less than 10 miles across; however, on the moon some craters are more than 150 miles in diameter. Perhaps such large craters are not recognizable on the Earth. Many of the smaller meteorites do not survive the heat generated by friction in passing through the Earth's atmosphere, producing meteors, or shooting or falling stars. It is estimated that between 1,000 and 1,000,000 tons of meteorites fall on the Earth each year.

Meteorites are believed to have formed at the same time as the rest of the solar system (discussed in the next section). Thus, they are used to date the formation of the solar system and as models of the Earth's structure. Meteorites are fragments of much larger bodies. This is shown by the coarse crystalline nature of many, suggesting slow cooling that is possible only in a large body. Many, however, have complex crystallization histories. As shown in the next section, the Earth developed its layered structure by melting after initial formation. Some meteorites are also differentiated bodies, and this melting can only occur in bodies large enough to generate the necessary heat, again suggesting that meteorites are fragments of large bodies. Meteorites are generally believed to be fragments from the asteroid belt. The asteroids are a group of about 30,000 small bodies ranging from about 500 miles to about one mile in diameter.

C.

Figure 18-19. (cont.) C. Rock returned by Apollo 16. This fragment was broken from a larger rock. Gas holes are prominent. Photos from NASA.

B.

Figure 18-20. Breccia from the moon. A. Two centimeters in length. B. Closeup photo showing glass-lined pits, probably caused by impact. Large pits are about 1.5 mm in diameter. Photos from NASA.

Billions more down to dust size are also thought to exist. The asteroids are believed to have formed by collisions between two or more small planets. The orbits of many meteors have been recorded by radar and are consistent with an asteroid origin.

However, some of the meteor swarms that the Earth passes through each year are very different in that they are predictable and are not associated with large meteorite falls. They are in the orbits of now dead comets. Small particles of meteorite material, called micrometeorites, probably originate from such meteor swarms and so have their origin in comets. Micrometeorites of this or other origin are believed to cause the erosion of the moon. Comets are described next.

Figure 18-21. Core sample of moon's surface collected by the crew of Apollo 12. The fine-grained nature of the material is well shown. Other cores from this area show distinct layering. Photo from NASA.

Cosmic ray bombardment in space can cause the formation of new elements, and the amounts of such elements record the time that the object has been in space. The various types of meteorites have different exposure ages, again suggesting that meteorites have complex histories.

Comets

Comets are among the most awe-inspiring spectacles in nature. (See Fig. 18-24.) In the past they were considered harbingers of evil times. We now know their make-up in general terms and can predict their reappearance, but we know almost nothing about how they form. Comets have very small masses. One passed through Jupiter's moons without noticeable effect on the moons, and the Earth has passed through the tail of a comet in historic time. However, in 1908 a great explosion, heard 500 miles away, occurred in Siberia that was believed to have been caused by a meteorite. When the area was visited in 1927, although the forest was knocked down for over ten miles, no large crater was found. It has been postulated that the explosion might have been caused by a gaseous comet rather than a solid meteorite. Comets are believed to be masses of frozen gas with some very small particles similar in composition to meteorites. This is sometimes called the "dirty snowball" theory. In the case of the Siberian explosion the ice of the comet is believed to have turned to gas very rapidly by frictional

Figure 18-22. Meteor Crater (Barringer Crater), Arizona. A similar crater would not be as well preserved in a more humid climate. Photo from Yerkes Observatory.

heating in the atmosphere and the resulting shock wave caused the explosion and damage.

When distant from the sun, a comet is very small, perhaps a half mile in diameter. At this distance it is seen only by the sunlight that it reflects. When closer to the sun, the sun's radiation melts some of the ice and ionizes some of the resulting gases, making the comet more visible. Thus, comets disintegrate by passing near the sun and die after a number of passes near the sun. During the process of disintegration, the spectra of water, carbon dioxide, methane, and ammonia can be seen. At this stage an average comet may have a head 80,000 miles in diameter, made mostly of very thin gas. The solar wind, to be discussed in the next section, pushes this gas outward from the head to form a tail up to 50 million miles long. This is the spectacular stage of a comet and occurs only when the comet is near the sun. One interesting point is that the solar wind pushes the tail away from the sun without regard to the direction in which the comet is moving. Most comets have very elliptical orbits, which explains why they reappear at intervals. As with the planets, the sun is at one focus of the ellipse, but when distant from the sun, comets are too small to see. Some comets have orbits that bring them back only after thousands of years; some may be parabolic and so never return. (See Fig. 18-25.)

The origin of comets is not known; they may form at the fringe of the solar system or may come from outside the solar system. Their composition is not too unlike the material from which the solar system is believed to have formed (discussed in the next section). Thus they may be fragments "left over" from the origin of the solar system.

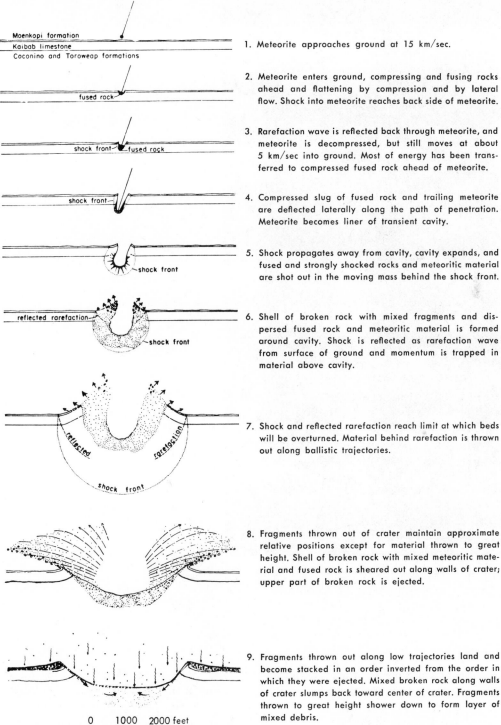

Moenkopi formation
Kaibab limestone
Coconino and Toroweap formations

fused rock

shock front — fused rock

shock front

shock front

reflected rarefaction
shock front

reflected   rarefaction

shock front

0   1000   2000 feet

1. Meteorite approaches ground at 15 km/sec.

2. Meteorite enters ground, compressing and fusing rocks ahead and flattening by compression and by lateral flow. Shock into meteorite reaches back side of meteorite.

3. Rarefaction wave is reflected back through meteorite, and meteorite is decompressed, but still moves at about 5 km/sec into ground. Most of energy has been transferred to compressed fused rock ahead of meteorite.

4. Compressed slug of fused rock and trailing meteorite are deflected laterally along the path of penetration. Meteorite becomes liner of transient cavity.

5. Shock propagates away from cavity, cavity expands, and fused and strongly shocked rocks and meteoritic material are shot out in the moving mass behind the shock front.

6. Shell of broken rock with mixed fragments and dispersed fused rock and meteoritic material is formed around cavity. Shock is reflected as rarefaction wave from surface of ground and momentum is trapped in material above cavity.

7. Shock and reflected rarefaction reach limit at which beds will be overturned. Material behind rarefaction is thrown out along ballistic trajectories.

8. Fragments thrown out of crater maintain approximate relative positions except for material thrown to great height. Shell of broken rock with mixed meteoritic material and fused rock is sheared out along walls of crater; upper part of broken rock is ejected.

9. Fragments thrown out along low trajectories land and become stacked in an order inverted from the order in which they were ejected. Mixed broken rock along walls of crater slumps back toward center of crater. Fragments thrown to great height shower down to form layer of mixed debris.

Figure 18-23. Diagrammatic sketches showing sequence of events in the formation of Meteor Crater, Arizona. From E. M. Shoemaker in Part 18, Report of XXI International Geological Congress, 1960.

Figure 18-24. Photo of Halley's
comet. Photo from Mount Wilson and
Palomar Observatories.

Tektites

Tektites are pieces of glass found on the Earth, both on the surface and in young
sediments. They are foreign to the places where they are found, occurring as rare objects
scattered in several distinct fields. (See Figs. 18-26 and 18-27.) Some have unusual
shapes that can be shown to be aerodynamic shapes that a melted object would assume
traveling at high speed in the atmosphere. Their chemical composition is similar to that
of the continental crust. Their origin is a problem, and several theories have been

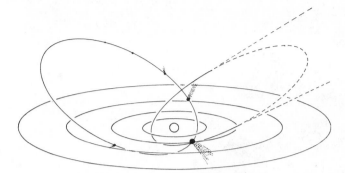

Figure 18-25. The solar system showing some possible comet orbits. The comet is shown on one orbit. Note that the tail always points away from the sun and that the comet gets larger where it is heated by the sun. The other orbit shows possible elliptical and parabolic orbits.

proposed. One theory suggests that they are fragments thrown out of the moon by meteorite impacts. The distribution pattern and other features of the 700,000-year-old tektites are consistent with an origin from one of the rays of the moon's crater Tycho. Another theory is that they are earth material dislodged by meteorite impacts. A third theory is that they may originate from comets because microtektites have been found near the site of the 1908 impact mentioned in the preceding section. Their shapes and distribution are consistent with any of these theories.

Recent studies suggest that at least some tektites may be of terrestrial origin. In each field the potassium 40-argon 40 dates are similar. Because argon, a gas, would be lost in melting, this is the date of melting. Rubidium-strontium dates for the fields studied have dates similar to the age of the bedrock at nearby meteorite-impact craters. These latter elements, unlike argon, would not be lost during melting. However, the distribution is not radial to the possible origin crater.

Recently microtektites were found in sediments in many parts of the Pacific and Indian Oceans by core sampling. Interestingly, the microtektites were found at the point in the cores where the Earth's magnetic field reversed. This suggests the possibility that whatever event caused the tektites might also have caused the magnetic reversal. The total amount of microtektites in the Indian and Pacific Oceans is estimated at a quarter of a million tons. The level at which the microtektites and the magnetic reversal are found also corresponds to the level at which a change in fossil life occurs.

Origin of the Solar System

The origin of the solar system is a problem that man has attempted to solve for many years. No completely satisfactory theory has yet been offered, and objections exist to all theories, including the one presented here. The best theory offered to date states that the planets and the other orbiting objects formed at the same time the sun formed. Other theories assumed that the planets formed from material pulled from the existing sun, generally by the gravitation of another star that happened to pass nearby. The many physical objections to this latter theory suggest that it is impossible. Only a brief outline of the newer theory can be given here.

A.

B.

Figure 18-26. Tektites. A. Typical tektites from Indochina. B. An actual tektite (right) compared with a synthetic tektite formed in a high speed wind tunnel that simulated the conditions of entry into the earth's atmosphere. The map in the background shows the area where the tektite was found. A. from Ward's Natural Science Establishment, Inc., Rochester, N.Y., B. from NASA.

Stars are believed to form by the condensation of a cosmic dust or gas cloud called a *nebula* (cloud), composed mainly of hydrogen and helium and perhaps frozen like snow. This condensation is caused largely by gravitational attraction. Condensation causes the temperature to rise; and when the temperature rises high enough, nuclear reactions begin and the new star begins to radiate energy.

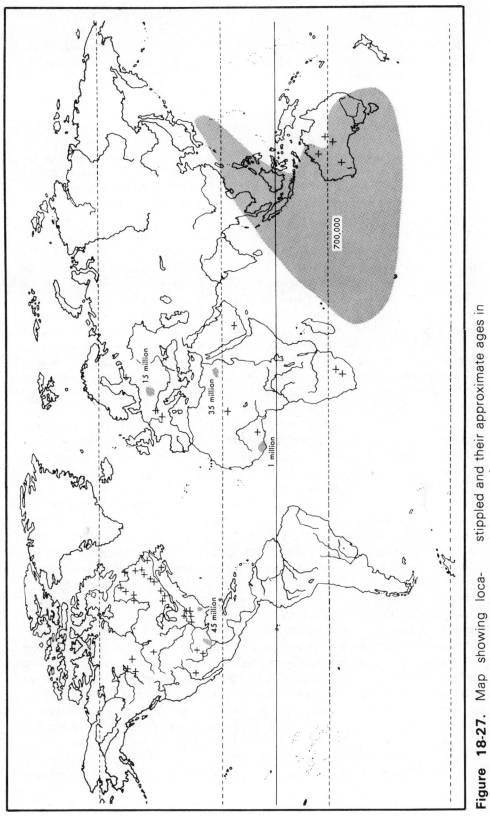

Figure 18-27. Map showing locations of impact craters and possible impact craters (crosses). Tektite fields are stippled and their approximate ages in years are given.

The planets are believed to have formed in a similar way during the origin of the sun. During condensation, such a gas cloud would almost certainly rotate because of the initial velocities of the particles brought together. As condensation continued, the rotation became more uniform as the random velocities cancelled out. Ultimately, the cloud flattened and became disk-shaped with most of the mass concentrated near the center. Further contraction and flattening made the disk unstable, and it broke into a number of smaller clouds. These smaller clouds were all in the same plane. Condensation occurred in each of these clouds, concentrating the heavier elements at the center. Because the nebula had about the same composition as the sun, the ancestral planets were surrounded by envelopes of hydrogen and helium. At about this stage, condensation had heated the ancestral sun to the point that nuclear reactions began. The sun's radiation ionized the remaining part of its nebula; and these ionized particles interacted with the sun's magnetic field, reducing the sun's angular momentum (spin).

Thus, the sun was slowed to its present slow rate of rotation, and the particles in the dust cloud gained in energy. Ionized particles radiated from the sun swept away the remaining gaseous dust surrounding the sun and the planets in the same way that the sun's radiation makes the tail of a comet point away from the sun, regardless of the direction of motion of the comet. These same particles from the sun form the cosmic rays and the Van Allen radiation belts discussed in Chapter 2. In this process the planets nearest the sun lost the most material. They are much smaller and denser than the more distant planets. As each planet condensed, it rotated faster, just as a spinning skater spins faster when she brings her arms in to her body. This theory explains the size and density of the planets, the distribution of momentum in the solar system, and why most of the planets revolve in the same direction.

The asteroid belt between Mars and Jupiter may have originated from a number of small centers of accumulation rather than from a single center, as did the other planets. Collisions among these bodies have increased their total number, and it is the resulting fragments from larger bodies that are the meteorites.

The moons formed close to the planets in the same way. When the planets lost much of their original masses because of the sun's radiation, the gravitational attraction between the planets and their moons was reduced. As a result, the moons moved further away from the planets. Some of these satellites may have escaped from their original planets to be captured by other planets. Such a transposition could account for the moons that revolve in the opposite direction. Pluto may once have been a moon of Neptune. Some of these erratic moons may even be captured asteroids.

This theory accounts for many other features of the solar system. The composition of the crust of the earth was our starting point in Chapter 2, and this can now be discussed further. The estimated composition of the universe is shown in Table 18-3; this is probably representative of the composition of the nebula from which the solar system condensed. Note that hydrogen predominates, that helium forms most of the rest, and that all the other elements constitute less than one per cent of the total. Most astronomers believe that originally the universe consisted entirely of hydrogen, the simplest atom, and that all other elements have been made by combining hydrogen atoms by a process called fusion. Such fusion can occur at the high temperatures, millions of degrees, inside stars. The development of the elements is believed to begin with the condensation of a gas cloud, or nebula, composed of hydrogen. The resulting star produces energy by atomic fusion of hydrogen to helium. Some of the possible reactions were described in Chapter 2. As the star ages, a stage is reached when the atomic "fuel" in the core has been nearly used up. At this time the temperature increases greatly and energy is produced by the formation of heavier elements from helium. At this stage, energy may be produced in the core faster than it can be radiated from the surface and an explosion results. The explosion may disintegrate the star, making it a *supernova* (a very bright, short-lived "star"); or the star may lose its material in less spectacular ways. The complete history of stars is complex and not completely known; however, it is known that the history of a star depends greatly on its mass, and that larger stars move through this cycle much more rapidly than do average-size stars, such as our sun. In any case, a star formed initially of hydrogen produces heavier elements and scatters these elements into the universe. The solar system formed from a dust cloud that contained heavy elements as well as hydrogen.

The correspondence between the composition of the universe and the composition of the earth's crust is quite apparent. The light elements, hydrogen and helium, are most abundant but were lost during the formation of the earth, as described above. The next most abundant element in the universe is oxygen, which is most abundant in the crust. The other elements are listed in Table 18-4.

Earth-Moon System

Interaction between the earth and the moon is believed to be slowing the earth's initially rapid rotation. This process is a continuing one, and the length of day is increasing. The year, the period of the earth's trip around the sun, is not affected because it is determined by the distance between the sun and the earth. Thus the number of days in the year is decreasing. The present average rate of slowing of the earth's rotation is estimated at between one second per day per 120,000 years to two seconds per day per 100,000 years. This is enough to produce apparent errors if the times of eclipses computed back in terms of our present day are compared with the ancient records. An ingenious check on this lengthening of the day in the geologic past has been devised. Some corals grow by adding layers much like tree rings. Thinner layers, apparently daily, occur within the annual layers. These daily layers are commonly not well preserved, but the few that can be counted suggest that in Middle Devonian time, about 350,000,000 years ago, the year had 400-410 days and in Pennsylvanian time, about 280,000,000 years ago, 390 days. (See Fig. 18-28.) The figures are in good agreement with the estimated slowing of the earth's rotation. Other types of fossils also show daily, monthly, tidal, and annual cycles, and it may soon be possible to determine the early history of the earth-moon system from such studies.

The slowing of the earth's rotation is caused by the tidal drag of the moon. The total rotational energy of the earth-moon system is conserved, so the energy lost by the earth is gained by the moon. The moon moves farther from the earth, and the length of the month, the moon's period around the earth, increases. Calculations show that ultimately the earth's rotation and the moon's orbital period will be the same, so that one part of the earth will always face the moon just as only one part of the moon now faces the earth. Similar calculations can be made to find the moon's past position and orbit. The results of such calculations depend to a large extent on the assumptions made. This is unfortunate, because otherwise the problem of the moon's origin and early history probably could be solved.

The origin of the moon cannot be determined, and all theories so far proposed have serious objections. The theory that the earth and the moon formed at about the same time at different centers of accumulations in a dust cloud was described above. The main objection to this is that both earth and moon should have the same composition if they formed in this way, but their average specific gravities are different. The average specific gravity of the earth is 5.5, and that of the moon is 3.3. Capture hypotheses were also mentioned in that earlier section, especially in regard to moons of other planets. Some of the recent calculations of the moon's early orbit strongly suggest a capture origin. However, if the earth captured the moon, the orbital energy (kinetic energy) of the moon would have heated the earth, probably to near its melting point, an event not

Table 18-3. Cosmic abundance of the chemical elements.

Element	Atomic number	Approx. atomic weight	Relative number of atoms*	Relative per cent of atoms
Hydrogen	1	1	25,000,000.0	86.64
Helium	2	4	3,800,000.0	13.16
Oxygen	8	16	25,000.0	0.09
Neon	10	20	14,000.0	0.05
Carbon	6	12	9,300.0	0.03
Nitrogen	7	14	2,400.0	0.008
Silicon	14	28	1,000.0	0.003
Magnesium	12	24	910.0	0.003
Sulfur	16	32	380.0	0.001
Argon	18	40	150.0	0.0005
Iron	26	56	150.0	0.0005
Aluminum	13	27	95.0	0.0003
Calcium	20	40	49.0	0.0002
Sodium	11	23	44.0	0.0002
Nickel	28	59	27.0	0.00009
Phosphorus	15	31	10.0	0.00003
Chromium	24	52	7.8	0.00003
Manganese	25	55	6.8	0.00002
Potassium	19	39	3.2	0.00001
Chlorine	17	35	2.6	0.000009
Cobalt	27	59	1.8	0.000006
Titanium	22	48	1.7	0.000006
Fluorine	9	19	1.6	0.000006
Vanadium	23	51	0.2	0.0000007
Copper	29	64	0.2	0.0000007
Zinc	30	65	0.2	0.0000007
Lithium	3	7	0.1	0.0000003
Strontium	38	88	0.06	0.0000002
Krypton	36	84	0.04	0.0000001
Scandium	21	45	0.03	0.0000001
Germanium	32	73	0.03	0.0000001
Beryllium	4	9	0.02	0.00000007
Boron	5	11	0.02	0.00000007
Selenium	34	79	0.02	0.00000007
Lead	82	207	0.02	0.00000007
Zirconium	40	91	0.014	
Gallium	31	70	0.009	
Yttrium	39	89	0.009	
Rubidium	37	85	0.006	
Bromine	35	80	0.004	
Barium	56	137	0.004	
Tellurium	52	128	0.003	
Xenon	54	131	0.003	
Arsenic	33	75	0.002	

Table 18-3. Cosmic abundance of the chemical elements (continued).

Element	Atomic number	Approx. atomic weight	Relative number of atoms*	Relative per cent of atoms
Molybdenum	42	96	0.002	
Tin	50	119	0.001	
Platinum	78	195	0.001	
Ruthenium	44	102	0.0009	
Cadmium	48	112	0.0009	
Neodymium	60	144	0.0009	
Niobium	41	93	0.0008	
Palladium	46	107	0.0007	
Dysprosium	66	162	0.0007	
Iodine	53	127	0.0006	
Cerium	58	140	0.0006	
Erbium	68	167	0.0006	
Osmium	76	190	0.0006	
Cesium	55	133	0.0005	
Lanthanum	57	139	0.0005	
Gadolinium	64	157	0.0005	
Iridium	77	193	0.0005	
Ytterbium	70	173	0.0004	
Mercury	80	201	0.0004	
Silver	47	108	0.0003	
Thallium	81	204	0.0003	
Bismuth	83	209	0.0003	
Rhodium	45	103	0.0002	
Antimony	51	122	0.0002	
Praseodymium	59	141	0.0002	
Samarium	62	150	0.0002	
Holmium	67	165	0.0002	
Indium	49	115	0.0001	
Europium	63	152	0.0001	
Hafnium	72	179	0.0001	
Tungsten	74	184	0.0001	
Gold	79	197	0.0001	
Terbium	65	159	0.00009	
Thulium	69	169	0.00009	
Rhenium	75	186	0.00005	
Lutetium	71	175	0.00004	
Thorium	90	232	0.00003	
Tantalum	73	181	0.00002	
Uranium	92	238	0.000008	
Technetium	43	99		
Promethium	61	147		
Polonium	84	210		
Astatine	85	211		
Radon	86	222		
Francium	87	223		
Actinium	89	227		
Protactinium	91	231		
Neptunium	93	237		
Plutonium	94	242		
Americium	95	243		
Curium	96	243		
Berkelium	97	245		
Californium	98	246		
Einsteinium	99	253		
Fermium	100	254		
Mendelevium	101	254		
Nobelium	102			
Lawrencium	103			

*The number of atoms, or the cosmic abundance, in the fourth column is relative to a value of 1,000 for silicon, atomic number 14. The figures in this column are from A. G. W. Cameron's revision of an earlier list by H. E. Suess and H. C. Urey.

suggested by the earth's thermal history. Many other ideas have been suggested, but all have physical objections. The exploration of the moon by man will probably provide much new evidence in the near future.

Figure 18-28. Growth lines in coral. The fine bands are daily growth layers. By counting the number of such bands within the yearly growth bands, it is possible to determine the number of days in the year at the time the coral lived. Photo courtesy of S. K. Runcorn.

Table 18-4. Comparison of the estimated composition of the crust, total solid earth, and universe. The figures for the total earth and the universe are much more uncertain than those for the crust. The compositions of the whole solid earth and the universe are discussed in Chapters 13 and 18 respectively.

Element	Atomic number	Approx. atomic weight	Abundance (weight per cent)		
			Crustal rocks[1]	Total solid earth[2]	Universe[3]
Hydrogen	1	1	0.14	—	61.0
Helium	2	4	0.0000003	—	36.8
Carbon	6	12	0.02	—	0.25
Nitrogen	7	14	0.002	—	0.08
Oxygen	8	16	46.60	30.0	1.0
Silicon	14	28	27.72	15.0	0.06
Aluminum	13	27	8.13	1.1	0.006
Iron	26	56	5.00	35.0	0.02
Calcium	20	40	3.63	1.1	0.006
Sodium	11	23	2.83	0.57	0.003
Potassium	19	39	2.59	0.07	0.0003
Magnesium	12	24	2.09	13.0	0.05
Titanium	22	48	0.44	0.05	0.0002
Phosphorus	15	31	0.105	0.10	0.0006
Manganese	25	55	0.095	0.22	0.0008
Nickel	28	59	0.0075	2.4	0.004
Sulfur	16	32	0.026	1.9	0.02
Chromium	24	52	0.010	0.26	0.001
Cobalt	27	59	0.0025	0.13	0.0002

[1] From Table 2-1.
[2] From Brian Mason, *Principles of Geochemistry*, John Wiley & Sons, 3rd ed., 1966, p. 53. (These data are estimates, and because only two significant figures are used the total is 101 per cent.)
[3] Calculated from Table 18-3.

QUESTIONS

18-1. List the components of the solar system.

18-2. How do the inner planets differ from the great planets?

18-3. In spite of the different sizes and densities of the planets, why is the origin of the earth inseparable from that of the solar system?

18-4. How can the different numbers of moons that each planet has be explained?

18-5. Which planets might contain life? Why?

18-6. What is the origin of the moon's surface features?

18-7. How do comets differ from meteorites?
18-8. How long would it take for meteorite falls to produce the earth? Is this a reasonable theory for the origin of the earth?
18-9. The uranium-lead and thorium-lead methods are used to date meteorites in spite of the problems in correcting for original lead. Why are not other radioactive methods used?

SUPPLEMENTARY READING

PLANETS

Ebbighausen, E. G., *Astronomy* (2nd Ed.) Columbus, Ohio: Charles E. Merrill Publishing Co., 1971, 160 pp. (paperback).
Faul, Henry, *Ages of Rocks, Planets, and Stars.* New York: McGraw-Hill Book Co., 1966, 109 pp. (paperback).
Ohring, George, *Weather on the Planets.* Garden City, New York: Doubleday and Co., Science Study Series, 1966, 144 pp. (paperback).
Sagan, Carl, and J. N. Leonard, *Planets.* New York: Life Science Library, Time, Inc., 1966, 200 pp.

MARS

James, J. N., "The Voyage of Mariner IV," *Scientific American* (March, 1966), Vol. 214, No. 3, pp. 42-52. (Mainly on the flight, not the results.)
Leighton, R. B., "The Photographs from Mariner IV," *Scientific American* (April, 1966), Vol. 214, No. 4, pp. 54-68.
Öpik, E. J., "The Martian Surface," *Science* (July 15, 1966), Vol. 153, No. 3733, pp. 255-265.
Sloan, R. K., "The Scientific Experiments of Mariner IV," *Scientific American* (May, 1966), Vol. 214, No. 5, pp. 62-72.

VENUS

James, J. N., "The Voyage of Mariner II," *Scientific American* (July, 1963), Vol. 209, No. 1, pp. 70-84.

MOON

Alfvén, Hannes, "Origin of the Moon," *Science* (April 23, 1965), Vol. 148, No. 3669, pp. 476-477.
Baldwin, R. B., *A Fundamental Survey of the Moon.* New York: McGraw-Hill Book Co., 1965, 149 pp. (paperback).
Dyal, Palmer, and C. W. Parkin, "The Magnetism of the Moon," *Scientific American* (August, 1971), Vol. 225, No. 2, pp. 63-73.
Hess, Wilmot, and others, "The Exploration of the Moon," *Scientific American* (October, 1969), Vol. 221, No. 4, pp. 54-72.
Hibbs, A. R., "The Surface of the Moon," *Scientific American* (March, 1967), Vol. 216, No. 3, pp. 60-74.
Mason, Brian, "The Lunar Rocks," *Scientific American* (October, 1971), Vol. 225, No. 4, pp. 48-58.

Runcorn, S. K., "Corals as Paleontological Clocks," *Scientific American* (October, 1966), Vol. 215, No. 4, pp. 26-33.

Schurmeier, H. M., *et al.,* "The Ranger Missions to the Moon," *Scientific American* (January, 1966), Vol. 214, No. 1, pp. 52-67.

Science, Apollo II issue, (30 January, 1970) Vol. 167, No. 3918.

Shoemaker, E. M., "The Geology of the Moon," *Scientific American* (December, 1964), Vol. 211, No. 6, pp. 38-47.

Wood, J. A., "The Lunar Soil," *Scientific American* (August, 1970), Vol. 223, No. 2, pp. 14-23.

METEORITES

Dietz, R. S., "Astroblemes," *Scientific American* (August, 1961), Vol. 205, No. 2, pp. 50-58. Reprint 801, W. H. Freeman and Co., San Francisco.

Heide, Fritz, *Meteorites.* Chicago: Phoenix Science Series, The University of Chicago Press, 1963, 144 pp. (paperback).

Mason, Brian, *Meteorites.* New York: John Wiley & Sons, Inc., 1962, 274 pp.

Urey, H. C., "Meteorites and the Moon," *Science* (March 12, 1965), Vol. 147, No. 3663, pp. 1262-1265.

TEKTITES

Barnes, V. E., "Tektites," *Scientific American* (November, 1961), Vol. 205, No. 5, pp. 58-64. Reprint 802, W. H. Freeman and Co., San Francisco.

Faul, Henry, "Tektites are Terrestrial," *Science* (June 3, 1966), Vol. 152, No. 3727, pp. 1341-1345.

Glass, B. P., and B. C. Heezen, "Tektites and Geomagnetic Reversals," *Scientific American* (July, 1967), Vol. 217, No. 1, pp. 32-38.

ORIGIN OF THE EARTH

Kuiper, G. P., "Origin, Age, and Possible Ultimate Fate of the Earth," in *The Earth and Its Atmosphere* (D. R. Bates, ed.), pp. 12-30. New York: Science Editions, Inc., 1961, 324 pp. (paperback). (This book was also published under the title, *The Planet Earth.* London: Pergamon Press, 1957.)

Urey, H. C., "The Origin of the Earth," in *Nuclear Geology* (Henry Faul, ed.), pp. 355-371. New York: John Wiley and Sons, Inc., 1954, 414 pp. This paper reprinted in J. F. White, (ed.), *Study of the Earth,* pp. 286-401. Englewood Cliffs, N.J.: Prentice-Hall, Inc., 1962 (paperback).

Chapter 19

Evolution and Fossils

In the following chapters the development of life is traced from its origin to the present. In this chapter the way that man recognized the meaning of the diversity of both fossils and present life is described. Indeed, it was not always recognized that fossils are the remains of once-living organisms; and even after this was generally conceded, separate creations were invoked to account for the diversity and for the numerous gaps in the record.

Evolution

Evolution is a biologic process, but it is also important to geology. The evidence for evolution comes both from fossils, the province of the geologist, and from living organisms, the province of the biologist. Biological evolution is the continuing change in populations of organisms that is now occurring and has occurred in the geologic past. In a sense, evolution is the biologic counterpart of the continuing changes, described in Parts I, II, and III, that have occurred and continue to occur on the earth. Because of its importance in man's intellectual development, evolution will be discussed historically.

At one time man considered himself at the apex of a static earth, located at the center of the universe. Copernicus indicated that the earth is dynamic and not at the center of the universe. Darwin, whose name is almost synonymous with evolution, showed

483

that in the broad view man occupies only a small, probably temporary, place on the earth. Such revolutionary changes in viewpoint, understandably, have not gone unchallenged.

An important step forward occurred in 1737 when Carolus Linnaeus published his book *Systema Naturae*. In this book he attempted to classify all living organisms in a logical system. He used two names, *genus* and *species,* for each kind of organism, and this system is still in use. The tenth edition of this book, published in 1758, is the generally accepted start of this system of classification. The value of this system is that it indicates that there are degrees of similarities and differences among organisms rather than random variations.

Although there is no completely satisfactory definition of species, it can be considered as a group of similar individuals that can interbreed to produce fertile offspring. A genus is a group of closely related species. Thus the cats belong to the genus *Felis;* the common house cat is *Felis catus,* and the ocelot is *Felis pardalis* (which is written *F. pardalis* if the genus has been mentioned nearby). The other terms in the system such as family, order, phylum, and so on are shown in Table 19-1. To the biologist working

Table 19-1. Classification of the more important fossil-forming organisms.

A. Classification scheme and an example.

Kingdom	Animalia
Phylum	Chordata
Class	Mammalia
Order	Primate
Family	Hominidae
Genus	*Homo*
Species	*sapiens*
Individual	John Doe

Most organisms identified by genus and species (in italics).

with living organisms, a species is a group that can interbreed, and the whole system attempts to show the blood relationships among organisms. The geologist working with fossils has the same goal, but he must work with dead remains of organisms. A typical fossil is a clam or a snail shell, so the geologist must define a species and deduce the relationships to other species on the basis of shell morphology alone. As will be described in later chapters, in rare cases he may have more evidence on which to base his conclusions. Because he works with form and structure alone, a geologist defines a species on the basis of a single fossil or a group of fossils called the type specimen(s). All other fossils referred to this species must be compared to the type(s). This is done by using descriptions, pictures, plaster casts, other specimens from the same location as the type, and direct comparison with type(s). Linnaeus laid this foundation for Darwin about one hundred years before Darwin left on the voyage of the *Beagle,* and in this intervening time the classification was refined and expanded.

Geology developed rapidly near the beginning of the 19th Century, when the geologic time scale was set up as was described in Chapter 17. Evolution was an idea mentioned by a few workers. In 1809, Jean-Baptiste Pierre Antione de Monet, Chevalier de Lamarck, described his theory of evolution. He believed that all organisms evolved from a single ancestor, and that the environment created the need for change or evolution. His mechanism, however, was not correct. Lamarck believed that the offspring inherited the acquired characteristics of the parents, and, in turn, passed them on to their children. For example, Lamarck believed that a blacksmith who developed large biceps would pass these muscles on to his children. We now know that traits are not inherited in this way, and many in Lamarck's day did not believe it either.

One of Lamarck's critics was Georges Léopold Chrétien Frédéric Dagobert Cuvier, who was mentioned earlier in connection with his studies in the Paris basin. Cuvier studied the vertebrate fossil bones found in the sedimentary beds of the Paris basin. He was an anti-evolutionist and attacked Lamarck on two grounds. Lamarck believed that organisms became more complex through evolution, but Cuvier noted that the fish that he found in ancient rocks were already very complex animals. He also could see no evidence of gradual evolution in the fossils that he studied. The bones in each successive bed were different from those in the underlying bed, and there was no trace of any of the intermediate steps necessary in evolution. For these reasons he believed in a separate creation for each bed, and that the change in rock type from bed to bed was caused by the revolution that killed the organisms and so set the stage for the next creation. Cuvier did some excellent work in his studies of fossil vertebrates. He founded modern comparative anatomy, and it is ironic that comparative anatomy is one of the more important lines of evidence for evolution. If the skeletons of vertebrates are compared, similar bones are found in each, although they may serve a somewhat different function in each. In some cases, the correspondence is very close (Fig. 19-1). It is highly unlikely that most vertebrates would have the same bones unless all evolved from a single common ancestor. Surely the most efficient design of a fin, a bird wing, and a human hand would not all use the same bones for such different functions.

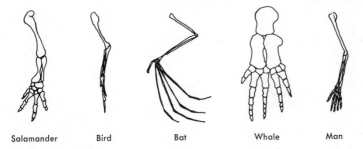

Salamander   Bird   Bat   Whale   Man

Figure 19-1. Similarity of bones in forelimbs of vertebrates. Similar bones serve different purposes in each of the examples. It seems very unlikely that such different limbs should have similar bones without evolution.

The stage is now set for Charles Robert Darwin, who was born on the same day as Abraham Lincoln. In 1831, when he was 22, he sailed on the *Beagle* as naturalist on a five-year voyage around the world. During the early days of the voyage, while fighting seasickness, he read Charles Lyell's *Principles of Geology* that had been published the preceding year. Lyell had gathered in his book much evidence for uniformitarianism, and showed that the earth is much older than was then currently thought. This book had a great influence on Darwin, and it started him thinking of slow biologic changes over an immense length of time.

During the voyage, Darwin made many observations, both biologic and geologic, but it is his study of finches in the Galápagos Islands that has become best known. The Galápagos Islands are 500 to 600 miles off the coast of Ecuador, too far for these birds to fly. Darwin found 13 species of finches that are unknown elsewhere in the world, although they are closely related to South American finches. The 13 species vary mainly in the size and shape of their beaks, and this reflects differences in their food. (See Fig. 19-2.) Darwin recognized the importance of his discovery at the time, but he did not

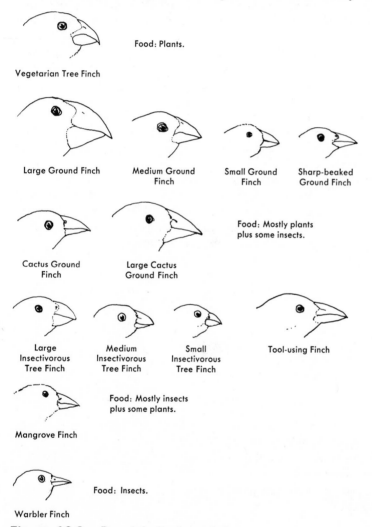

Figure 19-2. Darwin's finches of the Galápagos Islands. These finches apparently evolved from a common ancestor. Notice the differences in the shapes of their beaks and in the size of the birds.

develop his theory until later. Darwin believed that in some accidental way, such as being blown by a storm, a few mainland finches reached the islands, and from these birds the various species evolved, each species being adapted to a different food supply. The finches were able to take over the various food sources because no other birds except ocean birds are found in the Galápagos Islands—thus there was no competition.

Long after the voyage, in 1838, Darwin read Thomas Malthus' essay on population that had been published forty years before. Malthus' main point is that populations grow faster than their food supplies, and so food supply will ultimately control the world's population, an idea that is causing much concern at the present time. Malthus used the phrase "struggle for existence" and noted that most organisms produce more young than can easily survive. Darwin reasoned that those young with any slight advantage over the others would be the most likely to survive and produce offspring. He called this "Natural Selection" by survival of "the fittest."

As soon as Darwin realized that species are not fixed, he knew that he must answer all possible questions and produce an airtight case because he was sure to be attacked. He began to write, documenting the evidence for evolution. He produced a 35 page abstract in 1842, and in 1844 he allowed a few friends to read a 230 page essay. By 1858, after finishing his studies of the materials collected on the voyage of the *Beagle,* he had eleven chapters written of what would be *The Origin of Species* and was proceeding at a leisurely rate. At that point, he received an essay from a much younger man, Alfred Russel Wallace, that contained the main points of Darwin's work. Darwin was ready to abandon his work in favor of Wallace, but his friends urged a joint presentation. *The Origin of Species* was finally published late the next year. The attack that Darwin had expected came; it was bitter and still, in a few places, continues today. Up to Darwin's time, most geologists were trying to fit their geological history into Biblical history. Darwin finally broke this tradition. He was attacked because he believed that man descended from the apes, and because he believed that all species did not come from the Ark. Darwin's ideas on man were finally published in 1871 in the book *The Descent of Man, and Selection in Relation to Sex.*

One reason for the attacks on Darwin was that though he had documented the evidence for evolution, he had no process or mechanism to cause evolution to occur. He leaned toward Lamarckism, but he also recognized its weakness. Had Darwin known it, the mechanism had been discovered in 1865 and published in 1866 by Gregor Johann Mendel, an obscure Austrian monk. Mendel experimented over an eight-year period with garden peas. He used several characteristics, such as whether the peas were round or wrinkled. He discovered that if he cross-pollinated round and wrinkled peas, the first generation are all round peas. These hybrid round peas are not all the same, however, because the following generation is in the ratio of three round to one wrinkled pea. (See Fig. 19-3.) In succeeding generations, the wrinkled pea produces only wrinkled peas. In the same way one-third of the round peas produce only round peas in succeeding generations. The remaining two-thirds of the round peas produce in the next generation round and wrinkled peas in the ratio of three round to each one wrinkled. Mendel called characteristics such as roundness "dominant" and wrinkledness "recessive." He continued his experiments, using two characteristics such as color and roundness, and discovered again that the offspring occurred in fixed ratios. He showed that traits are inherited as discrete units. Mendel's genius was that he could apply mathematics to breeding. He was ahead of his time because no one else understood the importance of his discoveries. Darwin himself experimented with crossbreeding and

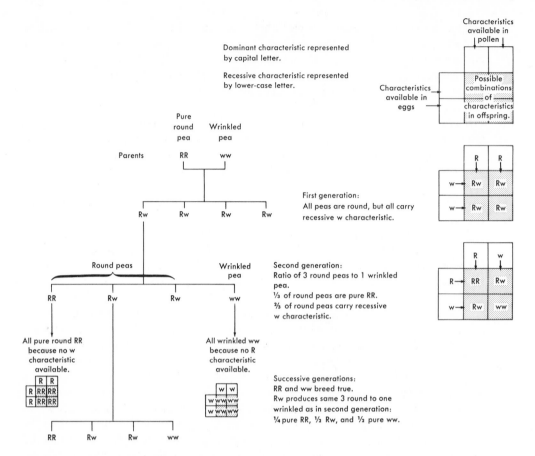

Figure 19-3. Mendel's inheritance law.

noticed the same three-to-one ratio, but the meaning and importance escaped him. Mendel's work lay buried until 1900 when three separate workers rediscovered both his findings and then his published work. The Darwin-Wallace simultaneous discovery and the triple rediscovery of Mendel's work are two examples of parallel discoveries that occur from time to time in science.

Mendel discovered that characteristics are passed to offspring by discrete units that are now called genes. At first, the mechanism was believed to be mutations that caused big steps. Darwin had referred to these as "sports." Further study and reflection show that most big changes are detrimental and the mutant generally dies before producing any offspring. Evolution is caused by mutation; but the changes are typically very small steps, and natural selection of these small changes is the mechanism of evolution. Genes change spontaneously, but, of course, at a very low rate. Changes are also caused by radiation. Natural radioactivity and cosmic rays can cause changes in genes, as can man-made radiation such as X rays. Many feel that such changes are the real danger from atomic bombs, and this is why atomic power plants are so carefully designed and monitored. Evolution can occur because in any species, especially if there are a large number of individuals, there is a large pool of genes. These genes are varied enough that favorable characteristics are available to meet almost any change in environment. Even if the mutation is only slightly favorable, it will, over a number of generations, replace the original characteristic and become the normal characteristic of the population. Thus, it is the population that undergoes evolution.

In the following chapters, the evolution of life from its origin to the present will be traced.

Adaptation, Radiation, and Extinction

As the history of the earth is outlined, it will be clear that some organisms have changed and others have not. Sponges, for instance, have not changed much during geologic history. They are very well suited to their form of life. Other organisms have changed or *adapted* to take advantage of new or changing conditions. These changes are generally a spreading out or *radiation* to different habitats. Radiation occurs again and again in the geologic record. A simple example is the fish that probably developed in fresh or brackish water and spread to every depth and water type. The rate of evolution has not been constant, but rather there have been times of rapid radiation.

A striking feature of evolution is the way that different organisms adapting to the same habitat develop similar characteristics. This is called *convergence,* and an example is the similarities between the present mammals and the marsupials of Australia. There is a marsupial in Australia to match almost every mammal habitat. For instance there is a marsupial "mouse," a marsupial "mole," and the Tasmanian "wolf." Other examples in reptiles of the Mesozoic Era will be described in Chapter 23.

Some of the most pronounced changes in life occurred at times when the environments changed, such as when the seas retreated from the continents. This could be more apparent than real, however, because of the unrecorded time represented by the resulting unconformity. When the environment changes, a species can do one of three things: it can migrate; it can adapt; or it can die out. The reasons for extinction are never simple and, generally, must be speculations because of the lack of evidence.

Disease epidemics have been suggested as the cause of the mass extinctions of the geologic past. The pathogenic organisms that cause disease generally only attack a single species so this cannot explain the extinction of many species, such as the dying out of all the dinosaurs at the end of the Cretaceous.

Climatic change has probably been the most popular explanation for mass extinctions. At many of the times that the mass extinctions of the geologic past occurred, mountain building also occurred and seas withdrew. These changes would, of course, drastically change the climate. Continental drift would also change climates. Many species would be unable to adapt or migrate and would die out, and new species, better adapted to the environment, would evolve. Close examination reveals flaws in this theory. At the times of great extinctions, the animal population changes drastically but not the plants, and plants seem to be at least as sensitive to climate as animals and probably more so. The last glacial events record a climate change, but the associated extinctions, which will be described later, occurred long after the main glacial advances.

Extinctions of some organisms and reversals of the earth's magnetic field correspond very closely. It was thought that radiation from space caused mutations that resulted in extinction. At the times when the earth's magnetic field reverses, the temporary lack of a magnetic field would allow high-energy particles from the sun to reach the earth.

If this effect did exist at times in the past, however, it would speed up evolution by causing more mutations. This effect should be most pronounced on terrestrial life. As noted before, plants are least affected at the times of mass extinction; and at most of these times marine life, somewhat protected from radiation by the sea, is most affected. The extinctions may occur because some organisms are adversely affected by a low-intensity magnetic field such as would exist at the time of reversal. Recent experiments have indicated that low-intensity magnetic fields have deleterious effects on a wide variety of organisms.

Another possibility is that many species must adapt or die if, in some way, an element on which they depend for food is destroyed or consumed by a more efficient competitor. This reduces the problem to finding a reason for the death of only one species instead of whole populations. Many of the theories offered above could explain the death of a single species. Many of the mass extinctions are associated with withdrawal of the sea, suggesting that such an event might shrink the environment of a key species, causing the demise of many, perhaps somewhat marginal, species. New species would develop when the seas again advanced, expanding the environment.

A variant of this theory is that most of the life of the ocean would be affected if the plankton died because the plankton, directly or indirectly, are the food for most forms of life in the ocean. The record suggests such a demise of at least some of the plankton. Chalk deposits of late Cretaceous age, composed almost entirely of the shells of plankton, are common in many parts of the world. These and other late Cretaceous deposits suggest that at many places this was a time of little erosion on the continents. It has been suggested that this lack of erosion reduced the amounts of inorganic nutrients in the ocean and so caused the dying of the plankton.

Changes in the amount of oxygen may have caused some extinctions. It was recently shown that the present-day representatives of organisms that underwent extinction need more oxygen than those that did not undergo extinction. The correlation is very high. Near the end of the Cretaceous Period, when a great many organisms died out, a new group of photosynthesizers expanded greatly. They are the coccoliths, and their remains form great chalk deposits, such as the White Cliffs of Dover, England.

In every case of extinction of a large group, some other group expands and so takes over the habitat. In most cases it is not possible to say why some groups died and others flourished.

Fossilization

Fossils are any evidence of past life, and this evidence may be preserved in many different ways. When most animals die, the body is rapidly destroyed; scavengers consume the body or bacteria cause it to decay. Only a very few escape this destruction and become fossils. Thus fossilization is a relatively rare event, and this is why the record of fossil life is so incomplete. Generally, if the body has hard parts, those parts are most likely to be preserved; and if the body is rapidly buried its preservation is more likely. For these reasons, the most common fossils are the shells of shallow-water marine animals such as clams.

The incompleteness of the fossil record makes reconstructions of past life very difficult. It has been estimated that only about one out of 44 species of the fossilizable marine invertebrates of the past is known. It is also estimated that of the 10,000 or so species found on a tropical river bank, only 10 to 15 are likely to be preserved.

Fossils form in the following ways:

Preservation of part of the animal, such as a clam shell.

Petrifaction by replacement, recrystallization or permineralization of the original shell or other part of the animal. This a very common type of fossilization, and most fossil shells are either recrystallized or replaced. Ground water is the agent that causes the replacement in most cases. Silica and calcite are the main replacing minerals, but many other minerals are also found as replacements. In some cases, very delicate internal structures may be preserved. Some of the best fossils with internal structures and surface ornamentations preserved are replacements by silica in limestone. In such cases, the fossils can be very easily recovered by dissolving the limestone in acid. (See Fig. 19-4H.) In permineralization the permineralizing mineral fills in the voids in the fossil, as silica does in petrified wood.

Molds of fossils form if the fossil is dissolved, leaving an empty space. (See Fig. 19-4A.)

Casts form if molds are filled with minerals or sediments. This is probably the most common way that marine fossils are preserved. (See Fig. 19-4A and B.)

Carbonization occurs if a fossil such as a leaf is squeezed, generally on a bedding plane, and the fluids are removed, leaving a thin carbon film. Worms and other animals without hard parts can be preserved in this way. Very tiny, delicate structures can be preserved by carbonization. (See Figs. 19-4C and I.)

Tracks and *burrows* are also evidence of life. (See Fig. 19-4J.)

Coprolites, or fossil excrement, can tell much about the eating habits of fossil life.

Frozen mammoths have been uncovered with every part of the animal intact. Such occurrences are, of course, very rare. (See Fig. 19-4E.)

Mummified animals are also quite rare. This could occur through the trapping and sinking of an animal in a tar seep; the tar would inhibit decay of the body. Peat bogs have yielded specimens with soft parts well preserved. Mummification can also occur in very dry climates.

A.

Figure 19-4. Types of fossilization. A. Mold and cast. Left is cast, right is mold of *Maclurites,* an Ordovician snail from New York (3 inches).

B.

C.

D.

Figure 19-4. (cont.). B. Internal cast of *Turritella,* a Cretaceous snail (length, 2 inches). C. A leaf preserved as a carbonaceous film. *Liquidambar* from Miocene of Oregon. D. Insect preserved in amber, Oligocene, East Prussia; insect preserved as a carbonaceous film.

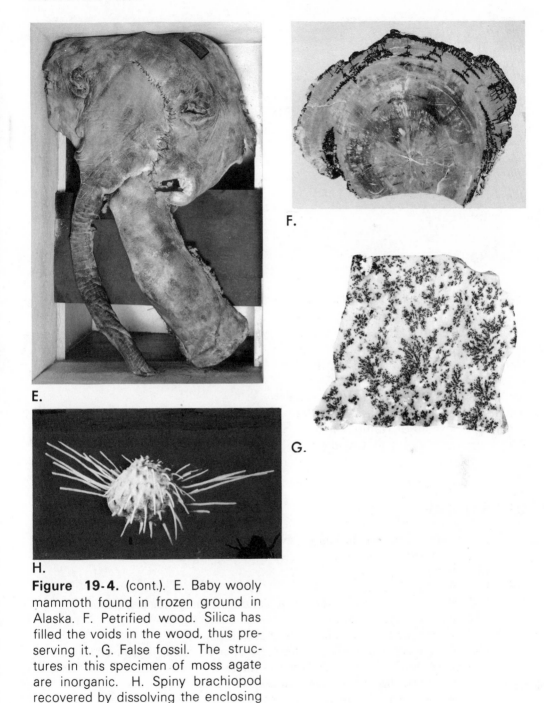

Figure 19-4. (cont.). E. Baby wooly mammoth found in frozen ground in Alaska. F. Petrified wood. Silica has filled the voids in the wood, thus preserving it. G. False fossil. The structures in this specimen of moss agate are inorganic. H. Spiny brachiopod recovered by dissolving the enclosing limestone with acid.

I.

Figure 19-4. (cont.). I. Delicate features are preserved in this worm-like animal by carbonization. J. Crawl marks. Parts A. and B. are from Ward's Natural Science Establishment, Inc., Rochester, N.Y. Parts C., D., F., G., H., and I. are courtesy of the Smithsonian Institution. Part E. is courtesy of the American Museum of Natural History. Part J. is courtesy of P. E. Cloud, Jr.

J.

QUESTIONS

19-1. List the evidence for biologic evolution.

19-2. Who first used genus and species in naming organisms? When?

19-3. What is a species?

19-4. What is a genus?

19-5. How does a geologist's use of species differ from a biologist's use?

19-6. What is a type specimen?

19-7. What were Lamarck's ideas about evolution?

19-8. What was Cuvier's evidence for separate creations? Is it good evidence?

19-9. What lead Darwin to his ideas on evolution?

19-10. What was the chief weakness in Darwin's *Origin of Species?*

19-11. What is Mendel's law of inheritance?

19-12. How do genes control evolution?

19-13. What is biologic radiation?

19-14. List and describe the types of fossilization.

19-15. Discuss the relationship between geologic changes and evolution of new species.

19-16. Describe several possible explanations for the extinction of species.

SUPPLEMENTARY READING

Bramlette, M. N., "Massive Extinctions in Biota at the End of Mesozoic Time," *Science* (June 25, 1965), Vol. 148, No. 3678, pp. 1696-1699.

Brues, C. T., "Insects in Amber," *Scientific American* (November, 1951), Vol. 185, No. 5, pp. 56-61. Reprint 838, W. H. Freeman and Co. San Francisco.

Collinson, C. W., *Guide for Beginning Fossil Hunters.* Illinois State Geological Survey, Education Series 4, 1959, 37 pp.

Dorf, Earling, "The Petrified Forests of Yellowstone Park," *Scientific American* (April, 1964), Vol. 210, No. 4, pp. 106-114.

Ericson, D. B., and Goesta Wollin, "Micropaleontology," *Scientific American* (July, 1962), Vol. 207, No. 1, pp. 96-106.

Hotton, Nicholas, 3rd., *The Evidence of Evolution.* New York: American Heritage Pub. Co., 1968, 160 pp.

Mathews, W. H., *Fossils, An Introduction to Prehistoric Life.* New York: Barnes and Noble, 1962, 337 pp. (paperback).

Moore, Ruth, *Evolution.* New York: Time, Inc., Life Nature Library, 1964, 192 pp.

Newell, N. D., "Crises in the History of Life," *Scientific American* (February, 1963), Vol. 208, No. 2, pp. 76-92. Reprint 867, W. H. Freeman and Co., San Francisco.

Simpson, G. C., *Life of the Past.* New Haven, Conn.: Yale University Press, 1961, 198 pp. (paperback).

Stirton, R. A., *Time, Life, and Man.* New York: Science Editions, 1959, 558 pp. (paperback).

Chapter 20

The Precambrian — The Oldest Rocks and the First Life

Anarchy is the law of nature, and order is the dream of man.

Henry Adams

One of the most important horizons in geologic history is the base of the Cambrian System where for the first time easily recognized fossils became abundant. This event occurred about 600 million years ago; and all of the time prior to that event, almost 90 per cent of geologic time, is called the Precambrian Era. The discussion of the origin of the earth in Chapter 18 was largely based on theoretical speculation and covered much of Precambrian time. This chapter is concerned with the classical meaning of Precambrian—that is, the history interpreted directly from study of old crustal rocks. All interpretation in geology is at least somewhat speculative, but Precambrian history is more speculative than that of younger parts of the geologic column. The reasons for this are many. The rocks are mainly metamorphic, making their origin difficult to decipher. They have been intruded by batholiths. They are generally deformed and may have been involved in more than one period of mountain making. (See Fig. 20-1.) At many places only fragments of once extensive rock units are preserved. Perhaps most important, dating is difficult, generally requiring extensive laboratory work. Thus, Precambrian history is even more sketchy, and more subject to major revision as new data become available, than the geologic history of later times.

Crystallization of the Interior of the Earth

The oldest crustal rocks so far dated by radioactive methods are 3,980 million years old. This is probably close to the time that the continental crust formed. The earth is

Figure 20-1. Deformed Precambrian metamorphic rocks in Northwest Territories, Canada. Such rocks are much harder to map than most sedimentary rocks. Photo by J. A. Donaldson, Geological Survey of Canada, Ottawa.

probably appreciably older than this. Meteorites yield dates of 4,550 million years; this is the time that they crystallized. As noted earlier, meteorites are dated using the uranium-lead and thorium-lead methods, and the correction for original lead is a problem. Meteorites are believed to have the same origin as the earth and to have the same composition as the interior of the earth. Thus the 4,550 million year date is perhaps the time that the interior of the earth crystallized. As will be seen shortly, this event is believed to have occurred sometime after the formation of the initially homogeneous earth. How much after is the problem because the size of the planet probably is important in determining the rate at which recrystallization occurred. At any rate, 4,500 million years is the time at which the bodies that became meteorites crystallized, and the solar system, including the earth, formed sometime earlier.

The earth, as originally formed, was probably nearly homogeneous and relatively cool. The contraction caused some heating, and radioactivity also must have caused heating. The radioactive heating must have been much greater than at present because much more radioactive material was present on the earth. (See Table 2-7.) This follows because after each half-life, half of the original radioactive element is gone. Meteorite-like material hitting the earth probably also caused local heating. The heating of the earth caused melting, and, under the influence of gravity, differentiation of the earth into core, mantle, and crust began. The heavier materials such as iron and nickel melted and moved toward the center, and the lighter silicates moved outward. The silicate mantle solidified from the bottom outward. The core, unable to lose its heat, remains liquid. The inner core may be crystallized. The radioactive elements that caused most of the heating—uranium, thorium, and potassium—all moved outward. These are all large atoms that do not fit into the iron-magnesium silicates, so they stayed with the melt as crystallization progressed from the core outward. The concentration of the radioactive elements near the surface enables their heat to be radiated at the surface. If they were still deep in the earth, their heat would be able to cause remelting.

The resulting layered structure of the earth has great geological significance. The fluid iron core makes the earth's magnetic field possible. The heat of the core may be responsible for convection currents in the mantle. As noted above, the other inner planets do not have magnetic fields, and may not have mountain ranges either. The moon also does not have folded mountains. Perhaps it is significant that the earth is the largest of the inner planets, and for this reason may be the only one with a liquid core. Surface geology on another planet would probably be much different from surface geology of the earth.

The actual formation of the earth is believed to have occurred 5,000 to 6,000 million years ago, perhaps closer to 6,000 million years. This estimate is reached from at least two lines of reasoning, but neither method gives more precise figures. The age of the sun is estimated to be about 6,000 million years. This figure is based on knowledge of stellar structure and rate of energy production, but it is not possible to discuss the complex physical reasoning further here. This does put an upper limit on the age of the earth, however, because the whole solar system is believed to have originated at the same time. The second line of reasoning is based on estimates of the time necessary for melting and recrystallization to occur. About 500 to 1,000 million years is a reasonable time for these processes, so this figure should be added to the 4,550 million years.

Origin of Oceans and Atmosphere

So far we have accounted for the main events between the origin of the earth, 5,000 to 6,000 million years ago, and 4,550 million years ago when the interior crystallized. Conventional geology begins with the oldest crustal rocks, 3,980 million years old. We know very little of the events in this interval between 4,550 and 3,980 million years ago. The continents had at least begun to form. After the initial continents formed in some unknown way, the crust was probably similar to that of the later times, but no details can yet be deciphered. Other events occurred during this time, however; the earth's initial rapid rotation was being slowed, as described earlier, and the atmosphere and the oceans began to evolve.

The present atmosphere and oceans have developed since the earth solidified, but the development is known only in general terms. Although the earth and the sun formed from dust clouds of probably the same composition, the earth's overall composition now differs from that of the sun. Therefore, it is clear that sometime after the initial formation of the earth by accumulation from a dust cloud, the earth lost much of its original thick gaseous envelope of mainly hydrogen and helium. We must account for the loss of these gases by processes not now operating because the earth is at present losing very little of its atmosphere. Much of the original dust cloud was lost when it was ionized by the sun's radiation and then swept away by moving ionized particles from the sun, as described above. The remaining gas nearer the earth was probably lost as a result of the rapid rotation of the earth, mentioned earlier, and as a result of the heating that melted the earth and formed the core and mantle.

It is difficult to reconstruct the next stage in the development of the early atmosphere; it may have consisted of such gases as ammonia, methane, and steam. This is the composition of the atmospheres of the great planets such as Jupiter (where the water is ice and hydrogen and helium are also present). In the case of the great planets, the source is believed to be the initial dust cloud. The earth, being nearer the sun, had lost by this stage the gases of the original dust cloud; however, frozen chunks of these gases left in the earth would have melted and been released during the later remelting of the earth to form this atmosphere. Evidence for the existence of this atmosphere is very weak. If such an atmosphere ever existed on earth, it probably was lost quickly because of the prevailing high temperature. Ammonia would be quickly decomposed by the sun's ultraviolet energy. Methane would cause fixation of carbon in the early sedimentary rocks, but there is no evidence that this occurred. The present atmosphere developed mainly from volcanic emanations; and the cooler, more slowly rotating earth retained these gases. Methane, ammonia, hydrogen, and water would form an atmosphere in which life could have begun; but as outlined here, the temperature was probably too high for life at the time the atmosphere had this composition. Life could also form in a volcanic-derived atmosphere. The origin of life will be discussed in more detail later. Returning to our main theme, volcanoes today emit mainly steam, carbon dioxide, and nitrogen, together with some sulfur gases; and there is no reason to believe that earlier volcanoes were different. Thus the atmosphere became steam, carbon dioxide, and nitrogen. With the cooling of the earth some of the steam condensed, and some oxygen may also have formed from the breakdown of steam. Such oxygen would recombine with hydrogen to form steam. The oxygen now in the atmosphere came from photosynthesis by plants after life began, and life probably could not form until the temperature was below 200°F. At the present time the atmosphere is 78 per cent nitrogen and 21 per cent oxygen, with all other gases forming about one per cent. The disposition of the volcanic gases probably follows this pattern: the steam has become water and is mainly in the oceans; the nitrogen is in the atmosphere; and the carbon dioxide has been partly broken down by photosynthesis to carbon in life and to oxygen in the atmosphere, and is partly in carbonate rocks.

This development of the atmosphere seems reasonable, and it is possible to work out a crude timetable. By about 4,000 million years, the crust had formed. By 3,375 million years, water was present because sedimentary rocks of this age have been found in southern Africa. Prior to that time, anaerobic life probably existed. Oxygen, however, was not present in the atmosphere because up to about 1,800 million years, clastic fragments of pyrite and uraninite are found in sedimentary rocks. If oxygen had been present, these minerals would have been oxidized. Another problem is that unless oxygen-mediating enzymes had preceded photosynthesizers, the oxygen would destroy the photosynthesizers. Banded iron formation may have kept the oxygen out of the atmosphere. Banded iron formation is found on every continent, but only in rocks 3,200 to 2,000 million years old and never in younger rocks. The iron could not be transported from the weathering site to the place of deposition in the oxidized state. Thus the iron must have been transported in solution in an unoxidized state, and oxidized at the place of deposition. It is believed that the photosynthesizers lived in the ocean, and that the iron combined with the biological oxygen and was precipitated to form banded iron formation. The banding or layering suggests a fluctuation between iron and oxygen.

Free oxygen appeared in the atmosphere at about 2,000 million years because from that time on, red beds—that is, clastic rocks in which the iron is oxidized—are common. Thick extensive red beds appear in the range 1,800 to 2,000 million years and so slightly overlap the last of the banded iron formation. Apparently at this time, oxygen-mediating enzymes formed, allowing life to expand rapidly. The life, probably algae, produced much oxygen; and as the oxygen accumulated in the atmosphere, iron was oxidized in weathering, preventing further development of banded iron formation.

At this stage, ultraviolet from the sun was lethal to life at the surface, so the organisms must have lived in water deep enough to shield them.

Some of the oxygen in the atmosphere was changed into ozone by ultraviolet radiation from the sun. Ordinary atmospheric oxygen is in the diatomic form, O_2, in which two oxygen atoms share electrons to form stable outer electron shells. In the ozone molecule, O_3, three oxygen atoms share electrons. Ozone is not stable and ultimately returns to ordinary diatomic oxygen. The formation of ozone absorbs much ultraviolet radiation that would otherwise reach the surface of the earth. Higher forms of life could not originate until the ozone layer had formed.

Precambrian Rocks and History

Precambrian time includes almost 90 per cent of geologic time. This section is about the rocks and the physical history that can be determined from them. This should be the longest and most interesting chapter in geologic history. However, the difficulties of dating these rocks, the repeated deformation many have had, and their fragmentary nature due to long erosion or younger covering rocks makes their study one of the most difficult and demanding aspects of geology. The story begins after the formation of the crust and extends to the base of the Cambrian System where extensive fossils begin and dating rocks becomes easier. All time correlation in Precambrian rocks must be done with radioactive dating. Locally, lithologic similarities or similar methods can be used to establish rock correlations. For these reasons, the decipherable history consists of scattered episodes.

From earlier chapters, it is obvious that geologic history, viewed very broadly, is a sequence of sea-floor spreading and the associated geosynclines. Such events have shaped the earth's surface. It seems likely that such events occurred in Precambrian time as well. However, the beginning had to be different. Somehow the first continent had to form. The record of these events is not clear, and so much speculation is involved in the following story.

The oldest rocks so far found are 3,980 million years old and are from western Greenland. Rocks 3,600 million years old are found at Barberton Mountain Land in southeast Africa and in Minnesota. At Barberton Mountain Land, the oldest rocks are volcanics, mainly andesite, and pyroclastic rocks with a few sedimentary beds. These rocks overlie peridotite and basalt that may be remnants of an ancient oceanic crust and Moho. The volcanic rocks are up to 6 miles thick and are overlain by two sedimentary sequences, each about two miles thick. All of these rocks have been intruded by granitic rocks, both during the emplacement of the layered rocks and afterward. The volcanic rocks are similar to those found on volcanic island arcs. A possible interpretation is that the volcanic rocks were erupted on an ancient oceanic-type crust. The volcanic rocks and the intruding granites formed a small continent, and erosion of this continent produced the upper layers of sedimentary rocks. Thus these rocks may record the formation of the continental crust.

Other Precambrian areas are similar to Barberton Mountain Land in that they consist of volcanic and sedimentary rocks surrounded by granitic rocks with most of the rocks metamorphosed. The amount of granitic rock appears to be greater than that associated with younger geosynclines. Thus it may be that in early Precambrian time, continents grew faster than later. Whether one or several continents formed in the Precambrian is not known, but Precambrian rocks are found on all continents. (See Fig. 20-2.)

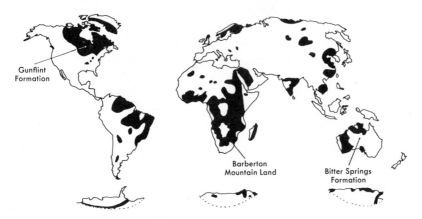

Figure 20-2. Areas where Precambrian rocks are exposed.

Some idea of the main episodes of Precambrian history can be obtained from a tabulation of radioactive dates. (See Fig. 20-3.) Such tabulation must be treated with caution because the rocks that have been dated are far from a random sample. However, the Precambrian rocks of North America have been fairly well studied so that the tabulation gives reasonable insight. Figs. 14-29 and 20-4 show the areal distribution of the different ages. Note in Fig. 20-4 that at places, younger rocks cut across the trends of older rocks. It is also apparent that Precambrian rocks underlie the central interior of North America where they are covered by younger rocks. This is shown by the Precambrian rocks exposed in the uplifted Rocky Mountains. Precambrian rocks also appear to underlie parts of the Appalachian Mountains.

The oldest rocks in North America appear to be about 3,600 million years and are in the Minnesota River Valley. The origin of these metamorphic rocks is not yet clear. Periods of extensive emplacement of granitic rocks occurred around 2,700, 1,800, 1,400, and 1,000 million years. The 2,700, 1,800, and 1,000 peaks may be significant because they mark times of granite emplacement on other continents.

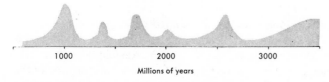

Figure 20-3. Ages of Precambrian granites of North America. After A. E. J. Engel, 1963.

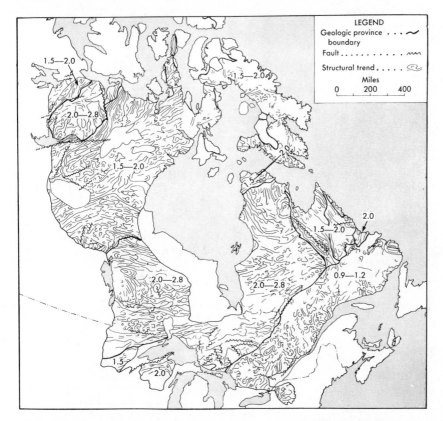

Figure 20-4. The Precambrian meta-morphic rocks of Canada. The structural trends (the foliation) of these much deformed rocks are shown. The numbers are the ages in billions of years. The extensions of the age provinces southward into the United States are shown in Figure 14-29 where these rocks are exposed mainly in the cores of the uplifted Rocky Mountains. Thus it appears that similar metamorphic rocks underlie much of the interior of the United States. The blank areas are places where sedimentary rocks are at the surface. Data from C. H. Stockwell, 1965.

Extensive exposures of Precambrian rocks are found in central, southern, and eastern Canada. (See Fig. 20-4.) Glaciation has exposed many of these rocks, and their mineral wealth has encouraged study. The region around the Great Lakes was the first to be studied. Four sequences of bedded rocks were distinguished. They were separated by unconformities and intrusive granites. Later radioactive dating altered some of the early interpretations. The early errors occurred because it had been assumed that the more highly metamorphosed rocks were older than the unmetamorphosed rocks. This is but one example of the pitfalls in the study of Precambrian rocks.

The growth of what is now North America is shown in Fig. 20-5. The extent of Precambrian rocks in both the Rocky Mountains and in the Appalachian Mountains can be determined in that figure. The size of the continent is only the minimum size at each time.

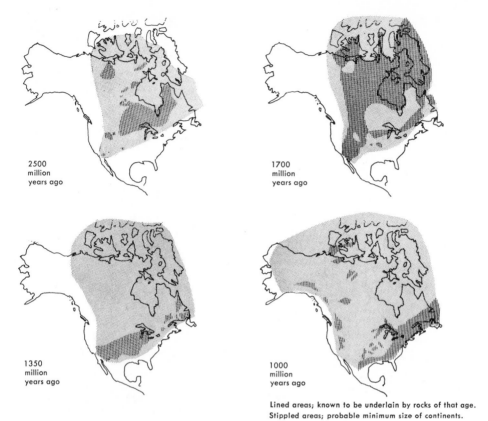

2500
million
years ago

1700
million
years ago

1350
million
years ago

1000
million
years ago

Lined areas; known to be underlain by rocks of that age.
Stippled areas; probable minimum size of continents.

Figure 20-5. Growth of North America in Precambrian time. The maps
show the minimum size based on radioactive dating of the rocks. After W. R.
Muehlberger, R. E. Denison, and E. G.
Lidiak, 1967.

So far, most of the rocks described have been metamorphic. Unmetamorphosed
Precambrian sedimentary rocks occur in both the Rocky Mountains and in the Appalachian Mountains. In the west these rocks are very thick. In the northern Rocky
Mountains, over 70,000 feet of sedimentary rocks were deposited in Idaho, Montana,
and Alberta. These rocks have been dated at between 850 and 1,450 million years. (See
Fig. 20-6.) Unconformably above these is another sedimentary sequence over 6,000 feet
thick. The second sequence is covered without notable unconformity by Cambrian
rocks at some places and unconformably at other places by rocks as young as Silurian.
Other thick sequences of Precambrian sediments occur in northern Utah in the Grand
Canyon and south of there in Arizona. At other places, Precambrian sediments grade
upward to Cambrian rocks without apparent unconformity and are described in the
next chapter with the Cambrian rocks with which they are more closely allied.

On all continents except Antarctica glaciation may have occurred in very late Precambrian time (750–850 million years). The evidence for this is rocks that resemble
glacial deposits in that they contain conglomerates and overlie striated rocks. Distinguishing between glacial deposits and other conglomerates is difficult. Similar rocks
2,000 to 2,400 million years old suggest a much earlier Precambrian glaciation.

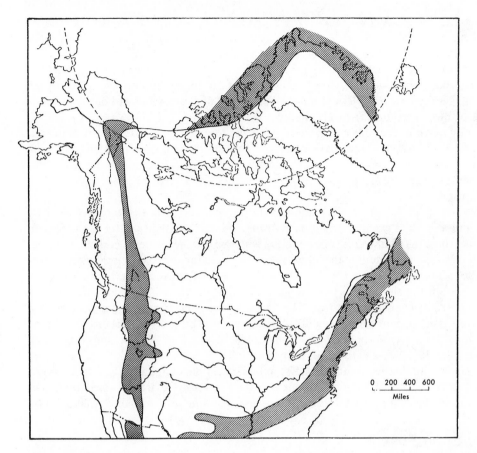

Figure 20-6. Late Precambrian sedimentary rocks were deposited in the areas shown. In parts of the western area, they accumulated to thicknesses up to 70,000 feet. This map was constructed from many scattered outcrops. These rocks may have been deposited elsewhere but subsequent erosion has removed them.

Origin of Life

All discussions on the origin of life are necessarily speculations. Life must have begun very early in the history of the earth, probably almost as soon as it became cool enough, perhaps 4,000 million years ago. The evidence of early life was reviewed when discussing the development of the atmosphere.

The steps involved in the formation of life are believed to start with the formation of amino acids. Next, these amino acids are brought together to form proteins, nucleic acids, and other compounds. Finally, the most difficult step is the transformation into living, self-replicating structures. The first of these steps is relatively easy. In a number of experiments, amino acids have been produced by passing an electrical discharge through a container of water, methane, ammonia, and hydrogen. This mixture was chosen because it contains the necessary elements and it is the composition of the atmospheres of some planets such as Jupiter. As discussed earlier, this mixture is believed to be one of the possibilities for the composition of the early atmosphere on the earth.

At the time life is believed to have formed on the earth, the atmosphere is believed to have consisted of nitrogen, carbon dioxide, carbon monoxide, and perhaps other volcanic gases. These gases dissolved in the ocean could produce amino acids using energy from ultraviolet radiation from the sun. Thus the early ocean would have been a very thin soup. The other steps necessary to form life could also take place in the ocean. The sun's ultraviolet radiation would probably destroy life without the protection of the ocean to absorb this energy.

Life could only have formed in an oxygen-free atmosphere or environment. If oxygen had been present, it would have destroyed the amino acids by oxidation. After life formed and oxygen was produced by photosynthesis, no new life could be formed. Thus all of the present life on the earth must have evolved from that first population of living organisms.

Clearly, if life formed in the manner described, it could form anywhere where these conditions exist. That is, it could have formed on another planet in the Solar System or elsewhere in the universe. The conditions needed are the elements carbon, nitrogen, and hydrogen, plus water in the liquid state. Such conditions must occur at many places in the vast universe, and so there must be other life in the universe. In the current space program, one of the goals is to search for life on Mars, the only planet with anything like the right temperature range. The materials returned from the moon have been carefully searched for evidence of life. Amino acids and other hydrocarbons were found in small amounts on moon rocks. They are believed to be of inorganic origin. On September 28, 1969, a carbonaceous meteorite fell at Murchison, Victoria, Australia. This meteorite that was dated as 4,500 million years old contains amino acids. These discoveries suggest that the conditions necessary to form life are not uncommon and that amino acids may have formed in pre-geologic time in the Solar System.

Precambrian Fossils

Until the late 1950's, very few Precambrian fossils were found; and many of the reported fossils found before that time are probably of inorganic origin. In recent years, microfossils have been found at several localities around the world, and larger fossils have been found in upper Precambrian rocks in a dozen or so places. The record is complete enough to be able to continue the story that began in the last section.

The first living organisms probably could not produce their own food but lived on material dissolved in the ocean. This situation could only exist for a short time; otherwise, the life would use all of the available nutrients. The development of photosynthesis probably made the first self-sustaining organism. The oldest sedimentary rocks so far found on the earth seem to confirm this theory. These rocks at Barberton Mountain Land in southern Africa are about 3,400 million years old. They contain tiny blue-green algae and bacteria, and chemical studies suggest that photosynthesis had developed.

The next younger abundant Precambrian fossils that have been studied are from the Gunflint Chert on the north shore of Lake Superior in Canada. (See Fig. 20-7.) Again, blue-green algae and bacteria are found, and these organisms were clearly photosynthetic. These rocks are about 1,700 million years old.

The organisms found at these two localities are primitive. Blue-green algae and bacteria are the only living things whose reproduction does not involve genetic material from two parents. They are essentially sexless, and so their form does not change. Thus these organisms in rocks over 3,000 million years old are much like living species.

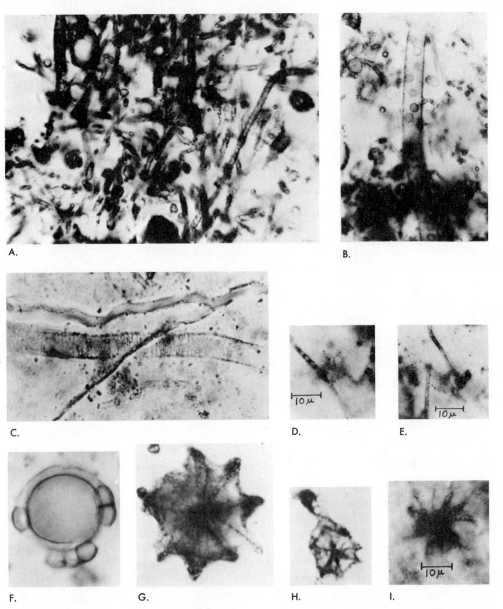

Figure 20-7. Microfossils from the Precambrian Gunflint Chert. A. Tangled filaments mainly *Gunflintia minuta,* and spore-like bodies *Huroniospora.* Diameter of field O. I mm. B. *Entosphaeroides amplus* with sporelike bodies within the filament. Diameter of field 40μ. C. *Animikiea septata* probable algal filament with transverse septae. Diameter of field 50μ. D. and E. Filaments with well-defined septae similar to iron-precipitating bacteria and blue-green algae. F. *Eosphaera tyleri.* Diameter of field 32μ. G. and H. *Kakabekia umbellata.* Diameter of field of G. is 20μ and H. is 12μ. I. Radiating structure similar to manganese- and iron-oxidizing colonial bacteria. One micron (μ) is one one-thousandth of a millimeter. Photos A, B, C, F, G, and H courtesy of E. S. Barghoorn; D, E, and I courtesy of P. E. Cloud, Jr.

Because genes are not involved, any mutation that occurs, in general, dies out in a few generations. This is not at all like the evolution described in Chapter 19.

The first organisms that have genes are found in rocks about 1,000 million years old at Bitter Springs in the Northern Territory of Australia. The organisms are green algae, and all green algae differ from blue-green algae in having sexual reproduction. Some of the fossils recovered from this chert show cell division. The development of these early forms of life is summarized in Fig. 20-8.

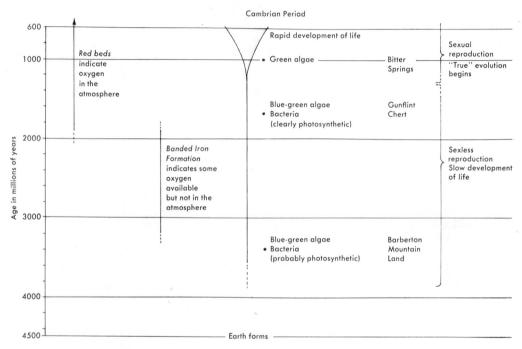

Figure 20-8. Development of Precambrian life and oxygen in the atmosphere.

These discoveries suggest that true evolution could not begin until about 1,000 million years ago. Up to that time, the organisms were too primitive and evolutionarily too conservative to change very much. After that time, evolution could proceed, and apparently did, because about 600 million years ago, life became abundant. Thus, in less than 500 million years, all of the many types of life found in Cambrian rocks may have evolved from green algae. Representatives of every phylum capable of fossilization are found in Cambrian or early Ordovician rocks.

Very little evidence of this evolution has yet been found. Perhaps the reason is that the organisms had few hard parts capable of fossilization. At the three localities just described, the algae were preserved in chert beds and were only found after searching with both optical and electron microscopes. Thus intensive search may unearth more evidence. Another possibility is that although the number of organisms was increasing rapidly during this time, only by the start of the Cambrian Period was life abundant enough to form abundant fossils. This is suggested by the discovery at ten places of a Precambrian fauna in rocks between 600 and 700 million years old. (See Figs. 20-9 and 20-10.) This suggests that by late Precambrian, life was becoming abundant but had few preservable hard parts.

Figure 20-9. Map showing locations where the unusual late Precambrian fauna shown in Figure 20-10 have been found. Other Precambrian faunas have been found elsewhere. After M. Glaessner.

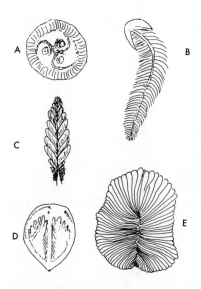

Figure 20-10. This unusual Precambrian fauna is found at the places shown in Figure 20-9. These organisms are found as impressions in sandstone. A. Circular form of unknown affinities, possibly with coiled arms. About ¾ inch. B. Probable worm. About 1 inch long. C. Leaf-like form of unknown affinities. About 7 inches. D. Oval form of unknown affinities. About ¾ inch. E. Worm-like form. About 1¾ inch.

The most common Precambrian fossils are algal mounds or columns similar to those made by living blue-green algae. (See Fig. 20-11.) Fig. 20-12 shows some unusual markings that may be worm tubes, or they may be of inorganic origin. Fig. 20-13 shows another type of marking found in Precambrian rocks. The example shown is thought by some to be a jellyfish. Crawl and scratch marks are found at many places in both young and old rocks. Fig. 20-14 shows both Precambrian and Cambrian crawl marks, and the similarities suggest that trilobites were present in some late Precambrian rocks.

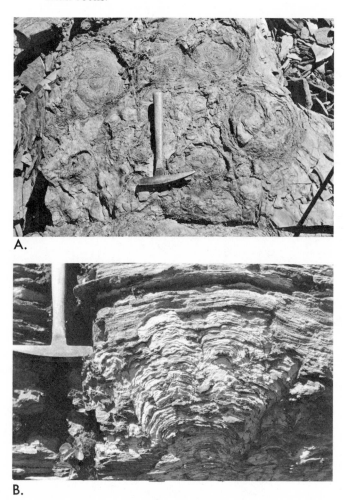

A.

B.

Figure 20-11. Algal structures in rocks about 1000 million years old in Glacier National Park, Montana. A. Top view, and B. side view of *Collenia undosa*. Photos by R. Rezak, U.S. Geological Survey.

Figure 20-12. Possible marks or tubes of a bottom-dwelling or burrowing organism. Some worms make similar markings. These markings are in sedimentary rocks 2000 to 2500 million years old. Scale in centimeters. Photo by Geological Survey of Canada, through the courtesy of H. J. Hofmann who first described these occurrences.

Figure 20-13. *Brooksella canyonensis,* a supposed jellyfish from Precambrian rocks in the Grand Canyon. Diameter of field is 3.5 inches. Photo from Smithsonian Institution.

Figure 20-14. Crawl and scratch marks, probably of trilobites. Marks such as these are commonly found both in Cambrian and younger rocks and in rocks well below (older than) the first trilobites. Parts A and C are from rocks of the Nama System, South West Africa, from well below the first trilobites. Parts B, E, F, and G are from the Deep Spring Formation, White Mountains, California, well below the first trilobites. Part D is from the Middle Cambrian Tapeats Sandstone, Grand Canyon, Arizona. Photos courtesy of P. E. Cloud, Jr.

QUESTIONS

20-1. Explain why the initial melting of the earth occurred sometime after the formation of the earth.

20-2. How did the earth's layered structure form?

20-3. When did the first life form?

20-4. Outline the steps by which the atmosphere is believed to have formed.

20-5. What is the significance of banded iron formation?

20-6. Can the beginning of Cambrian time be recognized at all places where sedimentary rocks of this age occur?

20-7. Where are the oldest rocks in North America found? What is their age?

20-8. What is the age of the oldest rocks so far found? Where are those rocks found?

20-9. Where are Precambrian metamorphic rocks exposed on North America?

20-10. Where are Precambrian sedimentary rocks exposed on North America? How thick are they?

20-11. How is life believed to have originated on the earth? Can this process account for the development of new types of life?

20-12. List the types and ages of the Precambrian fossils so far found.

20-13. What are some of the explanations for the sudden appearance of abundant fossils at the base of the Cambrian?

20-14. Why is Precambrian history so difficult to interpret?

SUPPLEMENTARY READING

Abelson, P. H. "Chemical Events on the Primitive Earth," *Proceedings of the National Academy of Sciences* (June, 1966), Vol. 55, No. 6, pp. 1365-1372.

Barghoorn, E. S., "Origin of Life," in *Treatise on Marine Ecology and Paleoecology*, Part 2 (H. S. Ladd, ed.), pp. 75-86. Geological Society of America, Memoir 67, 1957. This paper reprinted in J. F. White, (ed.), *Study of the Earth*, pp. 335-347. Englewood Cliffs, N.J.: Prentice-Hall, Inc., 1962 (paperback).

Barghoorn, E. S., "The Oldest Fossils," *Scientific American* (May, 1971), Vol. 224, No. 5, pp. 30-42.

Cloud, P. E., Jr., "Significance of the Gunflint (Precambrian) Microflora," *Science* (April 2, 1965), Vol. 148, No. 3666, pp. 27-35.

Eglinton, Geoffrey, and Melvin Calvin, "Chemical Fossils," *Scientific American* (January, 1967), Vol. 216, No. 1, pp. 32-43.

Engel, A. E. J., "Geologic Evolution of North America," *Science* (April 12, 1963), Vol. 140, No. 3563, pp. 143-152.

Engel, A. E. J., "The Barberton Mountain Land: Clues to the Differentiation of the Earth," Economic Geology Research Unit, University of the Witwatersrand, Information Circular No. 27, 1966. Revised, 1969. This paper reprinted in Preston Cloud, (ed.), *Adventures in Earth History,* pp. 431-445. San Francisco: W. H. Freeman and Company, 1970 (paperback).

Glaessner, M. F., "Pre-Cambrian Animals," *Scientific American* (March, 1961), Vol. 204, No. 3, pp. 72-78. Reprint 837, W. H. Freeman and Co., San Francisco.

Harland, W. B., and M. J. S. Rudwick, "The Infra-Cambrian Ice Age," *Scientific American* (August, 1964), Vol. 211, No. 2, pp. 28-36.

Keosian, John, *The Origin of Life.* New York: Reinhold Publishing Corp., 1964, 118 pp. (paperback).

Lansberg, H. E., "The Origin of the Atmosphere," *Scientific American* (August, 1953), Vol. 189, No. 2, pp. 82-86. Reprint 824, W. H. Freeman and Co., San Francisco.

Chapter 21

The Early Paleozoic – Cycles I and II

Mapping is the main tool of the geologist. Through mapping he learns the distribution and relationships of the various rock types exposed in the area under study. The rocks mapped are dated by comparison (correlation) with the type sections of the geologic column. The rocks under study are interpreted, using uniformitarianism. The resulting interpretations are integrated with similar interpretations of all other rocks of the same age to show the paleogeography (old geography) of that time. The interpretations of all parts of geologic time obtained in this manner are then synthesized to produce the earth's history. This is a very simplified sketch of how geologic history is interpreted. Note especially the many levels of interpretation. The evidence—the rocks—is commonly fragmentary, and establishing its contemporaneity may be very difficult.

Patterns in the Earth's History

The development of geology shows how the idea of patterns or cycles came about. All sciences had chaotic beginnings, and geology was more chaotic and slower to develop than were the other sciences because wide areas had to be studied before patterns began to emerge.

The first step was the establishment of the geologic time scale. The first systems to be defined were natural groupings; that is, they were rocks bounded by unconformities or other distinct horizons. It soon developed that the systems were, in general, natural subdivisions only at the places where they had been defined, and their boundaries transgress time as do all unconformities. At other places no natural boundary existed, and disputed zones were found where rocks had been deposited during the time of the unconformity at the type section. The type sections were not all in the same area, and this also led to correlation problems and other disputed zones. As a result of these problems, the type sections and extent of each system were defined arbitrarily. This established a workable time scale on which geology developed, but it is not in any sense a natural time scale. Because the time scale developed in Europe, it is even more unnatural in North America.

The recognition of geosynclines as important elements in geologic history was the next step. Geosynclines seem to provide a cyclic element in the history of continents, although they occurred at more or less different times on each continent. For the first time, though, a pattern emerged. The mountain-building episodes attracted the most attention in geosynclinal areas. Because of the batholithic intrusions, the metamorphism, and the later erosion, mountain building is difficult to date closely. However, it appeared that mountain building had occurred at roughly the same time at widely separated areas. When a closer look was taken, it became obvious that in many cases major mountain-building events at one place were being correlated with minor unconformities at other places. It also soon became apparent that at some place an unconformity could be found at almost any time in geologic history. Viewed on a world-wide scale, though, some parts of geologic time were very active and other times were fairly quiet. The quiet times were Cambrian and most of Triassic, and the more active times were Ordovician, Devonian, Pennsylvanian, and late Cretaceous. Thus a broad, fuzzy, world-wide pattern seemed to be emerging.

Another development has been the recognition of broad patterns of deposition and unconformity in the stable continental interior of at least North America. This pattern will be the framework of the brief outline of the geologic history of North America. Some of these unconformities can be traced into the geosynclines. It has been possible to recognize these cycles in the continental interior because there the rocks are relatively undeformed, are fossiliferous, and have been extensively studied in the quest for oil. Such a pattern requires a relative change in sea level. Deposition occurs when the continent is below sea level, and nondeposition and erosion occur when the continent is above sea level. Either the continents move up and down, or sea level changes, or both occur. If the continents rise and fall, one would expect tilting to occur rather than the overall changes observed, which suggests that sea level changes. Comparison with other continents will ultimately answer this question, but at present more mapping is necessary. The flat-topped seamounts and the atolls show changes in sea level in parts of the Pacific, suggesting that the shape of the ocean basins may change, causing a change in volume and thus a change in sea level. Such changes are probably associated with sea-floor spreading.

The pattern of geologic history of North America, then, is one of cycles of deposition followed by unconformity in the interior, and the geosynclines on the margins. This pattern has been ascertained for only the last 600 million years, the well-dated part of geologic time since fossil life became abundant. (See Fig. 21-1.) The present continental shelf, slope and rise may be geosynclines and Hudson Bay may be an example of a continental sea.

The study of fossils has also revealed some world-wide changes. Extinctions of large numbers of families occurred toward the end of the Cambrian, Devonian, Permian,

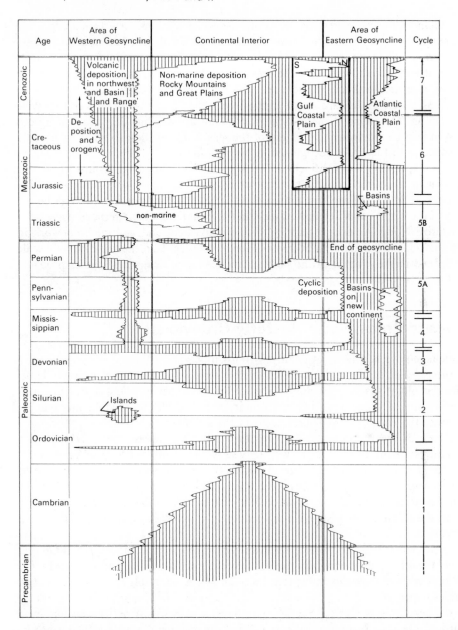

Figure 21-1. The geologic cycles of deposition in Central North America. This diagram shows in a general way where and when deposition occurred. Geosynclinal deposition in Texas and Oklahoma is omitted. Erosion and deformation cause many difficulties in constructing such a diagram. This is especially true in the lower Paleozoic of the far west. To gain a real understanding of the geologic history of this region, the student should add the rock types and their sources to the depositional sequences in the diagram. In general the times of non-deposition in the geosynclines correspond to orogenic periods. The vertical lines indicate non-deposition.

Triassic, and Cretaceous Periods. Soon after these extinctions, new forms appeared. The greatest extinctions occurred near the end of the Cambrian, Permian, and Creta- ceous Periods. The latter two mark the close of two of the eras, and it is these changes in life that were used to designate the eras. The Paleozoic (old life) Era ends with the Permian, and the Mesozoic (middle life) Era ends with the Cretaceous Period. The extinctions, although they are abrupt in the broad sense, took place over a considerable span of time. The cause of these extinctions is not known, but they may have been caused in part by changes in the extent of the seas. Some of these extinctions broadly correspond with the cycles recognized in North America.

Sea-floor spreading and continental drift seem to be the key to earth history. As we have seen, it is not yet possible to reconstruct the positions and movements of the continents in detail. When those data are available, it will be possible to describe the history of the whole earth in a unified way. In the meantime, the patterns or cycles that characterize North American geology form a convenient and natural framework in which to discuss the geological history of North America. Had the geologic time scale been defined in the interior of the North American continent, these cycles probably would have been the geologic periods.

Cycle I—Late Precambrian to Early Ordovician

Historical geology is concerned mainly with fossil-bearing sedimentary rocks because only these rocks enable reconstruction of the geography of an earlier time. Thus, historical geology is largely the story of the seas that have occupied the continents since Cambrian time. The nature of the sedimentary rocks enables us to make many infer- ences as to the nature of the source areas of the sediments.

The base of the Cambrian rocks with their abundant fossils is one of the most important points in geologic history. The base of the Cambrian System is generally defined as the first occurrence of certain fossils (trilobites); however, some place it at the first occurrences of easily recognized fossils. Thus, recognition of this important time line depends on chance preservation of fossils. In Cycle I the first Cambrian fossils occur in the midst of thick sedimentary sections so that dating the start of the cycle is impossible.

A general reconstruction of early Paleozoic geography is shown in Fig. 21-2. On this map and the other paleogeographic maps, the present outline of the continent and some of the present rivers are shown for reference. This does not imply that they existed at the times depicted.

At the beginning of Cycle I, the main framework of North American geology was established. Geosynclines occupied both sides of the continent as well as the far north. They probably record the existence of trenches associated with sea-floor spreading. As the cycle progressed, the seas spread over more and more of the continental interior; and by Early[1] Ordovician time, most of the interior was awash. The seas then with- drew, first from the interior, and finally from the geosynclines, ending the cycle. Cycle I is the simplest of the cycles.

---

[1]The terms *Early, Middle* and *Late* are used when referring to time. *Lower, Middle* and *Upper* are used when referring to rocks. All of these terms are capitalized when used in the formal sense.

Figure 21-2. Generalized paleogeography of the early Paleozoic. The Cambrian began with geosynclines occupying both sides of the continent, and the interior was dry. Precambrian sedimentary rocks are present in both geosynclines, indicating that they formed before the Cambrian began. As the Cambrian Period progressed the seas encroached on more and more of the continent until Late Cambrian time most of the interior was awash. The dashed lines indicate areas that were awash the least. During the Cambrian and much of the Ordovician the sediments were derived from the land areas interior of the continent. About the middle of the Ordovician and again in the Silurian, mountains were formed in the eastern part of the eastern geosyncline which were the source of much clastic debris. Ordovician, Silurian and Devonian sedimentary rocks near Hudson Bay record advances of the interior sea in that area. The crosses indicate the approximate position of the barrier that separates different trilobite faunas of Cambrian age.

Interior

Cycle I begins with the interior dry but surrounded by geosynclines. The seas gradually expanded over much of the southern, western, and far northern parts of the interior. (See Fig. 21-2.) Much of northeastern Canada and most of Greenland remained dry land, as did a broad arch from roughly Lake Superior to the Gulf of California. This pattern of land and sea was to be repeated but with a variety of minor variations in the later cycles. The land areas were undergoing erosion and provided the clastic sediments deposited in the seas. This is indicated by the fact that the sediments become coarser as these land areas are approached. The present continental interior is underlain by Precambrian metamorphic rocks, such as those presently exposed north of the Great Lakes. Erosion reduced these rocks to relatively smooth surfaces that underlie the Cambrian rocks. At some places, such as the Ozark Mountains of Missouri, Upper Cambrian rocks lap up against the flanks of the ancestral Ozark Mountains, showing that highlands did exist in the interior. The movement of the seas onto the interior was not a uniform advance, but probably had a number of short retreats and readvances. This is suggested by both rock types and changes in fossils.

At most places in the interior, the oldest rocks are sandstones. These rocks are generally less than a thousand feet thick, although they are much thicker at places. Except at the margins of the interior platform, the oldest rocks are Upper Cambrian in age. Lower and Middle Cambrian rocks are, for the most part, confined to the margins of the interior. The Lower Ordovician rocks conformably overlying these rocks are largely limestones with some shale. This change from clastic rocks to carbonates suggests that the land areas had been reduced to low relief by the end of Cycle I.

East

At the start of Cycle I, a geosyncline formed in the east. The oldest rocks are in the southern Appalachian Mountains. They consist of a sequence of 30,000 feet of clastic rocks. Overlying them is a 6,000-foot quartz-sandstone and shale unit, in which the first trilobites are found near the top. The two units are more or less conformable, although some workers consider the contact unconformable. At any rate, it seems clear that sedimentation began in Precambrian time. The source of these rocks was the interior of the continent. Overlying them are about 7,000 feet of Middle and Upper Cambrian shales and limestone. The amount of carbonate increased upward as the seas covered more and more of the interior. Above these are several thousand feet of Ordovician carbonates. The transition from Cambrian to Ordovician is in this conformable sequence of carbonates.

In the northern part of the geosyncline, volcanic rocks of Cambrian age are found with clastic rocks, suggesting that volcanic island arcs may have been present. Cambrian shale and sandstone up to 8,000 feet thick are found in Vermont. These are overlain by thick Lower Ordovician clastic rocks. Later mountain building has produced complex structures in New England and to the north.

In New England, the Cambrian trilobites are different on the two sides of the geosyncline, suggesting that a barrier may have separated them. (See Fig. 21-2.) The eastern trilobites are similar to European fossils, suggesting easier access to Europe than across Vermont. This distribution of fossils could have come about in many ways, and it is difficult to evaluate the various possibilities. It could mean a land barrier, a deep-water barrier, continental drift, or a difference in time—to name a few possibilities. Problems of this type occur throughout geologic history, showing another aspect of the difficulties in interpreting the geologic record.

West

The cycle also began in the West with the establishment of a geosyncline. The seas moved into the area from both north and south. A barrier existed between these two seas in Idaho until Middle Cambrian time. Although a much younger batholith obscures the evidence for the barrier, its presence is indicated by a difference in fauna between the two seas and by coarser clastic rocks near the old shoreline.

A very thick section of Cycle I rocks is found in the southwestern ranges of the Great Basin near the California-Nevada border. Here sedimentation began in Precambrian time. The oldest sedimentary rocks are an 8,000-foot sequence of clastics with some carbonates. Unconformably above those rocks are the Cycle I rocks. They begin with 1,500 feet of dolomite, followed by up to 10,000 feet of clastic sediments, largely quartz sandstones. Conformably above these rocks is another unit of similar clastic rocks about 5,000 feet thick. Near the top of this unit, the first Cambrian fossils are found. The Lower Cambrian rocks here are about 10,000 feet thick and are mainly sandstone, grading into shale near the top. The Middle and Upper Cambrian rocks are close to 8,000 feet thick and are mainly carbonates. They are overlain by similar thick Lower Ordovician carbonates.

The Cambrian clastic rocks had their source in the continental interior, as was the case with the other areas described. As the cycle progressed, the seas moved eastward onto the interior. As the seas transgressed onto the continent, the basal Cambrian deposits were a distinctive beach sand. This unit can be traced from southern Nevada through Utah and Montana to the midwest. This sandstone is progressively younger to the northeast, ranging from Precambrian to middle Cambrian in age. (See Fig. 21-3.)

Farther north in the western geosyncline in Canada, similar rocks accumulated in great thicknesses. The history is similar; except that at most places, the Lower Cambrian rocks appear to be unconformably on the late Precambrian sediments. The Precambrian sedimentary rocks are several tens of thousands of feet thick here.

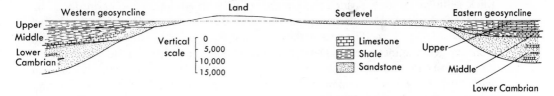

Figure 21-3. Generalized east-west restored section of Cambrian rocks. The section is drawn from the California-Nevada border to Virginia. Note that the basal Cambrian sandstone was deposited in a sea that gradually encroached on the continent so that the sandstone becomes younger as the interior of the continent is approached. In northern United States the late Cambrian sea was continuous across the continent. A section such as this is an interpretation from many fragmentary occurrences of deformed rocks.

Far North

Geosynclinal rocks are found in the Arctic islands of Canada. They may connect with similar rocks in Greenland and possibly even with northern Alaska. This is a very difficult area in which to study geology. Very few Cambrian rocks have been found. Ordovician rocks are widespread. To the north in the geosyncline, they are up to 10,000 feet thick and consist of volcanic and clastic rocks. Further south, they are thick shales; and still further to the south, on the edge of the Canadian Shield, they grade into carbonates up to 3,000 feet thick. In the geosyncline in East Greenland, about 40,000 feet of clastic rocks of Precambrian age are found. Overlying these are up to 8,000 feet of Cambrian and Ordovician carbonates. The Ordovician rocks are unconformably overlain by Devonian clastics.

Cycle II—Middle Ordovician to Early Devonian

This cycle began in Middle Ordovician time with the return of the seas. It differs from Cycle I in that mountain building occurred in the east, and the rest of the continent was less stable.

Interior

The seas returned in Middle Ordovician time, first to the geosynclines, and then across the interior. The pattern of land and sea was much like that of Cycle I. The new cycle began with the deposition of a remarkably widespread, thin, quartz sandstone over most of the interior. This sandstone is up to 300 feet thick at places and covers almost three-quarters of a million square miles. Its source was apparently the Canadian Shield. Overlying this sandstone are carbonates west of the Mississippi Valley and shale to the east. The carbonates are mainly dolomite that extends from western United States to Arctic Canada and is only a few hundred feet thick. The shale in the eastern interior grades into carbonate in the Silurian. The source of shale was a highland to the east in the area of the present Appalachian Mountains. Mountain building had begun in the geosyncline.

During Cycle II, the interior changed slowly from a flat lowland on which uniformly thick sediments were deposited to a number of very broad domes and basins (See Fig. 21-4). The domes and basins were pronounced by Devonian time, when they dominated the deposition pattern. The seas withdrew from the interior in late Silurian time and from the geosynclines in early Devonian time, ending the cycle. In Late Silurian time, a remnant of the sea was isolated in one of the basins in Michigan, Pennsylvania, and New York. Here a thick section of evaporites, especially rock salt (halite) and gypsum, was deposited. (See Fig. 21-4.)

East

The eastern geosyncline was an area of active mountain building throughout this cycle. This was the beginning of a new pattern here that culminated in the formation of the Appalachian Mountains in later cycles. During Cycle II, mountain building took place at different times and places throughout the geosyncline. None of the events described occurred simultaneously through the whole geosyncline. The mountain building and associated igneous and metamorphic events have destroyed much of the evidence, so reconstruction of the history is more difficult than for the simpler Cycle I.

Cycle II begins with clastic rocks of Middle Ordovician age. The source of these rocks was clearly the eastern part of the geosyncline because they thicken and become coarser to the east. Large fragments of Cambrian and Lower Ordovician limestones are found in thick clastic sections in the northern Appalachian Mountains, suggesting that uplifted and probably deformed rocks of Cycle I were eroded to form Cycle II deposits. This is shown diagrammatically in Fig. 21-5B. Volcanic rocks in eastern New England suggest volcanic island arcs there. Another middle Ordovician clastic fan had its source in eastern Tennessee.

Figure 21-4. Domes and basins of the interior. They began to form in the Ordovician and became pronounced by Devonian time. Lined area shows late Silurian evaporite basin. These features were not all active at all times.

Mountain building continued throughout the Ordovician, filling the geosyncline with clastics and spreading shale far to the west. From West Virginia to New York, Late Ordovician clastics unconformably overlie Middle Ordovician rocks. These clastics were shed so rapidly that the geosyncline was filled and the sediments accumulated above sea level. Some of these rocks are red beds, so called because their iron has been oxidized to hematite, which is believed to happen when the rocks are subjected to seasonal wetting and drying.

This orogeny reached a peak near the end of the Ordovician. Folding, faulting, and uplift were widespread. The Taconic Mountains in Massachusetts and Vermont were formed at this time. This range is a block of Cambrian and Ordovician shales 150 by 30 miles that were thrust at least 30 miles to the west. Middle and Upper Ordovician rocks were folded at this time.

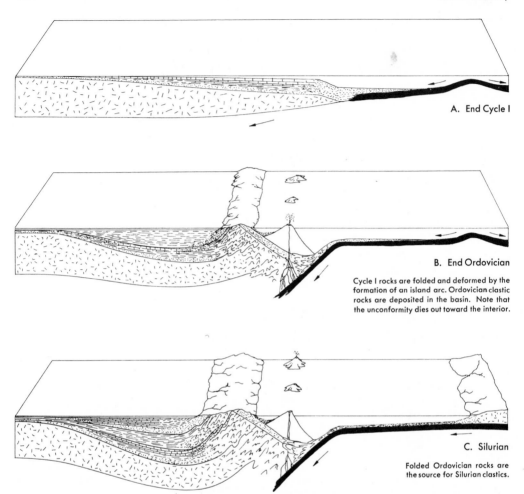

A. End Cycle I

B. End Ordovician

Cycle I rocks are folded and deformed by the
formation of an island arc. Ordovician clastic
rocks are deposited in the basin. Note that
the unconformity dies out toward the interior.

C. Silurian

Folded Ordovician rocks are
the source for Silurian clastics.

Figure 21-5. A. At the end of Cycle I, the geosyncline consisted of a continental shelf, slope, and rise. B. Cycle II began after a volcanic island arc formed and the Cycle I rocks were deformed. C. Renewed activity deformed the Ordovician rocks. This history is continued in Figure 22-7, and is also shown in Figure 16-21.

The Silurian record begins with a widespread quartz sandstone that is unconformable on Upper Ordovician rocks in the Appalachian Mountains. (See Fig. 21-5C.) Further to the west, Silurian carbonates overlie Upper Ordovician carbonates conformably. Uplifts in the highlands to the east occurred throughout the Silurian Period, but the resulting clastic sediments did not spread west of the geosyncline. Thus the mountain building was less than during the Ordovician. In parts of New England and in Maritime Canada, mountain building continued, resulting in thick sections of volcanic and clastic rocks. These rocks were folded and metamorphosed, and batholiths were emplaced. Elsewhere, the Silurian rocks were not affected by this mountain building.

The Lower Devonian rocks are a continuation of the Silurian patterns. Thick clastic deposits are found to the east, and thin limestones to the west. In the northeast, volcanic rocks occur with clastics, many of which are non-marine. The seas withdrew from the geosyncline in early Devonian time, ending the cycle.

West

Fewer rocks of Cycle II are found in the west, so the history is not so detailed. The pattern was apparently similar to Cycle I. In central Nevada, 20,000 feet of Ordovician coarse clastics with volcanic rocks are found. In Idaho, 10,000 feet of Ordovician clastics are found. In Alaska, thick Ordovician clastics with volcanic rocks occur. Late Ordovician batholiths and an unconformity between middle and upper Ordovician rocks suggest mountain building there. These western clastics grade eastward into the carbonates of the interior. In late Ordovician and early Silurian time, a chain of low islands in central Nevada may have separated these rock types.

Very few rocks of Silurian age are found in the west. About 4,000 feet of shale is found in central Nevada. Much thinner carbonate rocks are found toward the interior. In Alaska, thick clastic, carbonate, and volcanic sections are found.

This pattern continues to Early Devonian—that is, clastic rocks to the west, grading into thin carbonates in the interior. The cycle ends with the withdrawal of the sea in Early Devonian time.

Far North

The Arctic Islands contain thick Ordovician rocks as described in Cycle I. These rocks grade from volcanic and clastic rocks in the geosyncline to the north through shale to thin limestone to the south near the shield. The Silurian and Lower Devonian rocks are similar. In at least one place, deformation took place in the Silurian; and Late Silurian and Early Devonian rocks overlie younger Silurian rocks. At other places, Early Devonian rocks are folded.

In northern Greenland, dark limestone and coarse clastics of Silurian age are found.

Continental Drift—The Rest of the World

In these sections, the geological events on the other continents will be examined very briefly, and an attempt will be made to reconstruct the positions and movements of the continents. Given the present state of knowledge, this is difficult in the Paleozoic; but from Permian onward, the reconstructions are somewhat easier although disagreements do exist among those who have studied the problems closely. Much of what is said in these sections in each chapter will require revision as more data are gathered.

Although some have assumed that originally there was only one continent, recent work suggests that two or more continents may have existed (see Fig. 21-6). The northern continent was composed of what is now North America, Europe, and Asia, and the southern was composed of South America, Africa, India, Antarctica, and Australia.

Very little is known of the positions of the continents throughout most of the Precambrian. The evidence that parts of the southern supercontinent were together in the Precambrian is the matching of ages and structures such as shown in Fig. 16-20. Possible early Paleozoic locations are shown in Fig. 21-7. The locations of the equator and the south pole are from paleomagnetic data. Also shown are the locations where

Figure 21-6. A possible late Precambrian configuration of the continents. The shaded areas have ages greater than 1700 million years, and the two groupings may indicate that there were two early continents. After P. M. Hurley and J. R. Rand, *Science* (13 June, 1969), Vol. 164, p. 1238.

late Precambrian glacial deposits have been reported. Many of these reported glacial tills are merely conglomerates and not of glacial origin. Those near the pole may be glacial, but those near the equator most likely are not. The fold belt of Cambrian age shown is an excellent match between South America and Africa.

In the early Paleozoic, little is known of the southern continent, but what is known of paleogeography matches well. (See Fig. 21-8.) Europe and North America were in contact some of the time; this is shown by the similarity in fossils, match of the geosynclines, and match of the orogenic features produced by early Paleozoic folding. Development of a mid-ocean rise may have rotated North America relative to Europe in the early Paleozoic as suggested in Fig. 21-5. A small part of North Africa was also included in the European geosyncline, suggesting that the two supercontinents were in contact at that point. The geosyncline in central Eurasia may have connected with the one in northern North America. This geosyncline is on the site of the present Ural Mountains in Russia.

From late Precambrian to mid Paleozoic, the south pole remained in North Africa although it did move somewhat. Glacial deposits of Ordovician age have been found in the Sahara, confirming the location of the paleomagnetic pole. The equator passed

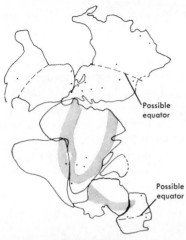

Figure 21-7. Late Precambrian-early Cambrian positions of continents based on magnetic and geologic data. The southern pole was in north Africa throughout the early Paleozoic. Late Precambrian mountain building in the southern continents is shown. The dots are places where glacial deposits of late Precambrian age have been reported. Those near the equator are probably conglomerates of non-glacial origin.

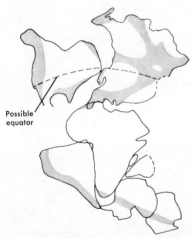

Figure 21-8. Reconstruction of continents in Silurian time. Geosynclinal areas are indicated.

through North America in a north-south direction in the Cambrian, and southwest-northeast direction in the Silurian.

Life of the Early Paleozoic

The first abundant fossils occur very abruptly at the base of the Cambrian System. Because well-developed, advanced forms of life appear suddenly at this horizon, the implication is clear that life must have existed for a long time previously. In the last chapter, some possible reasons for this were discussed. All of the phyla (the major subdivisions of life (see Appendix C), with the possible exception of Bryozoa, appear in the Cambrian. This suggests that a great radiation and diversification had occurred in the 500 million or so years since green algae had appeared. Accidents of preservation probably also bias our knowledge of life. This is suggested by the discovery of a number of excellently preserved, flattened carbonaceous films, even with fine hairs intact, of soft-bodied animals of Middle Cambrian age at a single locality on the crest of the Canadian Rockies near Field, B. C. This occurrence is the only fossil record in the whole world of some of these animals, and in most cases it is the only place where soft parts of the bodies are preserved. It took an unusual set of conditions to preserve these animals. (See Fig. 21-9.)

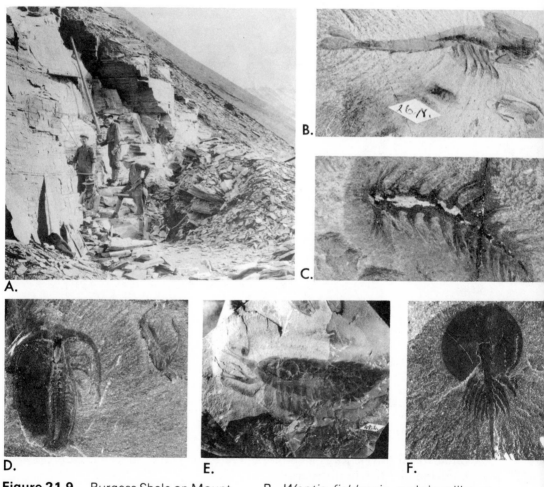

Figure 21-9. Burgess Shale on Mount Wapta, British Columbia. This Middle Cambrian formation has yielded extremely well preserved fossils of 130 species. A. Collecting fossils. The first fossils were found by chance in 1909. B. *Waptia fieldensis,* a shrimp-like arthropod. C. *Canadia setigera,* a worm. D. *Marrella splendens,* a trilobite-like arthropod. E. *Leanchoila superlata.* F. *Burgessia bella.* Photos from Smithsonian Institution.

Life began in the sea, and the first abundant fossils are marine. More than half of all Cambrian fossils are trilobites, and they and brachiopods together are about 90 per cent of Cambrian fossils. Trilobites were arthropods. (See Fig. 21-10.) Trilobites underwent a great adaptive radiation during the Cambrian, and they are good index fossils for that period. They were bottom dwellers and swimmers, and probably ate algae and soft-bodied creatures that are not preserved. They were probably scavengers, too. Trilobites became less abundant after Early Ordovician time as other invertebrates developed, and the trilobites became extinct near the end of the Paleozoic.

Figure   21-10. Trilobite.   *Elrathia kingi,* Middle Cambrian, about one inch long. Photo from Ward's Natural Science Establishment, Inc., Rochester, N.Y.

A reconstruction of a Cambrian sea bottom is shown in Fig. 21-11. Similar reconstructions are also shown for most of the other periods. The crowding together of so many organisms shown in all of these probably did not ever occur in nature but is necessary to show the varieties of life in each period. The reconstruction in Fig. 21-11 is based on the collections made at the site shown in Fig. 21-9. The fossils are the evidence; the reconstruction is an interpretation.

Near the end of Cycle I, in Early Ordovician time, the fauna changed somewhat. Brachiopods became more abundant (see Fig. 21-12), as did the other shell-bearing groups such as corals, bryozoa, crinoids, snails, and straight-shelled nautiloids. This change may have occurred in response to the development of predators. Few, if any, predators are found in Cambrian faunas, but they are found in Ordovician and later rocks. Straight-shelled nautiloids probably resembled present-day squids, and some could swim rapidly and catch prey with their tentacles. It may be that shells and

Figure 21-11. Reconstruction of Middle Cambrian (Cycle I) sea floor in western North America. The fauna shown include the soft-bodied animals in Figure 21-9. A. Sponge-like animal. B. *Marrella,* an arthropod. C. A worm. D. A jellyfish. Several different trilobites are also shown. From Field Museum of Natural History.

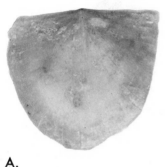

A.

B.

Figure 21-12. Ordovician brachiopods. Brachiopods are abundant in many Paleozoic rocks. They have two shells, and most were attached to the bottom by a stalk that came out through a hole near the hinge line. They are traditionally figured with the hinge line up. A. *Rafinesquina alternata,* from Ohio, about two inches. B. *Lepidocyclus capax,* also from Ohio, about one-half inch. Photos from Ward's Natural Science Establishment, Inc., Rochester, N.Y.

skeletons were developed mainly for support in the Cambrian, but evolved into protection in the Ordovician. The development of spiny shells in late Paleozoic adds credence to this idea. In any case, the change in the life occurred without a withdrawal of the seas, at least in North America.

The life of Cycle II is shown in Fig. 21-13, a typical Middle Ordovician sea floor. The many differences between this scene and the Cambrian one are obvious. In the Ordovician, corals and bryozoans expanded rapidly, and, to a lesser extent, snails, clams, straight-shelled nautiloids, and crinoids did also. This trend continued into the Silurian when corals became very abundant, forming reefs as shown in Figs. 21-14, 21-15, and 21-16.

Figure 21-13. Middle Ordovician (Cycle II) sea floor in north-central United States. A. Straight-shelled nautiloids (somewhat like squids). B. Trilobites. C. Snails. D. Colonial corals. E. Solitary corals (with tentacles). F. Brachiopods. G. Bryozoa. H. Seaweed. From Field Museum of Natural History.

Graptolites are important fossils in Cycle II, especially in the Ordovician. The floating graptolites evolved rapidly from their first appearance in Late Cambrian to their extinction near the end of the Silurian. They are good index fossils because they changed rapidly and are found in many environments because they floated. They tend to be found more commonly in black shale, probably because they sunk after death onto muddy bottoms where not much other life occurred. For this reason, it is common to

Figure 21-14. Bryozoa. These colonial animals build intricate structures on the sea floor in which a separate animal occupies each hole. Some build lacy structures, and others are stoney or twig-like. Fossils are generally fragmentary because the structure is broken after death of the colony. Silurian bryozoa, Rochester Shale, Lockport, N.Y. A few brachiopod shells can be seen in this slab with several types of bryozoa. Photo from Smithsonian Institution.

Figure 21-15. Middle Silurian (Cycle II) sea floor in north-central United States. A. Cystoid. B. *Favosites,* honeycomb coral. C. *Halysites,* chain coral. D. *Syringopora,* tube coral. E. Several types of nautiloids. F. Two types of brachiopods. G. Trilobite, several other types are also in the view. H. Two solitary corals. From Field Museum of Natural History.

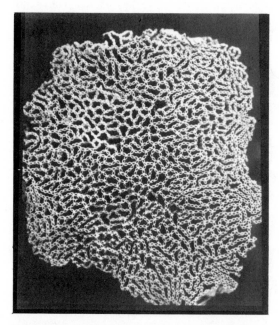

Figure 21-16. Chain coral. *Halysites,* Silurian, Louisville, Kentucky. A separate animal lived in each hole in this colonial coral. Photo from Smithsonian Institution.

speak of graptolite-bearing rocks versus shell-bearing rocks of Ordovician age. They were colonial animals, with many small individuals living in the stalks that hung down from the float. The fossils look like pencil marks on the enclosing shales (see Fig. 21-17), hence their name. In late Cambrian and early Ordovician, they were multi-branched; in middle and upper Ordovician, they had two branches; and in Silurian, they had one branch.

Vertebrates also first appear in Cycle II. They, too, began in water. The first evidence of their existence is scales and bone fragments found in a Late Ordovician sandstone in Colorado. Very few other fish fossils are found until Late Silurian, although in Cycle III they are dominant. Thus, there is little evidence for the stages in the development of the fish. A possible reason is that the early fish developed in fresh water rather than in the ocean, and very few nonmarine rocks of early Paleozoic age are preserved. As soon as marine life developed, it seems likely that some organisms would colonize the

Figure 21-17. Graptolites. The fossils of these curious animals look like pencil marks on rocks. They were colonial animals, and a separate animal lived in each of the saw teeth. The stalks hung down from a float, and because they floated the fossils are found in all environments. They evolved very rapidly; in Late Cambrian and Early Ordovician they had multi branches, in Middle and Upper Ordovician they had two branches, and in Silurian they had one branch. They became extinct at the close of the Silurian. Shown is *Didymograptus*, Middle and Upper Ordovician. Photo from Ward's Natural Science Establishment, Inc., Rochester, N.Y.

fresh water environments where they would have less competition. The early fish may have moved into fresh water because some organisms on which they could feed were already there and they could escape from their predators. Their predators may have been straight-shelled nautiloids or, probably more likely, the eurypterids. In any case, the first fish were armored, suggesting that they had predators.

Eurypterids became abundant in Silurian rocks. (See Figs. 21-18 and 21-19.) They are not found in normal marine rocks but in what probably were brackish waters or perhaps restricted saline waters. Some were as much as nine feet long, and they must have ruled their environment.

Scorpions are also found in Silurian rocks. (See Fig. 21-20.) They may have been the first animals to breathe air. The oldest definitely terrestrial scorpion known is Early Devonian. Scorpions, eurypterids, and trilobites were all arthropods.

Plants on the land must have preceded any animal life. The earliest land plants are Late Silurian in age. Possible land plants as old as Middle Cambrian have been reported from Siberia and India. Plants started from algae, and the first larger plants were marine. The widespread lowlands of Late Silurian time probably were a favorable environment for the development of plants. Plants were abundant from Devonian on. Beginning in Late Devonian, coal appears in every geologic period. The only earlier coal is in the Precambrian in Michigan; and although it may be of organic origin, it probably was not formed by land plants.

Figure 21-18. Late Silurian (Cycle II) sea floor in New York. Several types of eurypterids (see Figure 21-19) are shown with snails. Eurypterids probably lived in somewhat unusual environments such as restricted, saline, shallow seas. A. *Carcinosoma,* a scorpion. From Field Museum of Natural History.

Figure 21-19. Silurian eurypterid. *Eurypterus remipes.* This specimen came from the same formation as the scorpion shown in Figure 21-20, but from a different part of New York. Photo from Smithsonian Institution.

Figure 21-20. Silurian scorpion. *Carcinosoma scorpionis.* This specimen came from the same formation as the eurypterid shown in Figure 21-19, but from a different part of New York. Photo from Smithsonian Institution.

Figure 21-21 shows the occurrence of the invertebrates mentioned in this chapter, as well as those to be discussed in later chapters. In part, this diagram shows which kinds of life had the qualities that make good index fossils and in part which kinds of fossils have been well enough studied to be used for dating. This diagram will be useful in reading the next three chapters and should be referred to when needed.

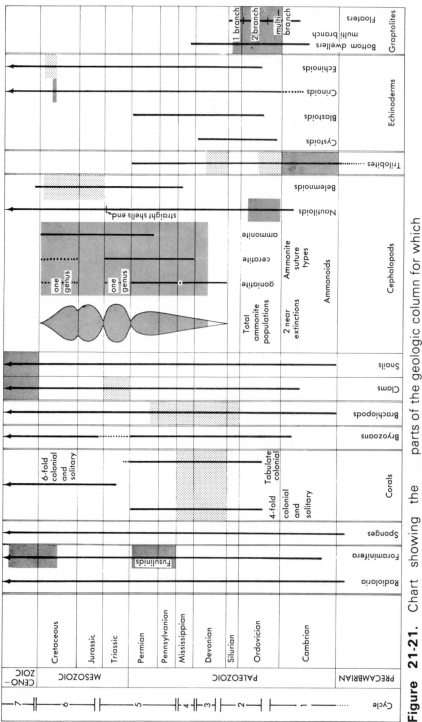

Figure 21-21. Chart showing the geologic occurrence of the more important fossil-forming invertebrate animals. The heavy lines show the time span of each animal group. Within some groups some simple distinctions are indicated that are of value in age determination. Also shown are the parts of the geologic column for which these fossils are useful for world-wide correlation (dark pattern) and for correlation of more restricted regions (light pattern). Correlation data after Curt Teichert, *Bulletin,* Geological Society of America, Vol. 69, 1958.

QUESTIONS

21-1. List as many of the assumptions made in interpreting geologic history as you can.

21-2. Describe the various patterns recognized or assumed in geologic history.

21-3. Is the Cambrian-to-Ordovician boundary a natural one in North America?

21-4. What events separate Cycle I from Cycle II?

21-5. What occurred in the eastern part of North America during the Silurian Period?

21-6. In what continent was the "south" pole in early Paleozoic time?

21-7. Where was the equator in North America in early Paleozoic time?

21-8. What is the importance of the Burgess Shale fauna?

21-9. What are typical Cambrian fossils?

21-10. In what ways do Cycle II fossils differ from those of Cycle I?

21-11. Why are graptolites important?

21-12. What were the first vertebrates, and when did they appear?

21-13. When did the first land plants appear?

SUPPLEMENTARY READING

GENERAL

Cloud, Preston, *Adventures in Earth History.* San Francisco: W. H. Freeman and Co., 1970, 992 pp. (A collection of papers by many authors.)

Dott, R. H., Jr., and R. L. Batten, *Evolution of the Earth.* New York: McGraw-Hill Book Co., 1971, 649 pp.

Engel, A. E. J., "Geologic Evolution of North America," *Science* (April 12, 1963), Vol. 140, No. 3563, pp. 143-152.

Kay, Marshall, and E. H. Colbert, *Stratigraphy and Life History.* New York: John Wiley & Sons, Inc., 1965, 736 pp.

PATTERNS IN GEOLOGIC HISTORY

Fairbridge, R. W., "The Changing Level of the Sea," *Scientific American* (May, 1960), Vol. 202, No. 5, pp. 70-79. Reprint 805, W. H. Freeman and Co., San Francisco.

Schleh, E. E., "Review of Sub-Tamaroa Unconformity in Cordilleran Region," *Bulletin,* American Association of Petroleum Geologists (February, 1966), Vol. 50, No. 2, pp. 269-282.

Sloss, L. L., "Sequences in the Cratonic Interior of North America," *Bulletin,* Geological Society of America (February, 1963), Vol. 74, No. 2, pp. 93-114.

Wheeler, H. E., "Post-Sauk and Pre-Absaroka Paleozoic Stratigraphic Patterns in North America," *Bulletin,* American Association of Petroleum Geologists (August, 1963), Vol. 47, No. 8, pp. 1497-1526.

Chapter 22

The Late Paleozoic – Cycles III, IV, and V (A)

Cycle III—Middle Devonian to Late Devonian

Interior

This relatively short cycle is confined to within the Devonian Period. The seas returned to the interior in Middle Devonian time, and at many places rocks of that age overlie Middle Silurian beds. In the eastern part of the interior, shale was the main rock type deposited. Its main source was highlands in the eastern geosyncline. Near the domes of the interior and the highlands in the geosyncline some sand was deposited. The western interior, extending to the Arctic, was the site of limestone and some shale deposition. Reefs developed in this limy sea; and in Alberta, these reefs are reservoirs for important oil fields. The domes and basins of the interior became more pronounced. In Montana and Alberta and at other places in Canada, evaporites formed in some of these basins that were cut off from the seas. The seas withdrew in the late Devonian, ending this short cycle.

East

The important events of this cycle occurred in the eastern geosyncline. In Middle and Late Devonian time, more orogeny occurred in the eastern part of the eastern geosyncline, and this area was transformed from geosyncline to continent by intrusive and metamorphic processes. This change altered the pattern of deposition. (See Fig. 22-1.)

541

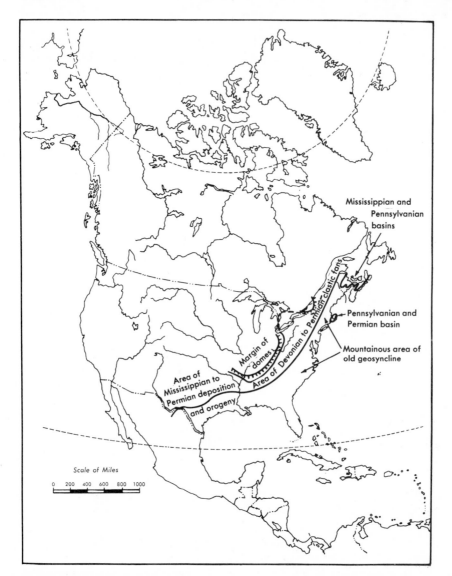

Figure 22-1. Middle Devonian to Permian sedimentation in the east occurred largely in basins that developed between the domes and arches of the interior and the mountains formed in the area of the old geosyncline. Compare with Figure 14-32.

Starting in Mississippian time in the north a few basins developed in the area of the old geosyncline. In Triassic time many basins developed throughout the area of the old geosyncline.

From Mississippian to Permian time, deposition and orogeny occurred in Texas and Oklahoma.

The area of the old geosyncline became an upland. This upland was rapidly eroded, and the resulting clastic sediments were deposited in the area just west of the old geosyncline. In this depositional area first one place and then another would downsink to trap the debris. Thus, in Middle and Late Devonian, over 10,000 feet of deltaic rocks were deposited in Pennsylvania and adjacent New York (Catskill Mountains) and West Virginia. These deposits were similar to the Ordovician and Silurian deltas but were much thicker and more widespread, and this type of deposition extended into the next two cycles in the east. In Mississippian time the clastic fan spread further south along the Appalachian Mountains, suggesting that orogeny spread southward. This pattern, as we shall see, continued throughout the remainder of the Paleozoic with the addition of smaller basins and adjacent highlands in the area of the old geosyncline. The seas again withdrew during part of Late Devonian time, ending the cycle.

West

In the far west, few Devonian rocks are found, but the history can be reasonably deciphered. A volcanic island arc is believed to have existed in the vicinity of the California-Nevada border. The evidence for this is that thick sections of clastic and volcanic rocks are found in western Nevada. Further east, these rocks interfinger with carbonates that, in turn, extend onto the western interior. Similar sequences are found in the Canadian Rockies in Alberta, where the carbonate rocks are very thick and contain both reefs and evaporites. In latest Devonian time, orogeny began in Nevada. This orogeny will be described in Cycle IV.

Far North

Cycle III begins quietly in the far North. The carbonate deposition of the interior extended to the Arctic. In the geosyncline, a few to several thousand feet of carbonates with some shale and siltstone, especially near the top, make up the Middle Devonian rocks. These rocks thin to the south on the interior platform. The Upper Devonian rocks are sharply different. Uplift occurred to the north, forming a highland that shed clastics into the geosyncline. These nonmarine clastic rocks are up to 10,000 feet thick and contain some coal beds. They closely resemble the rocks of this same cycle in New York and Pennsylvania.

In East Greenland, thick deposits of nonmarine clastic rocks were also deposited in this cycle. The rocks are up to 20,000 feet thick and were deposited in basins adjacent to uplands, and so record activity here also. The rocks are red and gray in color and resemble the Old Red Sandstone, a formation of the same age in England.

Cycle IV—Latest Devonian to Late Mississippian

Interior

This cycle is also short and begins with the return of the sea to the interior in latest Devonian time. The interval between Cycles III and IV was short, and it may be that at places the seas did not completely withdraw. The pattern of this cycle in the interior is much like the last cycle, but the rocks are quite different. The Mississippian System was named for the exposures in the area drained by the upper Mississippi River. The Lower Mississippian rocks in that area are black organic-rich shales. The shale is not the same age everywhere and is rarely more than a few tens of feet thick but is generally present over a wide area. It is difficult to reconstruct the conditions under which this shale was deposited. The organic content suggests stagnant water, but the fossils and scour channels suggest shallow water. The shale contains enough uranium that it may in the future be an economic resource. The source of the shale was highlands that had formed in the geosyncline to the east and south.

Middle Mississippian limestones overlie the black shale. Some of these limestones are composed largely of fossil fragments and some of spherical calcite grains (oölites). They are quarried for building stone. In Late Mississippian time, limestone and sandstone layers alternate. The source of the sands was highlands to the east. These are the first sandstones in the interior since Ordovician time.

The western interior was the site of limestone deposition through this cycle. These carbonates extend well into Canada. There the amount of shale increases to the north and in the later rocks.

The domes and basins of the interior were active and, to a large extent, controlled the thicknesses of all of the units described. The basin in Montana-Dakota was the site of evaporite deposition in Middle Mississippian time.

The cycle ended with the withdrawal of the seas in Late Mississippian time. Erosion followed over most of the continent. The erosion was more pronounced in the west than the east. The east was probably near sea level, setting the stage for the Pennsylvanian coal swamps.

East

The area of the old geosyncline was again uplifted, producing another clastic fan like that of Devonian time. The center was further south in Pennsylvania, and the total extent was less than in the Devonian. In Texas, Oklahoma, and Alabama, thick deposits of clastic rocks developed in Middle and Late Mississippian time, suggesting orogeny in that area. At the same time, in the area of the old geosyncline in New England and northeastern Canada, a number of basins developed. These basins persisted from Mississippian to Permian time and developed great thicknesses of clastic rocks eroded from the surrounding highlands, together with some carbonates, evaporites, and red beds.

West

In Cycle IV, the pattern of the far west changed. Mountain building occurred in central Nevada, and the resulting upland was to persist in some form throughout most of the later periods. The first evidence of the orogeny is found in late Devonian clastic rocks in the eastern half of Nevada. Prior to that time, clastic and volcanic rocks accumulated farther west, and the source of these rocks was apparently a volcanic island arc near the California-Nevada border. The orogeny began in Late Devonian time and extended into Pennsylvanian. The new mountain structures formed in the area where the older clastic rocks of western Nevada graded into the carbonate rocks of eastern Nevada. (See Fig. 22-2.) The orogeny thrust the western clastic rocks over the eastern carbonates. The rocks are estimated to have moved almost 100 miles. The resulting mountain range extended all the way across Nevada and into Idaho where younger igneous rocks obscure its extent. The mountains were never very high and underwent erosion as they were uplifted, shedding clastics both east and west. They were probably a series of islands connected by shallow water. To the east were deposited several thousand feet of conglomerate and sandstone that grade into the thinner limestones of the interior. To the west geosynclinal sediments accumulated. This pattern continued into Cycle V.

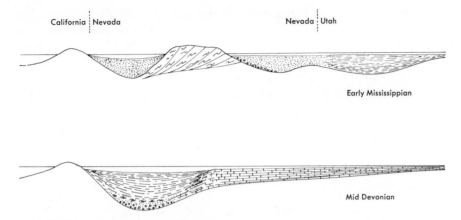

Figure 22-2. Devonian and Mississippian events in Nevada. Thrust faulting in Late Devonian and Early Mississippian time created two areas where clastic sediments were deposited. After R. J. Roberts.

Far North

In Alaska, uplift caused deposition of thick clastic rocks that grade into the thin limestones of the interior. In Arctic Canada, mountain building apparently took place because Middle Pennsylvanian rocks overlie Upper Devonian sediments on a major angular unconformity. The lack of Mississippian rocks makes it impossible to decipher the details. In East Greenland, the deposition of nonmarine clastics that had begun in Devonian time continued.

Cycle V(A)—Latest Mississippian to Triassic

Cycle V began with the return of the seas to parts of the interior in very latest Mississippian time and extended to early Jurassic time. The sequence, because it covers such a long period of time, lacks the continuity of the older cycles and is described in two parts in this and the following chapter. During this cycle mountain building occurred in many parts of North America, but complete withdrawal of the seas probably did not occur until early Jurassic time. The Triassic of the west will be covered in the next chapter.

Interior

Cycle V begins with deposition at the margin of the continental interior. The oldest rocks of this cycle are late Mississippian and early Pennsylvanian shale and sandstone that are found east of the Mississippi valley. These rocks were apparently deposited in the deltas of rivers flowing from highlands to the east and south. By Middle Pennsylvanian time, the seas again covered the interior, and great coal swamps had formed from the Mississippi valley eastward to the Appalachian Mountains. Thicker deposits formed in the basins of the interior. The climate was apparently warm, and the plants that grew in the swamp were buried before they decayed, becoming the very important coal deposits of eastern North America. (See Fig. 22-3.) Both the coal deposits and the

Figure 22-3. Generalized map of the United States in Middle Pennsylvanian time. After U.S. Geological Survey.

marine Pennsylvanian rocks of the interior are remarkable for the great number of repeated sequences of rocks called *cyclothems,* which are believed to be the result of periodic changes in sea level. (See Fig. 22-4.) The cause of these changes in sea level is not known, but they could be due to climatic changes or to vertical movements of the continents or the ocean bottoms. A typical cyclothem begins with coarse, nonmarine sediments that become finer grained as the sea approaches. Coal is deposited during a near-shore swampy time and is followed by marine sediments that grade upward from shale to limestone as the water depth increases. The sea then retreats; and then, generally after a short period of erosion, the sea readvances, starting a new cycle.

Permian rocks in the east are few and consist of clastic rocks in the area of the Pennsylvania-Ohio-West Virginia border. The upper Pennsylvanian rocks grade upward into the Permian. These non-marine Permian rocks, less than 1000 feet thick, are the youngest deposits associated with the eastern geosyncline. This will be discussed further in the next section.

There are no Triassic rocks in the eastern interior.

The Cycle V rocks of the western interior are very different from those of the east. In this cycle, the paleogeography of North America changed drastically. One of these changes was the development of a number of highlands and deep basins in the general area of the central Rocky Mountains and extending to Oklahoma and Texas. (See Fig.

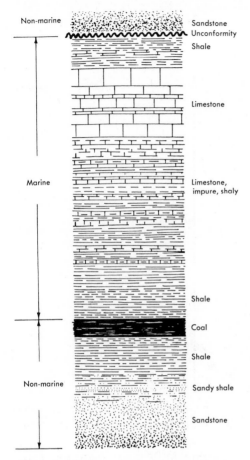

Non-marine — Sandstone
— Unconformity
Shale

Limestone

Marine — Limestone, impure, shaly

Shale

Coal

Shale

Non-marine — Sandy shale

Sandstone

Figure 22-4. A typical cyclothem begins with non-marine sandstone that becomes finer grained as the sea advances. Coal is deposited in a near-shore swamp. As the water depth increases the marine rocks grade from shale to limestone. The sea then retreats to start the cycle again.

22-5.) The Pennsylvanian rocks of the western interior are mainly quartz sandstone with some carbonates. In the deep basins adjacent to the highlands, thick clastic sections were deposited. The older broad basin in Dakota-Montana was the site of carbonate and evaporite deposition with some shale. Because of the changes in paleogeography, it will be more convenient to include the later history of the western interior under the heading West.

East

In this cycle, deposition in the eastern geosyncline ends. In the east the geosyncline was a highland at the start of the cycle, and rivers flowing from the highland brought sediments to the interior as was just described. Basins formed in Massachusetts-Rhode Island, and in eastern Canada. Thick clastic sections containing coal accumulated in these basins.

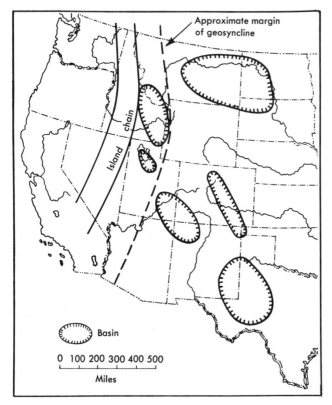

Figure 22-5. Late Paleozoic events in the west. In Late Devonian and Mississippian time orogenic events produced a chain of islands in the western geosyncline that was the source of clastic sediments. In the geosyncline the Early and Middle Pennsylvanian rocks are clastic west of the islands and limestone to the east. In late Pennsylvanian time renewed activity in the island chain resulted in clastic sediments east of the islands.

Starting in Mississippian and especially in Pennsylvanian and Permian a number of deep basins, some with adjacent highlands, developed both in the area of the geosyncline and in the interior sea.

Further to the south, thick clastic deposits indicate renewed orogeny. In the Texas-Oklahoma area, the geosyncline received thick clastic deposits, suggesting continual orogeny to the south. In this area, the geosyncline itself was deformed into a series of linear highlands and basins. These basins received sediments eroded from the adjacent highlands. Another orogeny affected this area, probably in latest Pennsylvanian time, when nonmarine clastic rocks were deposited and finally thrust plates moved northward over these rocks. In western Texas the thrusting was somewhat earlier, and the end of the Pennsylvanian was a time of erosion. (See Fig. 22-6.)

In Permian time the Appalachian geosyncline was the site of orogeny that formed the Appalachian Mountains. (See Fig. 22-7.) The only rocks of Permian age are Early Permian clastics, similar to the underlying Pennsylvanian rocks that were described in the last section. The exact age of this orogeny that turned the remaining parts of the old geosyncline into continental structure is not known. The Early Permian rocks were

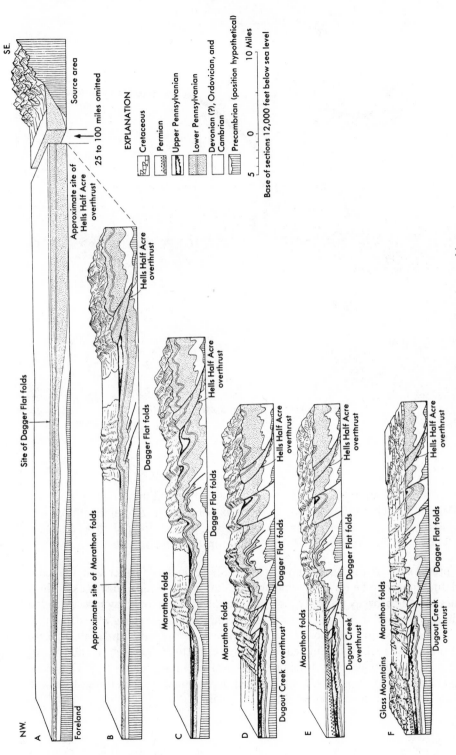

Figure 22-6. Hypothetical block diagrams showing the development of the Marathon Mountains in west Texas. Undeformed Permian rocks overlie deformed Upper Pennsylvanian rocks and so date the final orogenic phase. Note the inferred shortening of the rocks. From P. B. King, U.S. Geological Survey, *Professional Paper*, 187, 1938.

NW.

SE.

Foreland

Site of Dagger Flat folds

Source area

A

Approximate site of Marathon folds

Approximate site of Hells Half Acre overthrust

25 to 100 miles omitted

B

Marathon folds

Dagger Flat folds

Hells Half Acre overthrust

C

Marathon folds

Dugout Creek overthrust

Dagger Flat folds

Hells Half Acre overthrust

D

Marathon folds

Dugout Creek overthrust

Dagger Flat folds

Hells Half Acre overthrust

E

Glass Mountains   Marathon folds

Dugout Creek overthrust

Dagger Flat folds

Hells Half Acre overthrust

F

EXPLANATION

Cretaceous

Permian

Upper Pennsylvanian

Lower Pennsylvanian

Devonian (?), Ordovician, and Cambrian

Precambrian (position hypothetical)

5   0   10 Miles

Base of sections 12,000 feet below sea level

549

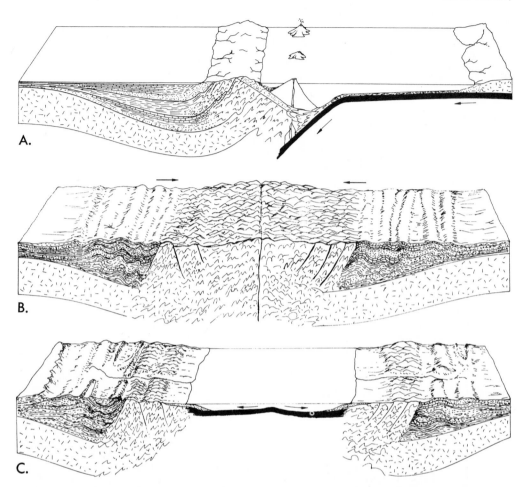

Figure 22-7. Formation of the Appalachian Mountains. This figure completes the history shown in Figures 16-21 and 21-5. A. Silurian Period. B. Collision of continents in late Paleozoic time produces the Appalachian Mountains. C. Reopening of the Atlantic Ocean in the Mesozoic.

folded, presumably at this time, and Late Triassic sediments further east were not affected and appear to lie on an erosion surface cut on the older folded rocks. Thus, it appears that the orogeny probably took place in the Permian and may have extended into Triassic time, and then, after a period of erosion, the Triassic rocks were deposited.

The Triassic rocks just mentioned are red clastics interbedded with basalt. In Late Triassic time these rocks accumulated to thicknesses up to 20,000 feet in a number of fault basins within and just east of the present Appalachian Mountains from Connecticut to North Carolina. (See Fig. 22-8.) The next depositional event in the east was the Cretaceous to Recent deposition on the coastal plain.

West

In the western geosyncline, to the east of the island chain that had appeared earlier, the Early and Middle Pennsylvanian rocks are limestones unlike the Mississippian

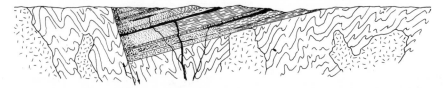

Figure 22-8. Triassic sediments formed in faulted basins in the metamorphic rocks east of the present Appalachian Mountains. The rocks are coarse clastic red beds that grade into finer clastics away from the faults, and basaltic volcanic rocks. Dinosaur footprints are found in these sedimentary rocks at some places. The Basin and Range fault structures are somewhat similar.

clastics. West of the islands, conglomerate and sandstone were deposited. In Pennsylvanian time the island chain was the site of orogeny and uplift. (See Fig. 22-9.) The Late Pennsylvanian deposits, in part eroded from this uplift, are sandstones and limestones and overlie earlier Pennsylvanian rocks with an angular unconformity. These accumulated in a very thick (40,000 feet) basin in eastern Utah and adjacent Colorado. (See Fig. 22-5.)

The area east of the geosyncline has, up to this point, been included with the interior; but after early Permian time, the seas never again covered the interior. Only the western interior was covered by seas, and because these seas were extensions of those covering the western geosyncline, it is better to include the closely related western interior with the western geosyncline.

The Pennsylvanian rocks of the western interior are quartz sandstones and carbonates. In southwestern Colorado, a highland developed, and thick accumulations of clastic rocks were shed into the adjacent basins. The basin on the Colorado-Utah border (see Fig. 22-5) received Early Pennsylvanian clastics, followed by thick evaporites and Middle and Late Pennsylvanian clastics and carbonates; the total thickness is 10,000 feet.

The Permian was an active time in the western geosyncline. In the far west, thick deposits of clastic and volcanic rocks are found. Orogeny occurred in western Nevada; and in Oregon, the Triassic overlies the Permian on an angular unconformity. In central Nevada, the orogenic island chain was uplifted and deformed and was again the site of thrust faulting. (See Fig. 22-9.) East of this area, the main deposits were quartz sandstone and carbonate. In the last half of Permian time, in parts of Idaho, Montana, Wyoming, and Utah, a restricted basin formed in which peculiar phosphate-rich rocks were deposited. They are an important economic source of phosphate. (See Fig. 22-10.) Further to the east, all of these marine rocks interfinger with clastic red beds whose source was the mountains in Colorado that had formed during the Pennsylvanian.

Parts of the Permian rocks of the western interior just mentioned have been studied very closely. At the start of the Permian Period, a shallow sea covered the area from Nebraska to west Texas. The main deposits were carbonates and shale along with clastics from the highlands of Colorado and Texas. This sea retreated to the south, so that by late Permian time, only west Texas was the site of marine deposition, and the highlands were largely eroded away by the end of the period. As the sea retreated, marine deposition was replaced by non-marine clastic red beds. These red beds interfinger with the marine deposits. They suggest that arid conditions prevailed there

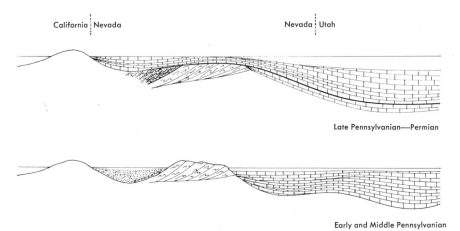

Late Pennsylvanian—Permian

Early and Middle Pennsylvanian

Figure 22-9. Pennsylvanian and Permian events in Nevada. These cross-sections continue the history shown in Figure 22-2.

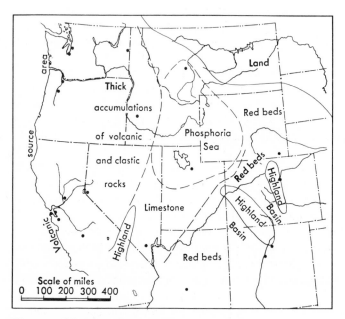

Figure 22-10. Generalized map of the Permian in western United States. All of the boundaries shown by dashed lines are gradational with much interfingering of rock types. The phosphatic shales that accumulated in the Phosphoria Sea are a valuable resource.

in Permian time, and similar Mesozoic rocks suggest that this climate persisted into later periods. Because of the retreating sea, almost all of the Permian System is represented by marine rocks in west Texas. It is the most complete Permian section in North America and would be studied for this reason alone; however, many environments are represented in these rocks, making their study even more valuable. This area had a number of basins separated from each other by shallower seas. Reefs developed, and evaporites formed in restricted basins, and all of these rocks interfingered with non-marine red beds. In very Late Permian time, non-marine red beds were deposited, ending the deposition.

At about the end of the Permian, the seas retreated elsewhere as well. This retreat was only partial in the western geosyncline. The Triassic rocks reflect a continuation of the Permian pattern, and so Cycle V is considered to continue into the Triassic. The Triassic of the west will be described in the next chapter because it sets the scene for the later Mesozoic periods, and because it is better to describe the life of the Triassic with the life of the rest of the Mesozoic.

Far North

The Devonian and Mississippian orogeny formed geosynclinal mountains to the north, composed of metamorphosed volcanic and clastic rocks. To the south of these is a basin in which most of the later rocks were deposited. To the south of the basin is a fold-mountain belt. South of the fold belt is the northern interior on which basins and arches formed just as in the interior further south in the United States.

In the basin just described, Pennsylvanian rocks were deposited. They are mainly limestone and chert, and are found at widely scattered points. Clastic rocks are also found, and they may be more or less continuous with similar rocks in the western geosyncline. Few Permian rocks have been found. Clastics, carbonates, and 5,000 feet of volcanics have been reported, but their relationships are not known. In Triassic time, shale, siltstone, and sandstone were deposited. Their source was to the south.

In Greenland, the Triassic begins with marine rocks and ends with continental deposits, all of which accumulated in faulted basins.

Continental Drift—The Rest of the World

By the end of the Paleozoic, the two supercontinents had joined to set the scene for the great breakup and drifting apart that formed the present earth. On the junction between Europe and Africa, a geosyncline formed that was active throughout the Mesozoic.

In the northern supercontinent, the mountains formed by the Silurian orogeny shed similar red clastics in Europe, eastern North America, and northern North America. Late Paleozoic folding produced mountains in both Eastern North America and Europe that were probably a continuous mountain system. (See Fig. 22-11.) The folding may have been the result of the collision of the continents. The Ural geosyncline was folded in the late Paleozoic, and this may have been caused by Europe and Asia joining.

A major change that occurred in late Paleozoic is the movement of the south pole from North Africa to near its present position in Antarctica. This means that the

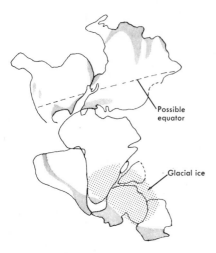

Figure 22-11. Reconstruction of continents in Late Paleozoic time. Although the positions of the continents relative to each other are similar to Figure 21-8, the pole has moved from North Africa to Antarctica. Mountains and areas of glaciation are indicated.

southern continent moved about 90 degrees north, assuming that the geographic pole remained in the same place. The Permian glacial deposits described and figured in Chapter 16 resulted from this drift. North America rotated about 90 degrees in this move, so that the equator ran east-west instead of north-south as it had earlier. The equator passed through southeastern United States, and this may account for the aridity indicated by the late Paleozoic and Mesozoic rocks.

The evidence that the southern continents were together in late Paleozoic time was presented in Chapter 16 as part of the evidence for continental drift and will not be repeated here.

Life of the Late Paleozoic

The marine invertebrates continued to evolve rapidly so that the life of the late Paleozoic differed from that of the early Paleozoic. Brachiopods (see Fig. 22-12), corals, and bryozoans (see Figs. 22-13 and 22-14) were abundant, and snails and clams were common. Graptolites died out, and trilobites declined, eventually becoming extinct at the end of the Permian. Fig. 22-15 shows a late Devonian sea bottom. Note the spines on the trilobite. Fish ruled the Devonian sea, as we will see, and spines on trilobites and brachiopods developed, probably to protect them from the fish. The first spined brachiopods appeared in the Early Silurian. Also shown in this Devonian scene is the coiled cephalopod, which developed by the coiling of a straight-shelled nautiloid. The coiled cephalopods developed very rapidly and are the most important fossils for dating the late Paleozoic and Mesozoic rocks. Both the external ornamentation and the shape of the margins of the partitions between the chambers changed rapidly. The cephalopods will be described further in the next chapter.

A.

B.

Figure 22-12. Paleozoic brachio-
pods. A. *Athyris spiriferoides,*
Devonian, New York, about one inch. B.
Mucrospirifer mucronatus, Devonian.
Photos from Ward's Natural Science
Establishment, Inc., Rochester, N.Y.

The Mississippian scene in Fig. 22-16 shows another aspect of late Paleozoic life. Crinoids and blastoids (See Fig. 22-17) expanded greatly in the Mississippian after some other orders of echinoderms died out in the mid-Paleozoic. At most places, however, the fauna was more balanced.

Pennsylvanian and Permian sea bottoms are shown in Figs. 22-18 and 22-19. The development of the cephalopods and spiny brachiopods shows clearly. Some excellently preserved examples of the Permian spiny brachiopods, on which the reconstruction in Fig. 22-19 is based, are shown in Fig. 22-20. Another aspect of the Pennsylvanian and

Permian is the development of fusulinids. (See Fig. 22-21.) They are a type of foraminifera and were very abundant at that time. They are good index fossils in Pennsylvanian and Permian rocks.

Figure 22-13. Honeycomb coral, *Favosites,* Devonian, Arkona, Ontario. Photo from Smithsonian Institution.

Figure 22-14. Permian bryozoa from Glass Mountains, Texas. Photo from Smithsonian Institution.

Figure 22-15. Late Devonian (Cycle III) sea bottom in western New York. A. Bryozoa. B. Crinoid. C. Colonial corals. Several types are shown, and corals are the main element in the fauna. D. Solitary coral. Several types are shown. E. Coiled cephalopod. F. Straight nautiloid. G. Spiny trilobite. Spines probably developed as a defense against fish that had developed by this time. Two smaller trilobites are also in the view. H. Brachiopods. Other types are attached to the coral. J. Sponge to the left of the letter. Also shown are snails. From Field Museum of Natural History.

Figure 22-16. Mississippian (Cycle IV) sea bottom in central United States. Crinoids and blastoids form a local "sea lily" garden. A starfish is also shown. At other places a more balanced fauna is present. From Field Museum of Natural History.

Figure 22-17. Blastoid *Pentremites,* Mississippian of Illinois, one-half inch. Members of the phylum Echinodermata are recognized by their five-fold symme- try. Blastoids are attached to the sea floor by a stalk. Photo from Ward's Nat- ural Science Establishment, Inc., Roch- ester, N.Y.

Figure 22-18. Pennsylvanian (Cycle V) sea bottom in north-central Texas. A. Sponges, both solitary and colonial. B. Crinoid. C. Solitary coral. D. Snail. Sev- eral other types are also shown. E. Brachiopods and clams. F. Cephalopod. Several other types are also shown. From Smithsonian Institution.

Near the end of the Paleozoic, many groups became extinct. No completely satisfac- tory explanation is available for this great dying. Trilobites and blastoids became extinct. Almost no corals or foraminifera survived, and several types each of bryozoans, brachiopods, and crinoids died out. The survivors, however, populated the Mesozoic seas. The only reasonable explanation is that many groups died when the seas retreated. An interesting point is that some organisms that died out at most places may be preserved somewhere. The Permian rocks of Timor contain echinoderms that became extinct elsewhere in the Mississippian.

Near the end of the Paleozoic, there was a world-wide time of retreat of the seas. Thus it appears that the major sub-division into eras may be a natural one. The great change in life was obvious to the early workers who defined the eras. In southwestern United States, deposition continued across this important boundary; however, the seas were very restricted, and most of the rocks are non-marine.

The Devonian is called the age of fish because they apparently ruled the seas. As mentioned in the last chapter, the first fish are found in Ordovician rocks, and they first had any abundance in Late Silurian. The fish apparently developed in fresh water and returned to the seas at that time. The first fish were not at all like those of today. They were less than a foot long, jawless and finless, and were armored with bone. (See Fig. 22-22.) Without jaws, they apparently fed in the bottom mud. Without fins they were poor swimmers. They probably could rise from the bottom by wiggling their tails like a tadpole, but without fins they could not change direction. All of these reasons, plus the armor, suggest that they were essentially bottom dwellers. Almost all were extinct by the end of the Devonian, although some of Pennsylvanian age are known and the living lampreys and hagfish are close relatives.

The jawless fish were replaced by the jawed fish (placoderms) in the Devonian. They, too, were armored, but they had paired fins and so were good swimmers. Some were quite large, and this, together with their jaws and swimming ability, made them fierce predators. (See Fig. 22-23.) The eurypterids that had been the predators declined in the

Figure 22-19. Permian (Cycle V) sea bottom, Glass Mountains, Texas. A. Several varieties of sponges. B. Two types of cephalopods. C. Spiny brachio- pods. Many other brachiopods and clams can be seen. Corals with tenta- cles can also be seen. From Field Mu- seum of Natural History.

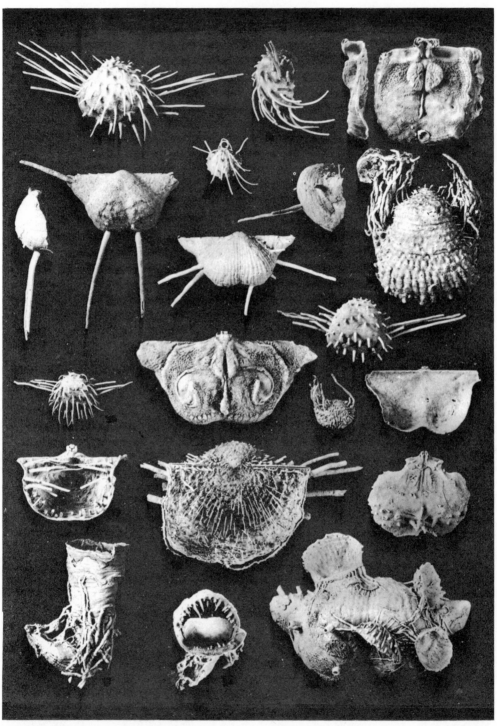

Figure 22-20. Spiny brachiopods from Permian limestones in Glass Mountains, Texas. These fossils were replaced by silica and were recovered, with the spines and other delicate features intact, by dissolving the enclosing limestone with acid. Photo from Smithsonian Institution.

Devonian as the jawed fish developed. The jawed fish spread rapidly because they were so well adapted to predation. Sharks, spiny fish, bony fish, as well as other types of fish that will be discussed later, all appeared in the Silurian or Devonian. Sharks and bony fish have continued to the present. The placoderms died in the Mississippian, and the spiny fish in the Permian. The relationships are shown in Fig. 22-24.

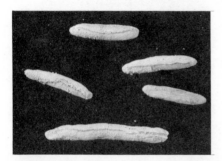

Figure 22-21. Fusulinid. Permian *Parafusulina*, Marathon, Texas. About ⅜ inch long. Fusulinids are very large foraminifera that are excellent index fossils in the Pennsylvanian and Permian. They died out near the end of the Permian. Photo from Smithsonian Institution.

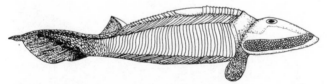

Figure 22-22. Early fish. The first fish were the Ostracoderms or jawless fish. The lamprey is the only living close relative. *Hemicyclaspis* of Devonian age is shown here (about one foot long).

Figure 22-23. Devonian jawed, armored fish. *Dunkleosteus intermedius* from northern Ohio. This fish is about four feet high. Other members of this group that died out in the Mississippian were up to 30 feet long with jaws over six feet high. Photo from Smithsonian Institution.

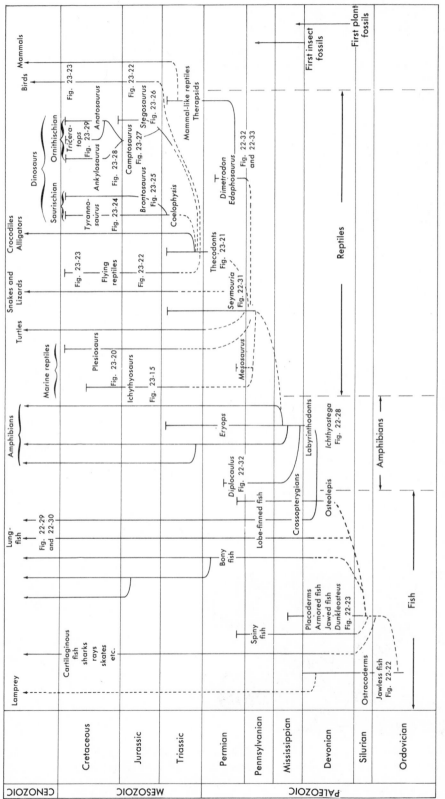

Figure 22-24. Evolution of the vertebrates. Although this chart appears formidable, the main theme can be seen along the bottom where progressive developments form a diagonal from primitive fish in the lower left to mammals in the upper right. Only the more important vertebrates are shown. Figures showing these animals are indicated.

The plant life on the land also developed rapidly in the late Paleozoic. From a few Late Silurian and some questionable older occurrences came the abundant flora of the Devonian. The Early Devonian plants were small and primitive and seem to have grown in marshes or swamps. The plants were successful and evolved rapidly. By Middle Devonian time, forests existed. (See Fig. 22-25.) A Pennsylvanian coal swamp

Figure 22-25. Middle Devonian forest in western New York. The Early Devonian plants were generally small, primitive marsh- or swamp-dwelling plants. A. *Psilophyton* was one of the larger that carried over into the first true forests of Middle Devonian age. B. An early tree fern that grew up to 50 feet high. C. Horsetail rushes. D. *Archaeosigillaria*. Painting by C. R. Knight, Field Museum of Natural History.

is shown in Fig. 22-26, and Fig. 22-27 shows typical Pennsylvanian plant fossils. The lush Pennsylvanian coal swamps disappeared in the Permian. This was the result of the climatic change to arid or semiarid conditions that is recorded by the red bed and evaporite deposits and the wind-deposited sandstones. Continental drift may have caused the climatic change. The southern supercontinent had the distinctive *Glossopteris* flora in late Paleozoic; and in the northern supercontinent, three floras are found in the late Paleozoic.

The first insects appeared in the Devonian, and they became large and relatively abundant during the Pennsylvanian. (See Fig. 22-26.) As noted earlier, arthropods may have been the first animals to breathe air. They have an impervious surface and so were probably among the first to move in fresh water and finally out of water altogether.

As soon as plants grew on the lands, terrestrial plant eaters could develop. This, too, occurred in the Devonian. The first land vertebrates developed from the lobe-finned fish. This probably occurred in a fresh-water environment because the change from ocean to fresh water solved some of the problems of living out of the water. The body fluids of most animals contain about the same amount and type of salts as sea water. Some way to enclose these fluids to prevent dilution was necessary for the animals to move into fresh water. Once that was accomplished, only a way to get oxygen from the air was necessary. It seems likely that the ability to travel out of the water was a great advantage if the pool a fish was in dried out. It could then crawl to the next pool. In the course of moving from pool to pool, it might find something tempting to eat and so begin to live more and more out of the water. In this way, perhaps, the first amphibians developed.

Figure 22-26. Reconstruction of a Pennsylvanian coal swamp. A. *Lepido-dendron.* B. *Sigillaria.* Note the large cockroach. C. *Calamites.* D. Large dragonfly. From Field Museum of Natural History.

The oldest amphibian fossils so far found are in the upper Devonian rocks of Greenland and eastern Canada. (See Fig. 22-28.) They developed from the lobe-finned fish. (See Figs. 22-29 and 22-24.) The fins of these fish developed into clumsy legs. Lungs developed from swim bladders that are found in Devonian fish. Swim bladders enable a fish to adjust its buoyancy and so are a help in swimming. The lobe-finned fish also developed another adaptation that enabled them to live to the next rainy season if their pool dried up. The lung fish (see Fig. 22-30) can burrow in the mud and breathe air while in a hibernation-like state.

The development of amphibians was a great step toward inhabitation of the continents, but they were still tied to the water. They had to return to the water to lay their eggs. In spite of this, they expanded very rapidly. The reptiles were the next step, and they first appeared in the Pennsylvanian. The reptiles are characterized by the amniote egg that can hatch on land. It is difficult to distinguish between reptiles and amphibians on the basis of skeletons alone because they are so similar. *Seymouria* (see Fig. 22-31), found in Lower Permian rocks in Texas, is known to be a reptile only because the fossil eggs are found with it. The reptiles must have evolved during the Mississippian Period, and *Seymouria's* line has now been traced back to the Pennsylvanian.

Many branches of reptiles developed in the Late Pennsylvanian and Permian. (See Fig. 22-24.) The Mesozoic is the age of reptiles, and they will be discussed in the next chapter. The Permian "sail lizards" did not continue into the Mesozoic. These unusual

A.

B.

C.

Figure 22-27. Pennsylvanian plant fossils. A. *Lepidodendron.* B. *Sigillaria.* Parts A. and B. are both parts of the trunk showing leaf scars. Photos from Field Museum of Natural History. C. *Asterotheca miltoni,* a fern leaf. Photo from Ward's Natural Science Establishment, Inc., Rochester, N.Y.

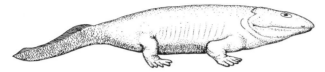

Figure 22-28. *Ichthyostega,* one of the earliest amphibians, from upper Devonian of Greenland. About two feet long.

Figure 22-29. *Latimeria.* A "living fossil" found near Madagascar. Lobe-finned fish similar to this developed into the early amphibians. About five feet long. Photo from Field Museum of Natural History.

Figure 22-30. Australian lungfish. The ability to breathe air shown by lungfish was probably a step in the development of amphibians from fish. Up to 40 inches long. Photo from Field Museum of Natural History.

Figure 22-31. *Seymouria* closely resembles the amphibians and would be classified as an amphibian if eggs were not found in the same rocks. From this stem reptile all other reptiles and higher vertebrates evolved. About 2.5 feet long. Photo from Field Museum of Natural History.

animals (see Fig. 22-32 and 22-33) had large fins. The purpose of the fins is not known. They may have been some sort of protection, or they may have been used to regulate their temperature. Reptiles are so called "cold-blooded" animals, and temperature regulation is a problem for them.

The land dwellers continued on from the Permian into the Mesozoic and expanded rapidly. Thus, whatever caused the great dying among the invertebrates had little effect on the vertebrates.

Figure 22-32. Permian "sail lizards." The one showing teeth is *Dimetrodon,* a carnivore. The smaller headed one is *Edaphosaurus,* a vegetarian. Both were about nine feet long. The fin may have been used to control body temperature. In the right foreground is *Diplocaulus,* a two-foot-long, bottom-dwelling amphibian. Painting by C. R. Knight, Field Museum of Natural History.

Figure 22-33. Skeleton of *Dimetrodon* shown in Figure 22-32. Photo from Field Museum of Natural History.

QUESTIONS

22-1. Where is the thickest section of Devonian rocks in North America? What is the evidence as to their source area?

22-2. Sketch a hypothetical cross-section through a dome of the type found in the interior during Paleozoic time. Pay particular attention to the changes in thickness of the sedimentary rocks and contrast these thicknesses with those in a dome formed by a much younger domal uplift.

22-3. What occurred in the far west during Cycle III?

22-4. What do the Cycle III rocks in the far north resemble?

22-5. Describe the Cycle IV rocks in the east.

22-6. What occurred in Nevada during Cycle IV?

22-7. How is the orogeny that produced the Appalachian Mountains dated?

22-8. What events were occurring in the west during this time?

22-9. What was the climate in western North America during Permian time? What is the evidence?

22-10. Where is the most complete section of Permian marine rocks in North America?

22-11. What are the dates of the important unconformities in the far north in late Paleozoic time?

22-12. Describe the movement of the southern continent in the late Paleozoic.

22-13. Which of the present continents were in contact at the end of the Paleozoic?

22-14. Why do we know so little about pre-Devonian plants?

22-15. When did the amphibians first appear? From what did they evolve?

22-16. When did the reptiles first appear?

22-17. What are some differences between reptiles and amphibians?

22-18. When did the first insects appear?

22-19. In what ways did late Paleozoic life differ from early Paleozoic life?

22-20. Name some organisms that became extinct near the end of the Paleozoic.

22-21. Were the jawless fish good swimmers?

SUPPLEMENTARY READING

Colbert, E. H., "The Ancestors of Mammals," *Scientific American* (March, 1949), Vol. 180, No. 3, pp. 40-43. Reprint 806, W. H. Freeman and Co., San Francisco.

Colbert, E. H., *Evolution of the Vertebrates.* New York: John Wiley & Sons, 1969, 542 pp.

Romer, A. S., "Major Steps in Vertebrate Evolution," *Science* (December 29, 1967), Vol. 158, No. 3809, pp. 1629-1637.

See also general references at the end of Chapter 21.

Chapter 23

The Mesozoic — Cycles V (B) and VI

Mesozoic means "middle life," and at the time it was named, it was thought to be the middle part of earth history. We now know the great age of the earth and that the Mesozoic is but a part of the late history of the earth. The life of the Mesozoic differs markedly from both the Paleozoic and the Cenozoic and was an obvious subdivision of geologic time to the early workers. In North America it was a very active time, as it was throughout the world.

In this chapter the Mesozoic history of western North America is completed. The Triassic history of the eastern and northern geosynclines was included in the last chapter. The reason for this division is that Cycle V extends from latest Mississippian to early Jurassic time; however, an almost complete withdrawal of the seas occurred near the end of the Permian Period. This made it convenient to include the eastern Triassic with the discussion of the formation of the Appalachian Mountains with which it is related, and to include the western Triassic with the rest of the Mesozoic with which it is more closely related.

Cycle V(B)—The Triassic of Western North America

At the beginning of the Triassic, the seas had retreated from the interior and a large part of the geosyncline as well. In part of the geosyncline in Idaho, Lower Triassic marine rocks overlie the Permian conformably. The geosyncline had changed somewhat from its configuration in the Permian. The island chain in central Nevada that

571

had been the site of Paleozoic orogeny was not present. In the far west, great thicknesses of marine clastic and volcanic rocks were deposited, at some places unconformably on Permian rocks. Thus the Paleozoic ended with orogeny at places in the active part of the geosyncline. Further east in the less active part of the geosyncline, marine rocks were deposited. In western Nevada, both clastic and carbonate sections tens of thousands of feet thick are found. At the edge of the interior, the marine rocks interfinger with non-marine red clastics.

Lower Triassic marine rocks are found only in the geosyncline but are known from southern Nevada to the Arctic, although in the United States, outcrops are limited. In Southern Nevada, they are clastic rocks about 8,000 feet thick, and in Idaho, they are limestone and shale 5,000 feet thick. These rocks interfinger with red sandstone and siltstone that in the northern Rocky Mountains are rarely over 1,000 feet thick. The Triassic sea in that area advanced and retreated a number of times, and apparently had an irregular shoreline with large embayments. Highlands in the interior were the source of some clastics. The non-marine red beds overlie similar Permian red beds, and it is difficult to place the boundary between the two systems because of the paucity of fossils.

In middle Triassic time, the seas retreated further west so that only the active part of the geosyncline was marine except in Canada where the whole of the geosyncline accumulated marine rocks. Evaporites formed in a restricted basin in Montana-Dakota. With retreat of the sea, some erosion apparently occurred in the areas of earlier deposition as well as elsewhere.

The area of maximum deposition of non-marine rocks was in Arizona and New Mexico, and here, Lower Triassic rocks are unconformably covered by Upper Triassic rocks with conglomerate at the base. This suggests uplift to the north. In this area, some of the Upper Triassic and Lower Jurassic sandstones appear to be wind deposited. These rocks are well displayed at Zion National Park. Because of the lack of fossils in these rocks, the boundary between Triassic and Jurassic is difficult to place. These wind-deposited rocks and the many red beds suggest that arid conditions prevailed over much of the United States.

In Middle and Late Triassic time, batholiths were emplaced in the Sierra Nevada Mountains according to radioactive dates. This activity was the forerunner of later intrusions and metamorphism.

The Jurassic began as a continuation of the Triassic. In the far west thick volcanic and clastic rocks were deposited, especially in a deep trough in the geosyncline in western Nevada. In Oregon, Jurassic rocks lie with an angular unconformity on Upper Triassic rocks, indicating orogeny in that area. The pattern, however, was soon to change, and the trough in western Nevada was the site of thrust faulting during deposition while still in Early Jurassic time. (See Fig. 23-1.) To the east of here, an uplifted area developed as the seas withdrew.

Cycle VI—Middle Jurassic Through Cretaceous

West

When the seas returned in Middle Jurassic time to start Cycle VI, the shape of the continent had changed. Orogenic events that formed an upland area had occurred in Nevada. (See Fig. 23-2.) This upland underwent erosion during most of this cycle and

NNW. 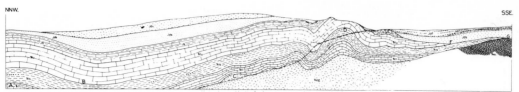 SSE.

1. Folding near margin of Luning embayment and deposition of conglomerate and fanglomerate of the Dunlap Formation in the marginal trough.

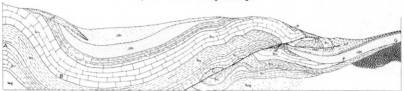

2. Development of Mac thrust on which the folded rocks to the north are carried toward the trough. Deposition of coarse material and folding and warping of Mac thrust impede further movement on this thrust plane. Folding spreads northward.

3. Further folding to the north accompanied by formation of the Spearmint thrust. Movement toward the trough was along an erosion surface cut on the upper plateau of the Mac thrust (D on this plate).

4. Folding of the Mac and Spearmint thrusts, as further southward movement is impeded. Major fold to the north was compressed into a recumbent syncline.

5. Involution of the fold to the north and its rupture along the Northwest thrust. Further folding of the Mac and Spearmint thrusts.

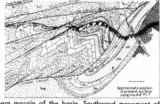

6. Development of West Ridge thrust on southern margin of the basin. Southward movement of the folded mass along the Cinnabar Canyon thrust, which is assumed to have used, in part, the folded plane of the Spearmint thrust. Folds in the area to the north are carried south along the Dunlap Canyon and South Fork thrusts above the Northwest thrust.

The sections are not drawn closely to scale. The assumed shortening below the Dunlap Canyon and South Fork thrusts is indicated by the changing positions of points A, B, C, D, E, F, G.

Approximate scale

1   0   1   2   3 miles

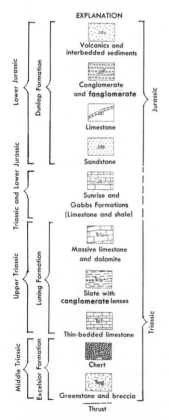

EXPLANATION

Lower Jurassic / Triassic and Lower Jurassic / Upper Triassic / Middle Triassic

Dunlap Formation / Luning Formation / Excelsior Formation

Jurassic / Triassic

Jdv
Volcanics and interbedded sediments

Jdf
Conglomerate and fanglomerate

Jdl
Limestone

Jds
Sandstone

Js
Sunrise and Gabbs Formations (Limestone and shale)

Tlu
Massive limestone and dolomite

Tls
Slate with conglomerate lenses

Tll
Thin-bedded limestone

Chert

Teg
Greenstone and breccia

Thrust

Figure 23-1. Diagrammatic sections showing progressive development in Early Jurassic time of complex structure in the northwestern part of Pilot Mountains in southwestern Nevada. From H. G. Ferguson and S. W. Muller, U.S. Geological Survey, *Professional Paper 216,* 1949.

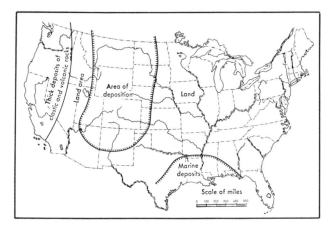

Figure 23-2. Middle and Upper Jurassic deposition. In Early Jurassic the Jurassic rocks were folded and faulted in the far west and a land mass formed. The area east of the land mass sunk and the sea entered from the north. Late in the Jurassic the sea withdrew to the north and non-marine deposits were laid down in the same area. Deposition in the Gulf of Mexico began in the Jurassic also.

shed clastics into seaways on both sides. Seas occupied the area east of the upland and covered the region of the present Rocky Mountains and Great Plains. This sea finally withdrew almost completely from the continent near the end of the Cretaceous Period, ending the cycle. This was the last time that seas occupied the interior of North America.

The cycle begins with seas returning to the northern Rocky Mountain area in Middle Jurassic time. The seas came from the north and at the margin of the Canadian Rockies, Early as well as Middle Jurassic sedimentary rocks were deposited. The shallow sea at first came as far south as southern Utah and covered most of Montana and North Dakota. Sand, silt, and clay, with some red beds and evaporites, were laid down, and the greatest thicknesses are found in Wyoming. At the maximum, the seas extended to northern Arizona and New Mexico, and possibly to the Gulf of Mexico. In Late Jurassic time the sea retreated to the north. As the sea withdrew, lake- and river-deposited rocks accumulated. These rocks are only a few hundred feet thick, but are of great interest because they contain dinosaur fossils and some rich uranium deposits.

On the western side of the upland, and probably on the upland itself, much activity took place during this cycle. Jurassic rocks are found at many scattered localities in Oregon, Nevada, northern Washington, and especially in California. These rocks are clastics and volcanics for the most part, and some have been metamorphosed. At some places, such as north-central California, the California-Oregon border, central Oregon, and southern British Columbia, the thicknesses are measured in tens of thousands of feet. In northern California, at least 16,000 feet of sandstone and shale were deposited in Late Jurassic time, and these rocks are covered by up to 25,000 feet of similar Cretaceous sediments. Further west is another sequence up to 50,000 feet thick that may be of the same age.

It is difficult to date orogenic events, and this is well illustrated in the Mesozoic of western United States. As we go westward from the sea in the Rocky Mountains, it becomes more difficult to date events the further west we go. This is because few fossiliferous rocks of Mesozoic age are found, especially in critical areas. Events in the eastern part of this area are, however, reflected in sediments deposited in the sea, enabling some dating. From western Utah all across Nevada is the Basin and Range country of isolated mountain ranges. The rocks in these mountains have been deformed, and the time of orogeny can be told at a few places from scraps of Mesozoic rocks. Detailed field mapping, in some cases of several ranges, may be necessary to date a single event. In California, in the Sierra Nevada Range, only igneous and metamorphic rocks are found, making study more difficult. Dating the granitic batholiths by radioactive methods has helped; and this method, together with detailed field mapping, has enabled the recognition of several belts of Mesozoic orogeny in the Great Basin, but it is not yet possible to reconstruct all the events in the Pacific Coast states.

The mountain building so far recognized in the Great Basin begins in the late Paleozoic in central Nevada as has already been described. In Triassic or Early Jurassic, thrusting occurred in south-central Nevada and southeastern California and extended to west-central Nevada. The evidence is radioactive dating in the south and stratigraphic dating to the north. In western Nevada and eastern Utah, uplift and deformation occurred mainly in Cretaceous time. The evidence is the Cretaceous rocks deposited in the sea to the east. In Late Cretaceous and Early Cenozoic time, orogeny occurred, both along the Pacific Coast and in the Rocky Mountains, and this will be described in Cycle VII.

In California some of the events are well dated, but the details remain obscure. The Sierra Nevada mountains are a vast complex of batholiths. The individual plutons have been closely studied, and hundreds of rocks have been dated by radioactive methods. Five distinct peaks in igneous activity have been found. They are Middle and Late Triassic, Early and Middle Jurassic, Late Jurassic, Early Cretaceous, and Late Cretaceous. In addition, Late Jurassic rocks are folded and intruded in the Sierra Nevada. As noted earlier, in central California almost on the flanks of the Sierra Nevada, sedimentation was continuous from Late Jurassic through the Cretaceous. (See Fig. 23-3.) These rocks are largely sandstone, and their source was probably an exposed batholith. Thus it appears that some batholiths were being emplaced as nearby batholiths underwent erosion. (See Fig. 23-4.)

The Mesozoic was a time of batholith emplacement in western North America. Similarly dated batholiths extend from Baja California to Alaska. The Coast Ranges of British Columbia are largely composed of granitic rocks. (See Fig. 23-5.)

Along the Pacific coast, a number of embayments existed, especially during the Late Cretaceous. The pattern was probably much like the Cenozoic history of that area. Cenozoic rocks and deformation obscure much of the Cretaceous story.

The seas returned to the area of the Rocky Mountains in Early Cretaceous time. This was the last great inundation of the sea on the North American continent. (See Fig. 23-6.) The seas extended from the Arctic to the Gulf of Mexico, and even along the Atlantic coast, sediments were deposited. In the Rocky Mountains and Great Plains areas, great thicknesses of clastic sediments were deposited. Up to 20,000 feet of clastics are found at some places. The source was the orogenic area in western Utah and eastern Nevada. The geosyncline sank as these sediments were rapidly deposited, and at times, the sediments accumulated above sea level. The source area was repeatedly uplifted, causing floods of conglomerate and sandstone to be deposited near the western margin of the basin. These rocks gave way to shale further east, and near the eastern margin,

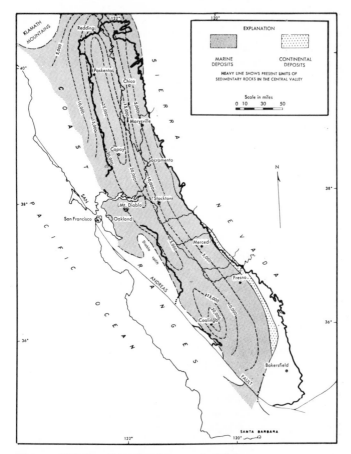

Figure 23-3. Distribution and thickness of Cretaceous sedimentary rocks in the Great Valley of California. The lines indicate the thickness that is more than 25,000 feet in places. From Otto Hackel in California Division of Mines and Geology, *Bulletin* 190, 1966.

to thin limestone layers. (See Fig. 23-7.) In Late Cretaceous time, the seas withdrew and non-marine clastic rocks were deposited. These rocks grade into the Cenozoic deposits that will be described in the next chapter. A small, short-lived embayment of the sea in North Dakota lingered into the early Cenozoic. The withdrawal of the sea marked the end of Cycle VI and the last great inundation of the continent. Orogeny occurred in parts of the Rocky Mountain area near the end of the Cretaceous. This orogeny is more closely related to the Cenozoic history and will be described in the next chapter.

Gulf and Atlantic Coast

Deposition in the coastal plain areas of the Gulf of Mexico and the Atlantic coast began in Cycle VI. (See Fig. 24-1.) It began in Jurassic time near the Gulf. The Jurassic rocks

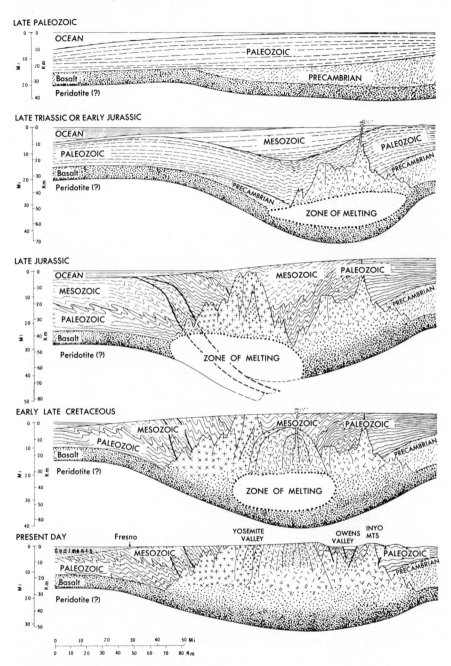

Figure 23-4. Hypothetical cross-sections showing the geologic development of California. The effects of sea-floor spreading are not shown. From P. C. Bateman and Clyde Wahrhaftig in California Division of Mines and Geology, *Bulletin 190,* 1966.

Figure 23-5. Mesozoic batholiths of
North America.

Figure 23-6. Generalized map of
Cretaceous time in the United States.
After U.S. Geological Survey.

do not crop out, but they are known from deep drilling for oil. The rocks encountered
include evaporites, carbonates, and shale. The salt domes of the Gulf Coast are believed
to have come from Jurassic salt deposits.

The Cretaceous deposits are much more extensive, and because of their overlap, the
Jurassic rocks are not exposed. In the Gulf area, the Cretaceous sediments are very
thick and consist mainly of shale and sandstone. Along the Atlantic coast, thin shale
and sandstone beds were the main deposits. The deposition on the Gulf and Atlantic
coasts continued into the Cenozoic, and thus the discussion will be continued in the
next chapter.

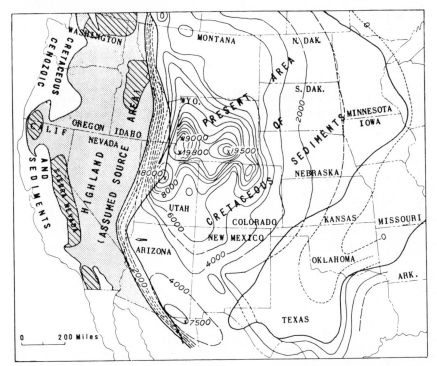

Figure 23-7. The distribution and thickness of Cretaceous sedimentary rocks in part of western United States. The lines show the thickness and are dashed where the rocks have been removed by erosion. The rocks are thicker and coarser near the source area. The great volume requires that about 5 miles was eroded from the source area and this does not include the Cretaceous rocks to the west of the highland. The batholiths in the source area are indicated by the diagonal lines. From P. C. Bateman and Clyde Wahrhaftig in California Division of Mines and Geology, *Bulletin 190,* 1966.

Far North

In the Arctic islands, the Jurassic is represented by a few hundred to a thousand feet of marine shale overlain by non-marine shale and sandstone. This is overlain by a shale unit of Late Jurassic–Early Cretaceous age that is up to 2,500 feet thick. In Cretaceous time, a clastic unit up to 4,500 feet thick was deposited. It is largely non-marine and contains some coal. Volcanic activity also occurred in the Cretaceous.

In eastern Greenland, Late Jurassic orogeny is recorded by thick clastic rocks of Early Cretaceous age. These are overlain by Cretaceous carbonates and red beds, suggesting a fluctuating shoreline.

Continental Drift—The Rest of the World

The Mesozoic was a time of drifting apart of the continents. The oldest rocks found so far on the sea floors are Jurassic, suggesting that in Triassic and Jurassic the present oceans, such as the Atlantic, may have begun to form. A probable sequence of drifting is shown in Fig. 23-8. The main geosynclines were along the west coasts of North America, South America, and Antarctica. These geosynclines probably formed on the border of oceanic plates. The geosyncline between Africa and Europe continued to develop.

During the Mesozoic, Africa and South America separated, as did Antarctica from Africa. India moved furthest during the Mesozoic.

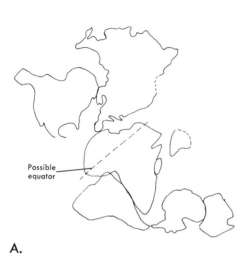

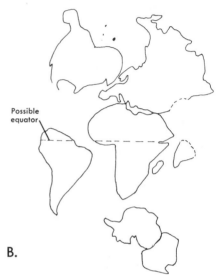

A.   B.

Figure 23-8. Continental drift during the Mesozoic. A. Shows positions at the end of the Jurassic Period. B. Shows positions at the end of the Cretaceous Period. Compare with Figure 22-11. Based on R. S. Dietz and J. C. Holden, 1970.

Life of the Mesozoic

The marine invertebrates of the Mesozoic are very different from those of the late Paleozoic. The extinctions, both in the Permian and at the end of that period, were probably caused by the withdrawal of the seas. When the seas returned in the Triassic, the main invertebrates were cephalopods and clams. Interestingly, no corals or foraminifera have ever been found in lower Triassic rocks, but they are present in Middle Triassic rocks so they did live somewhere. The Triassic corals and foraminifera are very different from those of the Paleozoic. (See Fig. 23-9.)

In the late Paleozoic and the Mesozoic, the cephalopods are almost ideal index fossils. They were abundant, are found in many environments, and they evolved rapidly. The stratigraphy of this interval is based on these animals. They were similar to the present-day nautiloids. (See Fig. 23-10.) They were swimming animals with coiled, chambered shells. They evolved rapidly, with differences between species being the ornamentation of the shell and the shape of the partitions between the chambers. (See Figs. 23-11 and 23-12.) Because they were swimmers, the fossils are found in all environments. It seems likely that they floated after death as well.

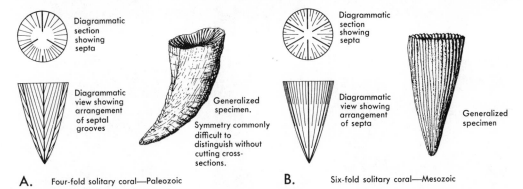

A.   Four-fold solitary coral—Paleozoic

B.   Six-fold solitary coral—Mesozoic

Figure 23-9. A. Paleozoic solitary or cup coral, showing four-fold symmetry. B. Mesozoic or Cenozoic solitary coral, showing six-fold symmetry.

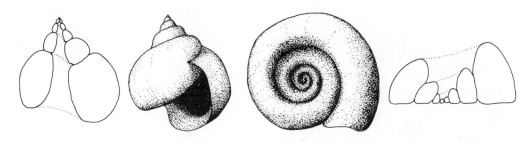

A. Shells and diagrammatic sections through coils of two generalized snails.

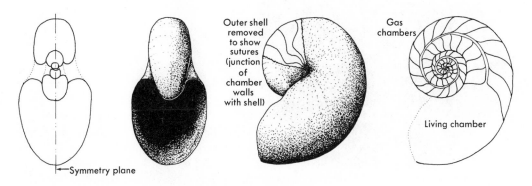

B.   Shell and diagrammatic sections of generalized cephalopod.

Figure 23-10. Differences between snail and cephalopod. A. Snail has no chambers and no symmetry. B. Cephalopod has chambers, some of complex pattern (see Figure 23-12), and is coiled in a plane, giving a symmetry plane. Some cephalopods are not coiled. The differences shown here are obvious morphology and there are other, more fundamental, differences.

A.

B.

C.

Figure 23-11. Cephalopods. A. *Metengonoceras,* Cretaceous with ceratite suture. B. *Baculites compressus.* A three-inch fragment of a straight-shelled ammonite. Note ammonite suture. C. *Scaphites.* Cretaceous ammonite. Length 1.5 inches. Photos from Ward's Natural Science Establishment, Inc., Rochester, N.Y.

A. Goniatitic

B. Ceratitic

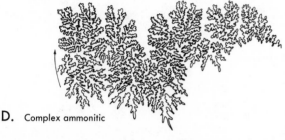

C. Ammonitic

D. Complex ammonitic

Figure 23-12. Cephalopod sutures. Arrows point toward the aperture. From *Treatise on Invertebrate Paleontology,* Courtesy of the Geological Society of America and the University of Kansas Press.

Fig. 23-13 shows the stratigraphic range of each of the cephalopod suture types. Perhaps one reason for their rapid evolution was that they nearly went extinct twice before finally dying out. At the end of the Permian, only one group survived; and again near the end of the Triassic, the same thing happened. At the end of the Cretaceous, they did die out. At each of these times, there was a withdrawal of the seas from the interior, but it seems likely that swimmers like these could have lived on in the oceans.

Why they developed such extremely complex partitions is not known, although they probably strengthened the shell much as corrugations strengthen cardboard. Another possibility is that the chambers were used by the animal to adjust buoyancy and for some reason the complex partitions were an advantage. In any case, the entire group died out at the end of the Cretaceous except the nautilus, which has simple, smooth partitions. Perhaps the more specialized forms could not adjust to the new conditions at the end of the Cretaceous, but this seems unlikely for such a successful group.

In the Jurassic, the seas again expanded, and many new forms appeared. One of these was the belemnoids which first appeared in the late Paleozoic. They were a type of cephalopod with an internal cigar-shaped shell. (See Fig. 23-14.) They were much like modern squids. (See Fig. 23-15.) Fig. 23-16 shows a somewhat more balanced sea floor in the Late Cretaceous.

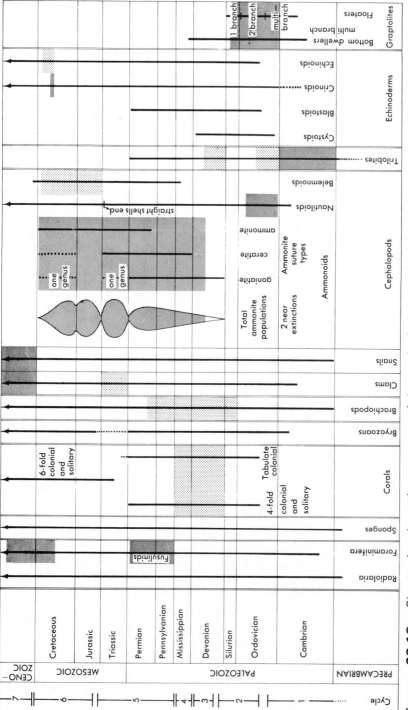

Figure 23-13. Chart showing the geologic occurrence of the more important fossil-forming invertebrate animals. The heavy lines show the time span of each animal group. Within some groups some simple distinctions are indicated that are of value in age determination. Also shown are the parts of the geologic column for which these fossils are useful for world-wide correlation (dark pattern) and for correlation of more restricted regions (light pattern). Correlation data after Curt Teichert, *Bulletin,* Geological Society of America, Vol. 69, 1958.

584

Figure 23-14. Cretaceous belemnoid. *Belemnitella americana* from New Jersey. Length, three inches. Photo from Ward's Natural Science Establishment, Inc., Rochester, N.Y.

Figure 23-15. Jurassic (Cycle VI) sea floor. Belemnoids (squid-like cephalopods) and oysters are the invertebrates shown. In the shadowy background the marine reptile ichthyosaur is shown. From Field Museum of Natural History.

Figure 23-16. Late Cretaceous (Cycle VI) sea bottom near Coon Creek, Tennessee. Coiled ammonite and straight-shelled ammonites *(Baculites)* and a number of snails and clams. From Field Museum of Natural History.

In both of the reconstructions, clams, including oysters, are prominent. In the Paleozoic seas, brachiopods were the main bivalves; but in the Mesozoic and Cenozoic, they are largely replaced by clams. The biologic differences between clams and brachiopods are great, and they can be distinguished easily from the symmetry of the shells. (See Fig. 23-17.) Clams live in many habitats, just as the brachiopods do. Some burrow into the bottom mud, and some are fastened to the bottom. One suggestion for why clams replaced brachiopods is that they burrowed deeper into the bottom mud and so developed a whole new habitat. They took over many habitats and expanded into many environments. They are very useful in determination of the paleoenvironment. The bottom dwellers attached themselves to the bottom by one shell, and some changed in appearance to resemble horn coral. Reefs of these coral-like clams formed at some places. Others evolved into oysters, also losing their symmetry. (See Fig. 23-18.)

The Mesozoic plants, too, were different from those of the Paleozoic, although the change is not apparent until Late Triassic. The lush swamps that produced the Pennsylvanian coal beds disappeared in the Permian. This was the result of the climatic change to arid or semiarid conditions that is recorded by the red bed and evaporite deposits and the wind-deposited sandstones. These desert conditions that persisted into Triassic

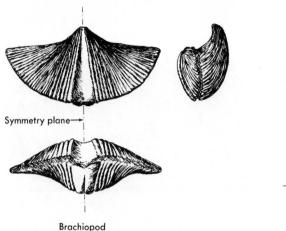

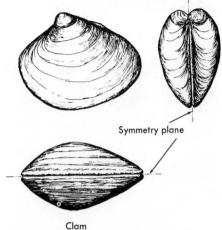

Figure 23-17. Differences between clam and brachiopod. A brachiopod has two dissimilar shells. A clam has two shells that are mirror images of each other because the clam has a symmetry plane between the two shells. The symmetry plane in a brachiopod cuts through both shells. Some clams (see Figure 23-18) do not have this symmetry.

time resulted in the preservation of very few members of the early Mesozoic flora. The few fossil plants, such as horsetails, found in early Mesozoic rocks are holdovers from the Paleozoic. In Late Triassic, the Mesozoic flora of ginkgoes, conifers, ferns, and especially cycads, all of which differed from Paleozoic forms of these plants, appeared. They appear in the background of Figs. 23-22 and 23-24 through 23-29. A few angiosperms, or flowering plants, also appear in Triassic rocks. The angiosperms, which are important elements of present floras, became abundant and spread to all continents in the Middle Cretaceous. They will be described in the next chapter.

The Mesozoic was the age of the reptiles, and, of course, the dinosaurs come to mind first. The dinosaurs were not the only reptiles. As noted in the last chapter, all of the reptiles have a *Seymouria*-like reptile of early Permian age as a common ancestor. (See Fig. 23-19.) In the Permian and most of the Triassic periods, the dominant group was the mammal-like reptiles. They became greatly reduced in Late Triassic time as the dinosaurs took over. The mammals evolved from the mammal-like reptiles, and the first mammal fossils are found in Upper Triassic rocks. The mammals were relatively unimportant until the Cenozoic, and so they will be discussed in the next chapter.

Seymouria's group lasted until near the end of the Triassic. The turtles evolved from them and are found from Early Triassic on. Snakes and lizards also had the same origin; the first fossil lizards are in Upper Triassic rocks, and the first snakes are Cretaceous.

The marine reptiles illustrate how a successful group like the reptiles moved into all environments. They returned to the sea, but they were better adapted than their amphibian ancestors were. (See Fig. 23-20.) They developed a system whereby the unhatched eggs were kept in the mother's body. Some Triassic ichthyosaurs have been found with unborn young in the body cavity.

A.

B.

C.

Figure 23-18. Clams. A. *Nuculana.* Pennsylvanian of Texas. Length 1.5 inches. A typical clam. B. *Trigonia thoracica,* a thick-shelled Cretaceous clam. C. *Exogyra arietina,* Cretaceous, 1.5 inches. A thick-shelled clam without the symmetry of most clams. Photos A. and B. from Smithsonian Institution, C. from Ward's Natural Science Establishment, Inc., Rochester, N.Y.

The thecodonts (Fig. 23-21) also evolved from *Seymouria's* group, and they gave rise to the dinosaurs, the crocodiles and alligators, the birds, and the flying reptiles. The thecodonts were a very successful group. They developed legs under the body instead of at the sides like the mammal-like reptiles and so improved their locomotion. Some developed a two-legged walk. In spite of this, they were generally small animals and were overshadowed by the mammal-like reptiles.

Birds are similar to reptiles in many ways. They differ in having feathers and being warm-blooded. The oldest bird is Jurassic and is known to be a bird rather than a reptile only because three almost complete specimens with feathers have been found in fine-grained limestone. This bird, *Archaeopteryx,* was about the size of a pheasant and had teeth in his beak. (See Figs. 23-22 and 23-23.) Birds are among the rarest of fossils. Their ecologic, but not evolutionary, predecessors, the toothed flying reptiles, are also found in Jurassic and Cretaceous rocks.

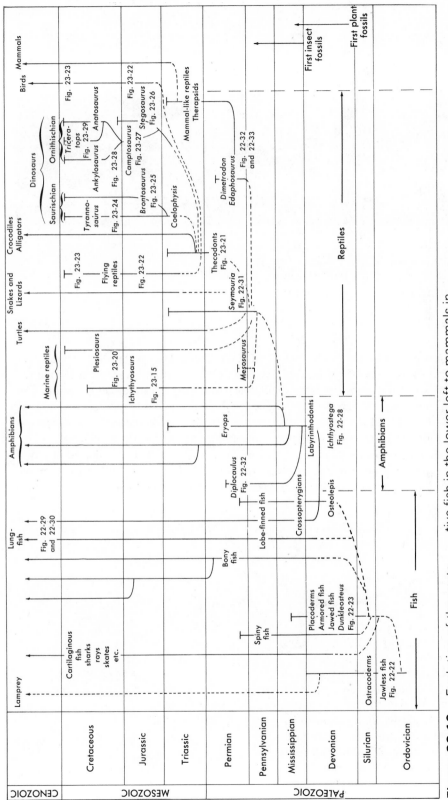

Figure 23-19. Evolution of the vertebrates. Although this chart appears formidable, the main theme can be seen along the bottom where progressive developments form a diagonal from primitive fish in the lower left to mammals in the upper right. Only the more important vertebrates are shown. Figures showing these animals are indicated.

589

Figure 23-20. Jurassic marine reptiles. Plesiosaurs are long-necked. Ichthyosaurs are porpoise-like and closely resemble fish. About six feet long. Painting by C. R. Knight, Field Museum of Natural History.

The first dinosaurs are found in Upper Triassic rocks. They were small animals, not at all like the huge beasts of Late Jurassic and Cretaceous that were soon to rule the lands. Many of the dinosaurs were small, but it is the big ones that get attention. The term *dinosaur* means "terrible lizard" and has been applied to two different groups of reptiles. It is an informal but useful term. They were very successful and evolved rapidly. Fig. 22-19 shows the relationships among the various dinosaurs.

The difference between the two types of dinosaurs is in their hip bones, one type having typical reptile hips (saurischians) and the other having bird-like hips (ornithischians). The saurischians were of two types, carnivores and herbivores. The carnivores

Figure 23-21. Thecodont. This group evolved from *Seymouria,* and the thecodonts gave rise to the dinosaurs, flying reptiles, birds, and crocodiles. Thecodonts had bodies similar to dinosaurs, but were all small, up to four feet, and were confined to the Triassic.

Figure 23-22. Jurassic scene. *Archaeopteryx,* the bird, has feathers. The flying reptile, *Rhamphorhynchus,* has no feathers and a wingspread of about three feet. Also shown are tiny dinosaurs and palm-like cycads. Painting by C. R. Knight, Field Museum of Natural History.

were mainly two-legged, with only small front legs; the largest flesh-eater of all times, *Tyrannosaurus* (20 feet tall, 50 feet long, weighing 8 to 10 tons), belongs to this group. (See Fig. 23-24.) The saurichian plant-eaters were mainly four-legged; some of the largest animals of all belong to this group (*Brontosaurus*). (See Fig. 23-25.) The ornithischians were apparently all plant eaters, and the bizarre dinosaurs, such as the duckbills, the horned dinosaurs, and the armored dinosaurs, belong to this group. (See Figs. 23-26, 23-27, 23-28, and 23-29.)

Most of the carnivores were two-legged, fast-running predators. The plant eaters developed many defenses. Some could run fast, some had armor, some had horns, some

A.

B.

Figure 23-23. A. *Pteranodon,* a
flying reptile. B. *Hesperonis,* a toothed
bird. Both from Upper Cretaceous of
Kansas. Photos from Smithsonian Institution.

were very large, and some lived in water. Much has been written about the small brains
of dinosaurs, but compared to other reptiles, their brains were near average. Compared
to mammals, their brains are small.

It is not clear why some of the dinosaurs became so large. It may be because reptiles
are cold blooded, and so it is an advantage to be large because a large animal needs
a much smaller percentage of its body weight of food each day to maintain its energy.

Most of the big dinosaurs were plant-eaters and so may have had a problem getting food. A disadvantage of being large is the difficulty of finding shelter in a storm. Size may also have been a defense against the predators.

The reptiles were very similar on all continents during most of the Mesozoic, suggesting that until the Cretaceous all continents were close enough to allow migration. In

A.

B.

Figure 23-24. Carnivorous dinosaurs. All carnivorous dinosaurs are saurischian. They could move rapidly on two feet and had only small front feet. A. *Ceratosaurus* from Upper Jurassic Morrison Formation near Canyon City, Colorado. About 24 feet long. Note the small horn on the top of the nose. Photo from Smithsonian Institution. B. *Tyrannosaurus,* the largest of the flesh-eating dinosaurs. About 50 feet long. Cretaceous. Painting by C. R. Knight, Field Museum of Natural History.

A.

B.

C.

Figure 23-25. Plant-eating saurischian dinosaurs. Mainly very large four-footed dinosaurs that lived in swampy water. A. *Brontosaurus,* Late Jurassic. About 70 feet long, weighing 35 tons. Eating enough to maintain such a size may have been a problem. Painting by C. R. Knight, Field Museum of Natural History. B. Skeleton of *Brontosaurus.* Photo from Field Museum of Natural History. C. *Diplodocus,* Upper Jurassic from Dinosaur National Monument, Utah. About 85 feet long, 13 feet high, weighing 13 tons. *Diplodocus* is found in Jurassic and Cretaceous rocks. Photo from Smithsonian Institution.

A.

B.

Figure 23-26. *Stegosaurus.* These plated dinosaurs lived in the Jurassic and Cretaceous. About 20 feet long, weighing two tons. A. Painting by C. R. Knight, Field Museum of Natural History. B. Skeleton. Photo from Smithsonian Institution.

the Cenozoic, there was less migration, and, interestingly, 30 orders of mammals evolved during the 65 million years of the Cenozoic, but only 20 orders of reptiles are found in the 200 million-year Mesozoic. This suggests that in the relative isolation of the Cenozoic, more orders formed because each isolated group evolved and radiated.

The orogenic activity that marked the Late Cretaceous and the Early Cenozoic produced a diverse topography that resulted in many climatic variations. These new climatic zones were soon occupied by plants, thus setting the stage for animals to evolve to take over these ecologic niches. The angiosperms, or flowering plants, that appeared in Early Cretaceous include the grains and grasses, and it is the rapid spreading of these to the dryer climates that created the new ecologic niches. The plant life of the Cenozoic

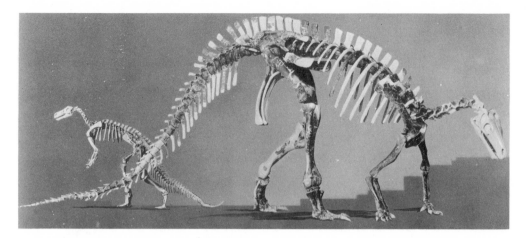

Figure 23-27. Duckbilled dinosaur. *Camptosaurus,* Upper Jurassic, Wyoming. *Camptosaurus* was one of the main stems of the ornithischian dinosaurs, and the ceratopsian (horned) dinosaurs evolved from them. *Camptosaurus* lived in the Jurassic and Cretaceous and was 15 feet long. Photo from Smithsonian Institution.

Figure 23-28. Cretaceous dinosaurs. On the right *Trachodon,* a duckbill. Other duckbills are the web-footed *Corythosaurus* in the water, and crested *Parasaurolophus* in the left background. In the foreground is *An-kylosaurus,* an armored dinosaur. In the center background is the ostrich-like *Struthiomimus,* a saurischian dinosaur. Painting by C. R. Knight, Field Museum of Natural History.

was essentially modern, consisting mainly of angiosperms and conifers. The many Cenozoic coal beds attest their development.

At the end of the Mesozoic, the dinosaurs, almost all of which had specialized to occupy the many different environments, died out. Some of the dinosaurs were big animals, making food gathering in times of changing climate and plant life a difficult task. At this time, the mammals were small and few in number. They apparently were better able to adapt to the new conditions and so replaced the dinosaurs. The mammals were a more successful group than the reptiles for several reasons. They had bigger brains. They had four-chambered hearts instead of the two-chambered hearts of reptiles, and had body hair rather than scales. Both of these features allowed them to maintain their body heat, rendering them less vulnerable to climatic changes. They were better able to preserve their young than the egg-laying reptiles. They also developed jaws and teeth better suited to their food. Their feet also underwent a series of changes that better equipped them to their ecology. In spite of all these reasons why mammals were better equipped than dinosaurs, the dinosaurs were not driven out by the mammals. Rather, for unknown reasons, the dinosaurs died out and the mammals took over.

A.

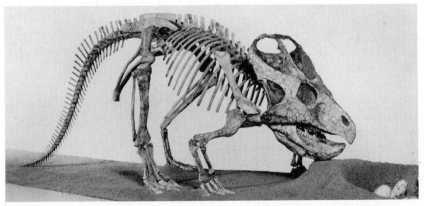

B.

Figure 23-29. *Ceratopsian* (horned) dinosaurs. Cretaceous. They evolved from *Camptosaurus.* A. *Protoceratops.* Eggs of this dinosaur were found in Mongolia. Painting by C. R. Knight, Field Museum of Natural History. B. *Protoceratops* skeleton. About six feet long. Photo from Field Museum of Natural History.

C.

Figure 23-29. (cont.) C. *Triceratops* skeleton. About 24 feet long. Photo from Smithsonian Institution.

QUESTIONS

23-1. Describe the Triassic rocks of western United States.

23-2. What was the extent of the Jurassic seas in North America?

23-3. Why is it difficult to date Cretaceous events between the Rocky Mountains and the West Coast?

23-4. Where were batholiths emplaced in North America in the Mesozoic?

23-5. Describe the Cretaceous rocks of the Rocky Mountains.

23-6. When did deposition start in the Gulf Coast area?

23-7. Which continents drifted apart during the Mesozoic?

23-8. Is there a relationship between continental drift and the locations of Mesozoic geosynclines?

23-9. Describe the fauna of the Triassic.

23-10. When did the first flying reptiles appear? The first birds?

23-11. How do Mesozoic corals differ from Paleozoic corals?

23-12. How are clams distinguished from brachiopods? Does this distinction hold true for all Mesozoic clams?

23-13. How are snails distinguished from cephalopods?

23-14. Outline the history of cephalopods.

23-15. How are Mesozoic plants different from Paleozoic plants? From Cenozoic plants?

23-16. What was the time of the dinosaurs?

23-17. Describe the various types of dinosaurs and other reptiles of the Mesozoic. Which ecologic areas or niches did they occupy?

23-18. What advantages do mammals have over reptiles?

SUPPLEMENTARY READING

Kurtén, Björn, "Continental Drift and Evolution," *Scientific American* (March, 1969), Vol. 220, No. 3, pp. 54-64.

See also references at the ends of Chapters 21 and 22.

Chapter 24

The Cenozoic

Cycle VII—Cenozoic

Following the withdrawal of the seas near the end of the Cretaceous, the seas never again occupied large areas of the interior, but they did cover most coastal areas during parts of the Cenozoic. Thus, the geologic history of the Cenozoic in most of North America is concerned with non-marine rocks, and so this cycle differs greatly from all of the previous cycles. During the Cenozoic, most of the present topographic features of the earth were formed, making its history of great interest to the traveler. It will be possible to describe only a few of the more important areas.

Atlantic and Gulf Coastal Plains

Marine sedimentation began in these areas in Mesozoic time and continued throughout most of the Cenozoic. The history is one of repeated advances and retreats of the sea. The Gulf area is one of the most closely studied areas in the world because of the great value of the oil recovered from these rocks. The rocks are penetrated by thousands of oil wells, and some of these are very deep. On the south Atlantic coastal plain, four major advances are recognized. They are mid-Cretaceous, Late Cretaceous to early Cenozoic, and two in the late Cenozoic—middle Miocene to early Pliocene, and Pliocene to Pleistocene. Florida was an area of limestone deposition much of the Cenozoic and separates the Atlantic from the Gulf coastal plain. On the Gulf, eight major advances and retreats are recognized, as well as many more minor ones. The maximum advance occurred in the Eocene when the Mississippi Valley was invaded as far as southern Illinois. This was the first advance of the Cenozoic, and the early Cenozoic rocks overlie the Cretaceous unconformably, indicating that the Cretaceous seas had retreated in late Cretaceous time. A typical advance of the sea is recorded by clastic rocks at the bottom of the sequence. The rocks grade from non-marine sand or clay to similar brackish water sediments to marine sandstone or shale that thickens greatly toward the Gulf. The advances and retreats of the sea are probably related to uplifts in the Appalachian Mountains, but it has not been possible to correlate these events. Much of the area drained by the Mississippi River was undergoing erosion throughout the Cenozoic and provided debris to the Gulf area. The sedimentary rocks on the Gulf coast are very thick, up to 50,000 feet with 15,000 feet of Pleistocene at some places. Fig. 24-1 shows a typical cross-section and shows the rise in the basement rocks offshore.

Through the Cenozoic, limestone was deposited in Florida. It is believed that this area was covered by shallow, warm water. Under such conditions, limestone is being deposited now in the Bahama Banks just east of Florida. Florida was not awash during all of Cenozoic, but at times islands were present.

Appalachian Mountains

The Appalachian Mountains were above sea level and undergoing erosion during the Cenozoic. Their history cannot be learned from sedimentary rocks except in a very general way because the uplifts of the mountains cannot be related to the coastal plain sediments. Thus we turn to study of the erosion surfaces to determine the history of the Appalachians. This was one of the first places that such studies were made. The culmination of these studies was the classical interpretation shown in Fig. 24-2. The history shown in the cross-section starts with high mountains in post-Triassic time (A). These mountains were eroded to a smooth peneplane before Cretaceous time (B); and in the Cretaceous, coastal plain sediments were deposited on the peneplane (C). This was followed by a broad up-arching (D). Erosion of the upwarped range formed a new peneplane (E). This peneplane was up-arched (F), and erosion developed a partial peneplane on the softer rocks (G). Another uplift caused erosional development of another partial peneplane on the softer rocks (H). Another uplift caused further dissection of the partial peneplanes and produced the present Appalachian Mountains.

More recent studies suggest that a simpler history could produce the same topography. In this case, the history would be the same up to block E. A single uplift followed by erosion could result in the ridge tops preserving remnants of the old peneplane at elevations of about 4,000 feet, and flat valleys cut on weak rocks.

Rocky Mountains and High Plains

Orogeny began in the Rocky Mountains near the end of the Cretaceous and extended to the Oligocene. The age and intensity of this folding vary from place to place. (See Figs. 24-3 and 14-38.) At some places Paleocene rocks unconformably overlie folded Late Cretaceous rocks, and at other places rocks as young as Oligocene are involved in the folding. In general, the more intense orogeny occurred near the west shore of the Cretaceous sea where thrust faults are found. The compressive phase was over by middle Eocene except in southern Colorado where late Eocene thrusts are found.

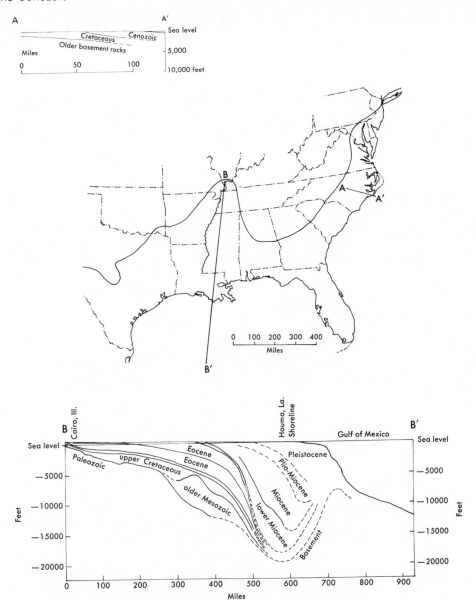

Figure 24-1. Cenozoic deposition in the Atlantic and Gulf areas. Note the very thick Cenozoic deposits in the Gulf Coast area.

Further east in Montana, Wyoming, and Colorado, most of the present mountain ranges were formed, with folding and faulting that was locally intense. Nearby on the high plains of eastern Montana, undeformed Late Cretaceous non-marine sediments are conformably overlain by similar Paleocene rocks, indicating that the orogeny was localized. This sequence of rocks is of great interest because it records the last of the dinosaurs and the rise of mammals. Further east in North Dakota and adjacent areas, a remnant of the Cretaceous sea remained until Paleocene.

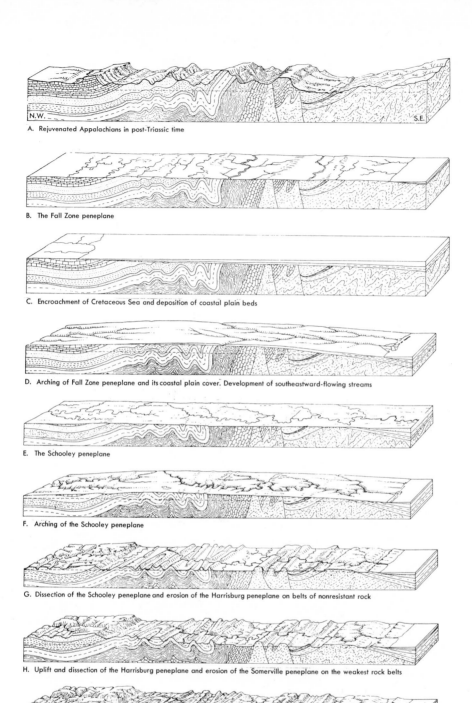

A. Rejuvenated Appalachians in post-Triassic time

B. The Fall Zone peneplane

C. Encroachment of Cretaceous Sea and deposition of coastal plain beds

D. Arching of Fall Zone peneplane and its coastal plain cover. Development of southeastward-flowing streams

E. The Schooley peneplane

F. Arching of the Schooley peneplane

G. Dissection of the Schooley peneplane and erosion of the Harrisburg peneplane on belts of nonresistant rock

H. Uplift and dissection of the Harrisburg peneplane and erosion of the Somerville peneplane on the weakest rock belts

Allegheny Front ⫶ Ridge and Valley Belt ⫶ Great Valley ⫶ Reading Prong ⫶ Trias lowld ⫶ Piedmont ⫶
APPALACHIAN PLATEAU ⫶ NEWER APPALACHIANS ⫶ OLDER APPALACHIANS ⫶ COASTAL PLAIN

I. Uplift and dissection of the Somerville peneplane to give present conditions

Figure 24-2. The development of the Appalachian Mountains. The history shown in the diagrams is the classical interpretation. New concepts suggest that the present form of the Appala-chian Mountains could have developed from a single uplift. From Douglas Johnson, *Stream Sculpture on the Atlantic Slope,* Columbia University Press, N.Y., 1931.

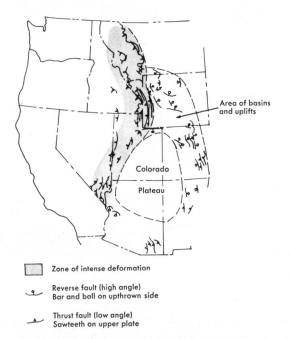

Area of basins
and uplifts

Colorado

Plateau

Zone of intense deformation

Reverse fault (high angle)
Bar and ball on upthrown side

Thrust fault (low angle)
Sawteeth on upper plate

Figure 24-3. Late Cretaceous and Early Cenozoic deformation in the Rocky Mountain area. In the western zone of intense deformation, folding and thrust faulting are common. In the eastern area of basins and uplifts, a number of basins sank and were filled by detritus from rising highlands. Many of these basins were folded with local thrust faulting during this interval. Very little deformation took place in the Colorado Plateau, an anomalous situation that has not been satisfactorily explained. The west coast was an area of active basins. Between the west coast and the intensely deformed zone, few Cretaceous rocks are present so deformation during this interval cannot be recognized. Structural details after Gilluly, 1963.

With the uplift of the mountain ranges in the Rockies, basins were formed between the ranges. (See Fig. 24-4.) These basins were filled with non-marine sediments during the early Cenozoic. Some volcanic sediments were deposited in some of the basins, and in the Eocene and Oligocene, great thicknesses of volcanic rocks were emplaced in northwestern Wyoming in the Absaroka Range. Volcanic rocks were deposited throughout most of the Cenozoic in the San Juan Mountains of Colorado. Erosion of the ranges filled the basins, mainly in Paleocene to Oligocene time.

From Oligocene to Pliocene time, the streams from the area of the present Rocky Mountains were spreading fine clastics over the Great Plains. Then, either due to a change in climate or to uplift of the whole Rocky Mountain area, erosion began. It is uncertain whether the uplift occurred at this time or earlier. This erosion removed much of the earlier deposits, forming the present topographic mountains. Prior to this downcutting, the rivers had been flowing on a relatively smooth surface, in part eroded and in part constructed by the rivers. The courses of the rivers on this surface bore no relationship to the rock types at depth, for they had developed their courses over tens of millions of years. Thus, the downcutting rivers at most places removed the soft basin-filling sediments, but at some places cut narrow, deep canyons through the

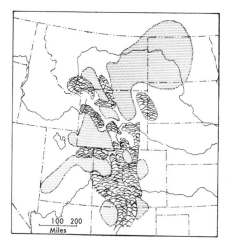

Figure 24-4. Early Cenozoic basins and highlands in the Rocky Mountain area.

mountain blocks composed of Precambrian metamorphic rocks. Examples of this include Big Horn River Canyon in the Big Horn Mountains, Laramie River in the Laramie Mountains, North Platte River, Sweetwater River, and many others. (See Fig. 24-5.)

Colorado Plateau

The events on the Colorado Plateau are related to those in the Rocky Mountains. In early Cenozoic time, the Colorado Plateau was at a low elevation, and rivers carrying sediments from the uplifted areas to the north and east deposited them on the plateau. By Eocene time, the plateau had downwarped, and a vast freshwater lake covered much of Utah. Organic-rich shale was deposited in this lake, and the resulting oil shales may someday be an important source of petroleum. Uplift and local volcanic activity occurred during the rest of the Cenozoic. During this time the Colorado River eroded the Grand Canyon. The rocks on the Colorado Plateau are, for the most part, horizontal, with some faulting and bending. All around this area, similar rocks are folded and deformed. Why the Colorado Plateau escaped this deformation is not known. (See Fig. 14-41.)

Basin and Range

This area is known variously as the Great Basin, the Basin-Range, or the Basin and Range area because, with the exception of the Colorado River, it is a basin with no external drainage to the ocean, and because the area is characterized by many fault-block mountain ranges with intervening basins. Like the Colorado Plateau, this is an unusual area. The deformation of the rocks that make up the ranges has been described in the preceding chapters. These deformed rocks have been uplifted in fault blocks to form the Basin and Range area. The faults appear to be normal faults, and so differ sharply with the compressional features displayed in the ranges. (See Fig. 24-6.) The faulting and uplifting began in the Miocene and are probably still going on, judging

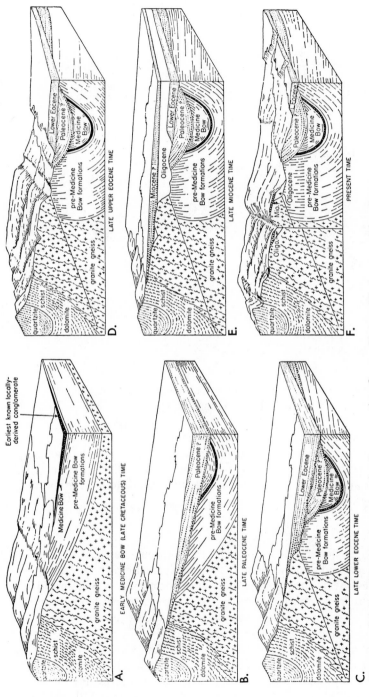

Figure 24-5. Development of the Rocky Mountains. The Medicine Bow Mountains shown here are typical of the Rocky Mountains in Wyoming. In early Cenozoic time deposition and deformation occurred at the same time (Parts A, B, and C). In late Cenozoic times the basins were filled to the level of the now-eroded highlands and the rivers spread sediments in both the area of the present Rocky Mountains and the Great Plains (E). In very late Cenozoic time uplift or climatic change caused the rivers to erode and downcut (F). The rivers removed much of the soft basin-filling sediments and in many cases cut deep canyons across previously buried ranges, producing the present mountain topography. From S. H. Knight, Wyoming Geological Association, 1953.

607

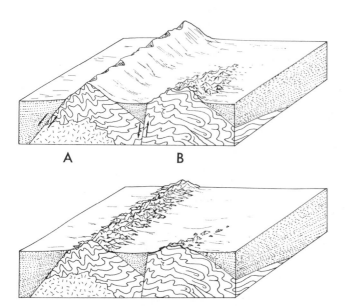

Figure 24-6. Typical Basin and Range structure. Both Ranges A and B are uplifted fault blocks with intervening basins filled by detritus eroded from the ranges. Range A is younger than range B. The rocks in both ranges were deformed prior to the present uplift. The lower view shows the ranges at a later time.

from the recent fault scarps and the number of earthquakes. The best example is probably the westernmost range, the Sierra Nevada in California. (See Figs. 24-7 and 24-8.)

The basins obviously have been sites of deposition since their formation. Prior to the faulting, Cenozoic rocks were deposited in part of the area, and these rocks are exposed in some of the ranges. Volcanic activity occurred at various times in different parts of the Great Basin. In some areas, thick, widespread welded tuffs can be traced for many miles from range to range. (See Fig. 14-42.)

Northwest United States

The northwest was an area dominated by volcanic rocks throughout much of the Cenozoic. Marine sediments and some volcanic rocks accumulated near the Pacific in a number of active embayments. A pre-late Miocene unconformity is widespread. Inland, the Cenozoic rocks are continental and volcanic. They were deposited in many basins, and the number of unconformities shows the crustal unrest. Further east, in much of eastern Washington and Oregon, these rocks are covered by the spectacular Columbia River Basalt to a depth of at least several thousand feet. Further east in the Snake River Valley are thick deposits of younger volcanic rocks. In Pliocene time the present Cascade Range was uplifted on a north-south axis that cuts across the earlier structures. The last step was the formation of the present volcanoes, such as Mt. Shasta, Mt. Hood, Mt. Rainier, and Mt. Baker, in late Pliocene and Pleistocene time. (See Figs. 24-9 and 14-44.)

Figure 24-7. The eastern fault scarp of the Sierra Nevada. The fault is covered by the alluvial fans. In the foreground is Owens River. Owens Valley was the site of a lake that is now dry, and some of the shoreline features can be seen. Photo from U. S. Geological Survey.

Figure 24-8. Aerial view of the west slope of the Sierra Nevada showing Yosemite Valley. Note the relatively smooth surface that was uplifted by the fault whose scarp is shown in Figure 24-7. The uplifted surface is incised by rivers whose valleys have been modified by glaciation producing features such as Yosemite Valley. At other places, especially on the skyline, erosion has produced rugged mountain topography. Photo from U. S. Geological Survey.

Figure 24-9. Oblique aerial view of
Mount Hood in northern Oregon. From
northern California through Washington the high Cascade volcanoes formed
on top of a mountain range composed
of deformed Cenozoic rocks. Photo
from U. S. Geological Survey.

California Coast

A brief description of the Cenozoic history of California will, together with the outline
of the northwest, give an indication of the types of events that occurred along the Pacific
coast. This was an extremely active area throughout the Cenozoic, and a detailed
history of the Pacific coast would be very long. In Paleocene and Eocene time, a sea
occupied the central and southern part of western California. Eastern California was
an upland, and its erosion shed sediments into the sea. Non-marine clastics were
deposited between the sea and the upland. In the area of the present Coast Ranges,
several large islands existed. At places, up to 10,000 feet of clastics were deposited in
this sea.

During the Oligocene, non-marine clastics and red beds were deposited inland with
some marine nearer to the coast. The seas returned in the Miocene, and there was much
volcanic activity. Orogeny also occurred at this time. In Pliocene many basins developed, and some of these were non-marine lakes. In the Los Angeles Basin, 15,000 feet
of sediments were deposited. Deposition continued into the Pleistocene. The San Andreas fault, which is associated with sea-floor spreading in the Pacific, was probably
active during much of the Cenozoic. In Middle Pleistocene, a major deformation
occurred. At this time the Coast Range Mountains were uplifted. Similar uplifts occurred all the way to Washington. As a result of this activity, the Pacific coast has
mountains along the coast (the Coast Ranges and Olympic Mountains), an inland
valley (the Great Valley in California, Willamette Valley-Puget Sound in Oregon and
Washington), and a mountain range (Sierra Nevada and Cascade Range). The only
exceptions are southern Washington where the coastal mountains are low hills, the
Oregon-California border where there is no valley, and southern California.

Arctic

The Cenozoic rocks of the Arctic Islands are mainly non-marine sandstone and shale with some coal. These rocks were apparently deposited in basins, and although most occurrences are only a few hundred feet thick, at one place they are 7,000 feet thick. These rocks were folded sometime in the Cenozoic.

Until the Cenozoic, the Arctic Islands and Greenland were part of the North American continent. The area apparently sunk to form the islands, as the waterways among the islands may be, at least in part, a drowned river system.

Pleistocene

The Pleistocene, although generally thought of as the glacial age, is defined by a type section like other periods. Glaciers only occupied a small part of the earth during the ice age, and they formed and melted at different times at different places. The Pleistocene is generally considered to have begun about one million years ago. Recent radioactive age determinations, however, indicate that glaciers began to form in Iceland about three million years ago.

Much of the Pleistocene history of North America was discussed in a general way in the chapter on glaciation, and this should be referred to at this time. The extent of the glacial advances is shown in Fig. 10-21.

The Pleistocene history of North America is a study worthy of a much longer book than this. Only an outline of events can be presented here because almost all of the present landforms were made or greatly modified during the Pleistocene. In Chapter 10, the geologic work of glaciers was described and the possible causes of glaciation were discussed. Here, only the effects of glaciation will be mentioned.

The principles involved in Pleistocene geology as outlined here are

1. Erosion and deposition by alpine and continental glaciers.
2. Drainage changes caused by glaciers, which include damming of rivers by the advance of the glaciers.
3. The effects of the generally rapidly released melt-water from retreating glaciers.
4. The effects of the weight of glacial ice lying on the continents.
5. The effects of the glacial climate in areas away from the ice.
6. The normal nonglacial geological processes, such as erosion, deposition and orogeny, that have been discussed in the history of the earlier geologic periods.

The most obvious effects of Pleistocene glaciation are seen in the areas actually glaciated. The continental glaciers covered much of the northern part of the United States. The glaciers made four major advances, and many of these advances had one or more smaller advances and retreats. The details of these oscillations have been deciphered by study of the moraines, as shown in Chapter 10. In north-central United States, the long, low morainal hills marking the farthest advance of the last glacier form conspicuous features. The soils of this area and of the midwest are largely glacial deposits scraped from the now almost bare areas to the north in Canada. In New England and New York, the low mountains were covered by the continental ice. The main effects were a rounding and smoothing of the mountains. Much of the material eroded by the glaciers was deposited nearby, producing the rocky soils and fields of this region. As in other places, the furthest advance of the glacier is marked by terminal moraines, of which Long Island is the most prominent.

At other places, rivers as well as moraines mark the limit of glaciation. In the mid-west, the Ohio River follows the approximate margin of the glacial ice. This river probably developed to carry the melt-water from the margin of the receding ice sheets to the Mississippi River. In the northern Great Plains in Montana, the Missouri River was diverted by the glacial ice. This river probably flowed northeastward toward Hudson Bay in preglacial time. The presence of the glacier forced it into a westerly and southwesterly course to join the Mississippi River. Much of the present course of the Missouri is within a few tens of miles of the limit of the glacial ice. (See Fig. 10-21.)

Mountain glaciers developed in the high mountains of western United States. These mountain glaciers sculptured the superb scenery of many of our ranges, such as at Yosemite National Park. (See Fig. 10-14.) At some places, so much snow accumulated on the mountains that only the highest peaks were not covered. At places, too, glacial tongues extended some distance out from the mountains. Generally this occurred where several coalescing glaciers extended into the lowland.

The area between the Cascade and the Northern Rocky Mountains was the scene of a remarkable event in glacial times. A wide path across the Columbia Plateau has been eroded by much rapidly flowing water. This is shown by the stripping off of most of the loose surficial material, eroded stream channels, and large sand bars. (See Fig. 24-10.) This erosion is believed to have been accomplished by glacial melt-water, but

Figure 24-10. Looking east across the Channeled Scablands in southeastern Washington toward farmlands where the soil was not removed by the glacial flood waters. Compare the many large-scale features of river erosion in the aptly named Channeled Scablands with the smooth farm lands. Photo by Bureau of Reclamation.

obviously some remarkable events must have occurred to channel enough water through the area to do this work. As the glacier retreated in the northern Rocky Mountains in Montana, melt-water filled the northward-sloping valleys, forming a large, irregularly shaped lake. The glacier to the north dammed this lake. When the water level of the lake overtopped the ice dam, the flowing water eroded the ice dam very rapidly. This is believed to be the source of the rapidly flowing, large amount of water. This water flowed across Idaho and met the Columbia Plateau near Spokane. The Channeled Scablands, as the area eroded in this unusual way is called, extend south from Spokane to Grand Coulee Dam. Apparently many floods of water reached the Columbia River in this way as the ice dam was reformed by forward-moving glacial ice and destroyed by the overtopping melt-waters again and again. Remember that even in a retreating glacier the ice moves forward. These flood-waters could not follow the Columbia River Valley because the Columbia was also dammed by glacial lobes extending south from the mountains north of the Columbia Plateau. These latter glacial lobes apparently blocked the Columbia River at different places as the various floods reached the Columbia, and they diverted the floods southwesterly across the Columbia Plateau. The diverted flood-waters eroded the scabland and rejoined the Columbia River in southern Washington. (See Fig. 24-11.)

The climatic changes that caused glaciation in high latitudes produced aridity near the equator. The evidence for this is both the type of sediment deposited in the oceans in low latitudes, and sand dunes in both northwestern Australia and west Africa that

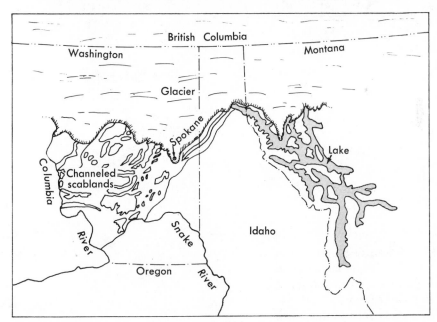

Figure 24-11. The development of the Channeled Scablands in eastern Washington. The lake formed from glacial melt-water in the north-sloping valleys in Montana as the glacier retreated. When the glacial ice that dammed the lake was removed by melting, or erosion by water that over-topped it, the lake drained rapidly along a broad valley south of the continental glacier. The water crossed the Columbia Plateau, eroding the Channeled Scablands, and then joined the Columbia River. Based on maps by J. H. Bretz in *Bulletin,* Geological Society of America, Vol. 67, 1956.

pass under the sea on parts of the shelf that were above sea level when glaciers occupied the northern lands. Nearer to the glaciers, the climate was wet.

In southwestern United States, during glacial times many large lakes, such as Great Salt Lake, formed in this region of interior drainage. These lakes were mentioned earlier in Chapter 10.

The Great Lakes of northern United States were also formed during the Pleistocene, but their origin was very different from that of Great Salt Lake. As long as the front of the continental glacier was south of the divide that separates the drainage of the Mississippi River from the St. Lawrence River, the melt-water all flowed out the Mississippi River. When the glacier retreated north of this divide, lakes were impounded between the glacier and the divide. As the water level rose and the glacier retreated, a number of different outlets were uncovered. If nothing else had happened, the lakes would all have been drained when the glacier melted away. However, the melting of the glacier reduced the weight pressing down on the crust and the crust began to rise, perhaps isostatically. This uplift acted as the glacier had to impound the Great Lakes. The development of these lakes is known in great detail as a result of much study. When the ice first melted from this area, the crust of the earth had been depressed by the weight of the ice so that it was below sea level, and marine conditions prevailed in part of the St. Lawrence Valley until the crustal rise forced the seas out. (See Fig. 24-12.) Other lakes also formed in this region, but were drained when the ice retreated and the crust rose. Lake Winnepeg is a small remnant of one of these other glacial lakes.

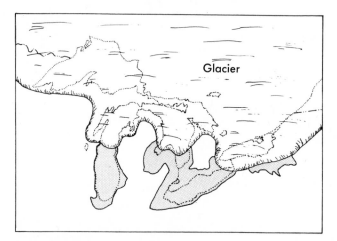

Figure 24-12. An early stage in the development of the Great Lakes. The glacier has melted back exposing large valleys that slope northward. Melt waters have filled the valleys that are dammed by the glacial ice.

Continental Drift—The Rest of the World

The present distribution of continents on the earth formed during the Cenozoic. Fig. 24-13 shows the drift that occurred. North America separated from Europe, and the

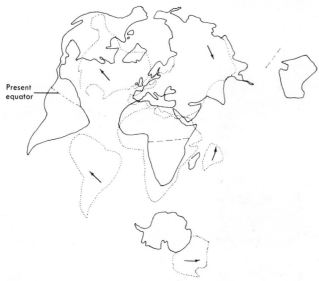

Present
equator

Figure 24-13. Continental drift during the Cenozoic. Dotted positions are
the end of the Cretaceous; solid line is
the present. Based on R. S. Dietz and J.
C. Holden, 1970.

Atlantic Ocean formed. Greenland separated from North America, and Australia left
Antarctica. The south Atlantic widened. India collided with Eurasia.

Orogeny occurred from Alaska to Antarctica, and apparently Central America
formed at this time. Africa and Europe may have drifted closer together because
orogeny occurred in the geosyncline between them. This mountain-building event
produced the Alps in Europe and a mountain chain across Asia. India collided with
Asia and may have caused the Himalayan Mountains to be folded.

North America moved somewhat northward, largely by rotation. This may have
caused a cooling of the climate reflected in some of the late Cenozoic life. Alaska and
Siberia may have collided, causing some deformation.

The Gulf of California and the Red Sea probably formed late in the Cenozoic. The
Bay of Biscay opened by the rotation of the Iberian Peninsula. Iceland formed by
volcanic activity on the Mid-Atlantic Ridge in late Cenozoic time.

Life of the Cenozoic

In the Cenozoic all types of life develop into the modern forms, making their study
especially interesting. It was an active time, with much mountain making and climatic
change. As a result, the changes in life, especially on the land, are great.

In the ocean, the biggest change is the absence of some groups of cephalopods that
were so abundant in the Mesozoic. Clams, oysters, and snails were the main elements
of Cenozoic seas, along with echinoids (see Fig. 24-14), bryozoans, and foraminifera.
The foraminifera underwent a new development and were very widespread. Because
they are small enough to be recovered intact in drill cuttings, they have been used
extensively to date the rocks encountered when drilling for oil. Limestones composed
of large foraminifera were deposited in Eurasia.

Figure 24-14. Echinoid *Eupatagus floridanus,* Eocene, Ocala limestone, Florida. Photo from Ward's Natural Science Establishment, Inc., Rochester, N. Y.

The plant life of the Cenozoic was essentially modern. (See Fig. 24-15.) The first fossil angiosperms are found in Jurassic rocks, but such fossils are relatively rare. In Early Cretaceous time, well-differentiated angiosperms became abundant, showing that they had developed somewhere. By the end of the Cretaceous, they replaced the Mesozoic plants. Their radiation continued throughout the Cenozoic.

The Cenozoic is the age of mammals. Mammals have many advantages over reptiles. They have a more efficient heart and hair, making them warm blooded. Their brain is larger, and the senses, such as smell and especially hearing, are better. Specialized teeth and jaws improve feeding, and the digestive system does not require inactive periods after eating as reptiles require. Teeth are easily preserved and are excellent mammal fossils. Growth of the young within the mother and a long period of nursing make the young more likely to survive. Just as the reptiles expanded to all habitats in the Mesozoic, so did the mammals in the Cenozoic. Bats are flying mammals; and seals, whales, and porpoises are marine mammals.

There are two main types of mammals: the marsupials, such as the kangaroo and opossum, that carry their young in an external pouch, and the placental mammals that carry their young internally. There is one other type of mammal, monotremes, living in Australia. Only two of these are known, the duck-billed platypus and the spiny anteater. They have hair and nurse their young but they lay eggs. There is almost no fossil record of these animals so they may be "living fossils," almost unchanged from the Cretaceous.

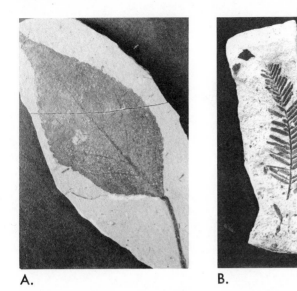

A.   B.   C.

Figure 24-15. Cretaceous and later leaf fossils have a modern aspect. A. *Alnus larseni,* Creede, Colorado. B. *Metasequoia,* Collawash River, Oregon. C. *Mahonia marginata,* Creede, Colorado. All are Miocene. Photos from Smithsonian Institution.

The oldest mammal fossils are small shrew-like skulls found in Upper Triassic rocks. Very few fossils are found until the Late Cretaceous. The largest Mesozoic mammals were the size of house cats. The Cenozoic placental mammals evolved from small, Late Cretaceous insectivores.

The marsupials were much less successful than the placentals. It is estimated that only about five per cent of all Cenozoic mammals were marsupials. The oldest marsupial fossils are Middle Cretaceous in age. The marsupials appear to have migrated to Australia and South America in the Late Cretaceous and then became isolated. As a result, the marsupials there did not have to compete with the placentals. They evolved into many types and occupied many habitats. Australia is noted for its marsupial fauna. In South America, the same thing happened, but here many of the marsupials became extinct at the end of the Pliocene. This occurred because about this time the Isthmus of Panama formed, and the placentals from North America migrated to South America. The marsupials could not compete with the placentals. The Cenozoic barriers and migrations are shown in Fig. 24-16. During most of the Cenozoic, there was some connection between Eurasia and North America.

In North America, the Paleocene was warm and temperate. The Paleocene mammals were archaic, and most were small. They lived in forests and near streams. Primates, rodents, insectivores, carnivores, and browsers were present. (See Fig. 24-17.) The browsers had hoofs and five toes. European and North American forms were similar, but the South American types were different. The first horses appeared in the Paleocene.

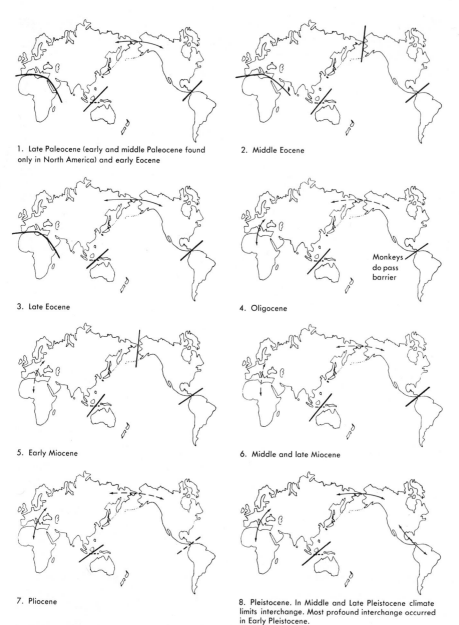

1. Late Paleocene (early and middle Paleocene found only in North America) and early Eocene

2. Middle Eocene

3. Late Eocene

4. Oligocene

Monkeys do pass barrier

5. Early Miocene

6. Middle and late Miocene

7. Pliocene

8. Pleistocene. In Middle and Late Pleistocene climate limits interchange. Most profound interchange occurred in Early Pleistocene.

Figure 24-16. Cenozoic vertebrate migrations. Barriers and migration routes are indicated diagrammatically. Continental drift probably caused many of the barriers and connections.

As the Eocene opened, the climate was subtropical and the animals were largely forest dwellers. All of the modern orders of mammals have been found although the species are different from those of the present, Rhinoceroses were present, and in the

Figure 24-17. Paleocene scene. *Barylambda,* a primitive hoofed animal with five toes. Western Colorado. About eight feet long. Painting by J. C. Hansen, Field Museum of Natural History.

late Eocene, the deer, pig and camel appeared. (See Fig. 24-18.) The largest land animal was the rhinoceros-like *Uintatherium.* Toothed whales appeared in the sea, indicating that this habitat was soon occupied after the extinction of the marine reptiles. (See Fig. 24-19.) In late Eocene time, more grasslands appeared as the climate became somewhat cooler and drier.

The Oligocene had a mild, temperate climate in North America. The Great Plains was a large flood plain with many rivers and some forests. Fig. 24-20 shows the differences in the climate and vegetation. The archaic forms disappeared, and the fauna took on a more modern aspect. The drier climate favored the faster, long-legged, hoofed grazers and browsers. Cats and dogs appeared. The titanotheres, which had begun in early Eocene as small rhinoceros-like animals, reached the size of elephants in the Oligocene and then died out. *Brontotherium* was the largest mammal ever to live in North America.

The drying and cooling trend continued into the Miocene in North America, and the grasslands expanded. This favored the rapid evolution of grazers such as horses, pigs, rhinoceroses, camels, antelopes, deer and mastodons. (See Fig. 24-21.) The large cats, bears, and weasels also appeared.

In the Pliocene, the climate cooled, remaining relatively dry. As a result of these changes, evolution continued rapidly. Many forms were weeded out, and those remaining became more specialized. The trend continued into the Pleistocene and produced our present life. Fig. 24-22 shows a scene on the high plains. Many of the animals resemble those in the area today. The differences are conspicuous, such as the shovel-tusked mastodon and the giant camel.

Figure 24-18. (Part 1)

620

Figure 24-18. (Part 2) Middle Eocene Bridger Formation of Wyoming. Although the plants and animals are reconstructed as accurately as possible, all of these creatures were probably never so close together. Painting by J. H. Matternes, Smithsonian Institution.

Smilodectes lemurlike monkey

Sciuravus squirrel-like rodent

Patriofelis large flesh eater

Hyopsodus clawed, plant-eating mammal

Crocodilus crocodile

Echmatemys turtle

Palaeosyops early titanothere

Stylinodon gnawing-toothed mammal

Helaletes primitive tapir

Orohippus ancestral horse

Homacodon even-toed hoofed mammal

Machaeroides saber-toothed mammal

Saniwa monitorlike lizard

Uintatherium 6-horned, saber-toothed plant eater

Trogosus gnawing-toothed mammal

Ischyrotomus marmotlike rodent

Hyrachyus fleet-footed rhinoceros

Helohyus even-toed hoofed mammal

Metacheiromys armadillolike edentate

Mesonyx hyenalike mammal

Sinopa small archaic flesh eater

621

Figure 24-19. *Basilosaurus cetoides*
Eocene whale from Alabama. These
toothed mammals quickly took the
place of the Mesozoic marine reptiles.
The jaws are three feet long. Painting
by C. R. Knight, Field Museum of Natural History.

The Pleistocene life of North America was dominated by large animals such as mastodons, mammoths, ground sloths, saber-toothed cats, bears, and giant beavers, as well as horses and camels. (See Figs. 24-23 and 24-24.) This was a time of great climatic change everywhere. Most of the animals just listed died out in North America about 8,000 years ago, when the last glacial stage was retreating. No one knows why this great extinction occurred, but hunting by early man has been suggested. It was not the climatic change because they had lived through four glacial advances. The present-day fauna of the African plains is somewhat similar to the late Cenozoic of North America.

Man

The development of man, the most successful of the mammals, is relatively poorly known. The man-like primates apparently evolved from the apes, but lack of fossils obscures the details. This lack of fossils in part reflects the cleverness of these animals in avoiding accidental deaths, and in part results from the fact that the apes evolved in the narrow transitional environment between the grasslands and the jungle or forest inhabited by other primates. This environment probably accounts for the development of two-legged locomotion, which freed the hands for other activities.

Fossil men are about the rarest fossils. In the past, each new discovery of a bone (complete skeletons are extremely rare) has been given a separate name because of disagreement as to where each fossil fits into the evolution of modern man. This disagreement, together with the scarcity of fossils and the problems of accurately dating the enclosing beds, has caused much confusion. Some of the more important discoveries and their more recent classification are shown in Table 24-1. In recent years much progress has been made in the understanding of fossil man. New fossil finds and reclassification of the many-named fossils into a few species have contributed to this advance.

Table 24-1. Some of the important early discoveries of fossil men and
their classification.

Date	Place	Fossils	Name	Recent classification
1848	Gibraltar	Skull	Neanderthal	*Homo sapiens*
1856	Germany	Skull and other bones	Neanderthal	*Homo sapiens*
1868	France	Five partial skeletons	Crô-Magnon	*Homo sapiens*
1894	Java	Skull cap, tooth, femur	Pithecanthropus	*Homo erectus*
1907	Heidelberg	Two lower jaws with teeth	*Homo heidelbergensis*	*Homo erectus*
1924	South Africa	Skull of child	*Australopithecus*	(subhuman)
1921	Rhodesia	Skull	*Homo rhodesiensis*	*Homo erectus?*
1929	Peking	Skull and molars	Pithecanthropus	*Homo erectus*
1931-1932	Solo River, Java	Several skulls	*Homo soloensis*	*Homo erectus?*

The primates, like other mammals, evolved from small insectivores in the Late Cretaceous. They adapted to life in the trees. The most important adaptations were the development of an opposable thumb, enabling grasping and swinging from tree limbs, and the shortening of the face, moving the eyes forward for stereoscopic vision. The latter enabled better judgment of distance—a necessity for a climbing animal. The size of the brain also increased. The higher apes came from the trees to the ground, perhaps because the cooling trend caused the forests to be replaced by grasslands. The gorilla and the chimpanzee live largely on the ground.

The oldest apes are early Oligocene, although other primates are found throughout the Cenozoic. Apes appear to have evolved rather rapidly in the Miocene, with scattered fossils found in Europe, Africa, and Asia, indicating differentiation into various types of apes. One, *Ramapithecus,* represented by jaws with teeth from India and Africa, appears to show the jaw distinctions leading to man rather than to the other apes; it has been dated about 14 million years ago, in late Miocene. Recently, earlier-discovered fossils were re-evaluated in the search for *Ramapithecus'* predecessors. These studies suggest that manlike jaw and tooth distinctions may have developed in Kenya, Africa, in the lower Miocene, 20 million years ago. (See Fig. 24-25.)

Australopithecus, the next known step, first appears in Kenya, Africa, in rocks that have been dated as approximately 3 million years old. This fossil is a recently discovered fragment of an arm bone. Over the next million years, he seems to have evolved rapidly. *Australopithecus* is found with crude stone and bone tools and the remains of increasingly larger animals that he apparently hunted. Early *Australopithecus* was small, but evolved into a much larger, more manlike creature; even the earliest forms walked erect —the first known to do so. The early discoveries led to great confusion, for he existed together with *Paranthropus,* a larger, more primitive-appearing, erect-walking vegetarian, who apparently evolved little over the same span of time—nearly a million years—until he disappeared. *Australopithecus* has been found only in Africa, with the possible exception of a disputed find in Java.

Olduvai Gorge, Tanzania, is a unique site in which a stratigraphic record going back two million years is exposed; its beds have been dated by the potassium-argon method.

Figure 24-20. (Part 1)

624

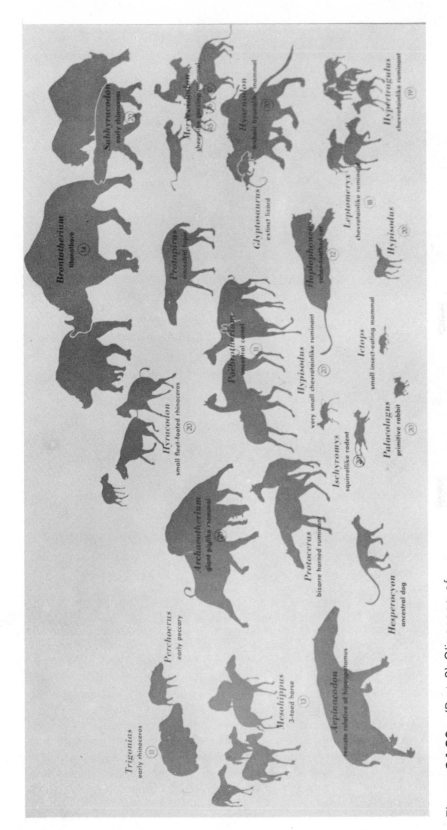

Figure 24-20. (Part 2) Oligocene of the White River Formation of South Dakota and Nebraska. Painting by J. H. Matternes. Smithsonian Institution.

625

Figure 24-21. (Part 1)

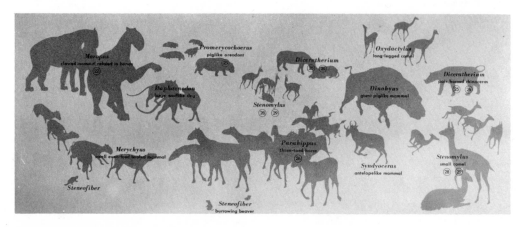

Figure 24-21. (Part 2) Lower Miocene of the Harrison Formation of western Nebraska. Painting by J. H. Matternes, Smithsonian Institution.

The lower and middle beds have revealed evolving *Australopithecus,* and the middle and upper beds his contemporary, the primitive *Paranthropus;* in a bed dated approximately one-half million years ago, *Homo erectus* has been found. (See Fig. 24-26.) In beds dated about a million years ago, a form appearing to be intermediate between *Australopithecus* and *Homo erectus* has recently been discovered, and may indicate a transition between the two. This latter form, temporarily called "Homo habilis," is the oldest known member of the genus *Homo.*

Homo erectus existed in the early middle Pleistocene from about one million to 500,000 years ago. Most of the fossil men that are not modern men are now classified in this species. The fossils are generally associated with stone tools, and some finds show evidence of the use of fire. The general characteristics used to distinguish *Homo erectus* are a thick, flat, narrow skull, with an average brain size of about 1,000 cubic centimeters (compared with averages of 1,400 cubic centimeters in modern man and 500 cubic centimeters in *Australopithecus* and a modern gorilla); protruding brow ridges; a marked angle for muscle attachment at the rear of the skull; and larger and more primitive teeth and jaw than modern man's. His legs are close to those of modern man, unlike the more primitive upright, *Australopithecus*. (See Fig. 24-27.)

The earliest and most primitive subspecies of *Homo erectus* have been found in Java, China, and South and East Africa. Later definitive fossils of *Homo erectus* come from these regions and Algeria. Tools similar to those associated with him have been found in Europe, but so far no fossils, with the exception of a jaw found near Heidelberg. This jaw is dated in the generally accepted mid-range of *Homo erectus,* but it is less primitive than the usual *Homo erectus* fossils, and skull portions which might give more positive assignment as to species are absent; its true position is greatly in doubt. (See Fig. 24-28.)

Although we are now approaching modern man and the record should begin to clear, in actuality it does not; there are too many anomalies in the idealized picture. *Homo sapiens* is the species to which modern man belongs. When and where he first appeared is now the question, and the answer is ambiguous. Some fossils seem to indicate that *Homo erectus* and *Homo sapiens* may have overlapped considerably in time, or possibly that the progress of evolution of *Homo erectus* varied greatly over its tremendous geographical range, affected locally by climatic conditions and migratory limitations.

The skull fragments found on the Solo River in Java (so-called Solo man) are extremely primitive in appearance, closely resembling *Homo erectus* and not *Homo sapiens;* however, the Solo fossils have been dated roughly 100,000 to 60,000 years ago, long after the generally accepted upper boundary date of *Homo erectus,* and contemporaneous with the *Homo sapien* Neanderthal man in Europe. Some authorities assign Solo man to *Homo erectus* on the basis of appearance, others to an extinct race of *Homo sapiens* on the basis of date. No lines of evidence leading to modern races of *Homo sapiens* have been found in the Far East to shed light on the Solo man anomaly.

Another find, a skull, in Rhodesia (Rhodesian man) is also in dispute. He is far less primitive than Solo man, but he is variously assigned to late *Homo erectus* or an extinct race of *Homo sapiens.* He existed very recently, 50,000 to 30,000 years ago, contemporaneous with late Neanderthal, and possibly early Crô-Magnon.

On the other end of the scale, in 1965, the rear portion of a skull was discovered at Vértesszöllös, Hungary. Its shape, curvature, and estimated brain size of 1,400 cubic centimeters have caused it to be classified as *Homo sapiens,* but its date of approximately 500,000 years ago, in the early middle Pleistocene, is well within the accepted range of *Homo erectus.* More complete skulls found in the 1930's in Swanscombe, England, and Steinheim, Germany, appear to substantiate this latest find. They are later in date (between 200,000 and 300,000 years ago), and they have the curvature and size of *Homo sapiens* and have been so classified. The Steinheim fossil includes a face which seems to indicate transitional brow ridges between *Homo erectus* and the Neanderthal man of later times. The Vértesszölös find may be a descendent of the earlier problematic Heidelberg jaw and an ancestor of the Swanscombe and Steinheim fossils.

Neanderthal man himself occurs near the boundary between middle and upper Pleistocene, and he is now classified as *Homo sapiens.* Extensive remains have been found in Europe and around the Mediterranean, dated from nearly 100,000 years ago until about 35,000 years ago. Neanderthal appears to have developed considerable

Figure 24-22. (Part 1)

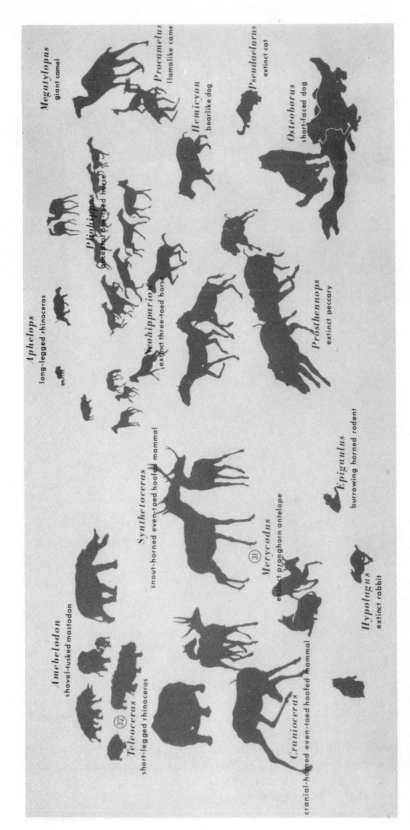

Figure 24-22. (Part 2) Early Pliocene of the high plains region. Painting by J. H. Matternes. Smithsonian Institution.

629

Figure 24-23. Pleistocene scene at La Brea tar seeps near Los Angeles, California. Saber-toothed cat, wolf, and vultures have gathered to feed on animals caught in the tar. Painting by C. R. Knight, Field Museum of Natural History.

Figure 24-24. Pleistocene scene in cool climate. Wooly mammoth and rhinoceros. Painting by C. R. Knight, Field Museum of Natural History.

variation. Those inhabiting western Europe, the best and earliest known, are the most different from modern man in limb and skull shape; and as time progressed they seem to have developed further from, instead of closer to, modern man until their disappearance. This seeming "regression" may simply represent an isolated race's adaptation to difficult climatic conditions. Finds in the Middle East indicate great variety and possible transitional forms between those with the extreme western-European characteristics and Crô-Magnon, the oldest modern man in the strict sense. This variety exists even in contemporaneous fossils in single locations.

Modern man himself emerged approximately 35,000 years ago, and the older forms disappeared. The most studied and best known is Crô-Magnon, whose remains are nearly indistinguishable from modern man's. He appears suddenly in Europe in beds of about the same age as late Neanderthal, and then Neanderthal disappears. Apparently he migrated into Europe, probably from the Middle East considering the possibly transitional forms recently found there. An important problem is whether Crô-Magnon evolved from Neanderthal or simply first intermingled with him in the Middle East and evolved from some other stock. He is presumed to be the ancestor of the modern European races. Nothing is known of the recent ancestors of the modern races of man outside Europe.

From this incomplete story it is possible to draw two versions of man's family tree. (See Fig. 24-29.)

The oldest fossils of man found in the New World are 11–13,000 years old on the Palouse River in Washington, and 11–12,000 years old at Tepexpan in Mexico. Sites

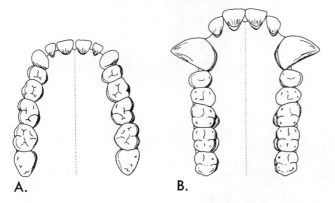

A.   B.

Figure 24-25. Upper jaw differences between *Australopithecus* A. and gorilla B.

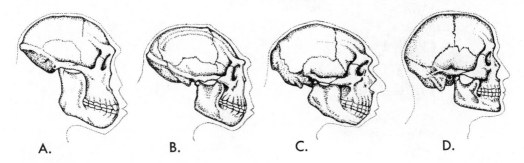

A.   B.   C.   D.

Figure 24-26. Comparison of skulls of A. *Australopithecus,* B. *Homo erectus,* C. *Homo sapiens neanderthalensis,* and D. *Homo sapiens.*

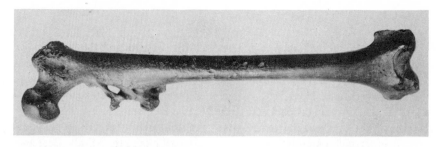

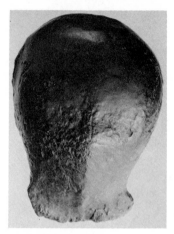

A.

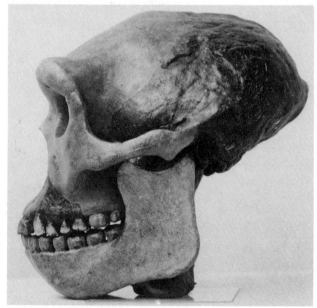

B.

Figure 24-27. *Homo erectus.* A. Typical fossils, teeth, femur, and skull cap. B. Restoration of skull by Weidenreich. Note the thick, flat skull with protruding brow ridges and a ridge at the back. The jaw has a receding chin, and the teeth are somewhat primitive. Compare with your skull by running your fingers above and beside your eyes and over the back of your skull and over your chin. Photos from Smithsonian Institution.

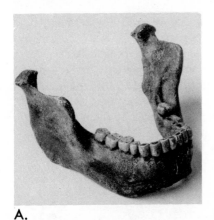

A.

Figure 24-28. Heidelberg Man (*Homo erectus*) is known only from the jaw, a cast of which is shown in A. The restoration by McGregor, shown in B., is based on that jaw. Photos from Smithsonian Institution.

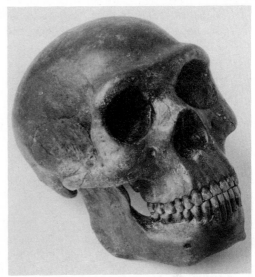

B.

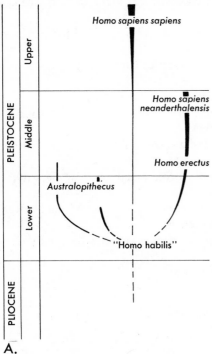

A.

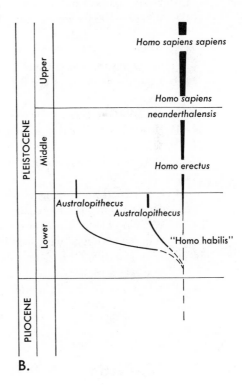

B.

Figure 24-29. Two possible versions of the evolution of man. The evidence is not complete enough to construct an accurate family tree for man. Modified from P. V. Tobias, *Science* (July 2, 1965), Vol. 149, pp. 31-32.

as old as 15,000 years in Alaska and South America, and up to 20,000 years in Mexico are known from artifacts; but these sites contain no human bones. Many anthropologists believe that human evolution can be described better from artifacts than from bones. Man must have reached the New World over the Bering land bridge between Siberia and Alaska. Men must have crossed the bridge about 35,000 years ago because the bridge was submerged between 35,000 years and 25,000 years ago, and at 25,000 years ago the corridor between the glacial ice sheets was closed by the meeting of the glaciers. Thus the search will continue for early man in the New World.

QUESTIONS

24-1. When did the four major advances of the sea occur on the Gulf coast?

24-2. Describe the erosional history of the Appalachian Mountains.

24-3. Describe the topography of the Pacific coast from Washington to California.

24-4. Outline the Cenozoic history of Arctic Canada.

24-5. Outline the development of the Rocky Mountains.

24-6. When and how did the Basin and Range structure of the Great Basin develop?

24-7. Where is the thickest section of marine Cenozoic rocks found in North America?

24-8. How much of North America was covered by ice during the Pleistocene? How many advances occurred?

24-9. Describe as many of the features of North America caused by glaciers as you can.

24-10. Describe the Pleistocene events in northwestern United States.

24-11. How does the marine life of the Cenozoic differ from that of the Mesozoic?

24-12. What are the two main types of mammals?

24-13. Outline the development of marsupials and account for their present distribution.

24-14. Describe the changing climates of the Cenozoic and how this affected the development of mammals.

24-15. What problems will future geologists have in correlating the present African "big game" fauna with North American fossils?

24-16. Why is it difficult to determine the development (i.e., describe the family tree) of modern man?

24-17. When and where did modern man first appear?

SUPPLEMENTARY READING

Broom, Robert, "The Ape-Men," *Scientific American* (November, 1949), Vol. 181, No. 5, pp. 20-24. Reprint 832, W. H. Freeman and Co., San Francisco.

Deevey, E. S., Jr., "Living Records of the Ice Age," *Scientific American* (May, 1949), Vol. 180, No. 5, pp. 48-51. Reprint 834, W. H. Freeman and Co., San Francisco.

Dorf, Earling, "The Petrified Forests of Yellowstone Park," *Scientific American* (April, 1964), Vol. 210, No. 4, pp. 106-114.

Eckhardt, R. B., "Population Genetics and Human Origins," *Scientific American* (January, 1972), Vol. 226, No. 1, pp. 94-103.

Howell, F. C., *Early Man*. New York: Life Nature Library, Time, Inc., 1965, 200 pp.

Howells, W. F., "Homo Erectus," *Scientific American* (November, 1966), Vol. 215, No. 5, pp. 46-53.

Krantz, G. S., "Human Activities and Megafaunal Extinctions," *American Scientist* (March-April, 1970), Vol. 58, No. 2, pp. 164-170.

See also references at the end of Chapter 21.

Chapter 25

The Future — Geology, Environment, and Man

In the West, our desire to conquer nature often means simply that we diminish the probability of small inconveniences at the cost of increasing the probability of very large disasters.

Kenneth E. Boulding
Human Values on the Spaceship Earth, 1966

The Geologic Future

One attribute of science is its ability to predict. A physicist uses his laws to predict the outcome when two masses collide, and the astronomer predicts eclipses many years in advance. Geologists can predict future geological events only in a general way. In this chapter, geologic aspects of the future will be considered.

To find the ultimate fate of the earth, we must look to astronomy, for it is the sun that gives life to the earth. The sun is believed to be about 5,500 million years old. It is slowly expanding and is now about 20 per cent larger than when it first formed. Ultimately it will become a giant star and will expand to Mercury's orbit. This will occur in about 3–4,000 million years. However, in about 2,000 million years, the sun will have expanded so much that its radiation will increase temperatures on the earth so that the oceans will boil. It seems unlikely that life, at least as we know it, can survive. After the giant phase, the sun will shrink and cool and become a white dwarf. At this stage, the earth will become very cold, again seemingly precluding any life. These predictions, of course, assume that the sun will not collide with another star, an event with a very small probability. Another possible interruption would be the collision of an asteroid with the earth, but, again, this is an unlikely event.

The moon will continue to recede, making both the day and the month longer. Much more must be known about the earth-moon system to make any more specific predictions. At present astronomers differ greatly on this question.

On a shorter time span, but still very long by human life spans, other predictions can be made, based on principles described earlier. The continents will probably be larger because they have apparently grown during geologic time. The configuration of the continents and their positions on the earth will change. As more is learned about sea-floor spreading, these changes will be predicted by future geologists. If radioactivity is an important source of energy for geologic processes, these processes will slow as more and more of our radioactive elements decay. In a much shorter time, erosion and sedimentation will modify the earth's surface in ways fairly easy to predict. This is about as far as the geologist can predict the future. In the following section, the environment will be considered, starting with geologic hazards.

Environment

Geologic Hazards

Where to Live? Most people answer this question pragmatically with near their work for the general location and, in detail, with a home that is to them subjectively pleasing and that they can afford. In the following pages a number of environmental factors will also be considered. Failure to take some geologic factors into account has cost some people much money and in some cases their lives.

Most people live in or near cities. The locations of cities are generally determined by natural factors such as good harbors, confluence of rivers, natural resources, and the like. For reasons such as these, California has become the most populous state in the union. However, many parts of California are geologically hazardous because of earthquake danger, but most Californians pay no heed because the odds are in their favor.

In this section some aspects of how geologic processes affect man will be considered. The principles involved in these processes were covered earlier. Also, earthquakes, a major geologic hazard, were covered in Chapter 13.

Trace Elements and Health. Some elements present in the environment in only trace amounts have a large influence on health. A few obvious examples of good and bad effects of trace elements are widely known. The relationship between dental caries and fluorine in the water supply, and selenium poisoning of cattle by "loco weed" (*Astragalus racemosus*) are known to almost everyone.

Although some trace elements, such as iodine, apparently move through the atmosphere, most come from the soil and enter the life cycle either through water supply or plants. Although climate is more important in soil formation than the type of bedrock, the bedrock does influence the trace elements in the soil. Because most of these elements are metals, mining areas are commonly enriched in one or more of them. Many other areas are also either enriched or deficient in some or many of the trace elements. The amount of chemical weathering seems important, and both arid and tropical areas are generally deficient. Not all trace elements are beneficial, as will be shown. Insecticides, sprays, and fertilizers also add elements to plants and soils.

Deciding which elements should be considered trace elements is not a simple decision. Most living tissue is composed of hydrogen, carbon, nitrogen, oxygen, sodium, magnesium, phosphorus, sulfur, chlorine, potassium, calcium, and in those animals with hemoglobin, iron. In general the trace elements regulate enzyme actions. The elements that do this are calcium, magnesium, iron, manganese, cobalt, copper, zinc, and molybdenum. Zinc seems to be necessary for growth and to reach maturity. Obviously there is overlap between the two lists. In addition, there is another group of elements called age elements because they accumulate during life. This group includes beryllium, aluminum, silicon, titanium, vanadium, chromium, nickel, arsenic, selenium, tin, lead, silver, cadmium, barium, gold, and mercury. The role of this last list is not clearly understood, and future studies may show that some of them belong in the other lists. Excess or deficiency of some of these can cause great changes or even death. Some can cause tumors and so may cause some cancers. The role of metals in both the cause and treatment of cancer is an area of current research. Some of the evidence showing the effect of trace elements on health is comparison of maps showing trace elements in soils with maps showing either mortality or morbidity of disease. Such evidence is subject to various interpretations and so is somewhat controversial. In the following paragraphs, a number of examples of the effects of trace elements will be described.

Fluorine aids in reducing dental caries, and many municipalities add it to drinking water for this reason. Areas of good teeth also differ from areas of poor teeth in having higher calcium, molybdenum, aluminum, titanium, and calcium-magnesium ratio, and lower copper, manganese, barium, and strontium. The evidence that fluorine reduces caries is excellent, but in spite of this, many people object to adding it to their water. Low-fluorine areas have higher rates of both osteoporosis and aortic calcification, so the benefits of fluorine may go far beyond sound teeth.

In Georgia, highland areas with soils containing many trace elements have lower cardiovascular death rates for middle-aged white males than coastal-plain areas where soils are deficient in trace elements. (See Fig. 25-1.) However, the amounts of trace elements in plants from these two areas are not too different, suggesting that if trace elements are the cause, they may enter the life cycle through the water supply or other means.

Goiter is well known to be caused by a deficiency in iodine. We avoid the enlarged neck of goiter by adding iodine to salt in places where iodine is low. Iodine is in sea water, so those living near the sea or eating much sea food generally get enough. Mountain areas and recently glaciated areas tend to be low in iodine. The distribution of iodine is erratic, so many exceptions can be found.

Cancer may also be related to trace elements. Low-iodine areas have a suspicious relationship to breast cancer. Cancer is less common near the Gulf of Mexico than in northern states, but this might be related more to climate than trace elements. Diabetes is less common in the tropics, and insulin doses are smaller there; again this may be climatic. In northeast Pennsylvania, coal contains chrome, nickel, and cobalt, which are known to cause lung cancer. The incidence of lung cancer and the amounts of these elements coincide. In another case, in southwest England near an old mining district, cancer mortality is high in an area of high trace element content.

Cattle fed forage low in selenium have a very high incidence of muscular dystrophy, which can be cured by selenium injection. The soils of part of the northern great plains are rich in selenium; some plants, such as loco weed, concentrate the selenium, and cattle eating it are poisoned. Apparently the selenium enrichment of this plant is seasonal or otherwise variable, because ranchers report conflicting observations. Thus too little and too much selenium are both detrimental to animal life.

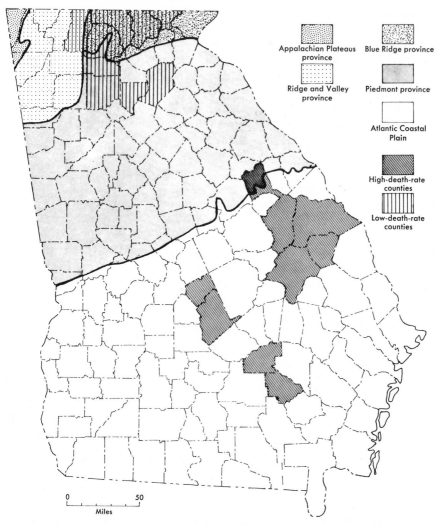

Figure 25-1. Relationship between cardiovascular disease and geologic provinces in Georgia. The areas with both high and low death rates for car-diovascular disease among white males of ages 35–74 are shown. From U. S. Geological Survey, *Professional Paper 574 C,* 1970.

In a Florida study, salt-sick disease in cattle and anemia in children were related to deficiencies in iron and copper in the soil. Addition of these two elements to diet promptly caused recovery. In addition, cobalt aided in copper retention, and molybdenum accentuated copper deficiency and also interfered with phosphorus metabolism.

Anthrax, another cattle disease, is controlled by soil type. Apparently the spores live in organic-rich calcium soils, and the disease reaches cattle after such infected soils dry out.

Some other studies suggest that some soils in western United States contain toxic amounts of vanadium, molybdenum, and selenium. Some of the peat used in gardens contains high amounts of uranium, and some natural fertilizers add cadmium and zinc to soils. Mercury is high in many soils. All of these points have a bearing on claims of the value of foods raised on organic versus chemical fertilizer, and apparently every case must be considered individually. Thus some food fads may be based unknowingly on trace elements. Medicinal plants in many cases also depend on trace elements for their value.

Lithium is sometimes used in the treatment of certain types of mental illness. Cities with trace amounts of lithium in their water supply have fewer admissions to mental hospitals than those that do not, according to one preliminary study.

Volcanic Hazards. The destruction caused by volcanoes is of two kinds—the direct destruction by material issuing from the volcano and the indirect damage caused by mudflows and floods. The former has already been described in the discussion of explosive volcanoes, such as Mt. Pelée and Krakatoa. There is little defense against such rapid destruction except evacuation at the first indication of activity by the volcanoes. The first eruption may be of the glowing cloud (nuée ardent) type and may occur anywhere on the volcano. In such cases there may be no escape. In the quieter eruption of basaltic volcanoes, such as those in the Hawaiian Islands, the fluid basalt moves at a few miles an hour and generally there is enough time to evacuate. This may seem of little value if your home and fields are inundated by either lava or air-borne ash, but at least you are alive and able to start over.

Indirect damage from volcanoes can also be swift and without warning. Many volcanoes are high mountains and even in the tropics are glacier covered. An eruption under a glacier can cause extensive melting that results in floods. Also many volcanoes, especially those that have been inactive for long periods, have large lakes in their depressed summit areas. Crater Lake in Oregon is an example. An eruption that breeches the lake, or an eruption into the lake, can cause the draining of the lake and so cause floods. The eruption of a submarine or shallow-island volcano such as Krakatoa can cause flooding by seismic sea waves.

Probably the largest areas inundated by volcanoes are those swept by mudflows. Although some mudflows are concurrent with the eruption, most occur at the next rainy season. Although the latter mudflows can be predicted, many lives are lost because many people refuse to leave their homes and fields. Mudflows are probably the most important way that pyroclastic volcanic rocks are distributed from volcanoes and so are a normal geologic process. A pyroclastic eruption can deposit ash and coarser material on the slopes of the volcano and on other slopes downwind. This loose material is not stable and at the next rain becomes mud that can flow even on very gentle slopes. Rain commonly accompanies volcanic eruption, and, although we don't know why, it may be related to the large amounts of steam that are part of most eruptions. Thus mudflows can accompany eruptions. Pyroclastic eruptions commonly kill vegetation by covering low plants, and the falling debris strips the leaves from trees. Without vegetation, the runoff is increased, and the possibility of mudflows is greater.

Mount Rainier was reported to have erupted 14 times between 1820 and 1894, and debris from some of these eruptions have been identified and dated. The recent history of Mt. Rainier shows both the potential hazards and gives some estimate of the frequency of such events. The main hazards at Mt. Rainier are eruptions of lava, eruptions of pyroclastics, and mud or debris flows. The youngest lava flows are probably those that formed the summit cone. Renewed activity there would melt snow, perhaps forming lakes in the two overlapping craters or causing floods or debris flows if the water moves down the mountain. Climbers have descended as deep as 450 feet in ice caves near the summit, and stones thrown from there caused splashes from deeper water. During the descent, many fumaroles and areas of hot rock were seen. Thus the main dangers of renewed lava eruptions are floods; and although the lavas themselves might cover some areas, they would move slowly enough to allow evacuation and probably would not flow far from the mountain.

Mudflows, or debris flows, which are similar to mudflows but contain much material coarser than mud, are common and are the most devastating type of activity at Mt. Rainier. Water and available rock material are needed to form mudflows. Therefore, they can result from either lava or pyroclastic eruptions, or heavy rainfall, or any other event that causes melting of glacial ice; because much rock material mantles the mountain, only water must be added. The largest postglacial mudflow, the Osceola Mudflow, occurred about 5000 years ago and had a volume of almost one-half cubic mile. It is believed to have originated by steam explosions near the summit. It traveled 40 miles down valley and spread up to 70 feet thick over 65 square miles in the Puget Lowlands. (See Fig. 25-2.) Remnants of the flow show that it was at least 500 feet thick

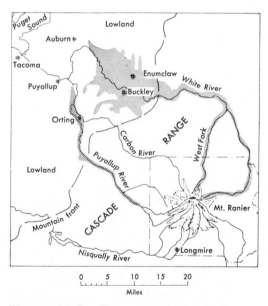

Figure 25-2. The extent of the Osceola and Electron mudflows from Mt. Rainier. The gray area is the Osceola mudflow, and the lined area the Electron mudflow. From U. S. Geological Survey, *Bulletin 1238,* 1967.

at places as it moved down the valley. The other large mudflow, the Electron Mudflow, occurred about 500 years ago. In 1947 heavy rainfall caused a debris flow at Kautz Creek that moved about 50 million cubic yards of debris. The banks of that creek expose six previous debris flows. In 1963 a rock avalanche from Little Tahoma Peak moved 4.3 miles and stopped only 2000 feet from White River Campground. It may have been caused by a small steam explosion.

The types of activity that have occurred at Mt. Rainier are listed in Table 25-1. Their effects, possible warning signs, and possible frequency, are also shown. Wise planning of land use and observation of the Cascade volcanoes can probably prevent a major disaster. The main problem is probably complacency in view of the infrequency of major eruptions.

Landslides. Landslides range from rapid, dramatic events that kill hundreds and cause millions of dollars damage all the way to slow, almost imperceptible creep. Landslides cost hundreds of millions of dollars each year in direct damage, and by delays because of blocked roads and railroads; but, perhaps more important to the average man, they commonly cause personal financial disaster if a new home is destroyed.

It is difficult to protect oneself from financial loss due to landslides. Landslides are typically excluded from homeowners' insurance, as is earthquake damage. Many policies also exclude damage due to subsidence. These coverages are available, but generally at high cost and should be investigated by the homeowner.

In many cases the sale of a home is voided if knowledge of hazardous conditions is concealed, but, in general, real estate sales are *caveat emptor*—let the buyer beware. Getting the opinion of a consulting geological engineer is an approach in some areas. The problem with this approach used to be finding a competent person or firm; but licensing laws in several states have made this easier, although many licensed engineers that can legally give such advice know very little about geology. The real problem is that consultants are vulnerable to law suits if their advice is in error; and with the recent rise in malpractice suits in all professional fields, consultants find it almost impossible to get insurance against such suits. Another problem the consultant faces is that he might also be sued by a landowner whose property loses value because of an inaccurate report that it has hazardous geologic conditions. It is difficult to give a geologic opinion on the stability of a single lot because much larger areas have to be studied to determine the geologic conditions. This means that consultants have to charge almost as much to examine a single lot as they do for a whole tract. Drilling and laboratory tests may be needed as well. Thus, the individual cannot do much to protect himself, but must look to government agencies to protect him.

A few cities have laws designed to protect the home buyer. The Los Angeles building codes are commonly cited as examples for the rest of the nation. Los Angeles is endowed with many natural attractions but also with many problems. Legislation dealing with earthquake hazards and air pollution also began in Los Angeles and also has served as a model for the rest of the country. The landslide problem became acute during the very heavy rains of 1952, in part because population pressures and the desire for property with a view led to building homes in the foothills surrounding the city. The January 1952 rains caused about $7,500,000 damage in Los Angeles. The building codes enacted before the following winter rains virtually eliminated high, steep cuts and fills, required permits for all hillside grading, and required inspection and geological reports. In addition, drainage has to be conducted to streets and away from cut and fill slopes, and erosion must be controlled by plantings or other devices.

Fills, if not properly compacted, may settle. Fig. 25-3 shows subsidence of an area in San Francisco that was built on filled ground over an old swamp. The city has raised the grade of the streets so that the streets are half way up what was the first floor of the houses.

Table 25-1. Types of eruptions that have occurred at Mount Rainier in postglacial time, anticipated effects and frequency of similar eruptions in the future, and possible warning signs of an impending eruption. From U.S. Geological Survey Bulletin 1238, 1967.

	Volcanic activity not necessarily associated with the rise of new magma	Types of volcanic activity associated with the rise of new magma
	Steam explosion (generally on small scale)	Steam explosion (may be on small to very large scale)
Direct effects	Formation of small-scale rockfalls and avalanches. Effects confined chiefly to flanks of volcano and valley floors immediately adjacent.	Formation of rockfalls and avalanches which grade down-valley into debris flows. Effects confined chiefly to flanks of volcano and valley floors. Formation of air-laid rubble deposits whose distribution would be limited to flanks of volcano and areas closely adjacent.
Indirect effects	Possible floods and (or) debris flows caused by damming of rivers by avalanche deposits. Principal effects probably would be limited to valley floors within and closely adjacent to park.	Very large avalanches of moist altered rock on flanks of volcano grading directly into debris flows on valley floors. Debris flows may be very long, extending tens of miles beyond park boundaries.
Possible warning signs of impending activity	Appearance of steam jets and clouds of water vapor, possibly accompanied by explosions and rockfalls. Abnormal melting of glaciers at hot spots; appearance of melt pits in glaciers.	
Indicated possible frequency	1 in 10 to 100 years - - - - - - -	1 in 2,000 years - - - - - - - - -

[The flanks of the volcano include an area within 4 or 5 miles from the summit]

The legal situation concerning landslides is not too clear because many cases are settled out of court. States and cities are involved, especially in slides caused by road construction, and in many cases they exercise their sovereign right of refusal to consent to be sued. Some states believe that warning signs on roads are sufficient protection for

Types of volcanic activity associated with the rise of new magma — cont.		
Pumice eruption	Eruption of bombs and (or) block-and-ash avalanches	Eruption of lava flows
Fall of pumice on flanks of volcano; widest distribution beyond flanks in a downwind direction, thus probably greatest to the northeast, east, or southeast. Thickness probably will be less than a foot beyond 5-mile radius of summit. Anticipate thin ash-fall over broad area without regard for topography.	Fall of hot to incandescent bombs, ash, and rock fragments. Distribution chiefly on flanks of volcano, but avalanches may extend into valleys.	Distribution probably limited to flanks of volcano. Major effects expectable only in immediate vicinity of flow.
Extensive melting of glaciers possible, caused by internal volcanic heat and by steam moving toward outside of volcano. Expected result: floods and debris flows on valley floors.		
Ejection of water from summit craters in early stage of eruption. Possible result: floods and debris flows on valley floors.		Ejection of water from craters if flows occur at summit of volcano.
Debris flows may result from extensive downslope movements of pumice onto valley floors.	Possible floods and debris flows caused by eruption of hot debris onto snow.	Extrusion of lava flow beneath or onto glacier might cause catastrophic floods.
Increase in frequency and magnitude of local earthquakes. Large increase in stream discharge unrelated to meteorological conditions. Appearance of clouds of water vapor associated with steam jets, possibly accompanied by small steam explosions and rockfalls. Increase in fumarolic activity and increased melting of snow at summit cone. Abnormal glacier melting at hot spots; appearance of melt pits in glaciers. Increase in temperature of fumaroles. Increase of sulfur and chlorine in fumarolic gases.		
1 in 2,500 years - - - -	1 in 5,000 years - - - -	1 in 10,000 years.

Figure 25-3. Subsidence. The street
has been regraded and raised so that
the garage of this house is unusable.

citizens. In some cases the question is whether the slide is an "act of God" or due to negligence.

Landslides can be stabilized in a number of ways. The most important point to make is that all of these methods are done much more economically during the initial construction rather than as a first-aid measure later on. In many cases it is possible, with very minor redesign of any engineering project, to avoid potential landslide areas. One very common misconception is that landslides can be stabilized with planting. Indeed, I have personally met a number of home owners in landslide areas in California who felt that growing either grass or ice plant would stabilize a landslide area. This is a complete misconception, and there are numerous examples of very large trees, even forests, moving as parts of landslide blocks. Many landslides move at depths greater than the normal root depth of most trees.

The main methods of stabilizing landslides are excavation, drainage, construction of retaining walls or similar structures, the use of rock bolts, and the injection of grout into the slide. Highways avoid landslides by bridging over the unstable areas. Most common of all these methods is to drain the landslide. Generally this is done by drilling holes into the slide and providing perforated pipes through which the water can escape. In some cases this is coupled with an impervious layer over part of the landslide to prevent the entrance of surface water into the slide. In a few cases the slides are actually triggered by blasting to stabilize the area. The quick clays (see Fig. 7-18) of Anchorage were subjected to both blasting and vibration in an effort to stabilize them after the 1964 earthquake.

Geologic Hazards of Running Water. "Floods are as much a part of the phenomena of the landscape as are hills and valleys; they are natural features to be lived with, features which require certain adjustments on our part." *Floods,* by W. G. Hoyt and W. B. Langbein, Princeton University Press, 1955.

Some primitive peoples adapted reasonably well to rivers and accepted seasonal or occasional floods as part of life. As man developed and built bigger and better structures for his comfort, he became less willing to accept the natural behavior of rivers.

A typical river valley consists of a channel, a flat area (called a flood plain) and sloping valley sides. (See Fig. 25-4.) Most of the time the flow is confined to the channel;

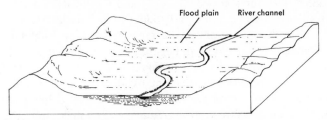

Figure 25-4. A typical river valley, showing the channel and the flood plain.

but at times of high flow, when the channel cannot carry the water, it spills onto the aptly named flood plain. The flood plain is the channel of the river during times of high water. This is the normal, expected behavior of a river. People who live or build structures on flood plains must expect these floods. Flood control is never one hundred per cent effective and can increase the actual damage as we will see later. Floods are not entirely bad events, however, because the silt deposited on the flood plains renews the soil of these generally very productive agricultural areas.

Rivers erode on the outside of curves and deposit on the inside of curves. (See Fig. 8-13.) This means that the location of a river is always changing. Rivers are commonly used as property boundaries as well as state and international boundaries. Such boundaries are not fixed and lead to many problems. In some cases the boundary is fixed at the position of the river on a certain date. Along the Mississippi, a farmer with land inside a meander loop may wake up some morning and find that his land is separated by the river from the rest of the state and county to which he pays taxes. He may have trouble getting police and fire protection. On the other hand, he now may have an ideal location for moonshining. Recently the United States and Mexico concluded a treaty fixing the international boundary near El Paso. (See Fig. 25-5.)

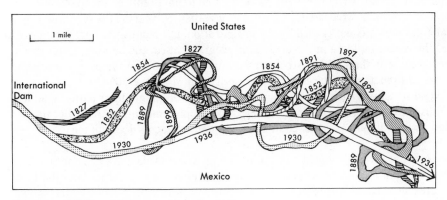

Figure 25-5. Changes in the Rio Grande River near El Paso from 1827 to 1936. After Boggs, 1940.

Storms and mudflows are geologically rapid but normal events. Intense storms that cause floods, landslides, or mudflows can occur almost anywhere. In the United States, such storms are common in the desert areas of the southwest where little note is taken of them because few people are affected. A few cities have expanded into hazardous parts of this region in Arizona and especially in southern California, and the resulting problems will be described later. The hurricanes of the Gulf and southeastern coasts of the United States cause similar problems, although flooding of the coastal areas is the usual result. Some of these storms cause problems as far north as New England. Hurricanes are generally accompanied by heavy rainfall that chokes the rivers, and strong onshore winds drive ocean waters onto the low coastal plain. The low barometric pressure associated with such storms increases the flooding by allowing much higher than normal tides. Hurricane Camille which hit the Gulf Coast in August, 1969, had winds up to 200 miles per hour, a low pressure of 905 millibars, and dumped up to 30 inches of rain in a day. The East Pakistan flood of December, 1970, killed between 250,000 and 300,000 people. This flood, erroneously described in the newspapers as a tidal wave, was caused by a hurricane (called a cyclone in that area) that had winds up to about 120 miles per hour and a low pressure of about 950 millibars. The wind and low pressure caused a storm surge about 18 feet high (normal tides are about six feet), which overran the low, flat area in the Bay of Bengal.

Mudflows caused by relatively rare but intense thunderstorms are not confined to desert areas. They can occur on any steep slope with enough overburden or loose material and not too much underbrush. Forested slopes, especially those with big trees whose shade prevents growth of underbrush, are susceptible to this type of erosion. The following example is from the Sierra Nevada Mountains near Lake Tahoe on the California-Nevada boundary. The Lake Tahoe area, because of its beauty, its nearness to San Francisco and Sacramento, and its climate that makes it an excellent skiing as well as summer resort area, is currently undergoing urbanization. In the late afternoon of August 25, 1967, more than an inch of rain fell in the drainage basin of Second Creek. This caused a mudflow of more than 50,000 cubic yards in the 1.5 square-mile basin. The mudflow caused extensive damage to homes, town houses, apartment houses, and roads. A seven-foot-diameter boulder moved in this flow, and at some places, the flow deposited up to 20 feet of debris. Similar events have been observed elsewhere in the Sierra Nevada Mountains, and the results of this type of erosion can be seen in many drainage basins. This particular event was estimated to have reduced the basin by about 0.02 feet and to have a recurrence interval of less than 50 years. This mudflow was a natural event and was not caused by the urbanization, although some conservationists blamed the recent building in the area. This event did affect the building programs in the area because it showed that structures should not be built in the path of potential mudflows and that much larger culverts and other drainage structures had to be installed on structures such as roads that must cross potential mudflow paths. Mudflow paths must be considered when purchasing or developing mountain areas. Because no mudflow or flash flood has occurred in the memory of residents does not mean that one will never occur.

Although not directly related to running water, snow avalanches are another geologic hazard in mountain resort areas. Avalanche chutes are also common in the Lake Tahoe area as well as any mountain area with moderate to heavy winter snows. Avalanche danger is controlled at ski resorts as a result of studies and regulations developed by the Forest Service and the National Ski Patrol System. In lift-served ski areas, the common method of controlling avalanches is to cause them to occur by blasting or other means at times when no one is skiing. Thus many small avalanches may fall harmlessly in a given chute. In uncontrolled areas an avalanche may not fall until a great depth of snow has accumulated. The writer has seen three- to five-foot-diameter trees broken by large avalanches in the Cascade Mountains of Washington. In general, avalanche chutes can be recognized by either a lack of vegetation where the slides occur at grass root level, or by underbrush-covered slopes (slide alder and the like) where sliding occurs after the brush is covered by snow. Slopes that avalanche frequently rarely allow trees to root on them, but it must also be remembered that at many places avalanches occur on forested slopes without disturbing fairly widely spaced trees of any size. Like mudflows and flash floods, avalanches must be considered in buying or developing mountain property. The writer has been offered attractive resort property with all of these hazards clearly shown in the landscape to anyone with an elementary understanding of geologic processes. Unfortunately, land sellers and developers rarely notice or understand these processes.

Floods—Dammed Rivers and Damned Rivers.
Any process that puts more water into a river than the river can carry will cause a flood. Intense precipitation and rapid melting of snow, especially in mountainous headwaters, are common causes. Other natural processes such as dams caused by landslides can cause upstream floods. In this latter case, commonly when the lake behind the slide overtops the slide-dam, the rapidly moving water quickly erodes the slide, emptying the lake rapidly and causing downstream flooding. When the ice breaks up each spring on northern rivers, the floating ice blocks themselves can cause damage to bridges and the like; and ice jams cause floods first upstream and then downstream in the same way as the slide dams just described. Except for landslides, the natural processes described are seasonal and are controlled by climatic processes.

Flood prediction requires much data and cooperation between hydrologists and meteorologists and even at best is not an exact science. The hydrologist must know such characteristics of the drainage basin as how long does it take for a rainfall of a given intensity to cause an increase in discharge of the river. This will occur after a low-intensity rain has saturated the ground, and almost at once in a high-intensity storm, but will also depend on how much moisture was in the soil at the start of the storm. He must also know the characteristics of the river. How much discharge can it carry before it floods? How much area will be flooded as the height of water rises above bank-full? How rapidly will the wave of high water move downstream? The meteorologist must predict the length and intensity of the storm. If melting snow is involved, he must consider the many factors involved in melting snow such as temperature, humidity, and wind, as well as the amount and condition of the snow. A number of stream gauging stations that report stream height and discharge, and weather stations that report precipitation and other weather elements are necessary to gather the needed data in a timely fashion.

Data on historical floods has been published by the U. S. Geological Survey in a series of publications called "Hydrologic Investigations Atlases." These atlases show areas flooded and stream gauge data and other data. Flood frequency data show the expected recurrence of floods of different height—that is the size flood that can be expected every 25 and 50 years, etc. If one of these atlases is available for your area, it should be consulted.

Accurate flood warnings can reduce damage by five to 15 per cent and save many lives. The forecasts must be accurate because if too conservative, people will be caught unprepared, and if unnecessary, will be disregarded. Flood preparations are costly and time consuming. Predictions are difficult because a small error in estimating the height of water as flood level or levee height is approached can make the prediction inaccurate. On the other hand, the more time between warning and flood, the better prepared the people.

A flood consists of several phases, each of which can be at least to some extent controlled. The first phase is the land phase and is characterized by flow of water on the surface. This occurs when the ground surface is either impermeable or saturated so the water must flow overland to the streams. Typically this occurs either during very intense storms or after the ground has been saturated by a long-duration storm. Typical damage caused by this phase is gullying. Control measures are revegetation and terracing. A gross example of how this phase leads to floods occurs when forest or brush fires destroy vegetation. In the following rainy season the water that was previously used by the plants runs off and floods occur downstream. This happens almost every year in southern California.

The second phase of a flood is the channel phase, and this, of course, occurs downstream from the land phase. The objective now is to keep the water within the channel. Dikes, dams, and channel improvements are the main methods of control. Dikes or levees may be temporary or permanent and will prevent flooding only if they are higher than the water level. They are brittle defenses in the sense that where overtopped and broken, the flooding is locally worse than if the water, in the absence of dikes, had spread over a larger area upstream. Channel improvements consist of dredging and straightening the channel to permit more efficient movement of the water. Dams hold back the water and release it at a rate that will not permit flooding downstream. Again, if the dam is too small, or the reservoir is allowed to fill at the wrong time, dams are brittle defenses, and once overtopped, the resulting flood is more destructive.

Dams are expensive and controversial. Many dams are multipurpose and produce electricity and irrigation water as well as flood control. Recreational use is also important. These sometimes conflicting uses lead to controversy. Large dams may affect several states or other countries which, of course, leads to other problems of a non-scientific nature. The scientific controversy that bears on these social and political problems is which is more efficient, big dams or little dams in flood control? Should floods be controlled by small dams in the small drainage basins in which they form, or is one large dam nearer to the place to be protected more efficient? There seems to be no clear-cut answer to this question. Probably each river system needs a carefully tailored approach using both methods of flood control. A larger question is: What other benefits can or should be gained from dams—power, irrigation, and recreation? This question has social, political, and scientific aspects.

Urbanization, Floods and Erosion. The main effect of urbanization is to increase the runoff, which, in turn, causes increased erosion and, of course, the eroded material is subsequently deposited. The resulting flooding, erosion, and deposition, as we have seen, may also occur in urbanized areas, compounding the problems. In a given drainage basin, a rainfall will, after some time period, cause an increase in the flow of the main stream. The length of the time lag and the amount of the increase in the flow will depend on the amount and the intensity of the rainfall, as well as the characteristics of the river. Fig. 25-6 shows an example. Urbanization of a drainage basin will increase

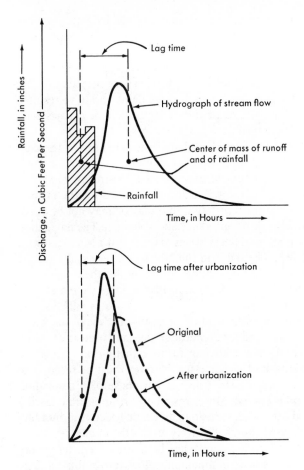

Figure 25-6. The effect of urbanization on stream flow. After urbanization, the peak flow is higher and occurs sooner. From U. S. Geological Survey, *Circular 554,* 1968.

both the peak runoff and the total runoff because buildings and paving reduce the infiltration.

The increase in discharge causes increased erosion. The first effect is a deepening of the channel and removal of accumulated debris. This increases the capacity of the river so that it can carry more water without flooding. As the main channel deepens, the tributary streams also deepen because their base level, which is the elevation at which they join the main stream, is lowered. Gullies may appear at many places in the drainage basin.

The increased runoff means less infiltration, so less recharge of ground water supplies. The water table will fall, and water wells must be drilled deeper.

A dam causes changes in a stream. Dams have finite lives because the sediment carried by the stream eventually fills the lake created by the dam. This deposition upstream from the dam occurs not only in the lake but all along the stream and its tributaries. This happens because the rising lake is the new base level of the river. This upstream deposition causes problems, especially for farmers whose irrigation ditches are silted up. Downstream the opposite problem, erosion, occurs. The clear water released from the dam no longer has the sediment load that it carried in the natural state, so it uses its excess energy to erode and deepen its channel. The tributaries also deepen to keep pace with the main stream, and gullies form where they can. These events have all happened to a greater or lesser extent at some dams and have resulted in many law suits. Climatic changes can also cause these effects, so some experts deny that dams are the whole cause of these troubles.

Other effects of dams are not always considered until after they are built. If the dam was built for water supply or irrigation, the amount of evaporation from the large surface area of a lake may reduce the amount of useful water. Upstream, the rising water table, due to the lake, may waterlog the soil, causing problems for the farmer. This may also cause a change in the natural vegetation that may or may not be good.

Another unexpected man-caused change occurred in Lake Michigan. The navigation channel built to bypass Niagara Falls allowed the alewives to reach Lake Michigan with unfortunate results for the native fish. The dams in the Northwest, especially on the Columbia River, prevented salmon from swimming upstream to their spawning grounds and so reduced the salmon in the Pacific Ocean. Fish ladders were added or built into new dams, reducing this problem.

The increased erosion caused either by urbanization or dams may make the water muddy, a type of pollution. To this may be added domestic sewage and industrial waste. The river is then no longer useful as a water source or for its esthetic beauty; and the long political, social, and economic task of cleaning up the river must be faced. The maximum amount of erosion occurs during the construction phase of urbanization when the natural slope of the ground is altered. More erosion occurs after urbanization than before, in spite of the reduced area where erosion can occur because of building and paving. This is because the increased runoff effectively erodes the uncovered areas.

Southern California, a semiarid area that recently has been urbanized, has many flood and erosion problems. The streams head in mountains, flow across alluvial fans, then across flat valleys into the Pacific Ocean. As described earlier, storms cause intense flows of water or mudflows. The debris is largely deposited on the alluvial fans at the foot of the mountains. As urbanization moved inland, first the valleys and then the fans were built upon. The rivers in many cases only rarely have water in them, so were disregarded in the building. Alluvial fans are built by the rivers changing their position on the fan from time to time. A mud or debris flow is deposited in the channel on the fan, choking it, and the next flow moves on the fan at a different, lower area. In this way, over a long time, the semicircular fan is built. The river valley and the fan are not distinctive features on the ground, and so the tendency to disregard them. They are, however, clear on topographic maps and air photos.

Thus mudflows, debris flows, or flooding cause damage on the fans. In the natural state, much of the water seeps into the ground on the fan. Where building prevents this, the water moves down the valley, flooding the urbanized areas there. Much work has been done to minimize these problems. Debris basins have been constructed behind dams to stop the sediment from reaching the fans. Flood conveyance channels have been constructed to carry the water to the ocean. This work was begun as long ago as 1915, or twenty years before the Federal Government began any flood control work. These measures have saved countless lives and probably billions of dollars' damage, but even so, lives and property are still lost. Zoning laws to control the use of the hazardous areas could prevent some of this loss. This simple measure is difficult to achieve because of political and economic complications. One facet of the problem is the large number of municipalities involved. In Los Angeles County alone, the number of incorporated areas has increased from 44 to 78 between 1935 and 1969.

Many of the ways of preventing flood damage have already been mentioned such as dams, debris basins, and zoning laws. Flood-proofing structures by making the lower floors waterproof, either permanently or during flood danger, can reduce damage. Accurate flood forecasting can make emergency flood proofing effective. Flood hazard maps, which are available for many areas, can be used for zoning and for predicting how often a given site will be flooded. (See Fig. 25-7.)

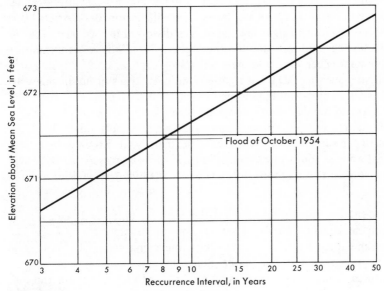

Figure 25-7. Frequency of floods on Salt Creek at Addison near Chicago. This example shows how often a flood of a given height can be expected on the average. From U. S. Geological Survey, *Circular 601 C*, 1970.

Flood-plain zoning is perhaps the best answer in spite of the relocation problems for those now using the flood plains. The resulting green belts could be parks that would improve urban life. The financial institutions that provide the money for buildings could protect themselves and their borrowers by consulting flood hazard maps before lending money. The Federal Housing Administration and the Veterans' Administration, as well as banks, could in this way prevent building in hazardous areas. Municipalities should not build streets or sewers in these areas, and utility companies should not provide service. All of these agencies would save themselves and their clients money by these simple measures.

Resources

Water, Population, and Food. Water and soil are the resources that ultimately support all life. Their availability will probably control the ultimate population of the earth. That food supply will limit population has been suggested since Thomas Malthus wrote his famous essay near the start of the 19th Century. Some have suggested that the sea is an important food source, but the best estimates are that the oceans can only feed a small fraction of our present population. It has also been suggested that food can be manufactured from fossil fuels such as petroleum, but no practical process has been developed yet.

Weathering processes form soil very slowly. Preventing soil erosion by both water and wind is important to food production. Replacing the materials in soil used by crops is also very important. The trace elements in soils that were discussed earlier are generally neglected in fertilizers, and this may be an important area for research. A current problem with fertilizers is that the runoff water from fertilized fields carries some of the fertilizer to rivers. In rivers and lakes the fertilizer provides nutrients that increase the growth of algae. The algae use the oxygen dissolved in the water, and the lack of oxygen causes the death of fish and other aquatic life. Phosphates in laundry detergents have the same effect. Pesticides used on crops get into rivers in this way, too. Long-lived insecticides such as DDT have gotten into most types of life in this way, and from animals such as birds eating poisoned insects. Thus, keeping soil productive without causing harm to the environment is a problem that requires much study.

In Chapter 12, figures were given suggesting that water supply might limit the population the earth can support. Since Malthus' time, many such estimates have been made and then cast aside as some improvement in agriculture increased yields. So will probably be the case with these figures, but it seems that ultimately Malthus will be right.

Mineral resources. The mineral resources of the earth are finite, and, once mined, no more new metal can be had. This is one aspect of the spaceship earth analogy. The day when any one metal runs out can be put off by improving mining technology so that poorer and poorer grade material can be mined, by recycling old metal, and by developing substitutes. Some metals, such as aluminum, will probably never run out, but some others, such as silver and mercury, may very well run out.

The distribution of metal mines is random, and no industrial country is self sufficient in all mineral resources. Obviously, the industrial countries are using mineral resources in much greater quantities than the undeveloped countries. The undeveloped or non-industrial countries are striving to become industrial countries and raise their standards of living. Thus the rate of use of minerals is bound to increase, and the increase will be rapid because most of the earth's population lives in the undeveloped countries.

Some examples of metal usage will show the magnitude of the problem. The per capita use of iron in the United States in 1967 was about one ton, and for the total world population the figure was .17 tons. The United States' usage was about six times that of the world average, or, said another way, iron production would have to be increased six times if the whole earth is to have the same standard of living that the United States had in 1967. It is estimated that the population of the earth will double by the year 2000. To keep the same world per capita use of iron, production will have to be doubled; or if everyone is to have the United States' 1967 standard of living, production must be increased 12 times. The world production of iron in 1967 was 550 million tons. Of course, not only iron but all other mineral resources will be needed in similar ratios to achieve these goals. See Table 25-2. These projections are probably too high for a number of reasons, such as export of manufactured goods by the United States. On the

Table 25-2. Current and predicted use of some metals. The amounts needed to bring world usage up to United States standards are enormous for the present population of the world, and staggering if population doubles by the year 2000. From C. F. Park, Jr., *Affluence in Jeopardy.*

Year 1967				
Metal	United States per capita	World per capita	World Production	Bring World to U.S. Standard
Iron	1 ton	.17 ton	550×10^6 tons	x 6
Copper	18 lbs.	3.2 lbs.	5.25×10^6 tons	x 5.5
Lead	12 lbs.	1.5 lbs.	2.44×10^6 tons	x 8

Year 2000		
Metal	Double Population As Today	Double Population World to U.S. Standard
Iron	x 2	x 12
Copper	x 2	x 11
Lead	x 2	x 16

other hand, per capita consumption in the United States and elsewhere has been rapidly accelerating, suggesting that the projections may be too low. It seems clear that the potential demand for metals in the year 2000 will be at least several times that of today. The question is, can this potential demand be met?

No clear answer can be made at present. The reserves of some metals, silver and mercury for example, appear to be too low, but history is full of examples of dire predictions that did not come true. New methods of prospecting and discovering deposits, and new technology using low-grade ore can change the gloomy picture.

The differences in mineral usage reveal a number of other differences among countries. The United States is a highly industrialized country. With only six per cent of the world's population and six per cent of the earth's land surface, in 1966 it used 16 per cent of the total earth's production of coal, 25 per cent of the petroleum, 26 per cent of the steel, 35 per cent of the copper, and 53 per cent of the aluminum. It must be remembered, too, that these amounts add to the already large quantities of these metals that are now in use in buildings, machinery, and the like. The amounts of mineral resources used are increasing in all countries; but although the amounts used per capita are increasing in the industrial countries, they are decreasing in many undeveloped countries. The rapid growth of population in the undeveloped countries is the cause. The industrial countries are increasing their standards of living, while in the undeveloped countries the people are becoming poorer, at least in mineral resources. See Fig. 25-8.

The United States, like most industrial countries, is not self sufficient in many mineral resources. It must import large quantities. A question of great importance is, what is a fair price to pay an undeveloped country for its unreplenishable mineral resources? All people look forward to the day when they will have a standard of living like ours.

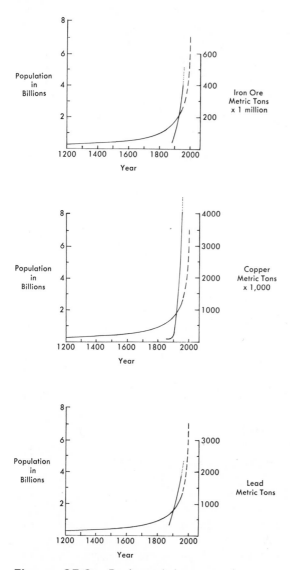

Figure 25-8. Projected increase in the use of copper, lead, and iron, compared to projected increase in population. Data from C. F. Park, Jr., *Affluence in Jeopardy*, 1968.

This will be difficult to achieve if their resources are gone. On the other hand, they cannot develop or even use their resources without capital, and their only ready way to get capital is to sell their resources. It takes many millions of dollars to find a resource, and many more to develop it, and, of course, much technical know-how. The historical background to the present world situation shows how undeveloped countries have been exploited in the past. The colonial system developed with the industrial revolution. The mother country was generally an industrial country, and the colonies supplied the mother country with raw materials. The political breakdown of this system began with the American Revolution and was pretty well completed after World War II with the formation of many new independent countries in Africa and Asia. Most of the colonialism of today is economic colonialism, and revolves around the question of the price paid for mineral resources. Prices may have to increase, and this may change the standards of living of the whole world. The constant upward revision of the royalties paid for oil in the Near East is an example of the kinds of change that seem likely to occur. Anyone who does not believe the importance of mineral resources in world politics needs only to study the history of Europe for the last 200 years or so. The territorial settlements after each war involved mainly the important mineral resource areas, such as the iron of Alsace-Lorraine, the coal of the Saar, and the lead-zinc of Poland.

Energy Resources. Modern civilization depends on energy. At first man's only use of energy was in the form of food. Later he used wood fires, and the industrial revolution required the use of coal. The fossil fuels, coal and petroleum, are the main energy sources at present, with nuclear power developing very rapidly.

The energy sources of the earth are:

> Solar radiation
>> Food
>> Fossil fuels
>> Water power
>> Direct solar energy
> Tidal energy
> Geothermal energy
> Nuclear energy

The amount of energy that the earth receives from the sun is many times more than the amount used by man. If a practical way to use it could be devised, no other source would be needed. It should be remembered that the earth re-radiates the energy received from the sun. If it did not, the temperature of the earth would increase, and geologic history shows that the average temperature has remained about the same for at least 500 million years. Solar energy is man's most important energy source. It is used by plants, providing the food that supports all of the life on the earth. In the past, plants in the form of wood provided an important source of energy. A very small amount of solar energy is preserved in the form of fossil fuels such as coal and petroleum.

Although the fossil fuels are our present main source of energy, their future is not bright. The earth's coal supplies are expected to last for two or three centuries, but petroleum will run out in 70 to 80 years. At the present time, about 60 per cent of the world's and 67 per cent of the United States' industrial energy comes from oil and gas. These figures suggest that many changes are due in the relatively near future.

Natural gas is the cleanest of the fossil fuels, and it will probably be the first to run out. Coal, which is the most widely used, releases sulfur and mercury. Sulfur is a major problem in some areas where it causes atmospheric pollution. Use of low-sulfur coal or removal of sulfur from the smoke is required in some cities. It is estimated that 3000 tons of mercury are also released from burning coal each year. This is about equal to the amount lost in industrial processes, and over ten times the amount released by natural weathering.

Water power is an attractive power source because of the lack of atmospheric pollution. Damming rivers, however, spoils some of their aesthetic value, and the lakes formed by the dams will ultimately fill with silt. Transporting the power from a hydroelectric plant requires transmission lines. In spite of these problems, water power will probably be important in Africa and South America because these continents lack large supplies of coal but have large unused water power. Water power could supply the earth with energy at the present rate of usage, but because of the silting of the lakes, it could only do this for one or two centuries.

Solar radiation as a source of energy is attractive because no fuel is used. The disadvantages are the large areas required and that only certain areas such as deserts receive enough sunlight to make such systems feasible. These disadvantages are offset somewhat because desert areas are largely unused for any other purpose, although many ecologists would disagree because they feel that deserts, like mountain, forest, and beach areas, should be preserved for recreation, ecologic and aesthetic reasons.

The relatively cloudless deserts of southwestern United States receive about 0.8 kilowatts per square meter for the middle six to eight hours of the day most of the year. A black surface can absorb most of this energy, but a black body is also an excellent radiator. Recently, surface coatings have been developed that allow the sun's radiation to pass through and be absorbed, but these surface films are reasonably opaque to the lower wavelength radiation from the collector. The collector would heat molten salts that would maintain a fairly constant temperature during overnight operation. A large-scale experiment will be necessary to prove the practicality of such a system.

Tidal energy can only supply a small amount of our energy needs, and no really practical method has been devised. The advantages are no fuel, no waste, no pollution, and minimum ecologic and scenic damage.

Geothermal energy is a new energy source in the United States, but has been in use since 1904 in Italy. The number of volcanic areas where geothermal steam is available is small. The life of a steam well is also limited. The best current estimates suggest that about one-fifth of our present energy needs could be met for about fifty years by geothermal power.

Nuclear energy seems most likely to be the energy source of the future. At present, its use is growing more rapidly than predicted. However, two problems must be overcome to insure its use. They are disposal of radioactive wastes, and development of reactors that use the more plentiful ores. Other less serious obstacles are public acceptance of nearby nuclear reactors and disposal of unused heat.

Nuclear energy can be released by fusion and fission. Fusion is the process used in the hydrogen bomb, and so far, no method has been found to release this energy slowly. It remains to be seen if this will ever become a practical power source. Fission reactors are in current use. Some of these reactors are burners in the sense that they consume their fuel; others are convertors or breeders. The difference is of supreme importance for the future. Uranium is the main fuel used in reactors. Natural uranium is a rare element and occurs in three isotopes. Uranium 238 is 99.283 per cent, uranium 235 is 0.711 per cent, and uranium 234 is 0.006 per cent of natural uranium. Of these, uranium 234 is so minor it can be neglected. Uranium 235 is the only naturally fissionable isotope and is used in burner reactors. Almost all of the present reactors are of this type, and they burn uranium 235. There is only enough uranium 235 to last about a century at the present rate of development. It is possible to build breeder reactors that will convert the abundant uranium 238 and thorium 232 into fissionable isotopes, but some uranium

235 is needed for these breeder reactors. Such breeders must be developed soon, or the uranium 235 will be used up and nuclear energy will not be available—and there seems to be no other substitute for the distant future.

Waste Disposal—"Garbage is a resource for which we have not yet found a use."

The amount of material discarded is alarming. In the United States it is about 5.5 pounds per person per day—almost twice what it was in 1920. Much of the increase is due to modern packaging and to disposable items. Most cities either bury or burn their garbage—processes that have been in use since before recorded history. Some cities have tried to make fertilizer out of garbage but have been unable to sell it. Several cities are compacting trash and either shipping it away or using it locally for land fill. Experiments are underway using trash as fuel to run electric generators and converting garbage into a nutritious protein.

One way to eliminate part of the problem is to recycle as much trash as possible. Bottles and aluminum cans are currently being recycled by consumers. Commercially reclaimed metals include about 40 per cent of our lead and 25 per cent of our copper. About three million tons of iron are discarded each year in the United States. Recycling trash can greatly reduce the need to find and mine metal ores, and so lessen one problem discussed earlier. In this sense, one can look at dumps and automobile wrecking yards as national resources although they still may be odorous eyesores.

Waste can be disposed of by releasing it into the atmosphere, rivers, the ocean, piling it on the surface, or burying it underground. Some of the geologic effects of each will be considered. Releasing gases into the atmosphere can produce haze and odor. Some gases, especially automobile exhaust, react under the influence of the ultraviolet in sunlight to produce damaging smog.

Rivers are used to carry away both sewage and industrial waste. Some of the resulting problems were described in Chapter 12. Excess heat has to be disposed of in many industrial processes and especially at power plants. It may seem strange that heat energy must be disposed of after the discussion of energy resources. The reason is that in any engine using heat energy, not all of the energy is available to do work. This is a well-known law of thermodynamics and is the reason why a perpetual-motion machine can not be invented. Steam boilers are less than 40 per cent efficient, nuclear burner reactors are about 30 per cent, and breeder reactors will be near 40 per cent efficient. The rest of the energy must go to the environment as heat. The current nuclear reactors must use water cooling. In other cases, cooling towers can be used so that the heat escapes into the atmosphere. The amount of water needed for cooling will increase. Raising the temperature of a river, lake, or the shallow nearshore ocean can have a deleterious effect on the aquatic life. Heating the atmosphere can change the climate, and the climate near cities is changing.

Surface disposal or burial both have the danger of contaminating water, especially underground water. In times of flood or heavy precipitation, surface waters can also be polluted. Decomposing garbage in a landfill produces methane, carbon dioxide, ammonia, and hydrogen sulfide. The carbon dioxide combines with water, forming carbonic acid that increases the amount of many materials that can be dissolved in the water. When such water seeps into surface or ground water, it generally pollutes it. It is possible to avoid pollution from disposal sites by paying attention to the ground water conditions as shown in Fig. 25-9.

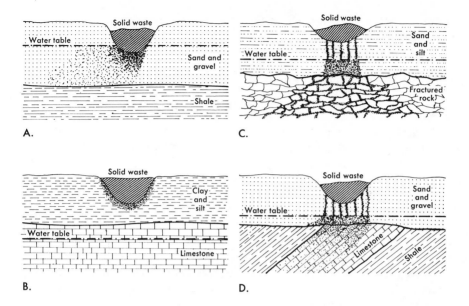

Figure 25-9. Effects of waste disposal on ground water. Only in B is contamination prevented by impermeable rocks. In all the other examples, contaminants from the waste move through permeable rock to the water table. From U. S. Geological Survey, *Circular 601 F,* 1970.

One special type of underground disposal is the pumping of fluids into deep rocks. This is a common way to dispose of oil-field brines. The earthquake problems at the Denver Arsenal suggest caution.

Radioactive waste products and similar very dangerous wastes present special problems. Some have been dumped at sea in special concrete and/or steel containers, but their potential danger to the whole ocean makes this practice not acceptable to at least some people. The waste from an atomic reactor is about the same amount as the total fuel. It must be held for about 600 years before it is safe. Salt is a very good material in which to bury atomic wastes. It is stable in earthquakes, has a high melting point, and is a good nuclear shield. For these reasons the Atomic Energy Commission hopes to bury radioactive wastes inside deep salt domes.

Thus waste disposal is a continuing problem. One recent suggestion that may solve it is to compact and encase the waste and then dump it into the deep sea trenches. Recycleable materials should be removed before dumping, or these resources would be lost. Sedimentation and sea-floor spreading will insure deep burial far below any depth affecting man. It might produce some unusual and puzzling rocks for future geologists. It is estimated that the United States each day produces enough waste to cover 400 acres to a depth of ten feet. The cost to government for collection and disposal is $3000 million per year, a sum only exceeded by funds for schools and roads. We may yet be "... standing knee-deep in garbage firing rockets at the moon. . ." (*Newsweek,* June 29, 1970, p. 77).

QUESTIONS

25-1. What will happen to the sun in the next few thousand million years, and how will this affect the earth?

25-2. List the trace elements that are harmful and those that are helpful to man.

25-3. What hazards do volcanoes pose?

25-4. In buying a hillside home, how can you protect yourself against possible landsliding?

25-5. Discuss whether building should be allowed on the flood plains of rivers.

25-6. Under what conditions are mudflows most apt to occur?

25-7. How can flood danger be reduced?

25-8. How does urbanization affect the runoff of streams?

25-9. Discuss whether the earth's mineral resources are sufficient for the next hundred years.

25-10. List the earth's energy sources, and discuss the future of each.

25-11. What are the problems facing us in disposing of our waste material?

SUPPLEMENTARY READING

GEOLOGIC HAZARDS

Bue, C. D., *Flood Information for Flood-Plain Planning.* U.S. Geological Survey Circular 539, Washington, D.C.: U.S. Government Printing Office, 1967, 10 pp.

Cannon, H. L., and D. F. Davidson, "Relation of Geology and Trace Elements to Nutrition," Special Paper No. 90, Geological Society of America, 1967, 68 pp.

Chorley, R. J., *Water, Earth, and Man.* London: Methuen and Co., Ltd., 1969, 588 pp.

Crandell, D. R., and D. R. Mullineaux. *Volcanic Hazards at Mount Rainier, Washington,* U.S. Geological Survey Bulletin 1238, 1967, 26 pp.

Gillie, R. B., "Endemic Goiter," *Scientific American* (June, 1971), Vol. 224, No. 6, pp. 92-101.

Glancy, P. A., *A Mudflow in the Second Creek Drainage, Lake Tahoe Basin, Nevada, and Its Relation to Sedimentation and Urbanization.* U.S. Geological Survey Professional Paper 650C, Washington, D.C.: U.S. Government Printing Office, 1969, pp. C195-C200.

Guy, H. P., *Sediment Problems in Urban Areas.* U.S. Geological Survey Circular 601-E, Washington, D.C.: U.S. Government Printing Office, 1970, pp. E1-E8.

Hoyt, W. G., and W. B. Langbein, *Floods.* Princeton, N.J.: Princeton University Press, 1955, 469 pp.

Leopold, L. B., *Hydrology for Urban Land Planning—A Guidebook on the Hydrologic Effects of Urban Land Use.* U.S. Geological Survey Circular 554, Washington, D.C.: U.S. Government Printing Office, 1968, 18 pp.

Rantz, S. E., *Urban Sprawl and Flooding in Southern California.* U.S. Geological Survey Circular 601-B, Washington, D.C.; U.S. Government Printing Office, 1970, pp. B1-B11.

Scherp, H. W., "Dental Caries: Prospects for Prevention," *Science* (September 24, 1971), Vol. 173, No. 4003, pp. 1199-1205.

Shacklette, H. T., H. I. Sauer, and A. T. Miesch, *Geochemical Environments and Cardiovascular Mortality Rates in Georgia.* U.S. Geological Survey Professional Paper 574C, Washington, D.C.: U.S. Government Printing Office, 1970, 39 pp.

Sheaffer, J. R., and others, *Flood-Hazard Mapping in Metropolitan Chicago.* U.S. Geological Survey Circular 601-C, Washington, D.C.: U.S. Government Printing Office, 1970, pp. C1-C14.

Thomas, H. E., and W. J. Schneider, *Water as an Urban Resource and Nuisance.* U.S. Geological Survey Circular 601-D, Washington, D.C.: U.S. Government Printing Office, 1970, pp. D1-D9.

U.S. Geological Survey, Hydrologic Atlases Portraying Areas Struck by Hurricane Camille on the Mississippi Gulf Coast. HA395 to HA408, 1969.

Van Ness, G. B., "Ecology of Anthrax," *Science* (June 25, 1971), Vol. 172, No. 3990, pp. 1303-1307.

Warren, H. V., "Medical Geology and Geography," *Science* (April 23, 1965), Vol. 148, No. 3669, pp. 534-536.

RESOURCES

Committee on Resources and Man, *Resources and Man.* San Francisco: 1969, 259 pp.

Flawn, P. T., *Mineral Resources.* Chicago: Rand McNally & Co., 1966, 405 pp.

Hubbert, M. K., "The Energy Resources of the Earth," *Scientific American* (September, 1971), Vol. 224, No. 3, pp. 60-70.

McKelvey, V. E., "Mineral Resource Estimates and Public Policy," *American Scientist* (January-February, 1972), Vol. 60, No. 1, pp. 32-40.

Park, C. F., Jr., *Affluence in Jeopardy.* San Francisco: Freeman, Cooper and Co., 1968, 367 pp.

See also supplementary reading for Chapter 6.

WASTE DISPOSAL

Clark, J. R., "Thermal Pollution and Aquatic Life," *Scientific American* (March, 1969), Vol. 220, No. 3, pp. 18-27.

Merriman, Daniel, "The Calefaction [warming] of a River," *Scientific American* (May, 1970), Vol. 222, No. 5, pp. 42-52.

Schneider, W. J., *Hydrologic Implications of Solid-Waste Disposal.* U.S. Geological Survey Circular 601-F, Washington, D.C.: U.S. Government Printing Office. 1970, 10 pp.

Appendices

Appendix A.— Minerals

Table A-1 helps to identify the more common minerals. To use the chart, first decide the sort of luster (metallic or nonmetallic) the mineral has; next, its color (light or dark); third, its hardness (compared to a knife blade); fourth, whether it has cleavage; and finally, read the brief descriptions to identify it.

Table A-1. Mineral identification key.

Dark-colored nonmetallic luster	Hard – not scratched by knife	Shows Cleavage	Black to dark green; cleavage 2 planes at nearly right angles; hardness, 5-6.	AUGITE
			Black to dark green; cleavage 2 planes about 60°; hardness, 5-6.	HORNBLENDE
		No cleavage	Red to red-brown; fracture resembles poor cleavage; brittle equidimensional crystals; hardness, 6.5-7.5.	GARNET
			Various shades of green and yellow; glassy luster; granular masses and crystals in rocks; hardness, 6.5-7 (apparent hardness may be much less).	OLIVINE
			White, clear, or any color; glassy luster; transparent to translucent; hexagonal (6-sided) crystals; hardness, 7; conchoidal fracture.	QUARTZ
			Any color or variegated; glassy luster; hardness, 5-6; conchoidal fracture.	OPAL
			Any color or variegated; waxy luster; hardness, 7; conchoidal fracture.	CHALCEDONY (AGATE)
			Red to brown; red streak; earthy appearance; hardness, 5.5-6.5 (apparent hardness may be less).	HEMATITE
			Yellow-brown to dark brown, may be almost black, streak yellow-brown; earthy; hardness, 5-5.5 (may have lower apparent hardness).	LIMONITE (GOETHITE)
	Soft – scratched by knife	Shows cleavage	Brown to black; cleavage, 1 direction; hardness, 2.5-3 (black mica).	BIOTITE
			Various shades of green; cleavage, 1 direction; hardness, 2-2.5 (green mica).	CHLORITE
			Yellow-brown, dark brown, or black; streak white to pale yellow; resinous luster; cleavage, 6 directions; hardness, 3.5-4.	SPHALERITE
		No cleavage	Red to brown; red streak; earthy appearance; hardness, 5.5-6.5 (apparent hardness may be less).	HEMATITE
			Scarlet to red-brown; scarlet streak; hardness, 2-2.5; high specific gravity.	CINNABAR
			Lead-pencil black, smudges fingers; hardness, 1; (one cleavage that is apparent only in large crystals).	GRAPHITE
			Yellow-brown to dark brown, may be almost black; streak yellow-brown; earthy; hardness, 5-5.5 (may have lower apparent hardness).	LIMONITE (GOETHITE)
			Dark to light green; greasy or waxy luster; some varieties are fibrous; hardness, 2-5, generally 4.	SERPENTINE

Table A-1. Mineral identification key (cont.).

Metallic luster			Black; strongly magnetic; hardness, 6.	MAGNETITE	
			Lead-pencil black; smudges fingers; hardness, 1; one cleavage that is apparent only in large crystals.	GRAPHITE	
			Brass yellow; black streak; cubic crystals, commonly with striations; hardness, 6-6.5.	PYRITE	
			Brass-yellow; may be tarnished; black streak; hardness, 3.5-4; massive.	CHALCO-PYRITE	
			Shiny gray; black streak; very heavy; cubic cleavage; hardness, 2.5.	GALENA	

Light-colored nonmetallic luster

Hard — not scratched by knife

Shows cleavage

White or flesh-colored; 2 cleavage planes at nearly right angles; hardness, 6. Large crystals which show irregular veining are PERTHITE. — ORTHOCLASE (POTASSIUM FELDSPAR)

White or green-gray; 2 cleavage planes at nearly right angles; hardness, 6; striations on one cleavage. — PLAGIOCLASE

No cleavage

White, clear, or any color; glassy luster; transparent to translucent; hexagonal (6-sided) crystals; hardness, 7; conchoidal fracture. — QUARTZ

Various shades of green and yellow; glassy luster; granular masses and crystals in rocks; hardness, 6.5-7 (apparent hardness may be much less). — OLIVINE

Any color or variegated; glassy luster; hardness, 5-6; conchoidal fracture. — OPAL

Any color or variegated; waxy luster; hardness, 7; conchoidal fracture. — CHALCEDONY (AGATE)

Soft — scratched by knife

Shows cleavage

Colorless to white; salty taste; cubic cleavage; hardness, 2.5. — HALITE

White, yellow to colorless; rhombohedral cleavage; hardness, 3; effervesces with dilute hydrochloric acid. — CALCITE

Pink, colorless, white, or dark; rhombohedral cleavage; hardness, 3.5-4; effervesces with dilute hydrochloric acid only if powdered. — DOLOMITE

White to transparent; 3 unequal cleavages; hardness, 2. — GYPSUM

Green to white; feels soapy; 1 cleavage; hardness, 1. — TALC

Colorless to light yellow or green; transparent in thin sheets which are very elastic; 1 cleavage; hardness, 2-2.5 (white mica). — MUSCOVITE

Green to white; fibrous cleavage; may form veins. — ASBESTOS

No cleavage

Green to white; feels soapy; hardness, 1. — TALC

White to transparent; hardness, 2. — GYPSUM

Yellow to greenish; resinous luster; hardness, 1.5-2.5. — SULFUR

Table A-2. Summary of common minerals.

Mineral	Composition	Color	Luster	Streak	Hardness	Cleavage	Other properties—Uses
Amphibole — *see hornblende*							
Asbestos	Hydrous magnesium silicate	Shades of green	Silky	None	Low apparent hardness	Fibrous	Fireproofing & insulation against heat & electricity
Augite	$(Ca,Na)(Mg,Fe^{II},Fe^{III},Al)$ $(Si,Al)_2O_6$ Calcium, sodium, magnesium, iron, aluminum silicate	Black— dark green	Vitreous	None	5–6	2 at nearly right angles	Rock-forming mineral
Biotite	$K(Mg,Fe)_3(AlSi_3O_{10})(OH)_2$ Hydrous potassium, magnesium, iron, aluminum silicate	Black— dark green	Vitreous	None	2.5–3	1 perfect	Rock-forming mineral
Calcite	$CaCO_3$ Calcium carbonate	Generally white or colorless	Vitreous to earthy	None	2.5–3	Rhombohedral, 3 not at right angles	Effervesces in cold dilute hydrochloric acid. Manufacture of cement
Chalcedony	SiO_2 Silicon dioxide	Any	Waxy	None	7	None	Conchoidal fracture
Chalcopyrite	$CuFeS_2$ Copper — iron sulfide	Brass-yellow— may be tarnished to bronze or iridescent	Metallic-dull	Greenish-black	3.5–4	None	Brittle

Table A-2. Summary of common minerals (cont.)

Mineral	Composition	Color	Luster	Streak	Hardness	Cleavage	Other properties— Uses
Chlorite	$(Mg, Fe, Al)_6(Al,Si)_4O_{10}(OH)_8$ Hydrous magnesium, iron, aluminum silicate	Green	Vitreous	None	2–2.5	1 perfect	Rock-forming mineral
Cinnabar	HgS Mercury sulfide	Vermillion-red to brownish-red	Adamantine to dull earthy	Scarlet	2.5	1	High specific gravity. Only important ore of mercury.
Clay	A family of hydrous aluminum silicates which may contain potassium, sodium, iron, magnesium, etc.	Generally light	Earthy	None	Very low apparent hardness	None	Rock-forming mineral
Dolomite	$CaMg(CO_3)_2$ Calcium-magnesium carbonate	Generally some shade of pink, flesh color, or white	Vitreous to pearly	None	3.5–4	Rhombohedral, 3 not at right angles	Powder effervesces in cold dilute hydrochloric acid
Feldspar — see plagioclase and orthoclase (potassium feldspar)							
Galena	PbS Lead sulfide	Lead-gray	Bright metallic	Lead-gray	2.5	Cubic, 3 at right angles	High specific gravity. Ore of lead
Garnet	A family of silicate minerals containing aluminum, iron, magnesium, calcium, etc.	Commonly red	Vitreous	None	6.5–7.5	None	Characteristic isometric crystals
Graphite	C Carbon	Black to steel-gray	Metallic or dull earthy	Black	1–2	1	Greasy feel— marks paper

Table A-2. Summary of common minerals (cont.)

Mineral	Composition	Color	Luster	Streak	Hardness	Cleavage	Other properties—Uses
Gypsum	$CaSO_4 \cdot 2H_2O$ Hydrous calcium sulfate	Colorless, white, gray	Vitreous, pearly, silky to earthy	None	2	1 perfect. Others poorer	Plaster of Paris
Halite	$NaCl$ Sodium chloride	Colorless to white	Vitreous	None	2.5	Cubic, 3 at right angles	Salty taste
Hematite	Fe_2O_3 Iron oxide	Reddish-brown to black or gray	Earthy to bright metallic	Light to dark Indian-red	5.5–6.5	None	Most important ore of iron
Hornblende	$Ca_2Na(Mg,Fe^{II})_4\,(Al,\,Fe^{III},Ti)$ $(Al,Si)_8O_{22}(O,OH)_2$ Hydrous calcium, sodium, magnesium, iron, aluminum silicate	Black– dark green	Vitreous	None	5–6	2 at 60°	Rock-forming mineral
Limonite (Goethite)	$FeO(OH) \cdot nH_2O$ Hydrous iron oxide	Yellow-brown to black	Generally earthy	Yellowish-brown	5–5.5	None	Low apparent hardness
Magnetite	Fe_3O_4 Iron oxide	Black	Metallic to submetallic	Black	6	None	Strongly magnetic. Important iron ore
Mica — see biotite, muscovite, and chlorite							
Muscovite	$KAl_2(AlSi_3O_{10})(OH)_2$ Hydrous potassium-aluminum silicate	Clear–light green	Vitreous	None	2–2.5	1 perfect	Rock-forming mineral
Olivine	$(Mg,Fe)_2SiO_4$ Magnesium-iron silicate	Green	Vitreous	None	6.5–7	None	Rock-forming mineral
Opal	$SiO_2 \cdot nH_2O$ Hydrous silicon dioxide	Any	Vitreous or resinous	None	5–6	None	Conchoidal fracture

Table A-2. Summary of common minerals (cont.)

Mineral	Composition	Color	Luster	Streak	Hardness	Cleavage	Other properties— Uses
Orthoclase (Potassium feldspar)	$KAlSi_3O_8$ Potassium-aluminum silicate	White to pink	Vitreous	None	6	2 at right angles	Rock-forming mineral
Plagioclase	$CaAl_2Si_2O_8$ & $NaAlSi_3O_8$ Calcium & sodium aluminum silicate	White to green	Vitreous	None	6	2 at right angles— striations on one cleavage direction	Rock-forming mineral
Pyrite	FeS_2 Iron sulfide	Pale brass-yellow	Metallic bright	Greenish or brownish-black	6–6.5	None	Brittle. Cubic crystals
Pyroxene — *see augite*							
Quartz	SiO_2 Silicon dioxide	Any	Vitreous	None	7	None	Rock-forming mineral
Serpentine	$Mg_6Si_4O_{10}(OH)_8$ Hydrous magnesium silicate	Dark to light green	Greasy to waxy	None	2–5	None	Metamorphic rocks. Fibrous variety is Asbestos
Sphalerite	ZnS Zinc sulfide	Commonly yellow-brown to black	Resinous to adamantine	White to yellow and brown	3.5–4	6	Zinc ore
Sulfur	S Sulfur	Yellow	Resinous	None	1.5–2.5	None	Fracture conchoidal to uneven. Brittle
Talc	$Mg_3(Si_4O_{10})(OH)_2$ Hydrous magnesium silicate	Apple-green, gray, white, or silver-white	Pearly to greasy	None	1	1	Cosmetics

Appendix B.— Topographic Maps

Topographic Maps

Maps are simply scale drawings of a part of the earth's surface. Thus they are similar to a blueprint of an object or to a dress pattern. Most maps are drawn on sheets of paper and thus show only the two horizontal dimensions. Geologists, together with civil engineers and many other users of maps, require that the third dimension, elevation, be shown on maps. Maps that indicate the shape of the land are called *topographic maps.* A number of different ways of showing topography are in current use, such as the familiar color tint that uses shades of green for low elevations and brown for higher elevations. The most accurate method, which will be described here, involves the use of contour lines.

Map Scales

Map scales are designated in several ways. (1) A scale can be stated as a specified unit of length on the map corresponding to a specified unit of length on the ground—for example, one inch equals one mile, meaning that one inch on the map represents one mile on the ground. (2) A scale can also be stated as a scale ratio or representative fraction, meaning that one unit (any unit) on the map equals a specified number of the same units on the ground. In the example used above, the scale would be given as 1:63,360, and one inch on the map would represent 63,360 inches on the ground (1 mile = 12 × 5280 = 63,360 inches). (3) Most maps contain a graphic scale, generally in the lower margin, for measuring distances.

Contour Lines

A contour line is an imaginary line, every part of which is at the same elevation. Thus a shore line is an example of a contour. Because a single point can only have one elevation, two contour lines can never cross (except on an overhanging cliff).

By convention, contour lines on each map are a fixed interval apart. This is called the contour interval. Typical contour intervals are 5, 10, 20, 40, 80, or 100 feet. The contour lines are always at an elevation that is a whole number times the contour interval. That is, the contour lines are drawn at 20, 40, 60, 80, etc., feet, never at 23, 43, or 63 feet.

A little imagination will reveal that if the contour lines are close together the surface is steeper than if they are far apart. (See Fig. B-1.)

If we think of an island and remember that the shore line is the zero contour, we can then imagine the successive contours, or shore lines, if the water level rises. From this we learn that a hill is represented on a contour map by a series of concentric closed lines, and, of course, a ridge would be similar, but elongate.

If we think of a valley, we see that the contour lines form V's pointing upstream. Remember a stream flows downhill at the lowest point in a valley; hence, if we start on one of the valley sides and walk in the upstream direction at a given elevation, we will eventually reach the stream.

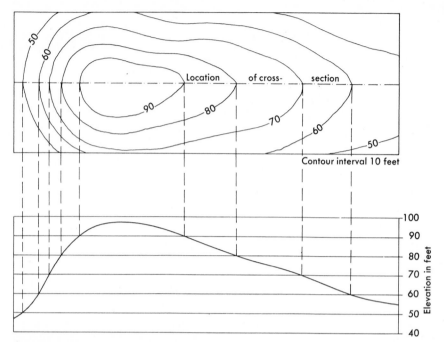

Figure B-1. Contour map of a simple hill. The cross-section shows that where the contour lines are closer together the hill is steeper.

A few minutes study of Fig. B-2 will reveal much information about reading contour maps.

Contour lines have the inherent limitation that the elevations of points between contours are not designated, although the approximate elevation can be determined by interpolation. In an effort to overcome this limitation, the elevations of hills, road junctions, and other features are commonly given on the maps.

Conventions Used on Topographic Maps

Because the United States Geological Survey produces most of the topographic maps in use in the United States, these are described here. Until recently these maps were made by actual field surveying, but are now made from aerial photographs with some field surveying to establish horizontal and vertical control points. The maps made since 1942 meet high accuracy standards. Ninety per cent of all well-defined features are within one-fiftieth of an inch of their true location of the published map, and the elevations of 90 per cent of the features are correct within one-half of the contour interval.

The maps, which are called quadrangles, are bounded by latitude and longitude lines. They are named for features within the map area, and because the names are not

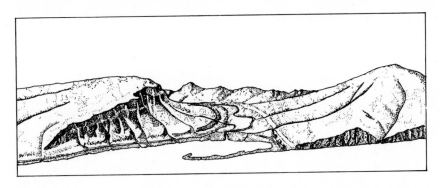

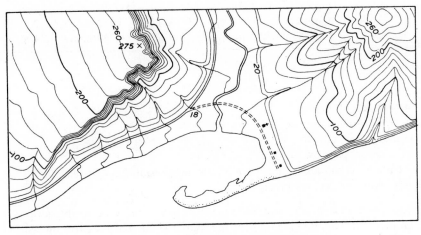

THE USE OF SYMBOLS IN MAPPING

These illustrations show how various features are depicted on a topographic map. The upper illustration is a perspective view of a river valley and the adjoining hills. The river flows into a bay which is partly enclosed by a hooked sandbar. On either side of the valley are terraces through which streams have cut gullies. The hill on the right has a smoothly eroded form and gradual slopes, whereas the one on the left rises abruptly in a sharp precipice from which it slopes gently, and forms an inclined tableland traversed by a few shallow gullies. A road provides access to a church and two houses situated across the river from a highway which follows the seacoast and curves up the river valley.

The lower illustration shows the same features represented by symbols on a topographic map. The contour interval (the vertical distance between adjacent contours) is 20 feet.

Figure B-2. Perspective view of an area and a contour map of the same area. After U. S. Geological Survey.

duplicated within a state, some are named for less important features. The sizes and scales used are shown in Table B-1. The number of square miles in a map of a given scale varies because of the convergence of longitude lines at the north pole.

Table B-1. Sizes and scales of
U. S. G. S. topographic maps.

Map designation	Scale		Quadrangle size (lat.-long.)	Quadrangle area (sq. miles)	Paper size (inches)
	Ratio	*One inch equals*			
7½ minute	1:24,000	exactly 2,000 feet	7½' × 7½'	49-70	22 × 27
					23 × 27
	1:31,680	exactly ½ mile	7½' × 7½'	49-68	17 × 21
15 minute	1:62,500	approximately 1 mile	15' × 15'	197-282	17 × 21
					19 × 21
30 minute	1:125,000	approximately 2 miles	30' × 30'	789-1,082	17 × 21
1 degree	1:250,000	approximately 4 miles	1° × 1°	3,173-4,335	17 × 21
1:250,000	1:250,000	approximately 4 miles	1° × 2°	6,349-8,669	24 × 34

Colors

Water features are in blue.

Works of man—roads, houses, etc.—are in black.

Contour lines are in brown.

In addition, some maps show

Important roads, urban areas, and public land subdivision lines in red.

Woodlands, orchards, etc. in green.

If you would like to obtain topographic maps of any area, write to the addresses below for an index to topographic maps for the state desired. They are free on request and show all of the maps available.

For states east of the Mississippi River:

United States Geological Survey
Washington, D.C. 20242

For states west of the Mississippi River:

United States Geological Survey
Federal Center
Denver, Colorado 80225

Appendix C.— Fossil-forming Organisms

Table C-1. Classification of the more important fossil-forming organisms.

    A. Classification scheme and an example.

Kingdom	Animalia
Phylum	Chordata
Class	Mammalia
Order	Primate
Family	Hominidae
Genus	*Homo*
Species	*sapiens*
Individual	John Doe

Most organisms identified by genus and species (in italics).

    B. Classification of more important fossil-forming organisms.

Kingdom Protista
    Phylum Schizomycetes—bacteria
    Phylum Sarcodina—foraminifers, radiolarians
Kingdom Animalia
    Phylum Porifera—sponges
    Phylum Coelenterata
      Class Anthozoa—corals
    Phylum Bryozoa
    Phylum Brachiopoda
    Phylum Mollusca
      Class Pelecypoda—clams, oysters
      Class Gastropoda—snails
      Class Cephalopoda
        Subclass Tetrabranchiata
          Order Nautiloidea—nautiloids
          Order Ammonoidea—ammonites, goniatites, ceratites
        Subclass Dibranchiata
          Order Belemnoidea
    Phylum Arthropoda
      Class Arachnida—scorpions, spiders, eurypterids
      Class Trilobita
      Class Insecta
    Phylum Echinodermata
      Subphylum Pelmatozoa (attached forms)
        Class Cystoidea
        Class Blastoidea
        Class Crinoidea
      Subphylum Eleutherozoa (free moving)
        Class Stelleroidea—starfish
        Class Echinoidea—echinoids

Table C-1B. continued

         Phylum Chordata
           Subphylum Hemichordata
             Class Graptolithina
               Order Dendroidea—dendroids
               Order Graptoloidea—graptolites
           Subphylum Vertebrata
             Superclass Pisces (fish)
               Class Amphibia
               Class Reptilia
               Class Aves (birds)
               Class Mammalia
      Kingdom Plantae
         Phylum Thallophyta—algae, fungi, lichens
         Phylum Bryophyta—liverworts and mosses
         Phylum Tracheophyta—vascular plants
           Subphylum Psilopsida
           Subphylum Lycopsida
           Subphylum Sphenopsida
           Subphylum Pteropsida
             Class Filicineae—ferns
             Class Gymnospermae—plants with naked seeds
               Order Coniferales
               Order Ginkgoales
               Order Cordaitales
               Order Cycadales
               Order Cycadofilicales
             Class Angiospermae—flowering plants
               Subclass Monocotyledonae
               Subclass Dicotyledonae

     In the following pages, the organisms in Table C-1B are described.

Kingdom Protista
    Phylum Schizomycetes

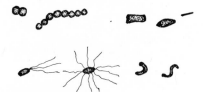

Figure C-1. *Bacteria* are rarely sought as fossils except in ancient Precambrian rocks. Bacteria are shown in Figure 20-7.

    Phylum Sarcodina
Many of the one-celled organisms in this phylum are important rock formers, and many are widely used to date rocks. Because these small organisms are recovered in drill cores, they are especially useful in correlating strata in oil wells.

Figure C-2. *Foraminifera* are mainly less than one millimeter in diameter, although a few types are much larger. Only a few of the many diverse forms are shown.

Figure C-3. *Fusilinids* are large foraminifers and were abundant in the late Paleozoic. Their size and shape are similar to wheat grains. See Figure 22-21.

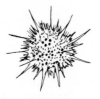

Figure C-4. *Radiolaria* are very small (less than one millimeter) and have shells composed of silica. They are seldom used in geologic dating in spite of their great diversity and numbers.

Kingdom Animalia
  Phylum Porifera

Figure C-5. *Sponges* are simple aquatic animals that feed by drawing water into their bodies through pores and expelling it through a central opening. The most common fossil remains are the spicules that help to maintain the body shape. See Figures 22-18 and 22-19.

Phylum Coelenterata

This phylum includes jellyfish, hydra, and corals. Of these, only the corals are important fossils. Coelenterates have a simple sack-like body with a mouth surrounded by tentacles at one end.

Class Anthozoa

Corals are both colonial and solitary. All are marine. Corals are widespread in Paleozoic rocks and are less important in Mesozoic and younger rocks. They are shown in many figures and the differences between some Paleozoic and Mesozoic corals are shown in Fig. 23-9.

Phylum Bryozoa

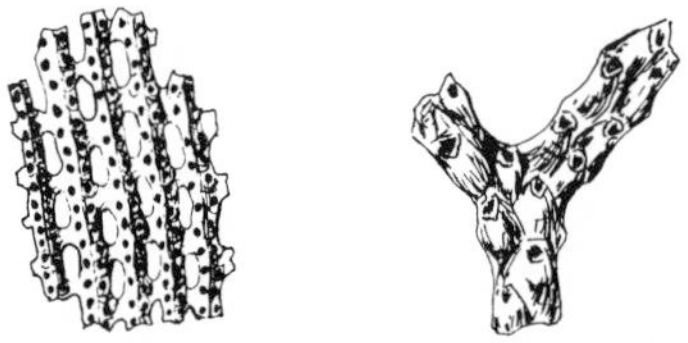

Figure C-6. *Bryozoans* are small colonial animals. Their fossil remains are generally lacy or stony fragments with many holes, or the framework that holds such fronds. See Figures 21-14, 22-14, and 22-15.

Phylum Brachiopoda

Brachiopods are bivalved animals. They were important in the Paleozoic and have declined in numbers since then. They are illustrated in many figures, especially Figs. 21-12, 22-12, 22-19, and 22-20. Fig. 23-17 shows the differences in symmetry between clams and brachiopods.

Phylum Mollusca

This diverse phylum includes clams, oysters, snails, and cephalopods, as well as other forms that are rarely fossilized.

Class Pelecypoda

Clam   Pecten   Oyster   Rudistid

Figure C-7. *Clams* and *oysters.* Clams have two shells that are mirror images. The forms are diverse. Most are bottom dwellers, but *Pecten,* a scallop, is a swimming type. The shells of oysters and rudistids do not have the bilateral symmetry of clams. See Figures 23-17 and 23-18.

Class Gastropoda

Figure C-8.   *Snails* have a single shell,
generally without symmetry. They are
the only terrestrial molluscs, although
most are aquatic. See Figures 19-4A
and B, and 23-10.

Class Cephalopoda
  Subclass Tetrabranchiata
    Order Nautiloids
    Order Ammonoidea
      Cephalopods are both straight and coiled. See Figs. 21-13, 21-15, 22-15, 22-18, and 23-11. They can be distinguished from snails by their symmetry as shown in Fig. 23-10. Nautiloids have simple sutures separating the chambers, and ammonites have more complex sutures. The distinctions among the ammonoids are shown in Fig. 23-12.
  Subclass Dibranchiata
    Order Belemnoidea
      Belemnoids are cigar-shaped shells that are internal. See Figs. 23-14 and 23-15.

Phylum Arthropoda — the joint-footed animals
  Class Arachnida
  The arachnids include spiders, scorpions and eurypterids. Eurypterids are shown in Figs. 21-18 and 21-19.
  Class Trilobita
  Trilobites were important in the Paleozoic, especially in the Cambrian Period. They are illustrated in Figs. 21-10, 21-11, 21-15, and 22-15.
  Class Insecta
  The *insects* are the largest group of animals. They are relatively uncommon fossils and so are of little stratigraphic value. See Fig. 19-4D.

Phylum Echinodermata
  The echinoderms are all marine and are "five-sided."
  Subphylum Pelmatozoa
    The pelmatozoans are all attached to the bottom by a flexible stem.
      Class Cystoidea   See Fig. 21-15.
      Class Blastoidea   See Figs. 22-16 and 22-17.
      Class Crinoidea   See Figs. 22-15, 22-16, and 22-18.

Cystoid

Blastoid

Crinoid

Figure C-9. *Cystoids* are less regular than other pelmatozoans. *Blastoids* are bud-like. *Crinoids* are sea lillies.

Subphylum Eleutherozoa
These echinoids are free moving.
Class Stelleroidea

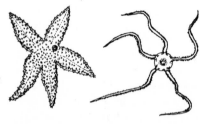

Figure C-10. *Starfish.* See also Figure 22-16.

Class Echinoidea

Figure C-11. *Echinoids* differ from starfish in their lack of arms. See Figure 24-14.

Phylum Chordata
Subphylum Hemichordata
Class Graptolithina
Order Dendroidea
Order Graptoloidea
The graptolites are extinct marine, colonial animals. They are too poorly known to classify accurately. The Graptoloidea are important Ordovician and Silurian fossils.

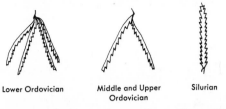

Lower Ordovician   Middle and Upper   Silurian
                    Ordovician

Figure C-12. *Graptolites.* See also
Figure 21-17.

Subphylum Vertebrata
    The vertebrates include all animals with backbones.
Superclass Pisces — fish
Class Agnatha

Figure C-13. The *agnathids* were
primitive jawless fish with bony armor.
See also Figure 22-22.

Class Placodermi

Figure C-14. The *placoderms* were
early jawed fish. Some were very large.
This fish, *Dinichthys,* reached 30 feet
long. See also Figure 22-23.

Class Chondrichthyes

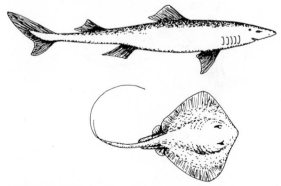

Figure C-15. *Sharks* and *rays* have
been predators since the Devonian.

Class Osteichthyes
This class includes all of the bony fish.
    Subclass Actinopterygii
       This subclass includes most of the modern fish.
    Subclass Sarcopterygii — lobe-finned fish
       Order Dipnoi — lungfish
       Order Crossopterygii — Crossopterigians
       These fish are shown in Figs. 22-29 and 22-30.
Class Amphibia
All amphibians must return to water to lay their eggs. Most live on land most
of their lives. Familiar amphibians include toads, frogs, and salamanders. See
Figs. 22-28 and 22-32.
Class Reptilia
    Order Cotylosauria
       The *stem reptiles* include *Seymouria* shown in Fig. 22-31.
    Orders Pelycosauria and Therapsida
       These orders include the *mammal-like reptiles.*

    Order Chelonia — *turtles.* See Fig. 24-18.
    Orders Sauropterygia and Ichthyosauria
       The *plesiosaurs* and *ichthyosaurs* were marine reptiles. See Fig. 23-20.
    Order Squamata — *lizards* and *snakes.* See Fig. 24-20.
    Order Thecodonta
       A *thecodont* is shown in Fig. 23-21.
    Order Crocodilla — *crocodiles.* See Fig. 24-18.
    Orders Saurischia and Ornithischia
       These orders include the *dinosaurs.* See Figs. 23-24 through 23-29.
Class Aves — *birds.* See Figs. 23-22, 23-23, and 24-23.
Class Mammalia
    Order Monotremata
       The *monotremes* are egg-laying mammals. The duckbilled platypus and the
       spiny anteater are examples. Monotremes are unimportant as fossils.
    Order Marsupialia

Figure C-16. *Marsupials* carry their
young in pouches.

    Subclass Eutheria — placental mammals
       Order Insectivora

Figure C-17. The *insectivores* include shrews, moles, and hedgehogs. See Figure 24-20.

Order Chiroptera — *bats*
Order Creodonta
    The *creodonts* were primitive carnivores from which the present carnivores evolved.
Order Carnivora

Figure C-18. The *carnivores* include dogs, cats, bears, weasels, hyenas, and civets. They are swift, smart hunters. See Figures 24-18, 24-20, 24-22, and 24-23.

Order Perissodactyla
    These are the "odd-toed" hoofed animals. They are characterized by having the axis of the foot go through the middle toe and by the loss of some of the other toes by straightening out of the foot.

Figure C-19. The "odd-toed" animals include horses, rhinoceroses, and tapirs. See Figures 24-17, 24-18, 24-20, 24-21, and 24-22.

Order Artiodactyla

These are the "even-toed" hoofed animals. They are more abundant than the "odd-toed" mammals. See Figs. 24-18, 24-20, 24-21, and 24-22.

Figure C-20. The "even-toed" mammals include camels, deer, and cattle.

Order Proboscidea.

This order includes the elephants and mammoths that have trunks. See Figs. 19-4E, 24-22, and 24-24.

Order Edentata

Figure C-21. Sloths and armadillos are modern edentates.

Order Cetacea — whales, porpoises. See Fig. 24-19.

Order Rodentia — mice, rats. See Figs. 24-18, 24-20, and 24-22.

Order Lagomorpha — hares, rabbits. See Figs. 24-20 and 24-22.

Order Primates

Suborder Prosimii

Figure C-22. The prosimians include lemurs, tarsiers, and lorises.

Suborder Anthropoidea
  Superfamily Ceboidea — new world monkeys
  Superfamily Cercopithecoidea — old world monkeys
  Superfamily Hominoidea — apes, men. See Figs. 24-25 through 24-28.

Figure C-23. Monkeys, apes, and man.

Kingdom Plantae
  Phylum Thallophyta

Figure C-24. The thallophytes include algae, fungi, and lichens. See also Figure 20-11.

Phylum Bryophyta

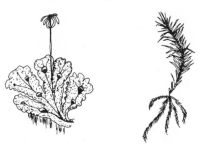

Figure C-25. The bryophytes include
liverworts and mosses.

Phylum Tracheophyta — vascular plants
  Subphylum Psilopsida
     These are the earliest vascular plants and they did not have true roots.

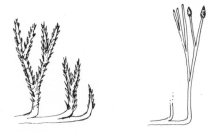

Figure C-26. Psilopsids. See Figure
22-25.

Subphylum Lycopsida

Lepidodendron

Sigillaria

Figure C-27. Lycopsids such as
Lepidodendron and *Sigillaria* were
spore-bearing plants with true roots.
See Figures 22-26 and 22-27.

Subphylum Sphenopsida

Figure C-28. *Equisetum,* a "scouring" rush, or horsetail. See Figure 22-25, and *Calamites,* Figure 22-26.

Subphylum Pteropsida — ferns and seed-bearing plants
      Class Filicineae

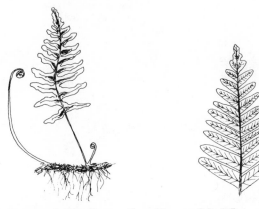

Figure C-29. Ferns. See Figure 22-27.

Class Gymnospermae — plants with naked seeds
      Order Cycadofilicales
            Glossopteris, shown in Fig. 16-6A is typical of this order.
      Order Cordaitales
            Cordaites was abundant in late Paleozoic.
      Order Coniferales.
            Pines, firs, cedars, junipers, and redwoods are common members of this
            order. See Fig. 24-15B.
      Order Ginkgoales

Figure C-30.   Ginkgo leaf.

Order Cycadales
    Cycads are shown in Fig. 23-22.
Class Angiospermae — flowering plants
    Subclass Monocotyledonae
    Subclass Dicotyledonae. See Figs. 19-4C and 24-15A and C.
    The angiosperms are predominant in modern floras.

Glossary

"When *I* use a word," Humpty Dumpty said, in rather a scornful tone, "it means just what I choose it to mean—neither more nor less."

"The question is," said Alice, "whether you *can* make words mean so many different things."

"The question is," said Humpty Dumpty, "which is to be master —that's all."

Through the Looking Glass
Chapter 6
Lewis Carroll

Aa lava   Hawaiian term for blocky surfaced basalt lava.

Abrasion   Erosion by friction—generally caused by rock particles carried by running water, ice, or wind.

Adamantine luster   Diamond-like luster.

Adaptation   The adjustment of organisms to the environment by improvements such as the development of new features.

Aftershock   An earthquake following a larger earthquake.

Agate   Chalcedony with colors in bands.

A horizon, soil   Uppermost layer of the soil. Topsoil. See Fig. 4-9.

Algae   A group of organisms that includes single-celled plants and sea weeds.

Alkali lake, flat   Desert areas where water or soil contains soluble alkalis. The alkalis, potassium, sodium, and calcium, come from weathering.

Alluvial fan   Low cone-shaped deposits built by a river where it issues from a mountain into a lowland.

Alpha decay   Radioactive decay in which an alpha particle is released.

Alpha particle   Helium nucleus. Consists of two protons and two neutrons.

Alpine glacier   Mountain glacier. A glacier in a mountain valley.

Amino acid   Nitrogenous organic compound that forms part of a protein.

Amniote egg   An egg in which the embryo is surrounded by membranes, one of which is the amnion, filled with watery amniotic fluid. This egg is an adaptation to protect the developing embryo against drying and physical shock.

Amphibian   A "cold-blooded" vertebrate that lives in water and breathes through gills in the early stages of development and breathes air through lungs in later stages.

Amphibole   A mineral group. The most important rock-forming amphibole is hornblende.

Amphibolite   A metamorphic rock composed mainly of amphibole, such as hornblende, and plagioclase feldspar.

Anaerobic   An adjective describing an organism that is able to live in the absence of oxygen.

Andesite   Fine-grained igneous rock. Intermediate in color and composition between rhyolite and basalt.

Andesite line   A line drawn on a map separating the oceanic areas of basaltic volcanism from the continental and island arc areas of andesitic volcanism.

Angiosperm   A plant with true flowers that bears seeds enclosed in ovaries.

Angstrom   A unit of length = 0.00000001 centimeters (approximately 0.000000004 inches).

Angular unconformity   An unconformity in which the older strata dip or slope at a different angle (generally steeper) from the younger strata.

Anorthosite   A coarse-grained rock composed almost entirely of plagioclase.

Anthracite   A type of coal formed by mild metamorphism of lower-grade coal.

Anticline   A type of fold in which the strata slope in opposite directions away from a central axis, like the roof of a house.

Apatite   A mineral. Number 5 on the hardness scale. $Ca_5(PO_4)_3F$

Aphanitic   A rock texture in which the minerals are too small to see with the naked eye. Generally the grains are smaller than about 1/8 millimeter.

Arkose   A sandstone containing much feldspar.

Artesian well   A well in which the water is under pressure so that it rises up the well bore. The water may or may not flow out onto the surface.

Arthropod   A member of the phylum Arthropoda. They have segmented bodies and jointed legs.

Asbestos   A name given to fibrous mineral material widely used as insulation. It may be composed of either an amphibole or a variety of serpentine.

Ash, volcanic   Unconsolidated pyroclastic material less than 4 millimeters in size.

Asteroid (astronomy)   One of a group of small planet-like bodies whose general orbit is between Mars and Jupiter.

Atoll   A ring-like coral island or islands with a central lagoon.

Atom   The smallest unit of an element. Composed of protons, electrons, and neutrons.

Atomic fission   A reaction in which an atom is split. The atomic bomb reaction is an example.

Atomic fusion   A reaction in which two or more atoms are combined to form a different element. The hydrogen bomb reaction is an example.

Atomic mass number   The total number of protons and neutrons in the nucleus of an atom.

Atomic mass unit   The unit used by physicists to express the weight of a nucleus. 1.66×10^{-24} grams (based on oxygen).

Atomic number   The number of protons, or positive charges, in the nucleus of an atom.

Augite   The common rock-forming pyroxene mineral. Generally black with two cleavages at right angles. $(Ca,Na) (Mg, Fe^{II}, Fe^{III}, Al) (Si, Al)_2O_6$

Aurora   Luminous bands or areas of colored light seen in the night sky near the poles. Northern or southern lights.

Autoradiograph   A "photograph" of radioactive material taken by placing film directly on a radioactive specimen. See Fig. 2-11.

Avalanche   The fall of a snow mass. The term is sometimes used for *landslide*.

Avogadro's number   The number of atoms in a molecular weight of a gas. 6.02×10^{23}

Banded iron formation   A sedimentary rock consisting of alternate layers of iron-rich and iron-poor silica.

Barchan dune   A crescent-shaped dune in which the cusps face downwind.

Barrier island   A low island between the ocean and the continent.

Basalt   A fine-grained volcanic rock. It is dark in color and composed mainly of plagioclase feldspar and pyroxene. It is one of the most common rock types.

Base level of river   The lowest level to which a river can erode. It is sea level for a river flowing into the ocean. The elevation of the mouth of a river.

Basement rocks   The rocks that underlie the sedimentary cover. The basement is generally a complex of igneous and metamorphic rocks.

Batholith   A very large discordant intrusive rock body, generally composed of granitic rocks. Some batholiths may be of metamorphic origin.

Bauxite   Aluminum ore. Bauxite is the term given to a weathering product composed of several aluminum-rich minerals.

Bedding   The layers or laminae of sedimentary rocks.

Bed load   The material moved along the bottom of a stream by the moving water.

Belemnoid   A squid-like animal with an internal shell. A member of the phylum Mollusca, which includes clams, snails, oysters, and ammonites.

Bench, surf-cut   A flat surface cut or eroded by wave action.

Beryl   A mineral, an ore of beryllium (a metal). $Be_3Al_2Si_6O_{18}$

Beta decay   Radioactive decay in which a beta particle is released.

Beta particle   In radioactivity, an electron moving at high velocity.

B horizon, soil   Second or middle layer of the soil. See Fig. 4-9.

Biotite   A common mineral, a member of the mica family. Recognized by its black, shiny color and perfect cleavage. $K(Mg,Fe)_3(AlSi_3O_{10})(OH)_2$

Bituminous coal   Soft coal.

Blastoid   A stalked echinoderm. Also called "sea buds."

Block fault   A fault that bounds an uplifted or down-dropped segment of the crust.

Block, volcanic   Pyroclastic fragment greater than 32 millimeters in size.

Blocky lava   Lava with a blocky surface. See *aa lava.*

Blowout   Hollow caused by wind erosion.

Bomb, volcanic   Pyroclastic material ejected from a volcano in a melted or plastic state. Its shape forms either in flight or upon impact.

Bornite   A mineral, an ore of copper. Cu_5FeS_4

Boulder   A rock fragment greater than 256 millimeters in size.

Bowen's reaction series   The sequence and reactions that take place in the crystallization of a basalt melt.

Brachiopod   A member of the phylum Brachiopoda. Marine animals with two unequal shells, each of which is bilaterally symmetrical. Also called "lamp shells."

Braided river   A river with many channels that divide and join.

Breccia   A clastic sedimentary rock with angular fragments, a sharpstone conglomerate. A volcanic breccia is composed of pyroclastic fragments greater than 32 millimeters in size.

Breeder reactor   An atomic reactor that produces more fissionable material than it uses.

Brown clay   A deep-sea deposit formed by fine material, mainly volcanic dust. The color is probably caused by oxidation. Also red clay.

Bryozoa   A phylum of colonial animals that build lacy, branching, or other types of structures with many small holes. Also called "moss animals."

Calcic plagioclase   Plagioclase feldspar with a high calcium content. Plagioclase is a continuous series between $CaAl_2Si_2O_8$ and $NaAlSi_3O_8$.

Calcite   A common rock-forming mineral, composes limestone and marble. Recognized by three directions of cleavage not at right angles, and reaction with dilute acid. $CaCo_3$

Caldera   A large volcanic depression containing volcanic vents.

Caliche   A whitish accumulation of calcite in soil in dry areas.

Capillary action   The rising of water in narrow passages, caused by surface tension. See Fig. 12-2.

Carbon 14   Radioactive isotope of carbon. Used in radioactive dating.

Carbonate   A rock or mineral containing carbonate (CO_3). The main minerals are calcite and dolomite, and the main rock types are limestone and dolostone (dolomite).

Carbonation   A reaction in chemical weathering in which carbon dioxide (CO_2) combines with the rock to form a carbonate (CO_3).

Carbonic acid   An acid important in weathering that is formed by the solution of carbon dioxide in water. H_2CO_3

Carboniferous Period   A part of the geologic time scale. It is so called because many coal deposits were formed at that time. It lasted from about 345 million years ago to about 280 million years ago.

Carbonization   A fossilization process in which the organism becomes a film of carbon, generally on a bedding surface of a sedimentary rock. Many leaf fossils form in this way.

Carnotite   An ore of uranium. $K(UO_2)_2(VO_4)_2 \cdot 3H_2O$

Cassiterite   An ore of tin. SnO_2

Cast, fossil   A fossil formed by the infilling of a cavity produced by the decay of an organism.

Cataclastic   A metamorphic texture in which at least some minerals have been broken and flattened.

Catastrophism   A doctrine that held most geologic changes were caused by sudden violent events.

Cementation   The process whereby binding material is precipitated between clastic grains, forming a rock.

Cenozoic   An era of the geologic time scale. It lasted from about 63 million years ago to the present.

Cephalopod   A marine invertebrate characterized by tentacles and a straight or coiled calcareous shell with chambers.

Chalcedony   An extremely fine-grained variety of quartz. Agate and chert are made of chalcedony.

Chalcocite   An ore of copper. Cu_2S

Chalcopyrite   An ore of copper. Fool's gold. $CuFeS_2$

Chalk   A soft white variety of limestone. Composed of shells of micro-organisms and very fine calcite.

Chemical weathering   The reactions that occur between a rock and its surroundings. The reactions tend to bring the minerals into equilibrium with the environment. Typically water, oxygen, and carbon dioxide are involved.

Chert   An extremely fine-grained variety of quartz. See *chalcedony*.

Chlorite   A compound containing chlorine.

C horizon, soil   The third layer of a soil, partially decomposed bedrock.

Chromite   An ore of chromium. $(Mg,Fe^{II})(Cr,Al,Fe^{III})_2O_4$

Cinder cone   A cone-shaped small volcano composed of pyroclastic material.

Cinder, volcanic   Pyroclastic material between 4 and 32 millimeters in size.

Cinnabar   An ore of mercury. HgS

Cirque   A glacial erosion feature. The steep amphitheater-like head of a glacial valley.

Clam   A bivalved animal, which has two shells that are mirror images. A member of the phylum Mollusca. Also called pelecypod.

Clastic   A textural term for rocks composed of fragments of some other pre-existing rock.

Clay   A mineral family. Most clays are very fine grained. Much of the material in soil and shales is clay.

Claystone   A rock composed largely of fine clay.

Cleavage   The tendency to split along planes of weakness.

Coal   A sedimentary rock formed of altered plant material.

Cobaltite   An ore of cobalt. $CoAsS$

Cobble   A rock fragment between 64 and 256 millimeters in size.

Colonial animal   Most of the colonial animals mentioned in this book live in groups or colonies that are organized so that there is a division of labor among the individuals. This term can also mean a type of animal that lives in close association with other similar individuals.

Columnar jointing   Jointing that breaks the rock into typically rough six-sided columns. Common in basalt and probably caused by shrinking due to cooling.

Compaction   Reduction in volume. Generally caused by pressure in sedimentary rocks.

Composite volcano   A volcano composed of both lava and pyroclastic material.

Conchoidal fracture   The rounded shell-like fracture typical of brittle materials such as glass.

Concordant intrusive rock   An intrusive igneous body whose boundaries are parallel to the enclosing sedimentary rocks.

Cone of depression   The cone-shaped depression of the water table caused by withdrawing water from a well faster than it can move through the rocks to the well.

Conglomerate   A coarse-grained clastic sedimentary rock. The fragments are larger than 2 millimeters.

Contact metamorphic rocks   Thermally metamorphosed rocks near the boundary or contact of an intrusive body such as a batholith.

Continental drift   The movement of continents on the earth's surface.

Continental glacier   A glacier caused by the buildup of an ice cap on a continent. Greenland has an example.

Continental rise   The transition between the continental slope and the deep ocean floor.

Continental shelf   The shallow ocean area that extends to about a 600-foot depth surrounding a continent. The underlying crust is generally continental.

Continental slope   The relatively steep slope between the continental shelf and the deep ocean basin.

Continuous mineral series   A mineral group such as plagioclase feldspar in which complete or continuous change in composition between end members is possible.

Contour line   A line on a map connecting points of equal elevation.

Convection current   Movement in a fluid caused by differences in density. The density differences are generally caused by heating.

Convergence   In evolution, the formation of similar structures in different types of organisms.

Coprolites   Fecal pellets or castings. Fossilized excrement or feces.

Coquina   A type of limestone composed largely of shells and shell fragments.

Coral   Bottom-dwelling marine animals. They are both solitary and colonial, and secrete external calcium carbonate skeletons.

Core of earth   The innermost layer of the earth. Consists of an inner core believed to be solid and an outer liquid core.

Correlation   The determination of the equivalence of two exposures of a rock unit.

Corundum   A mineral. Number 9 on the hardness scale. Al_2O_3

Cosmic ray   Very high-speed nuclei coming from space.

Covalent bonding   A bond formed by the sharing of electrons.

Crater, impact   A depression caused by impact of a falling object. Meteorites cause most large impact craters.

Crater, volcanic   A volcanic depression typically developed at the top of a volcano.

Creep   Slow downslope movement of surface material under the influence of gravity.

Cretaceous   A period in the geologic time scale. It lasted from about 135 million years ago to about 63 million years ago.

Crevasse   A crack in the surface of a glacier.

Crinoid   A type of echinoderm attached to the sea bottom by a stalk.

Cross-bedding   Beds or laminations at an oblique angle to the main bedding.

Crustal plate   A thin plate of the outer part of the earth. The depth is much less than the other two dimensions.

Crust of the earth   The uppermost layer of the earth. Above the Moho.

Crystal   A geometrical solid bounded by smooth planes.

Crystalline   A material with a crystal structure.

Crystallization   The processes of crystal formation.

Crystallography   The study of crystals.

Crystal settling   The sinking of early-formed crystals in a melt.

Crystal structure   An orderly internal (atomic) arrangement.

Crystal system   All crystals can be classified into one of the six crystal systems.

Curie point   The temperature above which a substance loses its magnetism.

Cycle of erosion   The sequence of changes that occur during erosion.

Daughter element An element formed from another element in radioactive decay.

Debris slide Landslide involving mainly unconsolidated material.

Decay, radioactive The spontaneous change of one element into another.

Deep-focus earthquake Earthquake with a focus below 180 miles.

Deflation Removal of material by wind erosion.

Delta The material deposited by a river at its mouth.

Density Denseness or compactness. The mass of a material in grams per cubic centimeter.

Deuterium An isotope of hydrogen.

Devonian A period in the geologic time scale. It lasted from about 405 to 345 million years ago.

Diamond A mineral composed of carbon. Number 10 on the hardness scale.

Diatomite A rock composed largely of the silicious shells of diatoms.

Differentiation The process by which different rock types are derived from a single magma.

Dike A tabular discordant intrusive body.

Dike swarm A number of more or less parallel dikes.

Dinosaur A group of extinct Mesozoic reptiles. They are characterized by the structures of their skulls and hips.

Diorite A coarse-grained igneous rock intermediate in composition between granite and gabbro.

Dipole magnetic field The type of magnetic field produced by a bar magnet. See Fig. 16-9.

Discharge The quantity of water flowing past a point in a river.

Discontinuity, Mohorovičić (M-discontinuity, Moho) The level at which a distinct change in seismic velocity occurs. It separates the crust from the mantle.

Discontinuous mineral series That part of the Bowen reaction series in which abrupt changes in mineral occur.

Discordant intrusive An intrusive rock body that cuts across the bedding of the enclosing sedimentary rocks.

Disseminated magmatic mineral deposit A mineral deposit in which the ore is scattered throughout an igneous body.

Dissolved load The material carried in solution by a river.

Dolomite A mineral $CaMg(Co_3)_2$, or a rock composed largely of dolomite (dolostone).

Dominant character A character inherited from one parent that appears in the offspring to the exclusion of the corresponding recessive character from the other parent.

Dreikanter A pebble faceted by sandblasting by wind-blown sand. Same as *ventifact.*

Drift, glacial A general term for glacier-deposited material. (See also *continental drift.*)

Drumlin A smooth hill of glacial drift with its long axis parallel to the direction of glacial movement.

Dune An accumulation of wind-blown sand.

Dunite A coarse igneous rock composed almost entirely of olivine.

Dynamic metamorphism Metamorphism caused by deformation of the rock.

Earthflow A slow flow of wet overburden.

Earthquake Vibrations of the earth caused by sudden internal movements in the earth.

Echinoderm A member of the phylum Echinodermata. Marine invertebrate animals, most of which have five-sided symmetry, and many of which have spines.

Eclogite A rock that forms under very high pressure, composed of garnet and pyroxene. Its chemical composition is similar to basalt.

Electron A fundamental sub-atomic particle. It has a very small mass and a negative charge.

Element A material that cannot be changed into another element by ordinary chemical means. Just over 100 elements are known. All atoms of an element have the same atomic number (number of protons).

Emergent coastline A coastline formed by either uplift of the continent or lowering of sea level.

Enargite   An ore of copper. Cu_3AsS_4

Eocene   A subdivision of the most recent era in the geologic time scale. It lasted from about 58 to 36 million years ago.

Eolian   A term applied to wind erosion or deposition. (Aeolian, obsolete)

Epicenter   The point on the earth's surface directly above the focus of an earthquake.

Epidote   A metamorphic mineral. $Ca_2(Al,Fe^{III})_3(SiO_4)_3OH$

Epoch   A subdivision of a period in the geologic time scale.

Era   A major subdivision of the geologic time scale.

Erosion   The wearing away and removal of material on the earth's surface.

Erratic   A glacially deposited boulder far from its source.

Esker   Curving ridges of glacially deposited material.

Estuary   Tidal inlet along a sea coast. Many form at the mouth of a river.

Eurypterid   An extinct arthropod.

Evaporite   A sedimentary rock formed by the evaporation of water, leaving the dissolved material behind.

Extrusive igneous rock   An igneous rock erupted on the earth's surface. A volcanic rock.

Fault   A break in the earth's crust along which movement has occurred.

Fauna   All of the animals of a given area.

Feldspar   The most abundant mineral family. Consists of plagioclase $CaAl_2Si_2O_8$ and $NaAlSi_3O_8$, and potassium feldspar $KAlSi_3O_8$ (orthoclase and sanidine). Mixtures of potassium and sodium feldspar are termed perthite.

Felsite   A light-colored, fine-grained igneous rock. Most have chemical compositions similar to rhyolite.

Ferromagnesian mineral   A mineral containing iron and magnesium. Most are dark colored.

Fissility   The ability to split easily along parallel planes.

Fission, atomic   A reaction in which an atom is split. The atomic bomb reaction is an example.

Flint   A variety of chalcedony generally dark in color. See *chalcedony.*

Flood basalt   Basalt that forms thick extensive flows that appear to have flowed rapidly (flooded).

Flood plain   The flat part of a river valley that is subject to floods at times.

Flora   All of the plants of a given area.

Fluorescence   The emission of light when exposed to ultraviolet light.

Fluorite   A mineral. Number 4 on the hardness scale. CaF_2

Fluvial   Pertaining to rivers.

Focus, earthquake   The point where an earthquake originates.

Folding   The bending of strata.

Foliation   A directional property of metamorphic rocks caused by the layering of minerals. Generally caused by crystallization under pressure.

Foraminifera   One-celled animals. Their generally microscopic shells are useful in dating some rocks.

Formation   A rock unit that can be mapped.

Fossil   Any evidence of a once-living organism.

Fracture zone   Long linear areas of apparent faulting in the Pacific Ocean.

Frost heaving   The lifting of the surface by the expansion of freezing water.

Frost polygons   Polygonal patterns of stones caused by repeated freezing.

Frost wedging   Mechanical weathering caused by expansion of freezing water.

Fusion, atomic   A reaction in which two or more atoms are combined to form a different element. The hydrogen bomb reaction is an example.

Fusulinid   A type of foraminifer shaped like a wheat grain. They are important fossils of the Pennsylvanian and Permian Periods.

Gabbro   A coarse-grained igneous rock composed mainly of plagioclase feldspar and augite or hornblende.

Galena   An ore of lead. PbS

Garnet   A mineral family.

Genes   The units of inheritance, transmitted in the sex cells of the parents, that control the characters of the offspring.

Genus   A group of species believed to have descended from a single ancestor.

Geode   A hollow space in a rock, generally filled later with another mineral such as chalcedony.

Geosyncline   A large elongate downsinking area in which many thousand feet of sedimentary rocks are deposited. They are later deformed to form mountain ranges.

Geyser   An intermittent hot spring or fountain from which hot water and/or steam issues.

Glacier   A mass of moving ice.

Glass   A supercooled liquid. A glass does not have an orderly internal structure like a crystal.

Glossopteris **flora**   A fossil assemblage characterized by the distinctive *Glossopteris* leaves. The flora occurs in rocks in the southern hemisphere and is an important line of evidence for continental drift.

Gneiss   A coarse-grained foliated metamorphic rock. This rock contains feldspar and is generally banded.

Goethite   A mineral. FeO(OH)

Gouge   The fine-grained ground rock in a fault zone. Mylonite.

Grade, metamorphic   A measure of the intensity of metamorphism.

Graded bedding   Bedding that has a gradation from coarse at the bottom to fine nearer the top.

Graded river   An equilibrium river that has just the slope and velocity to carry all the debris carried into it.

Gradient of a river   The slope of a river.

Granite   A coarse-grained rock containing quartz and feldspar. See Fig. 3-38.

Granodiorite   A coarse-grained rock containing quartz and feldspar. See Fig. 3-38.

Granule   A fragment between 2 and 4 millimeters in size.

Graphite   A mineral composed of carbon.

Graptolite   An extinct colonial animal. Found in early Paleozoic rocks.

Gravel   Unconsolidated fragments larger than 2 millimeters.

Gravity anomaly   An area where the attraction of gravity is larger or smaller than expected.

Gravity folding   Folding caused by beds sliding down a slope.

Gravity meter   An instrument for measuring the attraction of gravity.

Graywacke   A type of sandstone, generally poorly sorted with a clay or chloritic matrix.

Greenschist   A schist with much chlorite. Generally formed by the metamorphism of basaltic rocks.

Gypsum   A mineral. Number 2 on the hardness scale. $CaSO_4 \cdot 2H_2O$

Half-life   The time required for one half of the atoms of a radioactive element to decay.

Halite   A mineral. Common salt. NaCl

Hanging valley   A tributary valley whose floor is higher than that of the main stream valley at their junction. Commonly caused by glaciation.

Hardness   A mineral's resistance to scratching.

Hematite   An ore of iron. Fe_2O_3

Holocene   The most recent period in the geologic time scale. The Recent.

Horn   A pyramidal peak formed by glacial erosion.

Hornblende   A common rock-forming mineral. An amphibole. $Ca_2Na(Mg,Fe^{II})_4(Al,Fe^{III},Ti),(Al,Si)_8O_{22}(O,OH)_2$

Hornfels   A metamorphic rock formed by thermal or contact metamorphism of any rock type.

Humus   Dark organic material in soil.

Hydrolysis   A process in chemical weathering that involves the addition of water.

Hydrothermal deposit   A mineral deposit emplaced by hot water solutions.

Icecap glacier   (See *continental glacier.*)

Ichthyosaur   An extinct fish-like reptile.

Igneous rock   A rock that formed by the solidification of a magma. A rock that was once melted or partially melted.

Ilmenite   An ore of titanium. $FeTiO_3$

Index fossil   A fossil that can be used to date the enclosing rocks.

Insect   An arthropod of the class Insecta, which has a three-part body, three pairs of legs attached to the middle body part, one pair of antennae, and usually wings.

Intensity, earthquake   A measure of the surface effects of an earthquake. See Table 13-1.

Interlocking texture   A texture in igneous and metamorphic rocks in which the mineral grains grew together, filling all of the available space.

Intrusive igneous rock   A rock that solidified below the surface.

Invertebrate   An animal without a backbone.

Ion   An atom that has gained or lost electrons and so has an electrical charge.

Ionic bonding   The atoms are attracted to each other by electrical charges caused by concurrent gain and loss of electrons.

Island arc   A curving group of volcanic islands associated with deep-focus earthquakes, deep ocean trenches, and negative gravity anomalies.

Isostasy   The theory that the crust floats on a substratum in the mantle.

Isotope   A variety of an element having a different number of neutrons in its nucleus.

Jet piercing   Erosion by expanding gas bubbles.

Jointing   A fracture in a rock along which no appreciable movement has occurred.

Jurassic   A period in the geologic time scale. It lasted from about 181 to 135 million years ago.

Kettle   A depression in glacial deposits caused by the melting of a buried block of ice.

Laccolith   A concordant intrusive igneous rock body that domed up the overlying sedimentary rocks.

Lacustrine   Pertaining to lakes.

Lagoon   A shallow body of water with restricted connection to the ocean.

Landslide   Rapid downslope movement of bedrock and/or overburden.

Lateral fault   A fault with horizontal movement. Also called strike-slip fault, wrench fault, and rift.

Laterite   The red-brown soil formed by chemical weathering in the tropics.

Lava   Volcanic rocks that flow on the earth's surface.

Leaching   The dissolving of minerals by water passing through rocks.

Levee   The raised bank along a river.

Lignite   A brown-black coal that forms from the alteration of peat.

Limestone   A sedimentary rock composed largely of calcite.

Limonite   A general term for brown hydrous iron oxides.

Lithification   The process whereby a sediment is turned into a sedimentary rock.

Loess   Wind-deposited silt.

Longitudinal dune   Seif. An elongate sand dune parallel to the wind direction.

Magma   A natural hot melt composed of a mutual solution of rock-forming materials (mainly silicates) and some volatiles (mainly steam) that are held in solution by pressure.

Magnetite   A mineral. Fe_3O_4

Magnitude, earthquake   A measure of the amount of energy released by an earthquake.

Mammal   A warm-blooded vertebrate that bears live young and produces milk to feed them.

Mantle   The middle layer of the earth, between the crust and the core.

Marble   A metamorphic rock composed largely of calcite. Formed by the metamorphism of limestone.

Maria   The low, dark smooth areas on the moon.

Marsupial   One of a subclass of mammals in which the mother's nipples are located inside a pouch or marsupium. The relatively undeveloped young are carried in the pouch for several months after birth.

Mascon   From mass concentration. Areas on the moon with higher gravity, believed to be caused by concentrations of mass.

Mass number, mass unit   (See *atomic mass number, atomic mass unit.*)

Maturity   The stage between youth and old age in the cycle of erosion.

Meanders   A series of more or less regular loop-like bends on a river.

Mechanical weathering   The disintegration of a rock by physical processes.

Mesozoic   An era in the geologic time scale. It lasted from about 230 to 63 million years ago.

Metallic bonding   The atoms are closely packed and the electrons move freely among the atoms.

Metallic luster   The luster characteristic of metals.

Metamorphic grade   A measure of the intensity of metamorphism.

Metamorphic rock   A rock that has undergone change in texture or mineralogy.

Metamorphic zone   Essentially equivalent to metamorphic grade. Zones are generally thought of as related to depth at which metamorphism occurs. Generally recognized by index minerals.

Metaquartzite   Quartzite.

Meteor   A shooting star. A meteorite burning due to friction in the atmosphere.

Meteorite   Matter that has fallen on the earth from outer space.

Methane   Marsh gas. CH_4

Mica   A mineral family. Biotite, muscovite, and chlorite are common rock-forming micas.

Microfossil   Small fossils seen only by using a microscope.

Micrometeorite   A tiny meteorite that must be viewed under the microscope. (See *meteorite.*)

Microtektite   A tiny tektite. (See *tektite.*)

Millibar   A unit of pressure, one one-thousandth of a bar. One thousand dynes per square centimeter.

Mineral   A naturally occurring, crystalline, inorganic substance with a definite small range in chemical composition and physical properties.

Miocene   A subdivision of the Cenozoic Era of the geologic time scale. It lasted from about 25 to 13 million years ago.

Mississippian   A period in the geologic time scale.

Mohorovičić discontinuity (M-discontinuity, Moho)   The level at which a distinct change in seismic velocity occurs. It separates the crust from the mantle.

Mold, fossil   The impression left in the surrounding rock by the decay of organic material.

Molybdenite   An ore of molybdenum. MoS_2

Monocline   A flexure changing the level of any given bed. See Fig. 14-13.

Monotreme   One of a primitive subclass of mammals which lay reptile-like eggs, and which secrete milk through a number of non-united glands rather than through true nipples. Duck-billed platypus and spiny anteaters are monotremes.

Moraine   Material deposited by a glacier.

Mountain   An area elevated above the surrounding area.

Mountain glacier   Alpine glacier. A glacier in a mountain valley.

Mountain root   The low density rocks below a mountain range. The thickening of the crust below mountain ranges.

Mud cracks   The roughly six-sided cracks that form due to the shrinkage of mud when it dries.

Mudflow   The downslope movement of water-"lubricated" debris.

Mudstone   A fine-grained sedimentary rock formed from the consolidation of mud.

Muscovite   White mica. Recognized from its color and perfect cleavage. An important rock-forming mineral. $KAl_2(AlSi_3O_{10})(OH)_2$

Mutation   A spontaneous, inheritable change in an organism.

Mylonite   A fine-grained metamorphic rock formed by the milling of rocks on fault surfaces.

Nappe   A large plate of overthrust rocks.

Nautiloid   A cephalopod in which the septa separating the chambers are either straight or have simple curves.

Nebula   A cloud of gas or dust found in space.

Neutron   A fundamental particle with no electrical charge found in the nuclei of atoms.

Non-clastic sedimentary rock   A chemically or biologically precipitated sedimentary rock.

Normal fault   A fault caused by tension or uplift in which the hanging wall is depressed. See Fig. 14-14.

Nuées ardentes   A French term meaning glowing clouds. A volcanic pyroclastic eruption of hot ash and gas.

Obsidian   Volcanic glass.

Oceanic crust   That part of the earth above the Moho and under the oceans. Composed largely of basaltic rocks.

Old age   A stage in the cycle of erosion near its end.

Oligocene   A subdivision of the geologic time scale. It lasted from about 36 to 25 million years ago.

Olivine   A common rock-forming mineral. $(Mg,Fe)_2SiO_4$

Omphacite   A dense, green pyroxene found in the rock eclogite.

Oölite   A small spherical or ellipsoidal body, generally of calcite. Some rocks are composed of oölites.

Ooze   A fine-grained sediment found on deep ocean floors, composed of more than 30 per cent organic material.

Opal   An amorphous material with composition $SiO_2 \cdot nH_2O$. Loosely considered a mineral by most people.

Ordovician   A period in the geologic time scale. It lasted from about 500 to 425 million years ago.

Ore   Material that can be mined at a profit.

Ornithiscian   A dinosaur with pelvic bones similar in construction to those of birds.

Orogeny   The process of making mountains, especially folding, faulting, igneous and metamorphic processes.

Orthoclase   A common rock-forming mineral. Potassium feldspar. $KAlSi_3O_8$

Overburden   The surface material overlying bedrock.

Overturned fold   A fold at least one limb of which has been rotated more than 90 degrees so that the beds are overturned.

Ox-bow lake   A crescent-shaped lake formed in an abandoned river curve.

Oxidation   The process of combining with oxygen.

Oxide   A compound containing oxygen.

Pahoehoe lava   Hawaiian term for lava with a smooth ropy surface.

Paleocene   A subdivision of the geologic time scale. It lasted from about 63 to 58 million years ago.

Paleoenvironment   The environment of some time in the past.

Paleogeography   The geography of some time in the past.

Parabolic dune   A sand dune with the general shape of a parabola. The cusps point in the opposite direction to the movement of the wind.

Peat   Dark brown material formed by the partial decomposition of plant material.

Pebble   A fragment between 4 and 64 millimeters in size.

Pedalfer   A type of soil found extensively in eastern United States. It is enriched in iron and aluminum.

Pediment   An erosion surface cut in bedrock and superficially resembling an alluvial fan.

Pedocal   A type of soil found extensively in western United States. It is enriched in calcium.

Pegmatite   A very-coarse-grained igneous rock. The crystals are larger than 3 centimeters.

Peneplain   The smooth, rolling erosion surface that develops very late in the cycle of erosion.

Pennsylvanian   A period in the geologic time scale.

Percolation   Movement of water through rocks.

Peridotite   A coarse-grained igneous rock composed mainly of olivine and pyroxene.

Period, geologic time   A subdivision of the geologic time scale.

Period, ocean wave   The time it takes a wave to travel one wavelength.

Permafrost   Permanently frozen subsoil.

Permeability   The interconnection of pore space allowing fluids to move through a rock.

Permian   A period in the geologic time scale. It lasted from about 280 to 230 million years ago.

Permineralization   A process of fossilization in which material is deposited in pore spaces.

Perthite   A type of feldspar formed by the intergrowth of potassium feldspar and sodic plagioclase.

Petrifaction   The process of changing organic material into stone.

Phaneritic   A textural term meaning that the crystals or grains can be seen with the naked eye.

Phosphate rock   A sedimentary rock containing calcium phosphate.

Photosynthesis   Production of carbohydrates from water, carbon dioxide, and solar energy by chlorophyll in plants.

Phylum   One of the major divisions of the plant and animal kingdoms. A group of closely related classes.

Pillow lava   Lava extruded underwater or in wet mud develops a structure that resembles a pile of pillows.

Piracy, stream   The diversion of the headwaters of a stream by the headward erosion of another stream. Also stream capture.

Pitchblende   An ore of uranium.

Placental mammal   A mammal whose young develop within the mother's body. The embryo's nourishment and waste disposal take place through means of an organ called the placenta, which is a highly selective filter between the mother's bloodstream and that of the offspring.

Placer deposits   Accumulation of valuable heavy, durable minerals. Made by agents of transportation and deposition such as running water, wind, or waves.

Placoderm   Extinct primitive jawed fish.

Plagioclase   One of the feldspar minerals. A continuous series from $CaAl_2Si_2O_8$ to $NaAlSi_3O_8$. Recognized by its hardness, two cleavages, and striations on one of the cleavages.

Plastic   Can change shape permanently without rupture.

Pleistocene   A subdivision of the geologic time scale. The last glacial age.

Pliocene   A subdivision of the geologic time scale.

Plucking   The tearing out of blocks of rock by a glacier.

Plunging fold   A fold that is not horizontal. Compare Figs. 14-9 and 14-11.

Pluton   Any body of igneous rock that formed below the surface.

Polar wandering   Movement of the geographic poles (as opposed to continental drift).

Polymorph   Two or more minerals with the same composition but different form such as graphite and diamond.

Porosity   The open space in a rock.

Porphyritic texture   An igneous rock texture in which some of the crystals are much larger than the matrix.

Porphyry   A rock with a porphyritic texture.

Pothole   A hole worn into bedrock by moving water.

Precambrian   The oldest subdivision of the geologic time scale. It includes all the time from the origin of the earth to about 600 million years ago.

Precession of the equinoxes   A slow change from year to year of the position of the stars at the time of equinox.

Precipitate   Separate from a solution.

Precipitation   Rain or snowfall *OR* being separated from a solution.

Primate   One of an order of mammals that includes apes, monkeys, lemurs, and man.

Progressive metamorphism   Metamorphism of increasing grade.

Proton   A fundamental sub-atomic particle with a positive charge. Found in the nucleus of an atom.

Pumice   Very light cellular glassy lava.

P wave (push wave, primary wave)   One of the waves caused by earthquakes. The fastest traveling wave. A compression wave.

Pyrite   A common mineral. Fool's gold. FeS_2

Pyroclastic rock   A rock formed of volcanic ejecta.

Pyroxene   A mineral family. Augite is the common rock-forming member.

Quadrangle   Topographic map produced by the United States Geological Survey.

Quartz   A common rock-forming mineral, SiO_2. Recognized by its hardness, 7, and lack of cleavage.

Quartz diorite   A coarse-grained igneous rock. Composed largely of quartz, plagioclase, and dark minerals.

Quartz (quartzose) sandstone   A sandstone composed largely of quartz.

Quartzite   A metamorphic rock composed mainly of quartz. It forms from the metamorphism of sandstone.

Quick clay   A rock type that becomes fluid if vibrated or disturbed.

Radial stream pattern   A stream pattern like the spokes on a wheel.

Radiation, biologic   The branching into separate lines from a single ancestor.

Radiation halo   A colored zone or halo surrounding a tiny mineral grain. The color of the halo changes when the grain is rotated under a polarizing microscope. Also pleochroic halo.

Radioactivity   The property of some elements to change spontaneously into other elements by the emission of particles from the nucleus.

Rain shadow   Area of diminished rainfall on the lee side of a mountain range.

Reaction series   (See *Bowen's reaction series.*)

Recent   Holocene. Most recent subdivision of the geologic time scale.

Recessive character   A character inherited from one parent that is not developed in offspring when the corresponding dominant character is provided by the other parent.

Red beds   Red-colored sedimentary rocks.

Red clay   A deep-sea deposit formed by fine material, mainly volcanic dust. The color is probably caused by oxidation.

Reef   A rock built of organic remains.

Regional metamorphic rocks   Rocks such as schist and gneiss formed by regional metamorphism.

Remanent magnetism   Permanent magnetism induced in a rock by a magnetic field.

Reptile   A vertebrate which has a dry, hardened, usually scaled skin, breathes with lungs, lays shelled eggs, and whose temperature is dependent on environment.

Retrogressive metamorphism   Metamorphism at conditions of lower grade than those under which the original rock formed.

Reversed magnetism   A magnetic field with polarity reversed from that of the present field.

Reverse fault   A fault caused by compression. The hanging wall has been raised. See Fig. 14-14.

Rhyolite   A fine-grained igneous rock similar in composition to granite.

Rift fault   (See *lateral fault.*)

Rift valley   The valley formed along a rift fault or lateral fault.

Rille   A valley-like feature found on the moon.

Rip current   The return to the ocean of water carried landward by waves.

Ripple-mark   An undulating surface on a sediment caused by water movement.

Rock   A natural aggregate of one or more minerals *OR* any essential and appreciable part of the solid portion of the earth (or any other part of the solar system).

Rock cycle   The sequence through which rocks may pass when subjected to geological processes.

Rock flour   Very finely ground material from glacial abrasion.

Rockslide   Rapid downslope movement of bedrock.

Root of mountain   The low density rocks below a mountain range. The thickening of the crust below mountain ranges.

Ropy lava   Lava with a ropy surface. (See *pahoehoe lava.*)

Roscoelite   An ore of vanadium. $K(V,Al)_2(OH)_2AlSi_3O_{10}$

Runoff   The water that flows on the earth's surface.

Rutile   An ore of titanium. TiO_2

Salt dome   A structure caused by the upward movement of rock salt through overlying sedimentary rocks.

Sand   Fragments that range in size from 1/16 to 2 millimeters.

Sand bar   An accumulation of sand built up by waves or running water.

Sandblasting   Erosion by impact of wind-blown sand.

Sandstone   A clastic sedimentary rock with fragments between 1/16 and 2 to 4 millimeters in size.

Sandstorm   Strong winds carrying sand. Most of the sand moves close to ground level.

Sanidine   A potassium feldspar found in volcanic rocks. $KAlSi_3O_8$

Saurischian   A dinosaur with reptile-like pelvic bones.

Scarp   A steep slope.

Scheelite   An ore of tungsten (wolfram). $CaWO_4$

Schist   A foliated metamorphic rock generally containing conspicuous mica.

Scoria   Volcanic rock with coarse bubble holes.

Seamount   A submarine mountain. Generally basalt volcanoes.

Secondary wave (shake wave, S wave)   One of the waves generated by an earthquake. This transverse wave is second to arrive.

Sedimentary rock   Rocks formed near the earth's surface, generally in layers.

Seif dune   An elongate sand dune parallel to the wind direction. Longitudinal dune.

Seismic sea wave   Tsunami. A commonly destructive wave generated by submarine earthquakes. Commonly mistermed *tidal wave.*

Seismograph   A device that records earthquakes.

Semischist   A metamorphic rock produced by dynamic metamorphism.

Serpentine   A mineral family. $Mg_3Si_2O_5(OH)_4$

Serpentinite   A rock composed largely of serpentine.

Shake wave, shear wave, secondary wave   (See *S wave.*)

Shale   A fine-grained sedimentary rock composed largely of clay and characterized by its fissility.

Sharpstone conglomerate   A conglomerate composed of angular pebbles. (See also *breccia.*)

Shear   To force two adjacent parts of a body to slide past one another.

Shelf, continental   (See *continental shelf.*)

Shield   A continental area that has been relatively stable for a long time. They are mainly composed of ancient Precambrian rocks.

Shield volcano   A broad, gently sloping volcano.

Silicate mineral   A mineral containing SiO_4 tetrahedra.

Silicon-oxygen tetrahedron   The basic unit of the silicate minerals. Composed of four oxygen atoms in the shape of a tetrahedron surrounding a silicon atom.

Sill   A concordant igneous intrusive body.

Sillimanite   A mineral found in metamorphic rocks. Al_2SiO_5

Silt   A clastic sediment ranging in size between 1/256 and 1/16 millimeter.

Siltstone   A clastic sedimentary rock composed largely of silt.

Silurian   A period in the geologic time scale that lasted from about 425 to 405 million years ago.

Sink   A depression with no outlet.

Slate   A fine-grained metamorphic rock characterized by well-developed foliation.

Slope, continental   (See *continental slope.*)

Slump   A landslide that develops where strong, resistant rocks overlie weak rocks. A block slips down a curving plane.

Smaltite   An ore of cobalt. $(Co,Ni)As_{2-3}$

Soapstone   A metamorphic rock composed of talc.

Sodic plagioclase   The sodium-rich end member of the plagioclase feldspar series. $NaAlSi_3O_8$

Soil   The mixture of altered mineral and organic material at the earth's surface that supports plant life.

Solifluction   Downslope movement of overburden saturated with water.

Species   A group of similar organisms that can interbreed to produce fertile offspring.

Specific gravity   Ratio of the mass of a material to the mass of an equal volume of water.

Sphalerite   An ore of zinc. $(Zn,Fe)S$

Spodumene   An ore of lithium. $LiAlSi_2O_6$

Spring   A place where water issues from the ground.

Spur, truncated   A minor ridge that has been cut off.

Stalactite   A deposit formed on the roof of a cave.

Stalagmite   A deposit formed on the floor of a cave.

Staurolite   A metamorphic mineral. $FeAl_4Si_2O_{10}(OH)_2$

Stibnite   An antimony ore. Sb_2S_3

Stock   An igneous intrusive body smaller than a batholith.

Stoping   The sinking of blocks of heavier pre-existing rock through molten magma.

Stratigraphic correlation   (See *correlation.*)

Streak   The color of a powdered mineral.

Stress   Force per unit area. Force that causes strain.

Striation, glacial   Scratches on bedrock made by rocks carried by a glacier.

Striations, mineral   Parallel lines that result from intergrown crystals (twins). Commonly seen on one cleavage of plagioclase feldspar.

Submergent coastline   A coastline along which either the land has sunk or the ocean level has risen.

Subsoil   The middle layer or B horizon of a soil profile.

Supernova   A very luminous exploding star.

Surf   Nearshore wave activity.

Surface wave   One of the waves caused by an earthquake. They move slowly with large amplitude and cause much damage.

Suspended load   The material actually carried by wind or water, as opposed to bedload which is rolled along.

S wave (secondary, shake, or shear wave)   One of the waves generated by an earthquake. This transverse wave is second to arrive.

Syncline   A fold in which the beds slope inward toward a common axis.

System, geologic   The rocks deposited during a geologic period.

Talc   A metamorphic mineral. Number 1 on the hardness scale. $Mg_3Si_4O_{10}(OH)_2$

Talus   The slope at the foot of a cliff composed of fragments weathered from the cliff.

Tectonic   A term pertaining to deformation of the earth's crust, especially the rock structure and surface forms that result.

Tektite   A piece of glass commonly of unusual shape. Such pieces are believed to result from either meteorite impact or to have come from space.

Telluride   A mineral containing tellurium.

Terrace   A flat-topped surface.

Tertiary   A period in the geologic time scale.

Tetrahedron   A solid body with four sides, each of which is an equilateral triangle.

Texture   The size and arrangement of grains in rock.

Thecodont   An extinct group of reptiles. They were the ancestors of the dinosaurs, as well as several other groups of reptiles.

Thermal metamorphic rocks   Thermally metamorphosed rocks near the boundary or contact of an intrusive body such as batholith. Also contact metamorphic rocks.

Thrust fault   A reverse fault with a fairly flat fault plane.

Tidal current   A current caused by tidal movement of water.

Tidal wave   (See *seismic sea wave.*)

Tide   The periodic rise and fall of the ocean, caused by the gravitational attraction of the moon and to a lesser extent the sun.

Till   Material deposited directly by glacial ice.

Tillite   A sedimentary rock made of till (or a till-like sediment).

Titanothere   An extinct group of hoofed Cenozoic animals. Some of the later titanotheres were large.

Topaz   A mineral. Number 8 on the hardness scale. $Al_2(SiO_4)(F,OH)_2$

Topsoil   The upper or A horizon of a soil profile.

Tourmaline   A mineral.

Trachyte   A fine-grained igneous rock with composition similar to syenite.

Transform faults   Faults associated with mid-ocean ridges in which the actual motion is opposite to the offset of the mid-ocean ridge. See Fig. 16-5.

Transverse sand dune   An elongate dune with its long axis at a right angle to the wind direction.

Trench   A very deep area of the ocean, associated with volcanic island arcs.

Triassic   A period of the geologic time scale. It lasted from about 230 to 181 million years ago.

Trilobite   An extinct arthropod that lived in the Paleozoic and was most abundant in the Early Paleozoic. Characterized by segmented bodies with longitudinal grooves that divide the body into three segments.

Tritium   An isotope of hydrogen with two neutrons and a proton in its nucleus.

Troilite   A mineral. FeS

Truncated spur   A minor ridge that has been cut off.

Tsunami   (See *seismic sea wave.*)

Tuff   A pyroclastic volcanic rock with fragments smaller than 4 millimeters.

Tuff-breccia   A pyroclastic volcanic rock composed of coarse fragments in a tuff matrix.

Turbidity current   A current or flow caused by the movement downslope of water heavy because of its suspended material (muddy water).

Turbulence   A type of flow in which the fluid moves in a disorderly, irregular way.

Type area   The place at which a formation or system is well exposed and is defined.

Unconformity   A surface of erosion and/or non-deposition.

Undertow   The bottom, nearshore water returning to the ocean. Waves move the water onshore.

Uniformitarianism   The concept that ancient rocks can be understood in terms of the processes presently occurring on the earth.

Uraninite   An ore of uranium.

Van Allen radiation belt   A more or less doughnut-shaped zone of radiation in space high above the equator.

Varved clay   Fine, thinly bedded glacial sediments. Each bed is assumed to be a yearly deposit.

Ventifact   A pebble faceted by sandblasting by wind-blown sand. Same as *dreikanter.*

Vertebrate   An animal with a backbone.

Viscosity   The resistance to flowage of a liquid. Caused by internal friction.

Vitreous luster   Glassy luster. The shiny luster of broken glass.

Volatile   A material easily vaporized.

Volcanic sedimentary rocks (volcanic sandstone and volcanic conglomerate)   Sandstone and conglomerate composed of volcanic fragments.

Water table   The surface below which the pores are saturated with water.

Weathering   The alteration that occurs to a rock at the earth's surface.

Weight   The mutual attractive force between a mass and the earth.

Welded tuff   A tuff that was erupted when hot and whose fragments have fused together as a result of the action of heat and gases present.

Wolframite   An ore of tungsten (wolfram). $(Fe,Mn)WO_4$

Youth   The first stage in the cycle of erosion.

Zircon   A mineral. $ZrSiO_4$

Index